Converged Communications

Converged Communications

Evolution from Telephony to 5G Mobile Internet

Erkki Koivusalo
Advisor at Sofigate in Espoo, Finland

IEEE PRESS

WILEY

Published by John Wiley & Sons, Inc., Hoboken, New Jersey.
Published simultaneously in Canada.

For general information on our other products and services or for technical support, please contact our Customer Care Department within the United States at (800) 762-2974, outside the United States at (317) 572-3993 or fax (317) 572-4002.

Wiley also publishes its books in a variety of electronic formats. Some content that appears in print may not be available in electronic formats. For more information about Wiley products, visit our web site at www.wiley.com.

Library of Congress Cataloging-in-Publication Data is Applied for:
Hardback ISBN 9781119867500

Cover Design: Wiley
Cover Image: © zf L/Getty Images

Set in 9.5/12.5pt STIXTwoText by Straive, Pondicherry, India

Contents

Preface

This book has two basic goals. First of all, it explains to the reader how different fixed and mobile communications systems work when delivering data or media such as voice calls between remote communication parties. But another, perhaps not too obvious, point behind its story is to provide the reader with evolutionary understanding about why those systems look as they do. The book paints a broad picture of communication systems evolution from the early analog telephone networks with manually operated switches to the latest 5G mobile data networks.

To achieve its mission, this book is divided into five parts that together describe the pieces of a jigsaw puzzle of the evolution that has taken place in the world of telephony and wide area data communications:

- Part I: Fixed Telephone Systems
- Part II: Data Communications Systems
- Part III: Mobile Cellular Systems
- Part IV: IP Multimedia Systems
- Online Appendices - https://www.wiley.com/go/koivusalo/convergedcommunications

Part I describes the structure and protocols related to fixed telephone networks. After walking through the network architecture and its elements, the book describes the operation of a digital telephone network and its key building blocks – digital exchange and connecting trunk lines. Operation of a fixed digital telephone network is controlled by SS7 signaling protocol suite, which deserves some attention as the first protocol stack introduced by the book, also used in second- and third-generation mobile networks. Further on, Part I briefly explains the early approach to provide a fully digital data communication path over telephony network with ISDN. The final chapters of Part I provide insight on various transmission technologies used in fixed telephone access and trunk networks, with which both voice and data are transported between different elements of the network.

Part II is focused to data transmission and familiarizes the reader with commonly used link, network, and transport layer solutions. There are various compelling reasons to take data communications methods under study. First of all, the control mechanisms within any digital network are based on signaling data protocols. But what is even more significant from the perspective of this book is the evolution of networks from voice-only systems to systems carrying both voice and data and eventually to data-only systems where voice appears only as one small use case of all-IP data communications. Part II begins by introducing usage of telephony subscriber line to transport data with analog and digital modems. The next two topics are data link protocols and switching protocols that support virtual connections over the network. The last sections of Part II deal with IP network protocol and TCP/IP protocol suite, on which the current all-IP data networks are based. Due to the significance of the Internet protocol (IP) for modern Internet-based communications, the book describes two major versions of the protocol, IPv4 and IPv6, as well as the other main protocols of the TCP/IP suite.

Part III elaborates the evolution of mobile cellular networks from the first-generation analog cellular systems to the latest fifth-generation 5G cellular data system. For each generation of cellular systems, one specific system and its enhanced variants are explained. The focus of this part is to describe systems that are currently maintained or being developed within 3GPP standardization forum, as since the fourth-generation cellular technologies, only 3GPP solutions have been deployed all over the world. Part III starts with a generic description about what a cellular system is and what are the related concepts and functions. Thereafter, the book takes a brief look into NMT as one example of a first-generation analog cellular system. The main body of the part walks through technologies like GSM, GPRS, WCDMA UMTS, HSPA, LTE, and 5G. For each of those, the reader learns how the technology was specified, what were the goals and services to be provided, what does the system architecture look like, what kind of protocols does the system use, and how various system-wide procedures have been implemented.

Part IV eventually brings the topics of data and telephony together. The concept of convergence, providing voice or multimedia as a service of packet data network, is introduced. As practical implementations, the book explains how a basic SIP VoIP system works and how voice calls are supported in mobile operator domain with 3GPP IP Multimedia Subsystem (IMS) architecture. The final sections of Part III elaborate how the IMS mechanisms are bundled together with the underlying LTE or 5G New Radio access technologies as VoLTE and VoNR services.

Appendices cover telecommunications theory and technologies competing on complementary with those which have their roots in telephone network. Visit *The Online Appendix A* at https://www.wiley.com/go/koivusalo/convergedcommunications for access to online only materials which introduce the reader to the problem space of telecommunications. The reader gets familiar with the major challenges that the communications system designers face, to be addressed in the system design. For each challenge, standard solutions as known to the communications industry are briefly walked through. This part provides the reader with basic knowledge of telecommunications concepts and mechanisms needed to be able to comprehend the rest of the real-life systems as described in the rest of the book. The approach is qualitative, to introduce the reader with important terminology and the related functionality. Apart from the very basics, neither mathematical equations nor complex theoretical background are provided, as those might be difficult to grasp and are not necessary to understand functionality of system implementations. The reader is expected to have a basic understanding of physics related to electro-magnetic phenomena, to understand how electrical pulses propagate in cables and radio waves in space. Part I describes how such phenomena is applied for communications with different mechanisms as referred by the other four parts of the book. Other appendices describe technologies like WLAN, WiMAX cable modems, and Fiber-to-X.

Material of the book enables the reader to understand how each of the described technologies work, master the key terminology used with them, identify the similarities and differences between related systems, recognize the strengths and weaknesses of their technical choices, understand the context and limitations of the technologies coming from their historical perspective, and navigate in the specification jungles as available from various standardization forums. The book has some coverage of telecommunication theory and generic mechanisms but is focused to describe practical systems evolved over decades of engineering effort. The author of the book hopes the material will be useful for various interest groups including but not limited to students of telecommunications, data communications and mobile systems; staff working on network operators; members of different standardization forums; engineers designing new communications systems or specific device implementations. This book can be used as an introduction to wide area communications systems for anyone from newcomers to the topic to knowledgeable engineers who are specialized to some technologies but want to broaden their understanding about other related systems and technologies. The book is well suited for a university-level course of introduction to telecommunications technologies.

Acknowledgments

I would like to give special thanks to the members of my manuscript review team: Hannu Bergius, George Denissoff, Lauri Eerolainen, Jarkko Hellsten, Markus Isomäki, Petri Jarre, Mika Jokinen, Pasi Junttila, Mika Kasslin, Timo Lassila, Ari Laukkanen, Jussi Leppälä, Mika Liljeberg, Georg Mayer, Marko Ovaska, Arto Peltomäki, Antti Pihlajamäki, Jussi Silander, Ari Valve, and Jukka Vikstedt. These former colleagues gave many invaluable technical comments and suggestions of how to improve the structure and language of the book; they identified gaps in the covered areas, checked facts, and provided their insight about actual deployments. Big thanks also to Wiley team members who supported with crafting the book according to my vision. I am grateful to Sandra Grayson, Senior Commissioning Editor at Wiley, for successfully promoting my book during the proposal and contract stage. Thanks to Juliet Booker, Managing Editor, for actively guiding me through content development up to delivery of the final manuscript to Wiley. Ranjith Kumar Thanigasalam, Permissions Specialist, helped me check and correctly cite copyrighted material used from other organizations. My copy editor, Christine Sabooni, improved and polished the language throughout the book given I am a non-native writer of English. Additional thanks to Content Refinement Specialist, Ashok Ravi, who acted as my main contact during the typesetting and proofreading stages. Without their combined assistance the manuscript would have remained in a drawer, at my home. Last but not least, I would like to express the gratitude I have for my family members: my wife, Maarit, my daughters, Paula and Anni; and my son, Mikko. They have encouraged me to take the necessary steps and time to get this book completed and published. I have now done my part. Thank you for reading this book.

Acronyms

3GPP	3rd Generation Partnership Project
5GC	5G Core
5QI	5G QoS Identifier
AAL	ATM adaptation layer
ABM	Asynchronous balanced mode
ACELP	Algebraic code excited linear prediction
ACK	Acknowledgment (positive)
ADM	Add-drop multiplexer
ADSL	Asymmetric digital subscriber line
AF	Application function
AGCH	Access grant channel
AH	Authentication header
AICH	Acquisition indication channel
AIS	Alarm indication signal
AKA	Authentication and Key Agreement
AM	Acknowledged mode
AMF	Access and mobility management function
AMI	Alternating mark inversion
AMPS	Advanced Mobile Phone Service
AMR	Adaptive multi-rate
ANR	Automatic neighbor relation
AOR	Address of Record
AP	Access point
AP-AICH	Access preamble acquisition channel
APN	Access point name
APS	Automatic protection switching
ARM	Asynchronous response mode
ARP	Address resolution protocol
ARP	Allocation and retention priority
ARQ	Automatic repeat request
AS	Access stratum
AS	Application server
ASK	Amplitude shift keying
ATM	Asynchronous transfer mode
ATU	ADSL transmission unit

AuC	Authentication center
AUG	Administrative unit group
AUSF	Authenticating server function
AWG	American wire gauge
BC	Billing center
BCC	Bearer channel connection
BCCH	Broadcast control channel
BCH	Broadcast channel
BCM	Basic call model
BEC	Basic error correction
BER	Bit error ratio
BGCF	Breakout gateway control function
BICN	Bearer-independent core network
BLER	Block error rate
BPI+	Baseline privacy interface plus
BPKM	Baseline privacy key management
BS	Base station
BSC	Base station controller
BSR	Buffer status report
BSS	Base station subsystem
BSS	Basic service set
BSSGP	Base station subsystem GPRS protocol
BSSMAP	Base station subsystem management part
BTS	Base transceiver station
BWP	Bandwidth part
CA	Carrier aggregation
CA-ICH	Channel assignment indicator channel
CAP	Carrier-less amplitude/phase
CAS	Channel associated signaling
CBCH	Cell broadcast channel
CC	Component carrier
CCCH	Common control channel
CCE	Control channel element
CCF	Charging collection function
CCITT	Consultative Committee for International Telephony and Telegraphy
CCK	Complementary code keying
CCMP	Counter mode with CBC-MAC protocol
CCO	Cell change over
CCPCH	Common control physical channel
CCS	Common channel signaling
CD-ICH	Collision detection indicator channel
CDMA	Code division multiple access
CELP	Code excited linear prediction
CGI	Cell global identity
CIC	Circuit identification code
CIDR	Classless inter-domain routing
CM	Cable modem

CM	Communication management
CMTS	Cable modem termination system
CN	Core network
COO	Changeover order
CPC	Continuous packet connectivity
CPE	Customer premises equipment
CPCH	Common packet channel
CPICH	Common pilot channel
CPS	Coding and puncturing scheme
CQI	Channel quality indicator
CRC	Cyclic redundancy check
CRNC	Controlling radio network controller
CRS	Cell-specific reference signal
CS	Circuit switched
CSCF	Call state control function
CSD	Circuit switched data
CSFB	Circuit switched fallback
CSI	Channel state information
CSICH	CPCH status indication channel
CSMA/CD	Carrier sense multiple access with collision detection
CSPDN	Circuit switched public data network
CTCH	Common traffic channel
CUPS	Control and user plane separation
DAPS	Dual active protocol stack
DBA	Dynamic bandwidth assignment
DC	Dual connectivity
DCCH	Dedicated control channel
DCF	Distributed coordination function
DCH	Dedicated channel
DCI	Downlink control information
DCS	Digital communications system
DHCP	Dynamic host configuration protocol
DLCI	Data link connection identifier
DL-SCH	Downlink shared channel
DM-RS	Demodulation reference signal
DMT	Discrete multitone
DNN	Data network name
DNS	Domain name system
DP	Detection point
DOCSIS	Data over cable service interface specification
DPC	Destination point code
DPCCH	Dedicated physical control channel
DPDCH	Dedicated physical data channel
DPLL	Digital phase-locked loop
DRB	Data radio bearer
DRNC	Drifting radio network controller
DRX	Discontinuous reception

DS	Distribution system
DSCH	Downlink shared channel
DSL	Digital subscriber line
DSS	Digital signature standard
DSS	Dynamic spectrum sharing
DSSS	Direct sequence spread spectrum
DTAP	Direct transfer application part
DTCH	Dedicated traffic channel
DTMF	Dual-tone multifrequency
DTX	Discontinuous transmission
DXC	Digital cross-connect switch
DVA	Distance-vector algorithm
EAE	Early authentication and encryption
E-AGCH	E-DCH absolute grant channel
ECF	Event charging function
ECO	Emergency changeover
ECSD	Enhanced circuits switched data
E-DCH	Enhanced dedicated channel
EDFA	Erbium-doped fiber amplifier
EDGE	Enhanced data rates for global evolution
EDP	Event detection point
E-DPCCH	Enhanced dedicated physical control channel
E-DPDCH	Enhanced dedicated physical data channel
EDT	Early data transfer
EFM	Ethernet in the first mile
EFR	Enhanced full-rate
EGPRS	Enhanced general packet radio service
E-HICH	E-DCH HARQ indicator channel
EIR	Equipment identity register
eMBB	Enhanced mobile broadband
EMM	EPS mobility management
EOW	Engineering orderwire
EPC	Evolved packet core
ePDG	Evolved packet data gateway
EPS	E-UTRAN packet system
E-RGCH	E-DCH relative grant channel
ERP	Extended rate physical
ESM	EPS session management
ESP	Encapsulation security payload
ESS	Extended service set
ETSI	European Telecommunications Standards Institute
E-UTRAN	Evolved UMTS terrestrial radio access network
FACCH	Fast associated control channel
FACH	Forward access channel
FBSS	Fast BS switching
FCCH	Frequency correction channel
FCS	Frame check sequence

FDD	Frequency division duplex
FDM	Frequency division multiplexing
FDMA	Frequency division multiple access
FEC	Forward error correction
FEC	Forwarding equivalence class
FGI	Feature group indicator
FHSS	Frequency hopping spread spectrum
FISU	Fill-in signal unit
FP	Frame protocol
FR	Frequency range
FR	Frame relay
FSK	Frequency shift keying
FTTB	Fiber to the building
FTTC	Fiber to the curb
FTTH	Fiber to the home
FTTN	Fiber to the node
FTTx	Fiber To X
GBR	Guaranteed bit rate
GCID	GPRS charging identifier
GEM	G-PON encapsulation method
GERAN	GSM EDGE radio access network
GFSK	Gaussian frequency shift keying
GGSN	Gateway GPRS support node
GMM	GPRS mobility management
GMSC	Gateway mobile switching center
GMSK	Gaussian minimum shift keying
gNB	5G Node B
GPON	Gigabit-capable passive optical networks
GPRS	General packet radio service
GRUU	Globally routable UA URI
GSM	Global system for mobile communications
GSMA	GSM Association
GT	Global title
GTP	GPRS tunneling protocol
GUA	Global unicast address
GUTI	Global unique temporary identity
GW	Gateway
HARQ	Hybrid automatic repeat request
HDB3	high density bipolar 3
HDLC	High-level data link
HDSL	High-speed digital subscriber line
HFC	Hybrid fiber-coaxial
HLR	Home location register
HSCSD	High-speed circuit switched data
HSDPA	High-speed downlink packet access
HS-DPCCH	High-speed dedicated physical control channel
HS-DSCH	High-speed downlink shared channel

HSN	Hopping sequence number
HSPA	High-speed packet access
HSS	Home subscriber server
HS-SCCH	High-speed shared control channel
HSUPA	High-speed uplink packet access
HTTP	Hypertext transfer protocol
HTU	HDSL termination unit
IAM	Initial address message
IANA	Internet Assigned Numbers Authority
IAP	Internet access point
ICANN	Internet Corporation for Assigned Names and Numbers
ICIC	Inter-cell interference coordination
ICID	IMS charging identifier
ICMP	Internet control message protocol
ICSI	IMS communication service identifier
IMT	International mobile telecommunication
IUC	Interval usage code
IETF	Internet Engineering Task Force
IKE	Internet key exchange
IMEI	International mobile equipment identity
IMPI	IP multimedia private identity
IMPU	IP multimedia public identity
IMS	IP multimedia subsystem
IMSI	International mobile subscriber identity
IN	Intelligent network
IoT	Internet of things
IP	Internet protocol
IPSec	IP Security
IPX	IP roaming exchage
IR	Incremental redundancy
ISDN	Integrated Services Digital Network
ISI	Intersymbol interference
ISIM	IP multimedia services identity module
ISM	Industrial, scientific, and medical
ISO	International Standardization Organization
ISP	Internet service provider
ITU	International Telecommunications Union
ISUP	ISDN user part
IWF	Interworking function
Kbps	Kilobits per second
LA	Link adaptation
LAI	Location area identity
LAN	Local access network
LAPD	Link access procedure for channel D
LC	Link control
LCP	Link control protocol
LDP	Label distribution protocol

LDPC	Low-density-parity-check
LED	Light-emitting diodes
LLC	Logical link control
LPC	Linear prediction coding
LRF	Location retrieval function
LS	Link security
LSA	Link state algorithm
LSP	Link state packet
LSR	Label switched router
LSSU	Link status signal unit
LTE	Long-term evolution
LTP	Long-term prediction
LTU	Line termination unit
M3UA	MTP3 user adaptation layer
MAC	Medium access control
MAC	Message authentication code
MAIO	Mobile allocation index offset
Mbps	Megabits per second
MBR	Maximum bit rate
MCC	Mobile country code
MCCH	Multicast control channel
MCG	Master cell group
MCH	Multicast channel
MCS	Modulation and coding scheme
MDF	Main distribution frame
MDHO	Macro diversity handover
ME	Mobile equipment
MF	Multifrequency
MGCF	Media gateway control function
MGW	Media gateway
MIB	Master information block
MIMO	Multiple input, multiple output
MM	Mobility management
MME	Mobility management entity
MMS	Multimedia message
mMTC	Massive machine type communication
MMTel	Multimedia telephony
MNC	Mobile network code
MOS	Mean opinion score
MoU	Memorandum of Understanding
MPOA	Multi-protocol over ATM
MPLS	Multi-protocol label switching
MRFC	Multimedia resource function controller
MRFP	Multimedia resource function processor
MS	Mobile station
MS	Multiplex section
MSC	Mobile switching center

MSISDN	Mobile station ISDN number
MSP	Multiplex section protection
MSRN	Mobile station roaming number
MSRP	Message session relay protocol
MSU	Message signal unit
MTC	Machine type communication
MTCH	Multicast traffic channel
MTP	Message transfer part
MTRF	Mobile terminating roaming forwarding
MTU	Maximum transfer unit
MTX	Mobile telephone exchange
MUX	Multiplexer
NAI	Network access identifier
NACK	Negative acknowledgement
NAS	Non-access stratum
NAT	Network address translation
NBAP	Node B application protocol
NCP	Network control protocol
NEF	Network exposure function
NGAP	NG application protocol
NIC	Network information center
NID	Network interface device
NMT	Nordic Mobile Telephone
NR	New Radio
NRF	Network repository function
NRM	Normal response mode
NRZI	Non-return-to-zero-inverted
NSS	Network and switching subsystem
NSSAI	Network slice selection assistance information
NSSF	Network slice selection function
NTU	Network termination unit
NWDAF	Network data analytics function
OAM	Operation and maintenance
OCC	Orthogonal cover code
ODN	Optical distribution network
OFDM	Orthogonal frequency division multiplexing
OFDMA	Orthogonal frequency division multiple access
OH	Overhead
OLT	Optical line termination
OMCI	ONT management and control interface
ONT	Optical network termination
ONU	Optical network unit
OPC	Originating point code
OSI	Open systems interconnection
OSPF	Open shortest path first
OSS	Operations and support subsystem
PABX	Private branch exchange

PACCH	Packet associated control channel
PAD	Packet assembler and disassembler
PAGCH	Packet access grant channel
PBCCH	Packet broadcast control channel
PBCH	Physical broadcast channel
PC	Protection control
PCC	Policy and charging control
PCCCH	Packet common control channel
PCCH	Paging control channel
PCF	Policy and charging function
PCF	Point coordination function
PCFICH	Physical control format indicator channel
PCH	Paging channel
PCM	Pulse code modulation
PCPCH	Physical common packet channel
PCR	Preventive cyclic retransmission
PCRF	Policy and charging rule function
PCU	Packet control unit
PDCCH	Physical downlink control channel
PDCH	Packet data channel
PDCP	Packet data convergence protocol
PDH	Plesiochronous digital hierarchy
PDN	Packet data network
PDPC	Packet data protocol context
PDSCH	Physical downlink shared channel
PDTCH	Packet data traffic channel
PDU	Protocol data unit
PEI	Permanent equipment identifier
PFCP	Packet forwarding control protocol
P-GW	Packet data network gateway
PH	Packet handler
PHICH	Physical HARQ indicator channel
PIC	Point in call
PICH	Paging indication channel
PKI	Public key infrastructure
PLCP	Physical layer convergence procedure
PLMN	Public land mobile network
PLOAM	Physical layer operations, administration and maintenance
PMCH	Physical multicast channel
PMD	Physical medium dependent
PMI	Precoding matrix indicator
PMS-TC	Physical medium specific transmission convergence
PNCH	Packet notification channel
PON	Passive optical networking
POTS	Plain old telephone service
PPCH	Packet paging channel
PPP	Point-to-point protocol

PRACH	Packet random access channel
PRACH	Physical random access channel
PRC	Primary reference clock
PS	Packet switched
PSA	PDU session anchor
PSAP	Public safety answering point
PSC	Primary scrambling code
PSCH	Primary synchronization channel
PSD	Power spectral density
PSH	Payload header suppression
PSHO	Packet switched handover
PSK	Phase shift keying
PSPDN	Packet switched public data network
PSTN	Public switched telephone network
PSS	Primary synchronization signal
PTCCH	Packet timing control channel
P-TMSI	Packet temporary mobile subscriber identity
PT-RS	Phase tracking reference signal
PUCCH	Physical uplink control channel
PUSCH	Physical uplink shared channel
PVC	Permanent virtual circuit
QAM	Quadrature amplitude modulation
QoS	Quality of service
QFI	QoS flow identifier
QPSK	Quadrature phase shift keying
RA	Rate adapter
RA	Routing area
RACH	Random access channel
RAB	Radio access bearer
RAI	Release assistance indication
RAI	Routing area identifier
RAN	Radio access network
RANAP	Radio access network application protocol
RAND	Random number
RAR	Random access response
RAT	Radio access technology
REG	Regenerator
REG	Resource element group
RFC	Request for Comments
RI	Rank indicator
RIL	Radio interface layer
RIP	Routing information protocol
RLC	Radio link control
RNA	RAN-based notification area
RNC	Radio network controller
RNSAP	Radio network subsystem application protocol
RNTI	Radio network temporary identity

RPE-LTP	Regular pulse excitation-long-term prediction
RRC	Radio resource control
RRM	Radio resource management
RS	Regenerator section
RSPR	Reference signal received power
RTCP	Real time control protocol
RTP	Real time protocol
S1AP	S1 application protocol
SA	Security association
SACCH	Slow associated control channel
SAE	System architecture evolution
SAE	Simultaneous authentication of equals
SAPI	Service access point identifier
SCCP	Signaling connection control part
SCF	Session charging function
SC-FDMA	Single-carrier frequency division multiple access
SCG	Secondary cell group
SCH	Synchronization channel
SCP	Service control point
SCS	Subcarrier spacing
SCTP	Stream control transmission protocol
SD	Slice differentiator
SDAP	Service data adaptation protocol
SDCCH	Stand-alone dedicated control channel
SDH	Synchronous digital hierarchy
SDP	Session description protocol
SDU	Service data unit
SEC	SDH equipment clock
SEPP	Security edge protection proxy
SF	Single frequency
SF	Spreading factor
SFD	Start frame delimiter
SFID	Service flow identifier
SGsAP	SG application protocol
SGSN	Serving GPRS support node
S-GW	Serving gateway
SHDSL	Single-pair high-speed digital subscriber line
SIB	System information block
SID	Service ID
SigComp	Signaling compression
SIGTRAN	Signaling transport
SIM	Subscriber identity module
SIP	Session initiation protocol
SIR	Signal-to-interference ratio
SLF	Subscriber locator function
SLS	Signaling link selector
SM	Security management

SM	Session management
SM-CP	Short message control protocol
SMF	Session management function
SM-RL	Short message relay layer
SM-RP	Short message relay protocol
SMS	Short messaging service
SMSC	Short message center
SMSF	Short message service function
SMS-SC	Short message service serving center
SM-TL	Short message transfer layer
SNDCP	Subnetwork dependent convergence protocol
SNR	Signal-to-noise ratio
SPF	Shortest path first
SPI	Security parameter index
SRES	Expected response
SRNC	Serving radio network controller
SRS	Sounding reference signal
SRU	SHDSL regenerator unit
SRVCC	Single radio voice call continuity
SS	Subscriber station
SS7	Signaling System Number 7
SSB	SS/PCBH block
SSCH	Secondary synchronization channel
SSDT	Site selection diversity
SSID	Service set identity
SSM	Synchronization status message
SSP	Service switching point
SSRC	Synchronization source
SSS	Secondary synchronization signal
SST	Slice/service type
STM	Synchronous transport module
STP	Signaling transfer points
STU	SHDSL transceiver unit
STX	Start of text
SUCI	Subscription concealed identifier
SUL	Supplementary uplink
SUPI	Subscription permanent identifier
SVC	Switched virtual connection
SYN	Synchronization
TA	Terminal adapter
TACS	Total access communication system
TAU	Tracking area update
TBF	Temporary block flow
TC	Transmission convergence
TCAP	Transaction capabilities application part
TCH	Traffic channel
TCP	Transmission control protocol

TDD	Time division duplex
TDM	Time division multiplexing
TDMA	Time division multiple access
TDP	Trigger detection point
TE	Terminal equipment
TEID	Tunnel endpoint identifier
TFCI	Transport format combination identifier
TFI	Temporary flow identifier
TFT	Traffic flow template
TIM	Traffic indication map
TKIP	Temporal key integrity protocol
TLLI	Temporary logical link identifier
TLS	Transport layer security
TLV	Type-length-value
TM	Transmission mode
TM	Transparent mode
TMSI	Temporary mobile subscriber identity
TPC	Transmit power control
TPS-TC	Transmission protocol specific transmission convergence
TRAU	Transcoder and rate adapter unit
TRX	Transceiver
TS	Technical specification
TS	Timeslot
TTI	Transmission time interval
TUG	Tributary unit group
TUP	Telephone user part
UA	User agent
UDM	Unified data management
UDP	User datagram protocol
UDR	Unified data repository
UE	User equipment
UICC	Universal integrated circuit card
UL-SCH	Uplink shared channel
UM	Unacknowledged mode
UMTS	Universal Mobile Telecommunications System
UNI	User network interface
UPF	User plane function
UPS	Uninterruptible power supply
URA	UTRAN registration area
URI	Uniform resource identifiers
URLLC	Ultra reliable low latency communication
USB	Universal serial bus
USF	Uplink state flag
USIM	UMTS subscriber identity module
UTP	Unshielded twisted pair
UTRAN	UMTS terrestrial radio access network
VAD	Voice activity detection

VC	Virtual circuit
VC	Virtual container
VDSL	Very high-speed digital subscriber line
VLR	Visitor location register
VoIP	Voice over IP
VoLGA	Voice over LTE via generic access
VoLTE	Voice over LTE
VoNR	Voice over 5G New Radio
VoWiFi	Voice over WiFi
VPN	Virtual private network
WAN	Wide area network
WCDMA	Wideband CDMA
WDM	Wavelength division multiplexing
WEP	Wireless equivalent privacy
WiMAX	Worldwide Interoperability for Microwave Access
WLAN	Wireless local area network
WPA	WiFi protected access
X2AP	X2 application protocol
XnAP	Xn application protocol

About the Companion Website

From the website you can find the following online appendices:
Appendix A Challenges and solutions of communication systems
Appendix B Signaling System 7 and Intelligent Network call model
Appendix C Integrated Services Digital Network
Appendix D Fixed telephone access and transmission systems
Appendix E Digital subscriber line technologies
Appendix F Cable data access
Appendix G Wireless data access
Appendix H ATM systems
Appendix I Cellular systems
Appendix J Session Initiation Protocol suite
Appendix K Answers to questions

The companion website can be found at

www.wiley.com/go/koivusalo/convergedcommunications

Introduction – The Evolution

Since the emergence of spoken language, humans have always had a need to communicate remotely with peers located far away. Caravans and postal services have carried written letters, and American Indians used smoke as a method for quick communication over long distances. The discovery of electricity and radio waves made it possible to send signals over long distances with the speed of light, using wired or radio connections. Modern communications mechanisms, such as 5G radio access or VDSL Internet access, use sophisticated methods to provide the end users with stable, always on, high-speed connections with global reach to various services. It is a long way from smoke signals to 5G, so let's take a tour to see how all that happened.

The era of modern communications, powered with electricity, began during the 1800s along with various inventions related to electricity itself and later on the radio waves. In the early 1880s, a number of scientists made groundbreaking findings on electricity and magnetism. That eventually led Samuel Morse to create a telegraphy system, where letters were sent over a wire as morse code of short and long beeps. Around the same time, Alexander Bell created his telephone, which was able to capture and reproduce voice with help of a microphone and loudspeaker. The voice waveform was transmitted between two telephones over a set of wires in analog electrical form. Only a few years later, in 1892, Almon Strowger introduced a design for an automated telephone switch [1]. Sometime earlier in the 1860s James Maxwell was able to create a theory about electromagnetic radiation. The theory was verified a few years later. Just in the end of the century, Guglielmo Marconi created a wireless telegraphy system where morse code was sent over radio rather than wire. These early examples demonstrate how the development of technology was powered by scientific findings and innovations about how to apply those findings to communications.

In the first half of the 1900s, radio technology was developed further so that voice could be transmitted over radio and not only over wires. Electronic components, such as diodes and vacuum tubes, were invented, enabling mass market production of radio equipment. The period between World Wars made a leap for radio broadcast systems, and advances were made also for bidirectional radio communications devices, which could be used from vehicles or airplanes. Just before World War II, an important invention was done by Alec Reeves, who presented a way to represent voice in digital form by pulse code modulation (PCM) [2]. Technology was not yet available to implement PCM at war time. Another major invention was frequency hopping radio, which Hedy Lamarr and George Antheil had developed for torpedo guidance systems during wartime.

In the middle of the 1900s, the first steps toward digital communications were taken. Transistors were invented and time division multiplexing was applied for telephony. The first computers were built. In the 1960s, commercial production of integrated circuits started, automatic electronics telephone switches were put into service, and PCM was applied to voice trunks. The laser was invented in the 1960s. By the end of the 1960s, breakthroughs were made for optical transmission technologies with which it was possible to send signal over optical cable rather than electrical.

Digitalization of the telephone network was started and continued throughout the 1970s. From the end of the 1960s onwards, computers were used for specific purposes, such as business, defense, and science. First steps were taken to create packet switched protocols, to support data communications between computers. In the Arpanet project, Internet protocol was used to create resilient networks able to survive over loss of some nodes and links. Still in the business world, data was moved between companies over the telephone network in a totally different way, by scanning paper documents and sending them over to recipients as telefaxes. Telefax technology was adopted in the 1970s and was in common use throughout the following decade.

During the 1980s, businesses used analog modems for moving data over the telephone network between their different offices and business partners. The first analog cellular mobile systems were put into commercial service and standardization of second generation digital cellular systems was started. As telephone exchanges and trunk networks were already digitalized with the help of SS7 protocol suite, ISDN was specified to bring fully digital 64 kbps data channel up to the customer premises. At the same time, TCP/IP protocols came to common use by universities to support Internet use cases such as file transfer, newsgroups, and electronic mail.

In the 1990s, the pace of communications technology evolution increased even further. The first fully digital GSM cellular network was taken into commercial use in 1991. In roughly 10 years from the start, GSM had been taken into use in 200 countries by 600 operators and the number of GSM subscribers approached to 1 billion. This expansion was based on a few important factors. GSM was designed to be a scalable system and it performed well. Compared to digging new cables to ground, it was much easier to set up an antenna to cover a rather large area. From a subscriber point of view, GSM became attractive via the introduction of handheld and even pocket-size mobile phones, supporting short messages in addition to voice calls. Via economies of scale and increased competition, the prices of equipment and services came down.

Last, but not least, GSM came to the market at just the right time. After the World War II, the telephony business had been under tight regulation. Only the big national telephone companies were allowed to operate networks, but it all started to change in the political environment of the 1980s. Deregulation took place all over the world during the 1990s, which meant new business opportunities for new players. Challenger operators obtained licenses for radio spectrum and were allowed to build their own mobile networks.

The last decade of the century was disruptive also for data communications. While the Internet had been a playground of universities and US defense in the 1980s, something important happened in the end of the decade. While working for CERN, Tim Berners-Lee set up a project to share information in a networked environment as hypertext. Hyperlinks were used to point to referenced documents in remote computers. In a few years, the invention of the World Wide Web, or the Internet as we know it, was born. Early on, only a few academic and public organizations published any Web pages, but soon businesses found the potential of the new technology. The Internet boomed throughout the 1990s, and Internet service providers started to build Internet connections to homes, using new ADSL technology over existing telephone cabling. All this was enabled by the deregulation, especially in the US Internet consumer market, where incumbent operators were forced to open and lend their infrastructure for other challenger operators.

Very soon, it was found that access to the Internet would be desirable also from mobile terminals. Unfortunately, the rigid structure of circuit switched GSM made it difficult and expensive to support high-speed, asymmetrical, and variable bitrate Internet connections. In the beginning of the 2000s, GSM networks were enhanced with new GPRS technology, capable of allocating GSM timeslots for packet data traffic dynamically. Still the GPRS data rates stayed modest and latencies long, compared with what ADSL was able to deliver for fixed network customers.

As the need for mobile data access grew, third generation mobile networks, such as WCDMA UMTS and CDMA2000, were specified to support both circuit switched voice and packet switched data in an equal way. UMTS adopted its core network solution from GSM and GPRS, while the radio access technology was completely revamped. UMTS networks were deployed from 2001 onwards. In Europe, deployment was temporarily slowed down by operator economics. Many national states in Europe found out that their right of licensing radio spectrum was a valuable asset. They decided to arrange public auctions from 2000–2001 to grant licenses to operators for using radio spectrum allocated to third generation UMTS systems. Encouraged by the success of GSM, anticipating high returns for 3G investment, and being afraid of becoming locked out of the market, many operators ended up with rather high bids. However, just from 2000 onwards the global telecom boom cooled down. Operators had used high sums of money for 3G licenses and saw their business expectations declining just when they should have invested in building their networks. Based on these experiences, the pricing within later 4G auctions was much more conservative.

Initially, UMTS data rates were expected to support data rates up to 2 Mbps, but in the first networks only a few hundred bps were achieved. That initial disappointment was, however, resolved in a few years by introducing high-speed packet access (HSPA) technology as an enhancement to WCDMA networks. Smaller cell sizes were introduced to increase spectral efficiency over the network. At the same time, VDSL technology was developed to provide enhanced data rates to homes using interior telephony cabling, assuming that the last mile connection from the building to the network would be supported by optical fiber. During the first decade of the 2000s, mobile phones became so common that the number of fixed telephones started to decline. Eventually, fixed telephony became obsolete over the next 20 years for many developed markets. Old telephony subscriber lines were still used in digital data modem connections, but even the last remaining traditional types of table phones were gradually replaced with ones using cellular radio network rather than any cable, other than the one needed for power.

When the design of fourth-generation cellular technology was on the drawing board, a very important decision was made. Support for circuit switched connections would not be built at all for the new system. Instead, the 4G system was optimized only for packet switched data. It was seen that data consumption expanded so rapidly that the share of voice traffic became marginal. On the other hand, in the world of digital mobile communications, voice could be represented as data. It was deemed that voice would become a type of data application, just with very specific needs for stable and guaranteed Quality of Service. **Convergence** was the development where voice and data came together, sharing common mechanisms in the networks, rather than being two inherently different types of services relying on separate network designs.

4G LTE networks were in commercial use from 2010 onwards, and the very first commercial LTE network was launched on December 2009 in Scandinavia. Initially, no voice support was provided in LTE networks, and any 4G handsets had to fall back to using other 3G or 2G radio technologies for the duration of a voice call. Later, in 2015, operators started to open their VoLTE services as 3GPP compliant operator VoIP over LTE. VoLTE complemented the LTE networks to provide native voice support without switching over to other radio technologies. At the time of this writing, only a few operators provide VoLTE roaming service; thus, there is still demand for GSM or UMTS telephony by international travelers or in rural areas without LTE coverage [3]. LTE was successful in providing consumers with superior data service, with high bitrates between 10-100 Mbps and latencies of a few milliseconds over radio access.

At the same time, operators gradually lost part of their voice market share to Internet applications, such as Skype, Facetime, Hangouts, and WhatsApp supporting one-to-one calls or Zoom and Teams supporting multi-party multimedia conferences. All these applications also support instant messaging. Instead of being an operator core service, voice and messaging became a commodity supported by many different application communities. Operators were often no longer able to bill their customers with call minutes or by number of messages sent. Instead, they introduced billing models based on monthly flat fees, data volumes, or data rates provided.

But the world of communications is never ready. While operators were busy with building their LTE networks, the fifth-generation 5G cellular system technology was already on the drawing board in 3GPP. The LTE OFDMA radio technology uses radio spectrum already very efficiently, close to the theoretical maximum, but 5G New Radio essentially reused its method and structures. Higher bitrates could be provided by increased bandwidth and using very high, hitherto unused sub-6GHz and mmW frequencies above 24 GHz. In addition to the consumer broadband market, 5G was specified to support other use cases which either needed very low power consumption (IoT, sensor networks) or very high reliability and low latencies (self-driving cars, surgical operations, factory automation). The first 5G networks were in commercial use in 2019.

The following figure shows the overall timeline over which various wide area network technologies were introduced and rolled out since mid-1970s. The arrows depict evolution and impact from earlier technologies to the design of later ones.

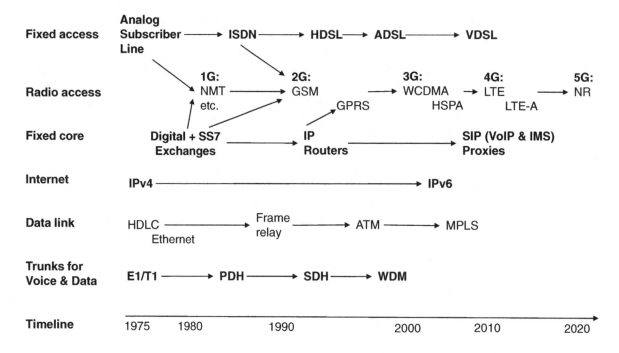

As can be understood from the earlier description, new telecommunications systems and technologies no longer emerge from a vacuum. New designs are not created from scratch. Instead, every new step of technology is built on top of previous technologies already deployed when the new technology was crafted. Communications technologies are developed in an evolutionary rather than revolutionary manner. To fully understand why a specific system was designed as it was, you need to understand the context in which the system specifications were created. The context has technical, political, and economic aspects. Knowing the virtues of prevailing technologies provides the background for setting the goals and making the technological choices for the next generation of technology.

In very many cases, the new technology uses some existing technologies as building blocks. To be able to understand how a specific technology works, you need to understand also those building blocks. In other cases, a new technology might emerge from a technical innovation, which is applied on top of an existing technology. That would mean an evolutionary step where the innovation adapts and enhances rather than replaces the existing systems. As communications systems are highly interconnected, a new technology has to be able to interoperate with relevant existing communications systems. The requirements for interoperability and compatibility set a number of constraints to the design of any new communications system.

To pick a few examples, let's consider the GSM system, which was a huge leap forward and created a global breakthrough for mobile phone deployments and accelerated the growth of mobile subscriber density. GSM emerged as a pan-European initiative to provide a common digital mobile telephony standard across a whole continent. When GSM specification started, there were country specific analog mobile networks being used, among those the Nordic Mobile Telephone (NMT) system being used in Scandinavian countries. The success of NMT paved the way toward the idea of having a standard mobile phone system in a much wider scale.

The GSM system was specified in Europe during the 1980s when telephony was perceived essentially as a fixed service. Both homes and the offices had fixed tabletop telephones cable-connected to wall sockets. Behind the scenes, the telephone exchanges and trunk networks were being digitalized. The digital telephone exchanges of fixed telephone networks used the new SS7 protocol suite for signaling purposes. With SS7, the exchanges were able to maintain the links between each other, over which SS7 protocol commands were used for various purposes, such as setting up and releasing calls. When GSM mobile networks also needed exchanges, the existing SS7 architecture was taken as a fundamental building block of GSM core network. The SS7 protocol suite was adopted as such for GSM apart from the fact that one new upper level SS7 protocol, Mobile Application Part (MAP), was developed for purposes specific to mobile GSM networks.

During the 1980s the Internet was not yet there for business use. When businesses had a need to communicate with each other in written format, they did not send emails. Instead, they sent telefaxes. After a secretary had written a letter with her typewriter, it was scanned by a telefax machine and sent over telephone connection as analog signal. The telefax machine on the remote end captured these signals and printed out the facsimile copy of the document. Naturally, telefax support was one of the services to be provided also by GSM system.

In the1980s, businesses had started to use slow analog modems over telephone networks to move data between their branch offices and head office or between the company and its bank. As digitalization developed, a brand-new idea was providing the digital interface to the network already at the customer premises. Digital telephony network was able to support 64 kbps connections optimized for emerged pulse code modulated voice. ISDN standardization took the goal to provide digital 64 kbps data interface to customer premises so that analog modems would no longer be needed. The complete end-to-end data path would be digital and for voice the PCM would be done by the phone at customer premises rather than by the exchange or the local multiplexer. When the GSM system was being specified, the goal was to build to be the first fully digital mobile phone system superior to all the existing mobile phone systems that were still based on analog technology. It was no wonder that one of the goals for GSM data transfer capabilities was to be compatible with ISDN. That would mean that you could easily connect your computer to the network either over an ISDN or a GSM connection, as both would provide similar service for the computer.

Regarding the capacity of GSM data connections, it must be recognized that GSM was not a standalone system. Initially, GSM was an access system enabling mobile users to connect with the users of fixed telephony networks. Only many years later, along with growing GSM subscriber density, GSM become the mainstream way of connecting calls between GSM users. When fixed networks used 64 kbps timeslots for the voice connections, it was evident that GSM systems would have to support the same. This constraint also affected the capacity which GSM would provide for data connections. GSM data service consisted effectively of 64 kbps circuit switched data so that the high-speed data option might use a few 64 kbps lines in parallel to provide improved data rates. The next step of evolution was to introduce GPRS, which reused the GSM radio interface but introduced a way to dynamically allocate multiple GSM slots for a single data connection with variable bitrate. Clearly, GPRS was not a new communications system, but an evolutionary step from GSM to improve its capabilities to support emerging packet data use cases, such as Internet access.

In summary, to be able to really understand a communication system, knowledge is needed about other related communications systems being used at the same time and together with the system being studied. This is what this book can provide to its reader. It contains a comprehensive but concise description about major wide area communications systems, every one of which has been the state-of-the art technology. The covered topics range from fixed telephony up to the latest 5G mobile data systems, VDSL modem links over the proven telephone subscriber lines, and VoLTE, which delivers traditional voice telephone service over 4G LTE mobile data connections.

References

1 Black, U. (1987). *Computer Networks: Protocols, Standards and Interfaces*. New Jersey: Prentice-Hall.

2 Anttalainen, T. and Jääskeläinen, V. (2015). *Introduction to Communications Networks*. Norwood: Artech House.

3 Sauter, M. (2021). *From GSM to LTE-Advanced Pro and 5G: An Introduction to Mobile Networks and Mobile Broadband*. West Sussex: Wiley.

Part I

Fixed Telephone Systems

1

Fixed Telephone Networks

This book starts its story from telephone networks, since the telephone network was the one and only globally available wide area communications network approximately over a hundred years. In the first half of the twentieth century, the communications systems supported analog voice with telephony and text with telegraph and telex systems. The era of data communications did not begin until after the Second World War, when the development of electronics and computer technology took giant steps decade by decade. At first, remote data traffic could only be transported over the telephone network with help of modems, but before long fully digital leased lines were also being used. The technology used for such leased data lines was originally developed for telephone network trunks to transport voice traffic converted into digital form. The first digital voice encoding method, **pulse code modulation (PCM),** had a profound effect on the design of the telephony network and transmission systems for which the basic bitrate 64 kbps equals the PCM codec bitrate.

Unless the reader is already familiar with basic telecommunications concepts such as amplitude, digital systems, encoding, modulation, protocols, frames, and so on, it is recommended to at first study *Online Appendix A*, which provides an introduction to various methods commonly used in telecommunications systems. The text within the book expects the reader to know those methods and also makes references to the Appendix, as appropriate.

1.1 Telephone Network

1.1.1 Analog and Digital Representation of Voice

Voice is the vibration of air caused by speech. The human ear detects sound waves since they make the eardrum vibrate by the alternating air pressure inside the ear. Figure 1.1 is a representation of a sample voice waveform, where time flows from left to right and the vertical peaks represent the intensity of the air pressure.

Telephones reproduce voice by converting this waveform, as captured from air, to an electric format with a microphone and then back to sound waves with a loudspeaker. The early telephone networks transported the analog electric waveform as such between the caller and the callee. The drawback of this scheme was rather low voice quality due to different sources of distortion, such as noise or echo, impacting the analog signal in the network. This problem was eventually solved by converting the analog waveform to a digital format, as transporting digital signals over long distances without errors became possible by the progress of electronics.

Pulse code modulation (PCM) was the first widely deployed method to convert voice into digital format in telephone networks. The PCM method globally used for the circuit switched telephony is specified in ITU-T Telecommunication Standard Recommendation G.711 [1]. Pulse code modulation works as follows [2]:

Converged Communications: Evolution from Telephony to 5G Mobile Internet, First Edition. Erkki Koivusalo.
© 2023 The Institute of Electrical and Electronics Engineers, Inc. Published 2023 by John Wiley & Sons, Inc.
Companion website: www.wiley.com/go/koivusalo/convergedcommunications

Figure 1.1 Time representation of voice waveform.

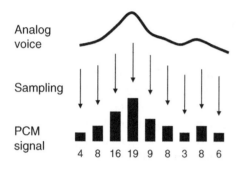

Figure 1.2 Concept of PCM sampling.

- Voice waves are converted with a microphone to an electrical analog signal, which has the same waveform as the sound waves had in the air.
- The amplitude of the electrical waveform is measured in regular short intervals. This **sampling** process is depicted in Figure 1.2.
- Each measured value of the amplitude is represented as a binary number.
- The sequence of the generated binary numbers is transported to the remote end. The numbers are sent with the same frequency as used for taking samples. When the numbers arrive to the receiver in a timely manner with constant latency, the receiver is able to reconstruct an analog electrical signal from these samples. By interpolating the waveform between the samples, the receiver can produce a waveform rather close to the one which was created by the microphone.
- The reconstructed electrical waveform is fed to a loudspeaker, which converts the signal to acoustic sound waves.

To convert an analog waveform to digital form accurately enough, the amplitude of the wave shall be sampled with a frequency at least twice as high as the maximum frequency of the analog signal. Majority of the power in a voice signal is conveyed within a sound frequency range between 400 and 3400 Hz, which was agreed to be the frequency range to be represented with PCM. Even if the low and high audio frequencies outside of this range would be filtered away, the reproduced speech can be well understood and the unique voice of the person speaking was also recognized. This fundamental agreement behind PCM is the reason why the highest voice frequencies transported over the fixed telephone network are below 4000 Hz. Consequently, PCM sampling frequency is 8000 Hz, which means samples are taken 8000 times per second and the interval between two samples is 125 microseconds.

The 8000 Hz sampling frequency is used with all the narrowband telephone network implementations around the world. Regardless of what the data rate of a digital link is, the frames of the telephone network always take 125 microseconds. On higher data rates, bigger frames are able to transport multiple voice channels simultaneously.

In the digital telephone network, the PCM modulation represents the value of measured amplitude as a binary number of 8 bits. Such a number may have 256 discrete values. As the actual value of the measured analog amplitude can be arbitrarily high or low, the PCM standards must specify how the available 256 values are mapped to a continuous range of amplitude value. This process is called **quantization**. One possibility would be to define the maximum and minimum values of the amplitude, which can be represented, and thereafter divide the range between those two limits to 256 equal parts, called **quantization intervals**. A measurement value which falls into one of the intervals is represented by the single discrete number assigned to that interval. Quantization introduces a rounding error, as every interval represented by a single number is in reality a range of amplitude values within the interval. In PCM, the difference between the actual sampled amplitude value and the amplitude represented as a single point defined by the interval number is called **quantization error**.

To divide the full range of voice waveform amplitudes to 256 equally spaced intervals is not a good approach. Its drawback is that on the intervals used for smallest amplitudes, the proportional quantization error would be largest, as measured in percentages of the actual value. As a consequence, quiet voices would be distorted most by the quantization process, when the loud voices would be represented reasonably well. While the human ear is most sensitive on quiet rather than loud sounds, the perceived voice quality would be low.

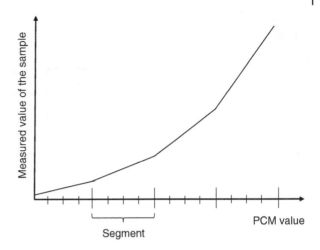

Figure 1.3 Operation of PCM quantization.

A better solution is to divide the range to quantization intervals or segments using logarithmic scale so that when the amplitude grows, the width of the interval also proportionally grows. This process is known as **companding**, with which the voice signal is compressed before quantization and expanded after decoding at the remote end. ITU-T has specified two companding schemes: A-law is used in European markets and μ-law used in North America and Japan. In practice, the companding approach can be approximated by splitting the whole range of amplitude to a few linear segments of different sizes and thereafter dividing those equally to smaller intervals, as shown in Figure 1.3.

For A-law, the eight PCM coding bits of a **PCM code word** are used as follows:

- Bit 1 means the polarity of the sample either as positive or negative.
- Bits 2–3 mean the linear segment within which the sample is located.
- Bits 5–8 mean the value of the sample within its segment.

The PCM circuits at first compress the sampled values for logarithmic scaling and thereafter apply the analog-to-digital conversion for producing the code words. The data rate of PCM coded signal is constant 64 kbps, as the sampling produces 8-bit code words for each of the 8000 samples taken in a second. This data rate is the basic rate of all digital fixed telephone networks.

1.1.2 Telephone Network Elements

The traditional fixed telephone network is a system which allows two persons to set up a voice call between each other, regardless of the location of their telephones. The network consists of telephones as voice terminals, exchanges or switches for connecting the calls, and links between all those devices to transmit voice and signaling necessary for call setup and release [3]. The endpoints of the call are identified with telephone numbers, each of them uniquely identifying one single terminal within the global telephone network.

Two terms are widely used for traditional fixed telephony:

- **Public switched telephone network (PSTN)** means the network itself.
- **Plain old telephone service (POTS)** means the telephony service provided by the fixed telephone network over a subscriber line to an analog telephone device.

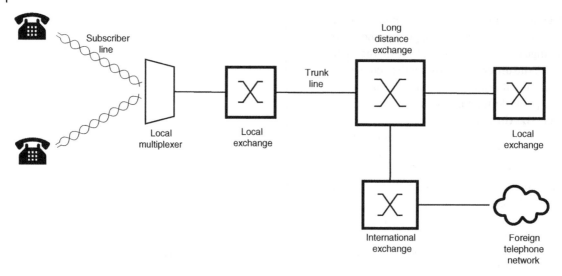

Figure 1.4 Telephone network structure.

The PSTN network consists of the following types of elements:

- Analog **telephones** at the subscriber premises
- **Subscriber lines** as a twisted pair of copper wires, connecting telephones to local multiplexers or in some cases directly to local exchanges
- **Local multiplexers** to terminate subscriber lines and to multiplex calls from them to trunk lines toward exchanges
- **Local telephone exchanges,** which connect calls within a region such as a town or city
- **Long-distance exchanges** connecting calls between different regions
- **International exchanges** to connect international calls
- **Trunk network lines** between exchanges and multiplexers

As can be seen from Figure 1.4, the global telephone network is a strongly hierarchical system. Logically, the telephone network can be divided into the access network and core network. The access network consists of subscriber lines and multiplexers. Its topology is a tree. The core network consists of exchanges, connected with trunks to form a mesh (or tree) network.

1.1.3 Evolution of the Fixed Telephony Network from Analog to Digital

Until the 1960s, the telephone signals were completely analog end-to-end on all the links and interconnected devices. All the related hardware (cables, exchanges, and multiplexers) just contributed to creating and maintaining the physical electric circuits between the two telephones which were engaged to a call and to amplifying the signal to support long-distance calls. The only exception were the trunks, which used frequency division multiplexing. For those trunks, the fundamental frequency of the analog voice signal was transposed to an available frequency division multiplexing (FDM) subband. Nevertheless, the nature of the signal was always analog at every part of the end-to-end circuit. Only the development of electronics and mass production of inexpensive components like transistors made it possible to start replacing analog transmission systems with more efficient **digital transmission systems** [4].

The evolution of the fixed telephone network from the 1800s to modern times took place over the following steps:

- The first subscriber lines had only a single iron wire so that the circuit consisted of that wire and the ground as the "return wire." Such a connection was very sensitive to external disturbances and the quality of the sound was awkward, except for very short connections. The first telephones were directly connected between each other, but when the number of phones grew, it was soon found that exchanges were needed to keep the number of lines reasonable.

- The earliest telephone exchanges introduced during the 1870s were operated manually. When someone wanted to initiate a call, the caller had to rotate a handle of the telephone to generate an electric signal toward the exchange. At the exchange, an indicator of the calling telephony line changed its position and a tone was produced to notify the operator to check which line had initiated the call attempt. After locating the line, based on its indicator, the operator connected her own phone to that line and asked the caller the identity of the callee. The electricity needed for the call was provided by the exchange to the connected circuit. Thereafter, the operator called the callee with a similar procedure to notify the callee to pick up the phone, which started ringing when receiving electric current from the exchange. When both the caller and callee had their phones ready for the conversation, the operator eventually interconnected the two lines to join the two originally independent calls. The operator then regularly checked from the line if the conversation was still going on. When no voices could be heard, the operator deemed the call to have ended and disconnected the lines. Later on, it became possible for the caller to notify the exchange about the call end by rotating the handle of the phone just like when initiating a call. With such manually operated exchanges, telephony was mainly a local service. The exchange was able to connect calls between phones directly connected to the local exchange, which was thus called the **central office** [2].

- In the 1880s, it was found that using two insulated wires for the subscriber line significantly reduced cross talk and increased the quality of the connection. A second improvement was to replace iron with copper for the wires. Compared to iron, copper had better conductive properties and was not subject to corrosion. Eventually it was found that twisting the copper wire pair made the wire more resistant to noise and decreased attenuation as the inductance between the wires grew. Progress was also made for exchanges, as the first electro-mechanical exchanges were introduced to replace the manual exchanges. Long-distance calls were introduced, but setting up such a call over multiple exchanges was still a time-consuming process. The caller had to wait for the operator to call back after the callee has been reached and all the necessary connections had been created for the call.

- In the early 1900s, the local exchanges were interconnected with automated long-distance exchanges and trunk lines. In the United States, the first transcontinental call, between New York and San Francisco, was placed in 1915. The structure and the operation of trunk lines in those days were similar to the subscriber lines. The trunk line was just a bunch of twisted copper wires, insulated with paper inside a protective rubber cover. When the distances were long enough, analog amplifiers were installed between trunk cables to keep the level of the signal good enough. Most typically, a single trunk line had two pairs of wires so that there was one wire pair to each of the directions. In that way, the amplifier could be unidirectional. If the city had multiple local exchanges, they were also interconnected with trunk lines. Since the number of calls between exchanges were just a small fraction of the number of local calls within an exchange, relatively few trunk lines were needed.

- Since installation of long trunk lines was expensive, creative new solutions were invented to transport multiple simultaneous calls over one cable. The bandwidth of a single analog voice signal over a telephone network is approximately 3 kHz. An analog trunk line made of copper provides at least 36 kHz of effective bandwidth. During the early 1900s, a method was used to modulate each separate voice signal on the trunk line with carriers separated from each other at 4 kHz intervals. In this way, one wire pair was able to transport 12 frequency-multiplexed calls on parallel. When using a coaxial cable instead of a twisted pair, the number of simultaneous calls could be increased even more, as the bandwidth supported by coaxial cable is much higher than that of a twisted pair.

- At the 1930s, the means to convert an analog voice signal into digital form with pulse code modulation (PCM) was invented. The invention was used to improve the quality of voice over long transmission links. At the time when the invention was made, the electronics that could be used for digitalization were so expensive that no wide PCM deployments were economically viable.

- At the 1950s, deployment of the automated exchanges was continued. Direct long-distance calling without operator intervention became possible in various regions in this period. Compared to the earlier manual approaches, direct calling significantly speeded up call connection time.

- At the 1960s, the development of new electronics and the falling prices of electronics made it finally feasible to introduce devices capable of multiplexing a number of voice signals into a single cable with PCM. By changing the multiplexing method on trunk lines from frequency multiplexing of analog signals to time division multiplexing of digital PCM signals, it was possible to increase the number of simultaneous calls over twisted pair from 12 to 24 or even 30. Such digital links were at first used on trunks between the long-distance exchanges but later on also between local exchanges. The digital signal was created and terminated at both ends of the trunk before connecting the individual calls to exchanges as analog signals in the traditional manner. Direct international and intercontinental calls between operators were also introduced in this period.

- A few years later, the digital line deployment reached also the trunks between exchanges and local multiplexers. The analog subscriber lines were terminated at the multiplexer, which took care of the pulse code modulation and passed the PCM signals toward the trunk. Consequently, the number of wires between exchanges and multiplexers could be reduced. Later on, coaxial cables replaced copper wire pairs on trunks.

- In the 1970s, the first digital exchanges were introduced for small towns, where the needed switching capacity was small. In those exchanges, the relays used by the old exchanges were replaced by microchips and signal buses. The digital switches were able to connect voice signals in digital PCM form between trunks used toward multiplexers and other exchanges, so the need to convert signal back to analog form for the exchange was avoided. Additionally, the digital exchanges were able to automatically process the digitally transported information about the state of the connected telephones and calls, such as the callee number and busy status.

- In the 1980s, the capacity of digital switches grew and the rate of increasing old analog exchanges with newer digital ones was increased. At the end of the decade, it become possible to transport the voice or data in digital form from the telephone device to the local multiplexer. To support that, the old analog telephone could be replaced with a newer ISDN phone capable of performing PCM process already at the customer premises rather than at the multiplexer. ISDN subscriber lines were able to support two voice channels and one data channel (2B+D). ISDN also supported using the voice channels to transport data. As the exchanges grew, the transmission capacity requirements increased between the exchanges. The capacity of trunk lines was increased at first with PDH trunks capable of multiplexing signals in a hierarchical manner and later on with optical SDH networks, to achieve even higher bitrates and direct access to sub-signals for demultiplexing.

- In the 1990s, the fixed telephone networks reached their peak as the cellular mobile phone technologies started to take off. During the first two decades of the twenty-first century, the mobile phones have nearly completely replaced fixed telephones so that we are currently living in the era of mobile telephony. Still, the telephone core networks continued to serve mobile phones, but their architecture was gradually upgraded toward packet switched rather than traditional circuit switched protocols. The twisted pair subscriber lines continued their life as carriers for DSL data connections while analog telephony was phased out. In enterprise contexts, table phones may still be used, but they typically use Voice over IP (VoIP) data connections instead of the traditional telephone network.

1.1.4 Telephone Numbering

Telephones are identified by their telephone numbers, unique within the network. As explained in *Online Appendix A.8.1*, a telephone number consists of the following parts [5]:

- Country code used for international calls
- Area code or national destination code, which defines the regional network within a country
- Subscriber number, which is unique within the region identified by the area code

When initiating a call in a fixed telephone network, the user has to provide the called number in one of the following ways:

- When calling a phone within the same area with the caller, only the subscriber number is needed.
- When calling a phone in a different area within the same country as the caller, both area code and subscriber number are needed.
- When calling a phone in a different country, all three parts are needed. An international prefix should be supplied before the country code to indicate to the exchange that the call is to another country.

Originally, the subscriber number was fixedly tied to a specific subscriber line. The house or apartment to which the subscriber line was connected owned the number. In the 1980s, when number portability was introduced in intelligent networks (see later this chapter, 1.2.2), this fixed mapping was broken. The semi-permanent relationship between the subscriber number and a subscriber line was recorded to a database rather than hardcoded to the exchange. This allowed the subscriber to keep the number even when moving to another apartment.

When mobile phones were introduced, the numbering system experienced a few changes:

- Providing the area code became necessary when calling from a mobile phone, since the mobile phones did not belong in a fixed way to any area even if they stayed being country-specific.
- Instead of the area code, mobile phones themselves had an operator code. Later on, with number portability, this fixed relationship was broken so that this code did not necessarily identify the serving operator.

1.1.5 Tasks and Roles of Telephone Exchanges

A telephone exchange has three major tasks in the network. First of all, with help of an exchange, the number of lines between all the phones in the network can be dramatically reduced. Instead of building a full mesh network where every pair of phones would have a direct connection between them, it is sufficient to connect a phone only to the exchange. Secondly, the exchange enables two phones to be connected just for the duration of the call so that the subscriber lines and trunk lines are released for other calls when the call is terminated. Since the early exchanges set up physical circuits for calls and released them after the calls were finished, the connections managed by exchanges are known as **circuit-switched** connections. Circuit-switched connections are created and released dynamically in the switching matrix of the exchange, based on control signals received from telephones [4]. Switching matrices of analog exchanges consisted of mechanical switches or relays used to connect and disconnect electrical circuits between the phones engaged to calls. With digital telephone networks, switching matrices become specialized integrated digital circuits, which were able to connect digital channels between each other. Finally, the exchange has, or is able to retrieve, the knowledge of how to route the calls from the caller to the callee, based on the telephone number provided at the call setup phase. Routing decisions may be static or depend on some dynamic conditions like the time of a day, as described in ITU-T E.170 [6]. In addition to transporting the end user traffic over circuits, exchanges collect and process charging data, make various measurements, and take care of timing and synchronization, as specified in ITU-T Q.521 [7].

Telephone exchanges can be categorized as follows:

- Local exchanges provide network connectivity to the subscriber lines within a certain geographical area, such as a town or suburb. The exchange is able to connect calls between phones in that area or forward the call attempts to another exchange, in case the callee resides in another region. In practice, there are local multiplexers between the subscriber lines and exchanges, so that the exchange receives the signals from subscribers within trunk lines from the multiplexers.
- Long-distance exchanges interconnect the local exchanges of different regions. A long-distance exchange may either be a separate standalone exchange or just a specialized function within a local exchange.
- An international telephone exchange interconnects national telephone networks over international trunk lines. An international exchange is connected either to the local or long-distance exchanges of the national networks.

1.1.6 The Subscriber Line

The traditional fixed telephones were analog devices. The microphone of a telephone vibrates from the voice air waves and converts the vibration into an electric signal with similar waveform. The telephone then passes the analog electric signal to the subscriber line leading to an exchange or a local multiplexer. After being carried over the telephone network to the telephone of the callee, the earpiece of the telephone converts the received electrical signal to sound, matching its waveform with the received electrical signal. The traditional telephone network has been designed to pass analog signals of frequencies within human voice with minimal distortion between two telephones.

Subscriber lines behind the telephone connectors on the walls are so called **unshielded twisted pairs (UTP)** of thin copper wires inside plastics insulation. The width of the cross-section of the wire is different in different countries and installations. In the United States, the options are defined within **American wire gauge (AWG)** standard, while in Europe the wires are according to the **international wire gauge (IWG)** standard maintained by the European Telecommunications Standards Institute (ETSI). When an analog voice call is set up, switches are closed at the telephone device and the exchange to form an electric circuit for the call. As this circuit is physically a loop, the subscriber line is also called **local loop** [8].

The subscriber lines were traditionally built (especially before the 1970s) so that a bunch of twisted pair cables were dug underground and run from a local exchange to a local multiplexer located in a cabinet along a street or road. The maximum span of such twisted pairs was a few kilometers. At the exchange, these wires were connected to a **main distribution frame (MDF),** from where each wire could be connected to a **subscriber line unit** of the exchange or to a data multiplexer serving leased lines. When new houses or apartments were built along with the street, spare wires were taken into use at the local multiplexers and connected to new twisted pair subscriber lines toward the new house or apartment. If a house was torn down and a new house built nearby, in some cases the unused subscriber line was left in place but a new branch connected to it. A joint was created to the middle of the wire at a location from the new wire that could be built to the new house. Such branches or wiretaps did not cause any harm to the analog voice signals below 4 kHz but would cause strong attenuation of certain higher frequencies used later for **digital subscriber line (DSL)** data connections [9].

The subscriber line attenuates the analog voice signal, which can travel in a twisted pair at maximum a few kilometers. If the subscriber premises were still farther away from the closest local multiplexer or exchange, special arrangements were needed to avoid the signal being faded out. A commonly used technique was to build the subscriber line from a chain of wires with different widths. In the United States, the subscriber lines in the rural areas were often extremely long, and loading coils were added to the line with regular intervals to keep the inductance of the line in the right level compared to the capacitance of the line. Such a solution minimized the attenuation on the analog voice frequencies but unfortunately greatly increased the attenuation for higher frequencies. On such lines, the digital subscriber line data technologies do not really work without removing the loading coils.

When relying only on twisted pairs, the local exchanges could not be located too far away from the subscriber premises. In large cities, the telephone operator had to build and maintain a large number of local exchanges. Digital transmission technologies, however, changed the game. A digital PCM trunk line extended the signal reach between a local multiplexer and an exchange to tens of kilometers, which allowed the operator to rely on much fewer local exchanges, still supporting a high number of subscribers. That was beneficial for the operators, as costs for exchange maintenance and rents paid for the central office premises could be decreased. As telephone service has recently been largely moved to mobile phones, the fixed subscriber lines have either been abandoned or reused as carriers of Internet data over DSL connections. With the progress of optical cable technology, operators have started to replace the copper wires leading to customer premises with optical trunk cables, so that the copper wires may be still used between an apartment and a local optical cable terminal in a street cabinet or in buildings' basements.

1.1.7 Telephony Signaling on the Analog Subscriber Line

Figure 1.5 represents an analog telephone system. An analog telephone gets its operating voltage via the subscriber line from the local multiplexer or exchange. Since the analog phone is not connected to a power socket, it continues operating as long as the exchange or the multiplexer is powered, even if the customer premises would otherwise have an electricity power outage. When the phone is unused, its earpiece disconnects the electrical circuit over the subscriber line. The phone is said to be in the **on-hook state**. When the user of the phone lifts the earpiece, the phone goes to **off-hook state** and closes the electrical circuit toward the multiplexer [3]. When the exchange finds out that the circuit is closed, it starts to send a continuous sound signal to the earpiece, informing the user that the exchange is ready for receiving the called number from the phone.

When the user thereafter presses the numeric keys or turns the rotary dialer of the telephone, the phone generates signals to the subscriber line to inform the exchange about the called number. Analog telephones used two different methods for encoding the number:

- Using **pulse dialing,** where the called number was expressed by a sequence of direct current pulses (100 ms) and idle break periods (60 ms) when the circuit was opened. The selected number (0–9) was indicated as the number of pulses transmitted. Old phones generated the pulses with a disc that rotated with help of a string. During its rotation, the disc connected the circuit for every pulse and disconnected it for every break. The newer phones were able to generate similar pulses according to the keypresses of the user.
- Using **dual-tone multifrequency (DTMF)** signal, which consists of two audible frequencies. These frequencies were selected so that they were within the 400–3400 Hz band supported by the telephone network, as shown in Table 1.1. DTMF tones could also be used to provide selections for automatic call processing machines and voice mail systems. The machines requested that the user press a certain key for a specific choice, such as selection between different types of services or the language that the customer wanted to speak. The frequencies used in the dual tones are given in the following table:

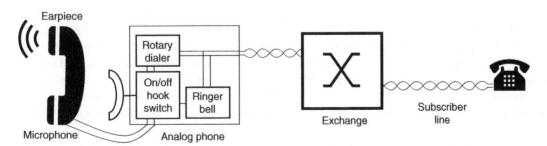

Figure 1.5 Analog phone system.

Table 1.1 DTMF dual frequency plan.

Number	1	2	3	4	5	6	7	8	9	0
Lower tone/Hz	697	697	697	770	770	770	852	852	852	941
Upper tone/Hz	1209	1336	1477	1209	1336	1477	1209	1336	1477	1336

After the called number is chosen, the exchange informs the user about the progress of connecting the call with different types of tones, as specified in ITU-T E.180 [10]. The tones used in a network are specific to the national conventions. Examples of such tones are the following:

- **Ringing tone** to tell that phone in the remote end is ringing. Exchange sends tones in a certain frequency over 0.67–1.5 seconds, after which there is a pause of 3–5 seconds.
- **Busy tone** to tell that the phone in the remote end is engaged to another call. Exchange sends tones followed by a pause, so that the duration of such a complete cycle is between 0.3–1.1 seconds.

To get the phone to start ringing at the callee, the exchange sends it a special **ringing signal,** which the phone connects to a ringer bell. The bell sound alerts the callee about the incoming call. Ringers are of two types, mechanical or electronic. Both types are activated by a 20-hertz, 75-V alternating current generated by the exchange. The ringer is commonly activated with 2-second pulses separated by a pause of 4 seconds.

1.1.8 Trunk Lines

Before digitalization of exchanges and trunks, end-to-end telephone signals were also analog. The analog voice signals were connected by exchanges to analog trunk lines. A single trunk line was able to carry multiple voice signals by using frequency division modulation. The bandwidth of the trunk cable was divided into subbands, each capable of carrying an analog voice signal with 3400 Hz bandwidth. The voice signal was used to modulate a carrier in the middle of the subband.

Since the 1970s, the trunk lines of a fixed telephone network were typically converted to digitally multiplexed links between exchanges and local multiplexers. To use a digital trunk, the multiplexer or the exchange itself terminates analog subscriber lines and multiplexes signals from them to a digital trunk line. PCM is used to encode the voice signal for the trunk. Any trunk line connected to a multiplexer is able to support tens of subscriber lines so that each of those has its own dedicated 64 kbps PCM channel on the trunk toward the telephone exchange. The higher order trunk lines between exchanges are able to support even thousands of simultaneous calls, depending on the type and capacity of the trunk.

The smallest capacity digital trunk lines were structured in slightly different ways in different continents. The basic American trunk was called the T1 link, supporting 24 voice channels and 1.5 Mbps total bit rate. The basic European trunk was called the E1 link, supporting 30 voice channels and 2.048 Mbps total bitrate. As described in Section 1.3.1, a 64 kbps voice channel is carried within a single timeslot of the time division multiplexed frame structures of T1 and E1 signals. As capacity needs increased, the digital technology was evolved to support further multiplexing of T1 and E1 signals to PDH or SDH/Sonet links of higher capacity. The physical media used for the trunk lines was typically twisted pair wire or coaxial cable for the smaller capacity links and coaxial or optical cable for the bigger links. Optical cables were preferred for long link spans. In some cases, microwave radios could be used instead of cable—for instance, when building a cable would be too expensive in difficult terrain. For further details about these digital transmission technologies, please refer to Section 1.3.

In addition to the actual voice channels, a fraction of the link capacity is reserved for the telephone network internal purposes such as transmission of notifications about remote error conditions or forwarding signaling and network management messages between the network devices. Such messages are processed within a signaling network, which consists of the following types of devices [11]:

- **Service switching point (SSP)**, which fundamentally is a telephone exchange creating and releasing the circuit switched connections between end user terminals.
- **Service control point (SCP)**, which is either a database storing information about subscribers or a centralized server with call processing logic.
- **Signaling transfer points (STP)**, which is a network device used for routing signaling messages between service switching points (SSPs) and service control points (SCPs).

Two different network architectures have been deployed for the transfer of signaling and management messages [8]:

- In the American model, the signaling and network management messages were transported in a dedicated signaling data network, which was separated from the voice network. SSP and STP were two separate types of network equipment.
- In other countries, the signaling and network management messages were sent directly between the telephone exchanges over the trunks used for voice traffic. A small part of the trunk frame structure was reserved for signaling traffic while the rest was used for the voice traffic. The telephone exchanges had both the roles of STP and SSP for signaling.

1.1.9 Telephone Networks and Data Communications

Traditionally, the capacity of trunk lines and exchanges has been dimensioned with the assumption that the length of an average call is between 3 and 5 minutes. When analog modems were placed into wide use during the first Internet boom of the 1990s, this old assumption become invalid. The duration of a data call could extend to hours, which caused overload for both trunk lines and telephone exchanges so that no new calls could be made, as all resources were already busy. To solve the problem, in theory three different approaches could have been used:

1) Increase the capacity of telephone exchanges and the interconnecting trunk lines. This solution was neither economically viable nor technically justified. Running packet switched data connections over a circuit switched infrastructure is fundamentally a bad idea. The telephony network lacks the flexibility needed for packet switched data. Telephony network was temporarily used for data connectivity due to the lack of better options, until data-centric network architectures became commonly available.
2) Separate the data traffic transported over subscriber lines into a dedicated data network and connect only the voice traffic to the telephone exchange. This solution become common in the early 2000s when asymmetric digital subscriber line (ADSL) and later on very high-speed digital subscriber line (VDSL) technologies were deployed. The basic idea of DSL was simple: the capacity of the subscriber line can be divided into two frequency bands to be used for different purposes. The low frequencies under 3400–4000 Hz were used for analog voice and any higher frequencies for DSL data carriers. In both ends of the subscriber line, there is a splitter which allows a single pair of wires to be connected to two devices, one for voice and another for data. In the customer premises, the devices would be an analog telephone and a DSL data modem. At the local multiplexer, the devices would be a voice multiplexer and a DSL data multiplexer, which would forward the traffic to different trunks and networks dedicated for voice and data.
3) Use totally separate networks for data traffic versus the voice traffic, already from customer premises. There are various ways of how to apply this approach to network configurations:
   - Using a cable modem over cable television network infrastructure for data and an analog telephone over the fixed telephone network for voice calls
   - Using DSL over the subscriber line or optical fiber to home for data and mobile telephony for voice calls

1.2 Telephone Exchange and Signaling Systems

1.2.1 Operation and Structure of a Telephone Exchange

A telephone exchange processes call attempts from analog subscriber lines with the following steps:

1) Detect a subscriber to have initiated a call attempt.
2) Collect the digits of the called telephone number.
3) Analyze the called number.
4) Route the call signaling to a trunk line toward another exchange or a local subscriber line of the callee.
5) Give an indication of an incoming call to the callee.
6) Create connections between digital 64 kbps channels (such as E1/T1 timeslots; see Section 1.3.1) used for the call at the switching matrix [4].
7) Provide the caller with ringing tone or other announcements depending on the progress of the call attempt.
8) Pass through voice media when the call is connected.
9) Release the call and collect charging records for it.

The specific functions of a telephone exchange are defined in ITU-T recommendation Q.521 [7]. A modern digital telephone exchange is a large system controlled by special computers or microprocessors embedded into the system. Although the exchange designs are vendor specific, this chapter describes a typical architecture of a digital telephone exchange.

The exchange consists of subracks fitted into a rack as shown in Figure 1.6. Each subrack contains a number of circuit boards or units, each of which has its own specific tasks to server the complete operation of the exchange. The units are able to communicate with each other over the communication channels available on the backplane of the subrack. The subracks are interconnected with cabling, providing data links internal to the exchange. The format of the data transported between the units and subracks may either be some vendor-specific proprietary format or any suitable standard format. Typically, the voice data is transported using the standard T1 and E1 links. In addition to the hardware, the exchange is controlled by a complex software system, which is distributed in one to the different units and the microprocessors running on those.

The parts of an exchange have been dimensioned in such a way that the exchange is capable of supporting simultaneous calls only from a subset of the connected subscriber lines, but not all of them. The capacity of internal links and the switching matrix of the exchange is just a fraction of the number of served subscribers. The capacity of the exchange should be sufficient for supporting the telephone traffic during typical rush hours. In exceptionally busy cases, it is possible that the exchange may run out of capacity so that any new call attempts are simply blocked.

The main functions of an exchange can be divided between units as shown in Figure 1.7:

- The units taking care of subscriber lines, when individual twisted pair lines are connected directly from the MDF to the exchange. The unit terminates the twister pair lines, converts the analog voice signal to digital PCM signal (and vice versa) and takes care of the analog signaling of the line. With ISDN subscribers, the unit terminates the ISDN 2B+D channels. One unit is typically able to terminate 30–120 subscriber lines. Note that typically, these kinds of units are part of the local multiplexer or concentrator rather than the exchange so that the exchange receives the traffic aggregated from subscriber lines within trunks from the multiplexer or concentrator. When using concentrators, only a subset of the subscriber lines can be connected simultaneously to the trunk toward the exchange.
- The units taking care of trunk lines from local multiplexers, concentrators, or other exchanges. The task of such a unit is to process calls and signaling between the subscribers and the rest of the telephone network. The unit typically processes the channel associated signaling or the lower layer protocols of the common channel

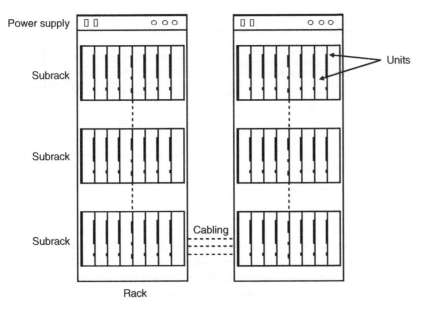

Figure 1.6 Physical structure of a telephone exchange.

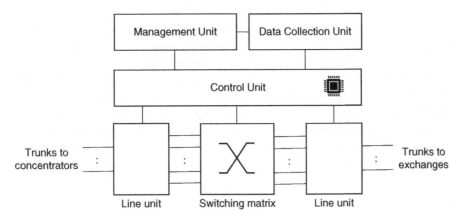

Figure 1.7 Logical structure of a telephone exchange.

signaling protocol stack (see Section 1.2.3). All the 64 kbps channels received from other exchanges can be simultaneously connected to the switching matrix of the exchange, while the same may not apply to the trunks received from local multiplexers.

- The cross-connect units hosting switching matrices where the calls are connected between the subscriber lines and/or 64 kbps channels of the trunk lines, as controlled by signaling. The switching matrix processes the voice channels in the same way, whether it gets them from a subscriber line unit or trunk line unit. Between the units, the signals are typically transported as T1 or E1 frames. When connecting a voice channel from a unit to another unit, the switching matrix does switching in two dimensions, space and time, as shown in Figure 1.8:
  - Space: to connect a timeslot between line units
  - Time: to rearrange the connected timeslots within a T1/E1 frame toward a line unit so that the timeslot is connected to the correct subscriber line at the line unit where the mapping between timeslots and subscriber lines is fixed

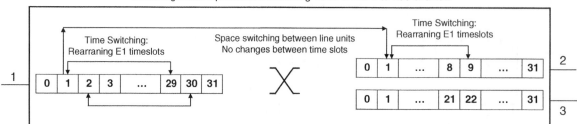

Figure 1.8 Operation of the switching matrix in time and space dimensions [8].

- The control unit, which processes the signaling data and controls the switching done on the switching matrix. The control unit may either be a separate unit or a function of the cross-connect unit in smaller exchanges. The control unit tells the switching matrix how the time slots of different line units should be connected. The control unit drives creating and releasing connections according to the signaling used to set up and clear calls. The control unit may also request the line units to perform line testing and will pass the results within signaling messages to other exchanges. The control unit may also take care of routing the signaling messages and the maintenance of the routing tables.
- The subscriber database unit is responsible for connecting the exchange to a separate database server. The database is used to help in routing calls to other exchanges and in taking care of different exceptional situations.
- The management unit of the telephone exchange. This unit monitors the whole telephone exchange, informs the network management system about fault conditions, and takes care of actions needed to recover from different types of faults. The management unit may also support a management console and a log printer.
- The data collection unit collects usage statistics needed for subscriber charging and network maintenance purposes. For each call, data is collected for the billing system. The unit provides statistics to support network planning and maintenance. From such statistics, network operators may find out that average load has increased to such a level that additional capacity would be needed.

1.2.2 Intelligent Networks

The earliest digital exchanges contained all the logic, databases, and programs needed for processing and connecting calls. During the 1980s, the technology was evolved toward distributed architectures where the logic, data, and connection processing were split to different pieces of equipment. The evolution was driven by the need to change and extend the call handling procedures due to the following reasons:

- The competitive situation between incumbent and challenger operators made the operators demand possibilities for quicker development of operator-specific services. Part of the call control logic was moved from the exchange to a separate computer system so that the operator was able to modify the logic and add new additional services used for service differentiation, such as
  - Directing calls toward a single nationwide hotline number to different call centers, depending on the time of a day.
  - Setting up a virtual private telephone network between different sites of an enterprise. Even if the calls would be connected over the public telephone network, the enterprise can define its own telephone number space for internal calls. Within a virtual network, the numbers can be freely distributed between different sites, regardless of the physical location of the sites and the related area codes of the public telephone network.

- Public authorities began to require operators to provide new regulated services, such as number portability. Number portability means that when a subscriber moves from one apartment to another, the telephone number provided by the fixed telephone network still stays the same for the subscriber. Consequently, the operators can no longer encode geographical areas such as suburbs into the telephone numbers. Instead, the mapping between numbers and individual subscriber lines must be defined in a separate subscriber database. When connecting calls, the exchange shall check this mapping from the database to route the call to the right destination.
- If the call control logic and related databases would be kept within the exchanges, any updates to the logic should be done simultaneously to all the exchanges of the network. But if the logic and databases are separated from exchanges, a single database server may be sufficient for the whole network and all its exchanges. The software and database updates would impact only the separate server, reducing the amount of maintenance effort needed and risks for having discrepancies in data stored in multiple locations. Naturally, for increased safety and reliability, the server should be duplicated so that if the server or its software upgrade process fails, a redundant server is readily available to support correct operation of the network.

The standardization of so-called **intelligent network (IN)** methods was started in the 1980s in the United States by Bellcore. In the 1990s, ITU took a role in the standardization work and published standards Q.1200 [12] –Q1400 [13] which closely follow the American standards. Eventually, ETSI defined their recommendations about how those ITU standards should be applied in European telephone networks. According to ITU-T Q.1219 [14] the intelligent network concept had the following targets:

- Efficient use of network resources
- Modularization of network functions
- Integrated service creation and implementation with reusable standard network functions
- Portability and flexible allocation of network functions to physical entities
- Standardized communication between network functions.

Various interrelated protocols were specified for the communication between intelligent network exchanges and external call control and database systems. The protocols IN/1, AIN, and IANP were used in the different intelligent network standard versions [8]. Each of these protocols specify what kind of data the exchange may query from other systems. To support standard patterns of communications, intelligent network (IN) call models were also standardized. Please refer to the *Online Appendix B.2* for a brief description of ITU-T CS-2 version of the IN call model defined in the ITU-T Q.1224 [15] specification. Since the standards were defined and enhanced over time, this common state model also evolved gradually as more detailed and complex. The unfortunate consequence was that in many cases the telephone exchange and call control server, which supported different versions of the standard, were not always able to interoperate correctly. IN protocols also specify the format of the query and response messages sent in well-defined steps of the call model. The messages compliant to these IN protocols were transported between devices with the TCAP protocol of the SS7 protocol suite (which is introduced in Section 1.2.3.2).

Support for IN features could be added to existing digital exchanges as a software upgrade. To complement such upgrades, new external call control and database systems were deployed for a complete solution. Interestingly, introduction of intelligent network technology took place during the 1980s when liberalization started to change the game for incumbent telephony operators. The 1980s and 1990s were a period of deregularization. Earlier, the telecom operators had a monopoly for providing telephony services within their countries. But from the 1980s onwards it become possible for challenger operators to establish and operate their own networks to compete with the incumbents. Together with the introduction of mobile telephony mass market, this caused the boom of the telecommunication industry during the 1990s and paved the way toward the mobile Internet era of today.

Intelligent network architecture made it possible for operators to specify additional supplementary services, supported by the logic in the external servers. Over time, those services proliferated. Eventually, ITU-T defined the various standard supplementary services, such as the following:

- Calling line identification presentation, ITU-T Q.731.3 [16]
- Calling line identification restriction, ITU-T Q.731.4 [17]
- Connected line identification presentation, ITU-T Q.731.5 [18]
- Connected line identification restriction, ITU-T Q.731.6 [19]
- Call diversion services: Call forwarding busy, Call forwarding no reply, Call forwarding unconditional, Call deflection, ITU-T Q.732.2 [20]
- Explicit call transfer, ITU-T Q.732.7 [21]
- Call waiting (CW), ITU-T Q.733.1 [22]
- Call hold (HOLD), ITU-T Q.733.2 [23]
- Terminal portability (TP), ITU-T Q.733.4 [24]
- Conference calling, ITU-T Q.734.1 [25]
- Closed user group (CUG), ITU-T Q.735.1 [26]

1.2.3 Signaling between Exchanges

As described in 1.1.7, an analog phone delivers the selected telephone numbers to the exchange either by cutting and closing the analog subscriber loop or transmitting special dual tone audio signals on a closed loop. The subscriber line is terminated either at the local multiplexer or the exchange. The signaling data such as the called number is transported in digital format between different types of digital telephone network equipment. Channel associated signaling and common channel signaling are methods used to transport signaling over digital trunks.

1.2.3.1 Channel Associated Signaling

The earliest method for transporting signaling over digital links was **channel associated signaling (CAS)** [8]. C5 signaling standards (CCITT Signaling System No. 5) defined the following CAS approaches:

- With C5 single frequency (SF) signaling, the telephone exchanges use the voice channels (T1/E1 timeslots) of a trunk to carry signaling tones while those channels are not used for calls. When the channel is not occupied, a certain constant audio tone is sent over it. Depending on the implementation, the tone is either within the range of 400–3400 Hz (used for carrying speech) or outside of it in the range of 3400–4000 Hz. In the former case, signaling cannot be transported on a channel when a call is connected to it, but in the latter case the signaling tones could be sent on parallel with a call. When the exchange requests to set up a call for a channel, it stops sending the signaling tone over the same channel. The remote exchange stops sending signaling tone to the same channel when it is ready for receiving the called number.
- With C5 multifrequency (MF) signaling, the telephone number selected by the caller is transported with 60-millisecond signals, which consist of two audible tones. Thus, the method is similar to DTMF tones, except for the strictly defined length of the signal and the frequencies used for the tones.

On digital PCM links, the channel associated signaling can be sent within the same channels as used for the calls or also over a separate signaling channel, if such exists in the frame structure. The former method is used on T1 links, while on E1 links the timeslot 16 is used for signaling. In the case of E1, the capacity of TS 16 is divided between the other 30 voice timeslots so that 16 consecutive E1 frames form a multiframe. To control voice channels, E1 multiframe provides 4 bits for signaling so that once within every multiframe half of the timeslot 16 is allocated for signaling associated to a specific voice timeslot. In this case, the signaling channel has lower capacity than audio channels and CAS signaling uses some binary indications instead of audio tones. The relation between one audio timeslot and the related four TS 16 bits of the multiframe used for CAS signaling is fixed. This makes the signaling still "channel associated." (For further details about T1 and E1 frame structures, please refer to 1.3.1.3.)

1.2.3.2 Common Channel Signaling and SS7

The drawback of CAS signaling is that its tone system allows only very limited types of states to be expressed. CAS can be used just to tell if a timeslot is occupied or not and to pass the called number between exchanges. When using voice frequencies for CAS tones, signaling is not available at all during the call. When different kinds of supplementary services were developed, new needs emerged for signaling, which CAS could not support. To facilitate the new use cases and to optimize the capacity used for signaling, a new signaling method **common channel signaling (CCS)** [8] was developed during the 1970s. CCS was introduced by the CCITT Signaling System No. 6 standard, published in the year 1972.

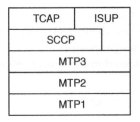

Figure 1.9 SS7 protocol stack.

Common channel signaling used an approach to transport signaling in digital rather than analog format. A set of CCS protocols were specified to be able to transport all the signaling information needed for various services of the telephone network. Common channel signaling protocols are members of the famous **SS7 protocol suite**. The **Signaling System Number 7 (SS7)** standards were specified and maintained during the 1980s and 1990s. The name SS7 originally referred to CCITT Signaling System No. 7 standard. Later, the SS7 protocol specifications were published within ITU-T Q.7xx standard series, after CCITT was renamed as ITU-T in 1993.

ITU-T standards define a number of SS7 protocol options, usage of which is determined within the national standardization organizations as profiles for local deployment. The SS7 protocol stack, as defined in ITU-T recommendation Q.700 [27], is depicted in Figure 1.9.

SS7 stack has the following protocol layers:

- **Message Transfer Part (MTP):** Set of protocols used to transport signaling messages in the right order and without errors from the source to the destination, such as between two telephone exchanges to set up a call. According to ITU-T Q.701 [28], MTP consists of three protocol layers on top of each other:
    - MTP level 1: Physical layer functions to transport bits with constant rate over a bidirectional link.
    - MTP level 2: Link layer functions such as alignment of the link, link monitoring, transport of SS7 frames, flow control, error correction, and retransmissions.
    - MTP level 3: Network layer functions such as routing of signaling messages, monitoring of links and routes, as well as route rearrangement in fault conditions. SS7 does not support dynamic routing protocols; thus, the network management system shall modify the contents of routing tables in each of the exchanges that perform routing.
- **Signaling Connection Control Part (SSCP):** Network layer protocol to complement features of MTP3. SCCP is used to transport such signaling messages which are not circuit related. The special task of SCCP protocol is to transport queries and notifications toward SCP network elements, such as intelligent network subscriber database servers.
- **Transaction Capabilities Application Part (TCAP):** Protocol used to deliver different kinds of commands and database queries related to telephone services. The specific commands and queries are defined in other national or international standards. TCAP is used to provide intelligent exchanges with information needed for call setup, while the actual call setup process is done with ISUP protocol.
- **ISDN User Part (ISUP):** An application layer protocol which is used to control the call setup and release processes between the exchanges. The predecessor of ISUP protocol in SS7 stack was called telephone user part (TUP). TUP did not yet support ISDN, while ISUP can and is used for both traditional voice call and ISDN call control.

In North American SS7 networks, the signaling messages use their own network, which is separated from the voice network. Totally separate links have been dedicated for signaling while voice traffic uses their own T1 links. Routing of signaling messages in the American model is done with dedicated equipment known as **signaling transfer point (STP)**. In the rest of the world, the signaling messages are transported between exchanges along the same links used for voice circuits [8]. On the E1 links, the timeslot 16 of the frame is reserved for common channel signaling. The exchanges deployed outside of the American market take care of routing the signaling messages.

More information of the above mentioned SS7 protocols can be found from the *Online Appendix B.*

1.2.4 ISDN

Narrowband Integrated Services Digital Network (ISDN) was developed during the 1970s and the 1980s after most of the trunk network and exchanges were digitalized [4]. ISDN call control relied on the SS7 protocol suite. The following goals were defined for the narrowband ISDN system:

- The system should be able to transport 64 kbps digital signals to subscribers over the existing telephone network cabling.
- Simultaneous usage of voice and data should be possible with a single ISDN terminal.
- The interface to the network should be such that it would enable connecting multiple terminals to a single cable connection, regardless of whether they all would use the same service or different ISDN services.
- The system should allow the end user to manage different supplementary services (such as call transfer, call hold, restriction of international calls etc.) with a numeric telephone keyboard.

The structure and functions of the ISDN system were defined within the I-series standards of CCITT. The specifications were published from 1984 onwards and later they were maintained by ITU-T. The concept and architecture of ISDN are specified in the following ITU-T recommendations:

- I.120 [29] Integrated services digital networks (ISDNs)
- I.310 [30] ISDN network functional principles
- I.324 [31] ISDN network architecture
- I.412 [32] ISDN user-network interfaces

For further details of narrowband ISDN, please refer to *Online Appendix C.1.*

The narrowband ISDN system was deployed at the first time in Great Britain in 1985. In the end of the 1980s, ISDN service was opened in the United States, France, and Germany. ISDN services were introduced in a slow pace for other countries so that the technology was in its widest use in the 1990s. In the early 2000s, ISDN became obsolete and ISDN connections were often replaced with DSL technologies.

Soon after introducing the narrowband ISDN, it was recognized during the 1990s that the narrowband 64 kbps data rate was too slow for high-performance use of the Internet. ITU-T started activities to specify **broadband ISDN** version to support higher bitrates. Broadband ISDN was designed to use a new asynchronous transfer mode protocol, described in 3.2.2 and *Online Appendix H.*

1.3 Transmission Networks

1.3.1 E1 and T1

1.3.1.1 Standardization of E1 and T1
In order to transport multiple 64 kbps PCM signals on a single twisted pair or coaxial cable link, two different standards based on synchronous time division multiplexing were developed at the 1960s:

- T1, where the TDM frame is divided into 24 timeslots to provide 1.544 Mbps total capacity for the link. T1 standard has been deployed in North America and Japan.
- E1, where the TDM frame is divided into 32 timeslots to provide 2.048 Mbps total capacity for the link. E1 standard has been deployed in the majority of the world, except in North America and Japan.

Since a 64 kbps channel (a single 8-bit timeslot of a T1/E1 frame) is the basic unit of transmission that may carry either PCM voice or data, it is called digital signal level 0 (DS0) user data channel [11]. The DS0 channel may also be referred as E0 (for E1 links) or T0 (for T1 links). T0 and E0 channels are typically used to carry PCM encoded voice. Due to the market specific requirements, T0 channels of T1 links typically carry PCM μ-law encoding while E0 channels of E1 links carry PCM voice encoded with A-law (see 1.1.3).

The frame structures of E1 and T1 as well as the functions of equipment processing those signals are specified in the following ITU-T standards:

- G.703 [33]: physical and electrical characteristics of PCM signals
- G.704 [34]: frame structures for signals with 1544 and 2048 kbps bitrates
- G.705 [35]: PDH equipment functional block model
- G.706 [36]: frame synchronization and CRC checksums of E1 and T1
- G.732 [37] and G.736 [38]: functions of E1 multiplexer generating the 2048 kbps signal
- G.796 [39]: functions of E1 cross-connect equipment

E1 and T1 links were designed to transport multiple 64 kbps PCM voice channels over trunk connections. While the technology was at first used for interconnecting exchanges, later it was used in various other kinds of contexts, such as connecting exchanges with local multiplexers, GSM base stations with base station controllers, as leased data lines to enterprises, or as data links within the early Internet core network. Later, such kinds of trunk lines were built with higher capacity link types, described in Section 1.3.

1.3.1.2 Endpoints of E1 and T1 Lines

The networks deploying T1 and E1 links use the following types of network equipment:

- **Regenerator (REG)** used on long transmission links to increase the length of the link. The regenerator interprets the received E1 or T1 signal and reproduces it to the next span of the link.
- **Multiplexer (MUX)**, which composes a single E1 or T1 signal from multiple DS0 channels and vice versa.
- **Add-drop multiplexer (ADM)**, which forwards one E1 or T1 signal from a link to another like a regenerator, but locally drops one or multiple DS0 channels from it and replaces them with other DS0 channels.
- **Digital cross-connect switch (DXC)**, which is able to interconnect the DS0 channels between different links and timeslots of multiple T1 and E1 links attached to the DXC.

Such pieces of equipment may either be standalone network devices or parts of a more complex telecom equipment, such as a mobile phone base station or microwave radio link.

1.3.1.3 Frame Structures

The structures of T1 and E1 signals are as follows (see Figure 1.10):

- T1: All the 24 timeslots are used for voice channels. In the beginning of a frame there is a framing bit which is used on every other frame to carry a signaling data channel. Further on, in the T1 multiframe structure every sixth frame has only 7 bits per timeslot used for user data or voice while 1 bit per timeslot is used for signaling.
- E1: Timeslot 0 is dedicated for frame synchronization and timeslot 16 for signaling. Other 30 timeslots are used for transporting voice channels.

Both types of frame structures are defined in ITU-T standard G.704 [34]. The length of one T1/E1 frame is 125 μs, which is the sampling interval used in PCM. When the PCM process produces 8 bits of information every 125 μs, it is possible to transport a PCM encoded voice signal over a single timeslot of the frame. Since the frame has either 24 or 30 voice timeslots, it is capable of transporting as many voice channels on parallel over a single link.

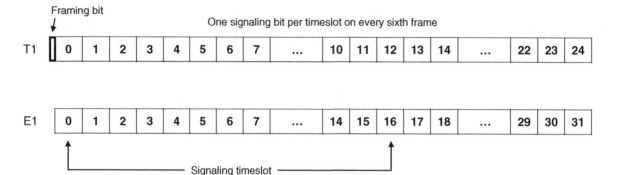

Figure 1.10 Structures of T1 and E1 frames.

The bitrates of the links in these two standards are as follows:

- T1: 1.544 Mbps = 24 × 64 kbps
- E1: 2.048 Mbps = 32 × 64 kbps

The length of T1 frame is 193 bits, transmitted in 125 microseconds, resulting 1544 kbps bitrate. The first bit of the frame is so called F bit, and the other 192 bits are used for 24 pcs of 64 kbps timeslots (TS); 24 consecutive T1 frames form a T1 multiframe.

The structure of T1 frame is as follows:

- F bit divided over the multiframe as follows:
  - The F bits of every other T1 frame (starting from the very first frame of the multiframe) belong to the 4-kbps network management channel. This channel is able to transport regular reports about the link status within LAPD protocol messages. Additionally, certain bit sequences have been defined, which are used to send commands or information related to reconnecting the link during the link test process, protecting the link with a redundant link or using the link as a synchronization source. The synchronization status messages (SSM) used for the latter purpose are similar to the SSM messages used in SDH, as described in *Online Appendix D.2.4*.
  - The F bits of every fourth T1 frame (starting from the second frame of the multiframe) belong to the CRC-6 checksum calculated from the information bits within the multiframe.
  - The F bits of every fourth T1 frame (starting from the fourth frame of the multiframe) are used to transport a constant frame alignment signal bit sequence 001011. This signal is used by the receiver to recognize the start of the multiframe.
- 24 timeslots as T0 channels. In the T1 multiframe structure, every sixth frame has only 7 bits per timeslot used for user data or PCM voice while 1 bit per timeslot is used for signaling.

The bits of a T1 signal are transmitted over the wire using AMI line coding [33]. T1 signal is transported over a coaxial cable or two unidirectional twisted pairs. The reason for using two pairs of wires is that the AMI line coding method used attenuates on twisted pair so quickly that unidirectional amplifiers are needed at every kilometer of the cable. Coaxial cable is a better option for long distances due to the lower attenuation.

The length of E1 frame is 256 bits, transmitted in 125 microseconds, resulting in 2048 kbps total bitrate. All the 256 bits of the frame are used for 32 pcs of 8-bit timeslots (TS). The first timeslot 0 is dedicated for frame synchronization and timeslot 16 for signaling. Other 30 timeslots are used as E0 channels to carry PCM voice or data. In GSM networks, a single E0 timeslot could be used to carry multiple voice channels encoded with GSM voice codecs. For instance, the GSM Full Rate voice codec produces a bitstream of 16 kbps, thus four GSM voice channels can be multiplexed into a single E0 channel. Sixteen consecutive E1 frames make up an E1 multiframe, which is divided into two submultiframes, both with eight E1 frames.

The structure of the E1 frame is as follows:

- Timeslot 0
  - In the eight frames of a submultiframe, the first bit of TS0 in every other frame may be used to carry a CRC-4 checksum calculated over the submultiframe. The usage of this checksum is optional. Bits 2–8 of the TS0 in those frames contain the constant bit sequence 0011011 used as a frame alignment word.
  - The bit 1 of TS0 of the other four frames also carry CRC-4 checksum while bit 2 has a constant value one. The bit 3 is used to pass information about link fault conditions detected in the remote end. The rest of the bits are reserved for other purposes, such as sending SSM or network management messages.
- Timeslot 16 is used as a signaling channel in one of these two ways:
  - As 30 separate CAS signaling channels, one for each voice/data timeslot. Each of the signaling channels use 4 bits per multiframe, so that TS16 of each frame carries two of the signaling channels.
  - As a single 64 kbps channel used for common channel signaling (CCS), which replaced CAS during the 1980s.
- Timeslots 1–15 and 17–31 are E0 channels with 64 kbps bitrate. It is possible to allocate multiple timeslots for a single N x 64 kbps data signal. Such aggregation of E0 channels has been used for leased data lines.

The T1 and E1 signals are synchronous so that the receiver shall constantly maintain the bit and frame synchronization with the received signal. Each network node processing T1 or E1 signals can derive its internal clock frequency either from an external precision clock signal or from the received T1 or E1 signal. In the older implementations, each link had to be synchronized separately, whereas newer implementations are able to forward the clock frequency from a link to another. In that way, all the devices of the network could be synchronized to one shared clock frequency.

1.3.2 V5

When the telephone exchange sites were close enough to subscribers, it was possible to build direct copper wire pairs from the exchange to telephones. Since end of the 1970s, the approach was changed and the copper wires were often terminated at a local multiplexer. T1 or E1 links were used to connect the local multiplexer with the exchange. Usage of the digital technology made it possible to decrease cabling costs as a single T1 or E1 cable was able to carry tens of simultaneous calls between the multiplexer and the exchange. On the other hand, the digital technology helped increasing the distance between telephone exchanges and subscribers, dramatically reducing the number of exchanges needed.

Initially, there was no standard specified as to how to connect subscriber lines via multiplexers to a digital trunk line. Instead, vendor-specific solutions were used. This typically caused a vendor lock and challenges for the operator to introduce new telephony services, which the vendor might not yet support. To ensure the compatibility between their systems, operators had to buy both their exchanges and multiplexers from the same vendor. To address these problems, during the 1990s, ETSI specified two different European interface standards, V5.1 and V5.2. While V5.1 specified a standard way of connecting subscriber links to trunks in a fixed manner, V5.2 supported dynamic connections allowing the operators to use statistical multiplexing to reduce the number of needed trunk lines. From the device perspective, V5.1 relied on multiplexers and V5.2 on concentrators. The first of these standards, V5.1, was widely deployed. The introduction of V5.2 took place so late that operators at that time were not too eager to make investments to their narrowband telephone networks. Instead, telephone operators started to invest in DSL to grow their Internet access business to complement the traditional voice business.

Please refer to *Online Appendix D.1* to find more information about V5 standards.

1.3.3 PDH

1.3.3.1 Standardization of PDH

Plesiochronous digital hierarchy (PDH) provided a way to multiplex many E1 or T1 signals to one cable and increase the transmission capacity over a trunk link for circuit switched connections. Plesiochronous refers to almost synchronous. PDH signals are not completely synchronous as the network does not have a common clock frequency.

In the trunk network, it was often necessary to transport a much larger number of voice channels as available on E1 and T1 links. The most straightforward way to solve this problem was to multiplex a number of E1 or T1 signals to one cable with TDM. In the next step, those signals could once again be multiplexed to higher-order signals. E1 and T1 signals themselves are products of TDM multiplexing, so the outcome of this process is nested TDM signals for the trunk lines. A single high-order signal may then consist of two to four nested TDM levels of lower-order signals. This plesiochronous digital hierarchy (PDH) solution was specified by CCITT in the following standards, which later became ITU-T documents:

- G.704 [34]: PDH frame structures with bit rates between 1.5–45 Mbps
- G.754 [40]: PDH frame structures with 140 Mbps bit rate

1.3.3.2 PDH Signal Hierarchy and Operation

The very basic dilemma of PDH comes from the lack of common clock frequency for individual signals multiplexed together. When building a new network or digitalizing old trunks, deployment of new digital links was done cable by cable. Since building of new cables was considered expensive, often the old cables were used and only the signal format within the cable was changed to increase bitrates. In early digital trunk networks, on each link the clock synchronization was done with help of a clock within the local receiver circuit. Since the accuracy of such local oscillators was limited to 50 ppm (i.e., 0.005%), the consequence was that when multiplexing different lower order signals into a higher order signal, there was no common clock frequency to be used for the combined signal. Because of this, the bitrate of the higher order signal was chosen to be slightly higher than the combined bitrates of the multiplexed lower order signals. This was done to ensure that the higher order signal could carry all the data from the multiplexed signals, even if their actual clock frequencies (and bitrate) would exceed the nominal one of the multiplexer. When the higher order signal had some extra capacity, the amount of which depended on the differences between the clock frequencies of the signals, the gap was filled with **justification bits** carrying no data. These justification bits were removed at the demultiplexing phase. As mentioned in the beginning of the chapter, the name plesiochronous digital hierarchy refers to almost synchronous hierarchy where the justification (or stuffing) bits are the exception.

Initially, only one level of multiplexing was used. In that case, the presence of justification bits did not cause any major issues. When the transmission technology and circuits were evolved and the need for capacity grew, higher bitrates were called for. Those needs were met by repeating the TDM multiplexing process to create higher order signals from a number of lower order signals. At each layer of multiplexing, justification bits had to be used as the network would not have a common clock frequency for any of the signals. The final hierarchical structure of PDH signals specified by ITU [41], with either E1 or T1 being the base rate, were as shown in Table 1.2:

The T/DS part of the hierarchy was used in American markets while the E part of the hierarchy was defined for the European market, but was deployed also on other continents.

Each level of the hierarchy has its own frame structure with a fixed length. PDH frames do not only contain the multiplexed sub-signals and justification bits, but also some amount of additional data bytes over which the link endpoints can provide information about detected fault situations or measured performance of the link [42]. Some of the extra bytes are used for a signaling channel used to transport LAPD protocol signaling messages. To organize the usage of those extra bytes, a multiframe structure has been defined over a number of consecutive PDH frames, similar to how multiframes are used for E1 and T1.

The type of cabling required to transport PDH signals depended on the distance and the type of the signal. Twisted pair wires could be used to carry E1/T1 signals over 1–2 km. For longer distances, repeaters or coaxial cables were needed. E2/T2 could also use twisted pairs, if the pairs would be properly insulated from each

Table 1.2 PDH signal multiplexing hierarchy.

	Multiplexing structure	Voice channels total	Bitrate
E1	32 x E0	30	2.048 Mbps
E2	4 x E1	120	8.448 Mbps
E3	4 x E2	480	34.368 Mbps
E4	4 x E3	1920	139.264 Mbps
T1/DS1	24 x T0	24	1.544 Mbps
T2/DS2	4 x T1	96	6.312 Mbps
T3/DS3	7 x T2	672	44.736 Mbps
T4/DS4	6 x T3	4032	274.176 Mbps

other in a cable bunch. Coaxial or optical cables were needed for higher order PDH signals, such as E3/T3 and E4/T4.

1.3.3.3 PDH Network Architecture

The only standard type of PDH equipment is multiplexer, which simply multiplexes four to seven lower order signals into one higher order PDH signal to be transmitted over one single cable. The basic structure of a PDH network is a tree of multiplexers, the depth of the tree depending on the type of the highest order signal transported over the trunk. The internal structure and functionality of PDH equipment is specified in ITU-T standard G.705 [43]. The following ITU-T standards define multiplexers for the different levels of PDH multiplexing hierarchy:

- G.733 [44], G.734 [45]: T1 multiplexers
- G.743 [46], G.752 [47]: T2 multiplexers
- G.732 [37], G.735 [48]: E1 multiplexers
- G.742 [49], G.744 [50], G.745 [51]: E2 multiplexers
- G.753 [52]: E3 multiplexers
- G.754 [40]: E4 multiplexers

Equipment vendors, however, produced other types of PDH devices, such as add-drop multiplexers and cross-connects. Those devices were typically able to demultiplex the input signal by one level, connect the demultiplexed signals as desired, and multiplex the signals back to the original level for transmission. For instance, an E4 add-drop multiplexer could demultiplex the E4 signal to four E3 signals, drop one of them and multiplex the rest into another E4 signal to be transmitted onwards within the network. Some devices were even able to support 1 + 1 protection of signals, as described for SDH in *Online Appendix D.2.6*.

The main problem with PDH is that each level of multiplexing and demultiplexing must be implemented with a separate piece of equipment. Because of the presence of stuffing bits and slightly different bit rates of the multiplexed signals, it is simply not possible to find a single E1 signal from an E4 signal, without demultiplexing the whole hierarchy of signals. If there is a need to drop one signal of low bitrate (like E1 or even an individual E0 voice channel) from a high-speed E4 or E3 signal, the whole signal must be demultiplexed to its E1 components. After dropping the needed low-order signals, the higher order signal must thereafter be hierarchically multiplexed once again before it can be transported over the next trunk. For every level, the stuffing bits are removed by the demultiplexing process and restored once again in the multiplexing process. The stack of equipment needed for this was known as PDH multiplexer mountain (see Figure 1.11).

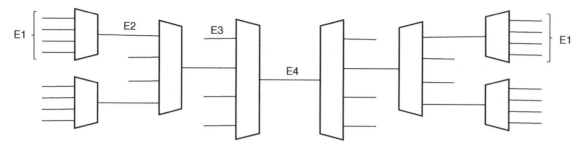

Figure 1.11 PDH multiplexer mountain.

It was very expensive for the network operators to acquire and maintain all the equipment needed by PDH transmission. In addition to the direct cost of the equipment, the building or rental costs of the premises were high as well when much space was needed for the equipment racks. Managing connections and cables between different multiplexers was complex and prone to human errors. Further problems were caused by the way PDH was specified. There was no global PDH market, since the American and European standards were not compatible. Even worse, the standards did not cover either PDH optical interfaces or network management aspects. During the 1980s, it became clear that for connections with higher bitrates than 140 Mbps, a better totally synchronous transmission standard was needed.

1.3.4 SDH

1.3.4.1 Standardization of SDH
Synchronous Digital Hierarchy (SDH) provides a way to multiplex many E1, T1, or PDH signals synchronously to one single electric or optical cable without additional justification bits. SDH increases the bit rates of a single signal even higher than the highest order PDH signals. SDH is an international standard, while it originated from an American SONET standard. SONET/SDH standardization had the following goals when the specifications were created in the end of the 1980s:

- SDH had to provide higher bit rates over an optical cable compared to what was available with PDH.
- SDH had to provide methods for adding and dropping individual lower order signals from a higher order signal, just using one single piece of equipment rather than a stack of multiplexers.
- SDH had to support multiplexing E1, T1, and any PDH signals.
- SDH was expected to provide methods for monitoring the SDH links as well as automatic reporting of faults detected.
- SDH was expected to provide methods for protecting links or individual lower order connections multiplexed to the links and taking the redundant capacity automatically into use in fault cases.
- SDH had to support centralized network management over a standard Q3 network management interface.

SDH multiplexing hierarchy, frame structures, functions and architecture are defined in the following ITU-T standards:

- G.703 [33]: Electric connectors of SDH equipment
- G.707 [53]: SDH multiplexing hierarchy and frame structures
- G.774 [54]: SDH management model
- G.781 [55]: SDH network synchronization
- G.783 [56]: SDH functional equipment model
- G.803 [57]: SDH network architecture
- G.957 [58]: Optical connectors of SDH equipment

ETSI has refined the SDH equipment model and functions in the specification ETS 300 147 [59], which is used as a reference for the latest versions of ITU-T standard G.783 [56].

This chapter provides high-level architectural overview of SDH, but further details of various SDH mechanisms can be found from the *Online Appendix D.2*.

1.3.4.2 Basic Principles of SDH Multiplexing

The very basic approach used for SDH is that all the SDH equipment within a single network is synchronized with one single **primary reference clock (PRC)** of the network. The clock frequency of the PRC is conveyed between the SDH equipment in the network over the SDH signals or with separate clock lines. The common clock frequency guarantees that SDH signals of certain order always have exactly same frame durations and bitrates, regardless from which link they are received. Because of this, the payload bit rate of a higher order SDH signal is exactly the sum of the multiplexed lower order SDH signals and no stuffing bits are needed. All that makes the SDH demultiplexing process straightforward even over multiple layers of nesting the signals.

An SDH frame is called **synchronous transport module (STM)**. When an external signal such as E1, T1, or any PDH signal is multiplexed into an STM, the clocking frequencies and bitrates of those external signals (based on their own clocking) must be adjusted to the common clock rate of the SDH network. Thus, while all the SDH signals of the SDH network have a common clock rate, the carried PDH signals do not follow it. Because of this, the transmission capacity reserved in SDH for a PDH signal is slightly higher than the nominal bitrate of the PDH signal. Extra additional bitrate is no longer needed for multiplexing SDH signals within the SDH multiplexing hierarchy. SDH uses stuffing bits only for adjusting the incoming PDH signals to the common SDH network bitrate. The transport capacity reserved within STM to carry PDH signals is called **container**. The received PDH signal and SDH stuffing bits in predefined positions are carried by the container. In addition to those bits, the container has a set of control bits telling if the stuffing bits carry any data or not. The number of stuffing bits depends on the ratio of the SDH primary clock frequency to the clock frequency of the carried PDH signal. As stuffing bits are no longer needed for multiplexing lower-order SDH frames into higher-order frames, it is possible to locate a single container from a nested SDH frame structure without demultiplexing the whole SDH signal. Figure 1.12 shows how higher-order SDH frame structures can be built from lower-order containers.

The purposes of containers within Figure 1.12 is as follows:

- C-11: Designed to carry T1 signals over SDH network
- C-12: Designed to carry E1 signals over SDH network
- C-2: Designed to carry T2 signals over SDH network
- C-3: Designed to carry T3 and E3 signals over SDH network
- C-4: Designed to carry E4 signals over SDH network
- C-4-4c: Transport capacity equaling to four concatenated C-4 containers to transport any very high bitrate data streams over SDH

ITU-T recommendation G.707 defines mappings for also other types of container payloads, such as stream of ATM cells and protocols such as DQDB and FDDI, which were used for local data transport while SDH was under specification.

A **virtual container (VC)** is a frame structure created by adding some **overhead (OH)** header bytes to the front of the **container (C)**. The location of a virtual container is not fixed within the SDH frame carrying the containers, but the start of a VC can be found with help of pointers located to fixed positions of the SDH frame. This mechanism is needed because there is no common frame synchronization at the edge of SDH network (due to processing of E1, T1, and PDH signals not synchronized with each other) even if the SDH network has a common clock frequency. While the length of each virtual container of the same order is exactly the same in the SDH network, the starting positions of the virtual containers as generated by different SDH edge equipment will differ. As an STM frame may multiplex virtual containers coming from different

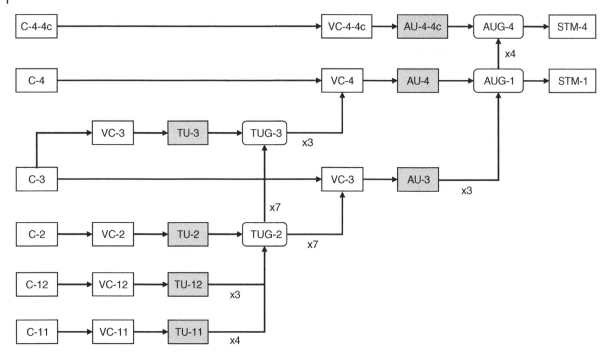

Figure 1.12 SDH multiplexing structure up to STM-4. *Source:* Adapted from ITU-T G.707 [53].

sources, the starting position of any virtual container may be arbitrary as compared to the start of the STM frame. Pointers are used to overcome this problem.

For each such SDH pointer there is a mechanism to adjust small differences between the clocking signals of the multiplexed signals. This mechanism is needed when virtual containers are transported between two different SDH networks, which are relying on different PRC clock sources. The starting positions of virtual containers multiplexed into one single higher-order SDH signal can be nudged by a few bits or bytes forwards or backwards. When moving the start position of VC backwards, a few bits can be taken from the pointer area to carry the virtual container data. When moving the start position of VC forwards, a few stuffing bits are added to the end of the pointer area.

Virtual containers are divided into two classes depending on the location of the VC in the SDH multiplexing hierarchy. The lowest-order virtual containers together with the pointers to VC frame start are called **tributary units (TU)**. A group of multiplexed tributary units are called **tributary unit group (TUG)**. TUG may carry either one high-capacity virtual container or multiple smaller capacity ones. The higher-order virtual containers together with the pointers to the VC frame start are called **administrative units (AU),** and the group of them are **administrative unit group (AUG)**, as shown in Figure 1.13.

SDH and SONET standards define frame structures for different bitrates. Since SONET is based on the PDH DS hierarchy used in American markets, it supports the related additional bitrates missing from the basic SDH. Table 1.3 describes the hierarchy of signals in the different standards:

The highest-order virtual container of SDH multiplexing hierarchy, VC-4 provides 140 Mbits bitrate for its payload to carry an E4 PDH signal. If bigger transmission capacity is needed for one connection, it is possible to concatenate multiple VC-4 virtual containers connected along with the same route as one logical VC-4-Xc channel, if all the equipment along with the path supports such concatenation. The value "X" is the number of concatenated VC-4 signals and is typically four, but larger values may also be used. The concatenation is done so that information about the concatenated VC-4 virtual containers is provided with the AU pointers. It must be noted that the path of the complete VC-4-Xc signal has to be kept always together end-to-end. If any of the VC-4 components or underlying transmission links break due to a fault then the whole concatenated VC-4-Xc signal is lost unless it can be moved to a redundant link.

1.3.4.3   SDH Network Architecture

It is possible to form topologically different types of networks from available SDH network element types (ITU-T G.803 [57]). Typically, at the network edge, a three-shaped structure is used. When moving toward the core of the network there is a need to protect links with redundant ones. Ring topology may be used to

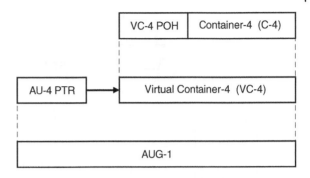

Figure 1.13   Building AUG-1 from C-4. *Source:* Adapted from ITU-T G.707 [53].

provide connection between two elements over two routes. In the very core of the network there is a need to build direct connections between all the elements as mesh, to support provisioning of both capacity allocation and link protection. In the network core, the transmission rates of the links are much higher compared to the network edge. See Figure 1.14 as an example of a small SDH network.

The following types of equipment are used in SDH networks:

- **Regenerator (REG)** is used on long trunks to extend their reach. The regenerator receives the attenuated optical signal, converts it to electric digital format, and regenerates the signal with its original transmission power to the next span of the link.
- **Multiplexer (MUX)** takes care of multiplexing lower-order signals to a higher- order signal to an optical (or an electric STM-1E) transmission link. Depending on its line units, SDH multiplexer is able to multiplex signals of different order and bitrates. For instance, an STM-4 multiplexer is able to compose an STM-4 signal by multiplexing 2 x STM-1, 1 x E4, and 63 x E1 signals into it.
- **Add-drop multiplexer (ADM)** forwards one STM signal from an SDH link to another but locally drops one or multiple lower-order sub-signals from the main STM signal and replaces them with other sub-signals.
- **Digital cross-connect switch (DXC)** is an equipment that is able to connect signals of SDH hierarchy (or the related virtual containers) received from different links to other links without any significant limitations, as depicted in Figure 1.15.

Table 1.3   SDH multiplexing hierarchy.

SDH signal	SONET electrical	SONET optical	Bitrate
STM-1	STS-3	OC-3	155.22 Mbps
STM-4	STS-12	OC-12	622.08 Mbps
STM-16	STS-48	OC-48	2488.32 Mbps
STM-64	STS-192	OC-192	9953.28 Mbps
STM-256	STS-768	OC-768	39 813.12 Mbps
–	STS-1	OC-1	51.84 Mbps
–	STS-9	OC-9	466.56 Mbps
–	STS-18	OC-18	933.12 Mbps
–	STS-24	OC-24	1244.16 Mbps

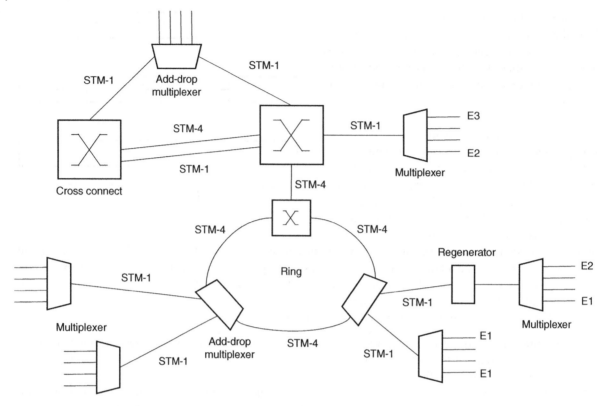

Figure 1.14   SDH network example.

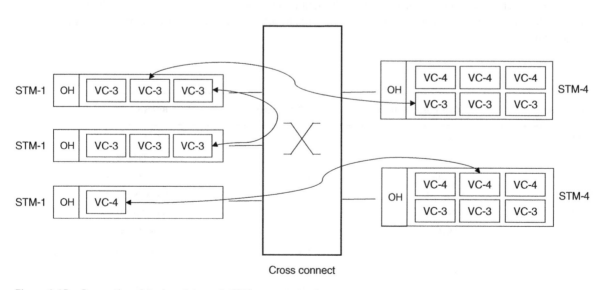

Figure 1.15   Connecting virtual containers in SDH cross-connect.

1.3.5 Microwave Links

1.3.5.1 Standardization of Microwave Link Systems

Microwave links can be used instead of cables on a point-to-point link, for instance, in the following cases:

- In an environment where installation of cables is difficult or expensive, such as rough terrain
- When setting up new connections quickly to areas where infrastructure is underdeveloped

Microwave links typically use a direct line-of-sight radio connection between two parabolic or horn antennas fixedly mounted to a tower or mast. The parabolic antenna has a dish shape, while horn antennas used for very high frequencies look like megaphones. Other types of antennas, such as microstrip patch or MIMO antennas, may also be used. The length of a radio link is from a few kilometers to 50 km. The microwave radio links are able to carry PDH or SDH frames over radio. They can be used as a part of a telecommunications network, for instance, to provide connectivity for a mobile phone base station antenna mast toward the core network.

Microwave links typically use licensed frequencies in frequency area 1–40 GHz. Additionally, the unlicensed 58 GHz band may be used. Each link connection uses a frequency band fixedly allocated for it. The directional antennas used at the end of the link have a direct line-of-sight radio connection between each other.

ETSI standard EN 302 217 [60] specifies some of the characteristics to be used for microwave radio links while it lacks complete standardization of signals transmitted.

1.3.5.2 Architecture of Microwave Radio Links

The architecture of a typical microwave radio link is as follows:

- An inside unit, which contains connectors for PDH or SDH trunks, an optional cross-connect unit, and a network management function.
- An outdoor unit, which contains the radio modem and equipment blocks for processing radio frequencies. The modem generates the frame sent over the radio link and takes care of line coding as well as carrier generation. The radio frequency processing blocks perform signal filtering and amplification.
- Antennas used for transmitting and receiving microwave radio signals over the link.
- Any cabling used between the units.

The signal to be transported is amplified both before transmission and after reception. Since the gain of the amplifier is typically constant, the transmit power is adjusted with adjustable attenuators.

1.3.6 Wavelength Division Multiplexing (WDM)

1.3.6.1 Standardization of WDM Systems

The Internet boom of the 1990s started exponential growth of global data consumption. The capacity needs for transporting data in access and core network rapidly increased. Supporting the requirements for ever higher link bitrates became more difficult than ever before. Extending the SDH signal hierarchy onwards from STM-64 was technically challenging due to the very large frequencies used on both the optical and electrical components. Since the problem was not the transmission rate of a single signal but the capacity available on optical cables, another approach was chosen to solve the issue. **Wavelength division multiplexing (WDM)** was the emerging solution to support multiplexing of different optical wavelengths to one optical cable [2]. Each wavelength was able to carry a single STM-16, STM-64, or Gigabit Ethernet data transport signal. This approach for the optical cables was similar to applying FDM to increase the capacity available over the electric cables. Both FDM and WDM allow transporting similar signals on parallel, using either different frequencies or wavelengths to carry the signals.

When using lasers with stable and narrow enough spectrum together with very precise optical filters, even 10–20 different wavelengths of light could be multiplexed to one single cable. In such a case, the cable could provide a total transmission capacity of tens or hundreds of Gigabits per second, depending on the bitrates of the individual signals multiplexed by WDM. One benefit of the WDM technology is that every separate wavelength could carry whatever upper layer protocol stack desired, making WDM solution a perfect platform for multiplexing different transport and link protocols.

WDM systems typically use bands that are 100 nm wide around the wavelength areas of 1300, 1530, 1550, and 1623 nm. In standards approximately 50 such wavelengths are defined. These wavelengths experience small enough attenuation in an optical cable to support spans with tens of kilometers for WDM links.

The functions of WDM systems are described in the following ITU-T standards:

- G.662 [61]: Characteristics of optical amplifiers and subsystems
- G.680 [62]: Physical functions of optical network elements
- G.694 [63]: Spectral grids for WDM lasers
- G.872 [64]: Architecture of optical transmission networks
- G.873.1 [65]: Linear protection in optical transmission networks
- G.874 [66]: Management of optical transmission networks

In addition to those individual documents, ITU-T has the following series of standards relevant for WDM networking:

- G.65x series recommendations define the characteristics of different types of optical fiber and cables.
- G.66x series covers various standards related to optical amplifiers, components, and subsystems.
- G.709.x [67] series defines architecture and multiplexing hierarchies for optical transport network (OTN). Following the SDH approach, OTN provides OTU frame structures with overhead information supporting forward error correction and performance monitoring. Success of OTN has yet been limited as many operators prefer to transport Gigabit Ethernet over raw WDM channels, instead of investing in the additional OTN layer to increase network reliability [2].

1.3.6.2 WDM System Building Blocks

WDM equipment can be built from the following types of optical components:

- Laser transmitters with narrow and adjustable wavelengths. Narrow enough wavelength can be produced by cooled distributed feedback lasers. The wavelength may be adjusted either mechanically by moving the mirrors of the transmitter or adjusting the filters used after the transmitter. The filters may be adjusted either electrically or with help of sound waves. Instead of adjustable transmitters, a matrix of fixed wavelength transmitters can be used so that the right transmitter instance is selected to produce the desired wavelength.
- Optical couplers, which are able to combine multiple optical input signals into one fiber. Couplers are used within WDM multiplexers and amplifiers.
- Optical receivers, which either convert the received optical power to electricity or modulate a locally generated laser with the received optical pulses so that the resulting microwave signal can be recognized.
- Optical filters and switches, which can be used to pick a specific wavelength from the received spectrum in an WDM add-drop multiplexer. The wavelength passing a filter may be fixedly configured or adjusted by changing the optical properties of the crystals in the filter electrically or acoustically.
- Optical amplifiers, which can amplify all the wavelengths received withinin a certain range of wavelengths. Operation of the amplifiers may be based on semiconductors or using erbium-doped fiber amplifier (EDFA), shown in Figure 1.16. In the latter case, the erbium in the fiber is driven with local laser pump to a state where

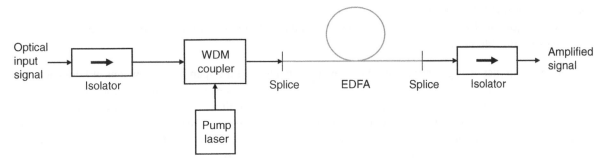

Figure 1.16 EDFA amplifier.

it is very close to emit light. When the amplifier receives light power also from the network, the total power absorbed by erbium makes it to emit light according to the received light pulses. The energy received from the input link is amplified with the energy erbium gets from the local unmodulated laser pump. Amplifiers may be used either at transmitters or receivers to amplify the signal before arriving to the receiver circuitry. It is also possible to use stand-alone amplifiers in the middle of a link to increase the span of the link.

- Wavelength converters, which may use only optical or both optical and electrical signals. The converter may change the received optical signal to an electrical one and thereafter back to an optical signal of another wavelength. Full optical converter may change the wavelength of the signal using a set of modulation, amplification, and optical filtering techniques.

The following techniques have been used for building optical cross-connects:

- It is possible to embed very small mirrors or hatches to a semiconductive board so that their positions can be adjusted electrically. With a very simple approach, such a hatch would either let or stop an optical signal to pass a hole in the board. A more complex implementation is able to adjust mirrors to different positions so that the mirrors can be used to guide the signal to a specific path. With modern technology it is possible to locate more than 100 mirrors to 1 square centimeter of the board.
- A bubble switch uses bubbles in liquid to connect and switch optical signals between optical waveguide fibers. The straight waveguides are attached to a circuit board and cut so that they are terminated at crossings. At the crossings, the waveguides meet with small angle between their paths. In each crossing there is a small pool of liquid. In basic configuration, the light passes the liquid directly so that the direction of the light ray is not changed. The light ray continues propagating over the crossing to a codirectional waveguide. With help of a bubble technique used for inkjet printers, bubbles can be created in the liquid. Such a bubble causes light to be reflected and the ray to change its path. In this way, the path of the light can be changed from one waveguide fiber to another one connected to the same crossing.
- Thermo-optical switch is able to switch the rays of light in the fiber crossings from one fiber to another with help of a thermo-optic phase shifter. When heating up the shifter electrically, the optical properties of the shifter can be adjusted so that the phase of the light wave changes within the shifter. Such a phase change causes the ray to change its path when passing through the shifter.
- Liquid crystal switch uses technique developed for liquid crystal displays. In the switch, the polarization state of a liquid crystal is electrically changed so that the ray is polarized between two directions. Based on its polarization state, the ray will pass one of the polarization filters used in front of the two fibers.
- Optical signal can be converted as an electrical signal before the switch and back to optical signal after the switch. Such a cross connect may be based, for instance, on an SDH cross-connect architecture.

1.3.6.3 WDM Network Architecture

The types of equipment used for WDM networks are similar to those used in SDH networks even if the actual technology is quite different:

- An optical amplifier or regenerator is used on long links to increase the span from a few tens of kilometers up to 100–200 km. A regenerator receives the attenuated optical signal from an upstream link, converts it to digital form, and re-creates the signal to the other downstream link. An amplifier just amplifies the optical signal, but it is much more inexpensive than a regenerator. However, spontaneous emissions from EFDA add extra noise to the signal, so the number of amplifiers that can be added to a link is limited. To reach very long distances, over 600–1000 km, typically regenerators are used instead of amplifiers.
- A multiplexer adds wavelengths from multiple sources to one optical cable.
- An add-drop multiplexer passes one WDM signal from a link to another, but locally drops and adds one or a few wavelengths.
- Optical cross-connect is able to interconnect wavelengths between links. The connections may be either permanent or they can be changed by the management system.

ITU-T recommendation G.872 [64] describes the architecture of optical transport network as a chain of links and connections between them. The links carry multiplexed wavelengths. The end-to-end path of the optical signal is called a trail. Network equipment functions, such as trail termination, adaptation, and connection functions similar to the SDH equipment, are defined in the specifications.

The simplest possible WDM network is one in which a wavelength is transported over only one single link between the source and destination. To build such a network, two WDM multiplexers are needed, one for both ends of the link. The topology of a WDM network may also be a ring so that the ring consists of WDM add-drop multiplexers (ADM) and interconnecting links between them. At each ADM, some of the multiplexed wavelengths exit and enter the ring. The dropped signals are connected to SDH or Gigabit Ethernet links. The other wavelengths continue their paths within the WDM ring. The WDM network can also be built as a partial mesh so that each of the WDM equipment within the network has a sufficient number of direct links with other endpoints.

When the topology of a WDM network becomes more complex, planning and building the network become more difficult. The planner will have cases where a path across many links cannot use a single wavelength, as none of the wavelengths are available on all the links along the path, even when every link may have some spare wavelengths. The solution to such problems is to use wavelength converters or transponders where necessary. When designing such networks, the following challenges must be solved:

- How to reserve necessary capacity and wavelengths from each link of the network?
- How to locate the expensive wavelength converters to the network?
- How to configure the connections of the network so that there are no loops in any unused wavelengths? This is a necessary condition, since loops would cause problems when the signals in the loop would be amplified forever.

1.4 Questions

1 What are the tasks of a telephone exchange?

2 How does the pulse code modulation work?

3 How does a telephone network benefit from signaling?

4   What were the reasons for introducing intelligent networks?

5   Which improvements did SDH bring over PDH?

6   What is meant by wave division multiplexing?

References

1   ITU-T Recommendation G.711, Pulse code modulation (PCM) of voice frequencies.

2   Anttalainen, T. and Jääskeläinen, V. (2015). *Introduction to Communications Networks*. Norwood: Artech House.

3   Anttalainen, T. (2003). *Telecommunications Network Engineering*. Norwood: Artech House.

4   Griffiths, J.M. and Explained, I.S.D.N. (1998). *Worldwide Network and Applications Technology*. West Sussex: Wiley.

5   ITU-T Recommendation E.164, The international public telecommunication numbering plan.

6   ITU-T Recommendation E.170, Traffic routing.

7   ITU-T Recommendation Q.521, Digital exchange functions.

8   Dryburgh, L. and Hewitt, J. (2005). *Signaling System No. 7 (SS7/C7) – Protocol, Architecture, and Services*. Indianapolis: Cisco Press.

9   Goralski, W. (2001). *ADSL & DSL Technologies*. New York: McGraw-Hill.

10   ITU-T Recommendation E.180, Technical characteristics of tones for the telephone service.

11   Sauter, M. (2021). *From GSM to LTE-Advanced pro and 5G: An Introduction to Mobile Networks and Mobile Broadband*. West Sussex: Wiley.

12   ITU-T Recommendation Q.1200, Principles of intelligent network architecture.

13   ITU-T Recommendation Q.1400 Architecture framework for the development of signalling and OA&M protocols using OSI concepts.

14   ITU-T Recommendation Q.1219. Intelligent network user's guide for Capability Set 1.

15   ITU-T Recommendation Q.1224, Distributed functional plane for intelligent network Capability Set 2.

16   ITU-T Recommendation Q.731.3 Stage 3 description for number identification supplementary services using Signalling System No.7 - Calling line identification presentation.

17   ITU-T Recommendation Q.731.4 Stage 3 description for number identification supplementary services using Signalling System No.7 - Calling line identification restriction.

18   ITU-T Recommendation Q.731.5 Stage 3 description for number identification supplementary services using Signalling System No.7 - Connected line identification presentation.

19   ITU-T Recommendation Q.731.6 Stage 3 description for number identification supplementary services using Signalling System no.7 - Connected line identification restriction.

20   ITU-T Recommendation Q.732.2 Stage 3 description for call offering supplementary services using Signalling System No. 7: Call diversion services: - Call forwarding busy - Call forwarding no reply - Call forwarding unconditional - Call deflection.

21   ITU-T Recommendation Q.732.7 Stage 3 description for call offering supplementary services using Signalling System No. 7: Explicit call transfer.

22   ITU-T Recommendation Q.733.1 Stage 3 description for call completion supplementary services using Signalling System No. 7: Call waiting (CW).

23   ITU-T Recommendation Q.733.2 Stage 3 description for call completion supplementary services using Signalling System No. 7: Call hold (HOLD).

24   ITU-T Recommendation Q.733.4 Stage 3 description for call completion supplementary services using Signalling System No. 7: Terminal portability (TP).

25 ITU-T Recommendation Q.734.1 Stage 3 description for multiparty supplementary services using Signalling System No. 7: Conference calling.

26 ITU-T Recommendation Q.735.1 Stage 3 description for community of interest supplementary services using Signalling System No. 7: Closed user group (CUG).

27 ITU-T Recommendation Q.700, Introduction to CCITT Signaling System No. 7.

28 ITU-T Recommendation Q.701, Functional description of the message transfer part (MTP) of Signalling System No. 7.

29 ITU-T Recommendation I.120, Integrated services digital networks (ISDNs).

30 ITU-T Recommendation I.310, ISDN - Network functional principles.

31 ITU-T Recommendation I.324, ISDN network architecture.

32 ITU-T Recommendation I.412, ISDN user-network interfaces - Interface structures and access capabilities.

33 ITU-T Recommendation G.703, Physical/electrical characteristics of hierarchical digital interfaces.

34 ITU-T Recommendation G.704, Synchronous frame structures used at 1544, 6312, 2048, 8448 and 44 736 kbit/s hierarchical levels.

35 ITU-T Recommendation G.705, Characteristics of plesiochronous digital hierarchy (PDH) equipment functional blocks.

36 ITU-T Recommendation G.706, Frame alignment and cyclic redundancy check (CRC) procedures relating to basic frame structures defined in Recommendation G.704.

37 ITU-T Recommendation G.732, Characteristics of primary PCM multiplex equipment operating at 2048 kbit/s.

38 ITU-T Recommendation G.736, Characteristics of a synchronous digital multiplex equipment operating at 2048 kbit/s.

39 ITU-T Recommendation G.796, Characteristics of a 64 kbit/s cross-connect equipment with 2048 kbit/s access ports.

40 ITU-T Recommendation G.754, Fourth order digital multiplex equipment operating at 139 264 kbit/s and using positive/zero/negative justification.

41 ITU-T Recommendation G.702, Digital hierarchy bit rates.

42 ITU-T Recommendation G.775, Loss of Signal (LOS), Alarm Indication Signal (AIS) and Remote Defect Indication (RDI) defect detection and clearance criteria for PDH signals.

43 ITU-T Recommendation Q.705, Signalling network structure.

44 ITU-T Recommendation G.733, Characteristics of primary PCM multiplex equipment operating at 1544 kbit/s.

45 ITU-T Recommendation G.734, Characteristics of synchronous digital multiplex equipment operating at 1544 kbit/s.

46 ITU-T Recommendation G.743, Second order digital multiplex equipment operating at 6312 kbit/s and using positive justification.

47 ITU-T Recommendation G.752, Characteristics of digital multiplex equipments based on a second order bit rate of 6312 kbit/s and using positive justification.

48 ITU-T recommendation G.735, Characteristics of primary PCM multiplex equipment operating at 2048 kbit/s and offering synchronous digital access at 384 kbit/s and/or 64 kbit/s.

49 ITU-T Recommendation G.742, Second order digital multiplex equipment operating at 8448 kbit/s and using positive justification.

50 ITU-T Recommendation G.744, Second order PCM multiplex equipment operating at 8448 kbit/s.

51 ITU-T Recommendation G.745, Second order digital multiplex equipment operating at 8448 kbit/s and using positive/zero/negative justification.

52 ITU-T Recommendation G.753, Third order digital multiplex equipment operating at 34 368 kbit/s and using positive/zero/negative justification.

53 ITU-T Recommendation G.707, Network node interface for the synchronous digital hierarchy (SDH).

54 ITU-T Recommendation G.774, Synchronous digital hierarchy (SDH) - Management information model for the network element view.

55 ITU-T Recommendation G.781, Synchronization layer functions for frequency synchronization based on the physical layer.

56 ITU-T Recommendation G.783, Characteristics of synchronous digital hierarchy (SDH) equipment functional blocks.

57 ITU-T Recommendation G.803, Architecture of transport networks based on the synchronous digital hierarchy (SDH).

58 ITU-T Recommendation G.957, Optical interfaces for equipments and systems relating to the synchronous digital hierarchy.

59 ETSI ETS 300 147, Transmission and Multiplexing (TM); Synchronous Digital Hierarchy (SDH); Multiplexing structure.

60 ETSI EN 302 217, Fixed Radio Systems; Characteristics and requirements for point-to-point equipment and antennas;.

61 ITU-T Recommendation G.662, Generic characteristics of optical amplifier devices and subsystems.

62 ITU-T G.680 Physical transfer functions of optical network elements.

63 ITU-T G.694 Spectral grids for WDM applications.

64 ITU-T Recommendation G.872, Architecture of the optical transport network.

65 ITU-T Recommendation G.873.1, Optical transport network: Linear protection.

66 ITU-T Recommendation G.874, Management aspects of optical transport network elements.

67 ITU-T Recommendation G.709, Interfaces for the optical transport network.

Part II

Data Communication Systems

2

Data over Telephony Line

While Part I of the book described the evolution of the telephone systems, Part II discusses transport of data rather than voice. The telephone network was designed for voice, but it was conveniently available to be used also for remote data communications. Initially, analog modems were used to adapt the properties of data streams to the end-to-end path over the telephone network. Later on, only the subscriber lines remained in use for data traffic. Those lines could be used to carry data with digital subscriber line technologies over the "last mile" to the customer premises. At the operator, the data from the subscriber line was redirected into a dedicated data network, and only the voice traffic was carried over the rest of the telephone network.

2.1 Subscriber Line Data Technologies

2.1.1 Narrowband Analog Modems

Before the Internet became a commercial network during the 1990s, long- distance data connections were provided only by telephone companies. The available options were **leased data lines** or **modem connections** over the traditional analog telephone network. As described in Section 1.1.6, initially the telephone network was designed to transport the frequencies of voice waves transformed to analog electric signals. Consequently, the bandwidth of the telephone network was in the range of 400–3400 Hz. The binary data from a computer could be transported over the telephone network only using signals in that very narrow frequency band. Later on, the pulse code modulation (PCM) process was applied at the operator to convert the analog waveform to a digital 64 kbps signal transported over digital trunks. Nevertheless, without ISDN the subscriber line stayed as an analog link, transporting signals under 4000 Hz frequency, suitable for the PCM process.

The goal of narrowband modem design was to introduce ways to modulate the sine wave carrier with binary information so that the modulated signal could be adapted to the narrow ~3 kHz bandwidth available over the telephone network. In order to do that, the resulting waveform should carry most of the binary information with one fundamental frequency and 1–2 harmonic frequencies, which would be very close to the fundamental frequency. In the local multiplexer or exchange, the PCM process would then be used to convert the analog signal into 64 kbps digital form. In the remote end of the circuit, PCM would convert the signal and back to analog format for the subscriber line. In the end of the line, an analog modem would demodulate the signal to retrieve the binary information from it, as shown in Figure 2.1. For bidirectional transmission, both the modulation and demodulation were done within one single device called a modem (modulator–demodulator), which connects the computer to the subscriber line.

Converged Communications: Evolution from Telephony to 5G Mobile Internet, First Edition. Erkki Koivusalo.
© 2023 The Institute of Electrical and Electronics Engineers, Inc. Published 2023 by John Wiley & Sons, Inc.
Companion website: www.wiley.com/go/koivusalo/convergedcommunications

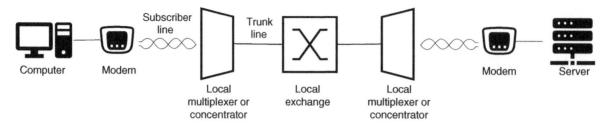

Figure 2.1 Analog modem connection over telephone network.

Table 2.1 Types of standardized narrowband modems.

Standard	Bitrate	Modulation method	Introduced
V.21 [1]	0.3 kbps	FSK	1964
V.22 [2]	1.2 kbps	PSK	1980
V.22bis [3]	2.4 kbps	QAM	1984
V.32 [4]	9.6 kbps	QAM	1984
V.32bis [5]	14.4 kbps	QAM + Trellis	1991
V.34 [6]	28.8 kbps	QAM + Trellis	1994
V.34 [6]	33.6 kbps	QAM + Trellis	1996
V.90 [7]	56 kbps (downstream) 33.6 kbps (upstream)	Digital from service provider to digital telephone network	1998

The narrowband modem essentially converts the binary data into an electric waveform, which has the same frequency characteristics as human voice. The telephone network would then transport those artificial sounds to the other end where another modem converted them back to digital data. If a signal coming from an analog modem was connected to an analog telephone in the remote end (rather than to a modem), the earpiece of the phone would convert the modem signals to audible sound.

Table 2.1 summarizes different types of analog modem signals, as specified by CCITT and later on by ITU-T. Analogy modems were commonly used for more than four decades to provide data connectivity over the telephone network. It is worth noting that half of these standards emerged during the 1990s, when the commercial use of the Internet took off. The last of these analog technologies supported 56 kbps bitrate, which is already quite close to the 64 kbps bitrate provided by DS0 signals, being the upper limit of data rate over a single telephone circuit. For higher bitrates, dedicated data network connections would be needed.

2.1.2 Digital Subscriber Line (DSL) Technologies

The basic idea behind **digital subscriber line (DSL)** technologies was to use the full bandwidth of the copper wire pair subscriber line to increase the bitrates as compared to the analog modem technologies [8]. The frequency area available over the end-to-end telephone network is between 400 and 3400 Hz, but the subscriber line alone is able to support much higher frequencies. DSL technologies use modulation and line coding methods, which take advantage of these otherwise unused frequencies to transmit digital data over the subscriber line. Just as analog modem technologies were improved over four decades, a number of different DSL technologies have been specified since the 1990s. Some of those technologies use the whole bandwidth of the subscriber line whereas others use only that part of the bandwidth that is not used for analog voice. In the latter case, the subscriber line could simultaneously support an analog voice call and a DSL data connection.

Since DSL technologies rely on the bandwidth of subscriber lines being higher than that of the rest of the telephone network, the data had to be separated from the voice signals at the end of the subscriber lines. Only the voice could be carried over the telephone network trunks, and high-speed data needed its own network infrastructure. Such a solution offloads the data connections from telephone exchanges and makes it possible to use higher data bitrates than the 64 kbps provided by the telephone network. When deploying a DSL variant, which supports simultaneous transport of analog voice and DSL data on the subscriber line, a filter device called a **splitter** had to be used on both the ends of the line. At the customer premises, the splitter passes the voice frequencies to an analog phone and the data frequencies to the DSL modem. At the central office or local multiplexer site, the splitter passes voice frequencies toward the telephone exchange and data frequencies to the DSL multiplexer.

Most of the DSL technologies use one pair of wires to transport the signal to both directions. The signals for different directions can be separated in two ways [9]:

- Using frequency multiplexing so that both the directions use their own frequency areas. The drawback of this approach is that either of the directions can use only part of the available bandwidth, which limits the achievable bitrates.
- Using the same frequency areas for both the directions. The drawback of this approach is that when part of the transmitted signal is reflected back from joints in the subscriber line, the echo interferes with the transmission from the remote end since both the signals arrive over same frequencies. Such interference would make reception of the transmission unreliable. The solution is to use echo cancellation techniques (see *Online Appendix A.3.1*) to eliminate the echo. In DSL lines, severe echo can easily be caused by a connection of a two-wire subscriber line to four wires, which was a technique used to extend the analog voice line over long distances between customer premises and the central office.

Table 2.2 provides a summary of different DSL technologies and their characteristics. The data rates and lengths given in the table are approximations, and actual values depend on the distance, quality of cable, and any local interference sources. The earliest DSL technologies were symmetric so that the bitrate is the same for both the directions. Asymmetric DSL technologies provide higher bitrates downstream from network to subscriber and slower rates to upstream. Asymmetric technologies were good for home data connections where the pattern of Internet browsing meant most of the data was being downloaded from the network to the home computer. This is still the mainstream pattern of using the Internet, even while social media use has introduced the need for larger amounts of upstream traffic (Table 2.2).

From these DSL technologies, ADSL was the one in widest use in the early 2000s; VDSL has largely replaced it within the latest decade. Deployment of VDSL has been expanded as operators have rolled out optical cables to customer premises, utility cabinets, or operator-owned premises close enough to customers, so that the twisted pair subscriber line has become short enough for high bitrates. In some cases, the same optical cables may be shared between DSL traffic and the traffic from/to small cellular base station sites close to customer premises. It is worth noting that VDSL can be used also over longer subscriber lines, similar to ADSL deployments, but then the data rates may also stay comparable to ADSL data rates. To increase bitrates from what could be provided by a single subscriber line, capacity of multiple DSL lines can be aggregated for an upper layer protocol by **bonding** the lines. ITU-T has defined DSL bonding for protocols (such as ATM and Ethernet) carried over DSL in the G.998 recommendation series. ADSL and VDSL are further described in the upcoming Sections 2.2 and 2.3, but interested readers can find brief descriptions of older HSDL and SHDSL technologies from the *Online Appendix D*.

When many early Internet service providers started to deploy ADSL over existing telephone subscriber lines, others took a different path and took television network cables into use for home Internet access. Cable modems

Table 2.2 Summary of DSL technologies.

	Bitrate	Subscriber line length	Number of wires	Application
HDSL	1.5/2.0 Mbps	4–5 km	2/3 pairs	The first DSL technology applied for business use
IDSL	0.128 Mbps	6 km	1 pair	DSL variant compatible with ISDN data
SDSL	0.1–1 Mbps	3–4 km	1 pair	Alternative for HDSL with one pair of wires only
SHDSL	1.5/2.0 Mbps	2–3 km	1 pair	Second-generation HDSL technology
ADSL	1.5–8 Mbps downstream 1 Mbps upstream	3–5 km	1 pair	Home data connections for Internet access on parallel with fixed voice telephony
ADSL2	4–12 Mbps downstream 1–3.5 Mbps upstream	3–5 km	1 pair	Like ADSL
ADSL2+	20–48 Mbps downstream 1.4–6 Mbps upstream	1–2 km but lower rates with distances up to 5 km	1/2 pair	Like ADSL
VDSL	13–52 Mbps downstream 1.5–16 Mbps upstream	0.3–1.5 km	1 pair	Home data connections for Internet access, requires optical cable connection close to subscriber premises
VDSL2	300 Mbps downstream 100 Mbps upstream	0.3 km but lower rates with longer distances up to 1 km	1 pair	Like VDSL

were developed to deliver data over coaxial cables installed for cable TV service. This book has its focus in technologies evolved from the telephony network, and the cable modem technology is described in *Online Appendix F.1* as an alternative to DSL.

In the early days of DSL, the data connectivity was provided between DSL multiplexers and the dedicated data network with leased lines. Before long, the Internet service providers started to build their own packet-switched IP networks to reach the subscriber lines. For such access, networks' passive optical networking technologies were often used, as described in *Online Appendix F.2.* The optical networks could be terminated either to homes, buildings, or separate utility cabinets near them. Because of that, such solutions are commonly known as **Fiber to X (FTTx)** technologies, where the X denotes the various types of termination points of the optical network.

2.2 Asymmetric Digital Subscriber Line

Asymmetric Digital Subscriber Line (ADSL) is a term that covers a range of DSL technologies that use different modulation methods (CAP and DMT) but share the following common properties:

- Transport of data over telephone subscriber lines on parallel with an analog voice signal.
- Asymmetric data rates up to 6 Mbps upstream and up to 48 Mbps downstream. The maximum bitrates depend on the specific ADSL variant.

ADSL systems have been standardized by ANSI and ITU-T as follows:

- ADSL specification ANSI T1.413 [10] was published in the year 1998.
- ADSL specification ITU-T G.992.1 [11] was published in the year 1999.

- ADSL2 specification ITU-T G.992.3 [12] was published in the year 2002.
- ADSL2+ specification ITU-T G.992.5 [13] was published in the year 2005.

When referring to ADSL within Section 2.2, the text in many cases applies to all the preceding detailed ADSL system versions. In certain cases, the description is scoped for "basic" ADSL compliant to T1.413 [10] or G.992.1 [11] to distinguish it from ADSL2 and ADSL2+ systems. The distinction between the systems is necessary when describing features that are not generic but specific to an ADSL version.

2.2.1 Architecture of ADSL System

An ADSL system consists of the following devices [9] (shown in Figure 2.2):

- **ADSL modem** or **ADSL Transmission Unit Remote (ATU-R)** is the customer premises equipment (CPE) or a device that converts data between the ADSL format used on the subscriber line and the format used on the local link to the computer. The modem may be an internal modem card installed into the case of a desktop computer, but typically it is an external device with integrated router functionality. In the latter case, the modem is connected to one or multiple computers over wired or wireless local area network links.
- **The ADSL multiplexer** is a network element in operator premises. It is collocated either with a local telephone multiplexer or the local exchange. The ADSL multiplexer terminates the ADSL link of the subscriber line. The **ADSL Transmission Unit Central office (ATU-C)** unit of the multiplexer is connected to the subscriber line with a splitter unit. ATU-C is an ADSL modem integrated to the multiplexer and thus managed by the operator. The multiplexer is connected toward a dedicated data network with one or multiple trunk lines to which the data traffic from subscriber lines is statistically multiplexed (see *Online Appendix A.9.2*). With statistical multiplexing, the capacity of trunk lines is not fixedly allocated between the subscriber lines. Instead, the multiplexer allocates the trunk capacity dynamically for packets transported over the subscriber lines.
- **A splitter** is used at customer premises to divide the frequencies of the subscriber line for the analog telephony and ADSL data. The former is delivered to the phone and the latter to the ADSL modem. At the operator premises, the splitter is used to provide voice frequencies to the local multiplexer or exchange and data frequencies to the ADSL multiplexer. The splitter has one connector for data and another for voice frequencies. For these connectors, the splitter filters out those frequencies that are not needed by the connected device. At the customer premises, the splitter is either integrated into the ADSL modem or is a separate accessory plugged to the telephone wall plug.
- **A network interface device (NID)** is a box that sits between the network managed by the operator and the customer premises wiring managed by the customer. A NID can be either an active or passive one. A passive NID does not need a power supply, while an active NID contains electronics and needs power.

Most often, the ADSL multiplexer, local telephone multiplexer, and splitter are co-located in a street utility cabinet set up for the multiplexers. In that case, the wiring inside of the cabinet can be managed easily, as shown in Figure 2.3.

If local telephone multiplexers are not used, the subscriber lines are terminated at the **main distribution frame (MDF)** within the central office site hosting the local exchange. MDF is used to manage wiring between

Figure 2.2 Structure of an ADSL system.

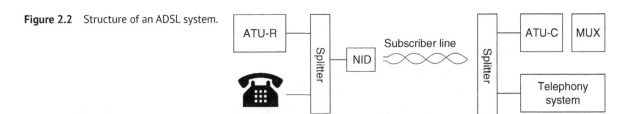

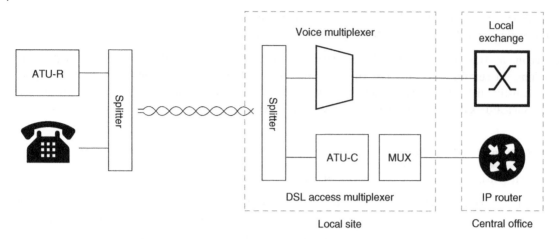

Figure 2.3 ADSL multiplexer located at a local multiplexer site.

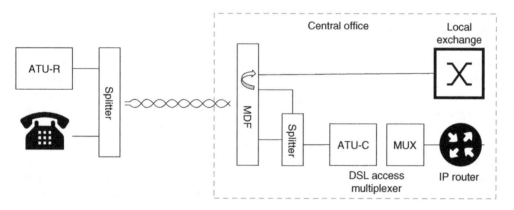

Figure 2.4 ADSL multiplexer in a central office – double wiring case.

the subscriber lines and subscriber line units of the exchange. Typically, the MDF is located close to the exchange so that the wires between the MDF and the exchange can be kept short. If ADSL multiplexers are installed to the central office site, operators must deploy splitters between the MDF and the exchange. The splitter filters ADSL frequencies out of the signal provided to the exchange and provides those ADSL frequencies to the ADSL multiplexer. Depending on the available space on the site, the ADSL multiplexer may be located in the same room as the exchange, in a different room in the same building, or in a separate building. In the latter case, if the splitters are co-located with the ADSL multiplexers in another building, double wiring may be needed between MDF and the splitters. One pair of wires would carry all the signals from the subscriber line via MDF to the splitter and another pair would bring the telephony frequencies back from the splitter to MDF where they are connected to the exchange, as shown in Figure 2.4.

2.2.2 ADSL Modulation Methods

Carrierless amplitude/phase (CAP) was a modulation technique used by the majority of early ADSL systems introduced before 1996. CAP was a proprietary solution of a major ADSL circuit vendor and was never standardized. Operation principles of CAP are similar to QAM modulation, described in *Online Appendix A.4.5*, but the simpler

CAP does not require use of local oscillators and phase locked loops, which add complexity of circuits. With the emerging ADSL standardization, CAP-based systems were soon dropped from the ADSL market.

Discrete multitone (DMT) is the modulation method defined in ANSI standard T1.413 [10] and used also in various ADSL standards of ITU-T from recommendation G.992.1 [11] onwards. DMT uses OFDM multiplexing described in *Online Appendix A.9.1*. In DMT, the 1.104 MHz bandwidth of the subscriber line is divided into 256 subchannels. The bandwidth of each subchannel is 4.3125 kHz. A subchannel is used to transmit an independent subcarrier modulated with QAM. In the context of ADSL, the subcarriers are also referred to as tones or bins. Bitrate of a subchannel can be adjusted by choosing the level of QAM used to modulate its subcarrier, which determines the number of data bits conveyed by a QAM symbol. The choice of the QAM level is done based on the properties of the subchannel, which are determined when initializing the ADSL connection for a subscriber line. The attenuation and noise of each subchannel are measured during the ADSL line startup. The measurement uses predefined training signals, which are transmitted to the subchannels within the line initialization sequence. Each of the modems has its own turn to transmit the training signals on the subchannels, and the modem in the other end compares the received signals to expected baseline values. With this procedure, the receiver may calculate the distortion of the signal on the subchannel. Based on the results, the ADSL modems will agree on the QAM levels to be used for each subchannel. The total bitrate of the ADSL line is the sum of bitrates of each subchannel. In this way, DMT modulation is robust against disturbances caused by different narrowband radio frequency noise sources and attenuation of certain frequencies, as caused by branching the subscriber line.

In Figure 2.5, the number of tones is much lower than the real ADSL deployment, with 265 tones. The figure demonstrates two phenomena typical for DMT:

- The attenuation increases for higher-frequency subchannels. Even if the transmit power of every tone would be the same, the received power is lower for the higher-frequency tones.
- One of the tones is distorted and its power reduced by a reflection of the signal wave from a wiretap. The phase of the weaker reflected wave is just opposite to the phase of the transmitted tone. The affected tone frequency depends on the length and location of the tap.

ADSL subchannels 1–6 occupy frequencies close to those used for analog voice. They may be used for ADSL if the subscriber line is not used for voice calls. The bandwidth of analog voice signals would fit to subchannel 1, but channels 2–6 are often used as a guard band to avoid any possibility of interference between voice and data signals. Either FDM or echo cancellation may be used to separate the DTM upstream and downstream traffic. DTM subchannels 7–39 are always used for upstream traffic, and downstream traffic uses channels 40–250 with FDM or channels 7–250 with echo cancellation. Downstream bitrates up to approximately 10 Mbps can be reached with 265 ADSL subchannels and the 1.104 MHz system bandwidth.

Figure 2.5 DMT modulated signal in the end of the subscriber line. *Source:* adapted from [8].

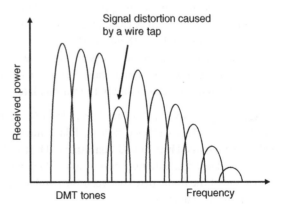

ADSL2+ specification (ITU-T G.992.3 [12]) extends the total downstream bitrates to 24 Mbps by doubling the downstream bandwidth to 2.208 MHz and the number of downstream subchannels to 512. Such high bitrates can be reached on lines shorter than a kilometer, as the higher frequencies used by ADSL2+ experience heavy attenuation on longer lines. Annex M of the specification shifts some of the downstream capacity to upstream by moving the upstream/downstream frequency split from 138 kHz up to 276 kHz. In this way, the maximum upstream bitrate can be increased to 3.3 Mbps while downstream bit rates decrease slightly by changing the direction of a few subcarriers. ADSL2+ supports also bonding the capacity of two subscriber lines to one high-speed ADSL transport path. In that way, even higher bit rates such as 48 Mbps can be achieved on short enough subscriber lines.

2.2.3 ADSL Latency Paths and Bearers

ADSL uses two ways of transporting the data depending on the Quality of Service requirements of the data streams. The two complementary approaches are known as **interleaved** and **fast data.**

- Interleaving is used for data streams that can tolerate delay better than transmission errors. Interleaving distributes a data block to multiple frames to limit the impact of noise burst only to a small fraction of bits rather than the whole block. When the bit error ratio is small enough for a block, forward error correction can be used to restore the correct bit values at the receiver. The operation of interleaved data transport in ADSL is as follows:
  - The user data stream is split to blocks. For each block, a Reed-Solomon forward error correction code (R-S FEC) is calculated and attached to the end of the block. If a small enough number of bytes are corrupted during the transmission of the block, the Reed-Solomon code can be used to correct the transported data, as explained in *Online Appendix A.6.4.*
  - Every protected block is divided into smaller parts, each of which is sent in a different ADSL frame. In this way, one frame contains just a fraction of every consecutive block; thus, the frame carries data interleaved from multiple blocks.
  - If a noise burst corrupts a part of a frame or a complete frame, it impacts only a subset of bytes of each block within the frame. When the complete block has been transported with those frames into which its data has been interleaved, the errors in the received block can in many cases be corrected with the Reed-Solomon forward error correction mechanism.
  - Interleaving causes extra delays for data transport since the bytes of the block are interleaved and transported over many frames. Only after receiving the last one of those frames can the complete data of the block be captured, verified, corrected, and passed to the upper layer protocol.
- Fast data is used to transport data that tolerates errors better than delay. Fast data does not use interleaving and works as follows:
  - The user data stream is split to blocks. For each block, a forward error correction code such as a Reed-Solomon code word is calculated and attached to the data block. The blocks are transmitted one by one in their original order without interleaving.
  - If there are individual bit errors impacting just a few bytes of the block, forward error correction can fix those. Any longer noise burst may cause errors that cannot be corrected. In this case, the application receives some of the data with errors, but with minimal latency.

According to ANSI T1.413 [10] and ITU-T G.992.1 [11] specifications, each ADSL frame is divided into two parts, one of which carries interleaved data and the other fast data, as shown in Figure 2.6. These two parts are called **interleaved buffer** and **fast buffer**. The two buffers introduce two **latency paths** for the data, as the non-interleaved fast data experiences only a small latency while the interleaving process introduces long latencies. The allocation of the frame bytes between the two latency paths and corresponding buffers depends on the configuration of the ATU devices.

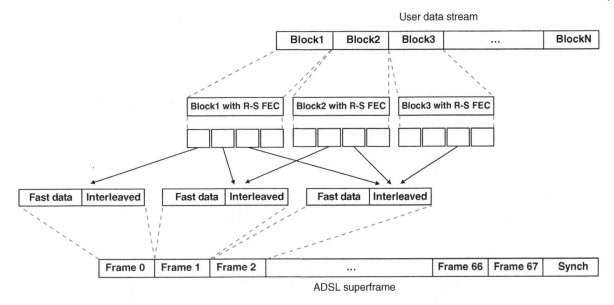

Figure 2.6 ADSL interleaving process.

The bit streams carried by an ADSL connection are divided to at most seven **bearer channels**. The exact number of bearer channels used depends on the transport class applied as described in Section 2.2.6. The four downstream bearer channels are called AS0 to AS3. The three bidirectional bearer channels are called LS0 to LS2. Each of the AS and LS bearer channels carries either only interleaved or fast data. In the ADSL line initialization sequence, the bearer channels taken into use are mapped one by one either to the interleaved or fast buffer area of the frame. The chosen latency path mappings of the bearer channels are indicated with the synchronization control bits of every ADSL frame.

The boundaries of the bearer channels can be found from received frames based on the information that ATU has got about the bitrates used for each channel. When counting bytes from the start of frame (or start of interleaved and fast data buffers), the bytes belong to channel AS0 until the channel A0 bit rate is reached. The following bytes belong to the channel AS1 and so on, for as many AS channels as used for the configuration. After the AS channels the next bytes of the frame belong to LS channels, starting from channel LS0.

While the original ADSL specifications had only two latency paths, ADSL2 specifications (including ADSL2+) generalize the concept of latency path. In ADSL2, the frame can transport one to four different latency paths, their latencies depending on an adjustable interleave depth used for the data blocks on each path. Figure 2.7 shows ADSL2 frame structure with two latency paths and the related mux data frames, subject to the interleaving process. Also, the bearer structure of ADSL2 is different from ADSL. Instead of using the AS and LS bearer channels, ADSL2 frame may carry a maximum of four frame bearers to both directions. Every frame bearer belongs to either an upstream or downstream frame so that upstream frames carry only low-speed frame bearers while downstream frames may carry either low-speed or high-speed frame bearers. The frame bearers used, their bit rates, and mappings to latency paths are all agreed on during the line startup sequence. Mapping rules are very flexible. Either multiple bearers may use one single latency path or a bearer may use its own latency path alone.

2.2.4 ADSL Modem Functional Block Model

The **ADSL transceiver unit (ATU)** functional block model of ADSL specification G.992.1 defines ATU as a chain of the following blocks: multiplexer of LS and AS channels, scrambler, forward error correction block,

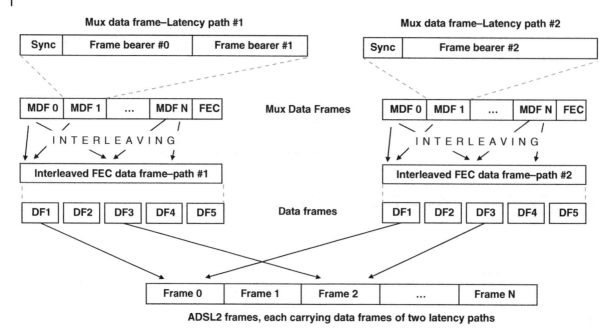

Figure 2.7 Mapping of ADSL2 frame bearers and two latency paths to frames.

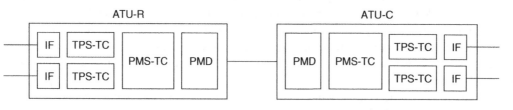

Figure 2.8 ADSL2 functional block model.

Figure 2.9 ADSL2 PMD transmitter block diagram.

interleaver, and blocks for generating QAM symbols to the DMT subcarriers. The ADSL2 specification G.992.3 provides a more structured model, which uses blocks very similar to the SHDSL spec G.991.2.

According to G.992.3, the ADSL2 ATU consists of the following functional blocks, shown in Figure 2.8, and PMD sub-blocks, shown in Figure 2.9:

- **The physical medium dependent (PMD)** ATU contains one PMD block for each pair of wires. PMD block has the following tasks:
  - Clock synchronization of the line needed for DMT QAM symbol timing
  - Data coding and encoding
  - Modulation and demodulation of DMT subcarriers

- – Adding cyclic prefixes between symbols as guard periods to avoid intersymbol interference
- – Echo cancellation and line equalization
- – Link startup and physical layer overhead
- **The physical medium specific transmission convergence (PMS-TC)** block has the following tasks:
  - – Multiplexing of bearer channels for different latency paths
  - – Scrambling and descrambling of the signal to guarantee that the transmitted signal has sufficient number of changes between bit values 0 and 1, to support the PMD clock synchronization
  - – Error detection and forward error correction
  - – ADSL frame structure and synchronization to carry both the user data and ADSL overhead data
- Transmission protocol specific transmission convergence (TPS-TC) block is responsible for adapting ADSL signal toward different payloads, such as ATM, STM, or packet data.
- Interface (I/F) block, with which an ADSL unit connects to the customer data terminal or the computer over a data connection such as LAN or SDH interface

2.2.5   ADSL Frame Structure

The transmission time of an ADSL (G.992.1) frame is 250 µs. An ADSL superframe consists of 68 consecutive ADSL frames and one synchronization symbol. The superframe structure repeats once every 17 ms. Since ADSL data rate is variable but the frame transmission rate is constant, the number of bits per frame depends on the bitrate used on the subscriber line. The bitrate of the line is determined by the number of bits per DMT symbol (i.e., the level of QAM modulation chosen for each of the DMT subchannels).

An ADSL frame consists of the fast and interleaved data buffers. Fast data buffer is transmitted at first and the interleaved buffer follows it, as shown in Figure 2.10.

The first byte of fast data buffer of the ADSL frame is called the fast byte, which is used to transport overhead data of the ADSL protocol, as generated and consumed by the ADSL system. Different overhead data items are mapped to the fast bytes of the 68 frames within an ADSL superframe as follows:

- Frame 0: CRC checksum of the fast buffer of the previous superframe
- Frame 1: Unused indicator bits reserved for future use

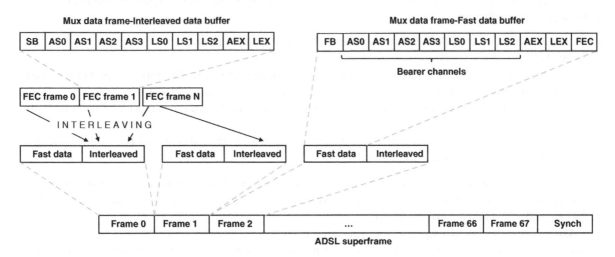

Figure 2.10   ADSL frame structure as specified within ITU-T G.992.1.

Figure 2.11 Structure of a message on the eoc channel.

- Frames 34 and 35 contain indicator bits, which are related to the **operation and maintenance (OAM)** function of the ADSL subscriber line:
  - Loss of signal (los) bit indicates whether the power level of received pilot signal is dropped under a predefined threshold.
  - Remote defect indicator (rdi) bit indicates a loss of frame synchronization. Synchronization is lost when the contents of the synchronization symbols in two consecutive superframes were not as expected.
  - Far end block error (febe) bits indicate if the receiver has detected any bit errors on the frame. This information is given separately for interleaved and fast data.
  - Forward error correction code (fecc) bits indicate whether the receiver has used forward error correction to fix bit errors from a received frame. This information is given separately for interleaved and fast data.
- Frames 2–33 and 36–67:
  - The fast bytes of these frames may be used either as synchronization control "sc" bits or as part of the embedded operation channel (eoc). The purpose of the fast byte in the frame is indicated by its first bit "sc0."
  - The sc (synchronization control) bits describe the used bearer channels and their allocation for fast data.
  - Embedded operation channel (eoc) is used to transport line management data and commands between ATUs, using the format shown in Figure 2.11. Examples of such data items are requests for self-checks, notifications of the device being powered off for power saving, or requests to read or write some data from/to the internal registers, such as
    - o Identifier of the equipment vendor and serial number of the ATU
    - o Results of performed self-tests
    - o Attenuation of the line and margin of its signal-to-noise ratio
    - o Channel configuration used

The first byte of interleaved data buffer of the ADSL frame is the sync byte, which is used within the frames of ADSL superframe as follows:

- Frame 0: CRC checksum of the interleaved buffer of the previous superframe
- Frames 1–67:
  - The Sync Byte of these frames may be used either as synchronization control "sc" bits or as part of the ADSL overhead channel (aoc). The purpose of the sync byte in the frame is indicated by its first bit "sc0."
  - The sc (synchronization control) bits describe the bearer channels used and the allocation of them for interleaved data.
  - The aoc channel is used by ATUs to negotiate the number of bits transported over each DMT subchannel and the corresponding bit rates per subchannel.

The aoc channel is carried over the sync byte only when none of the bearer channels are mapped to the interleaved data buffer. Otherwise, the sync byte is used for synchronization control and the aoc channel is carried over a LEX byte within the interleaved data buffer.

As ADSL2 does not support a separate fast data path, the mapping of the overhead channel to ADSL2 frame is organized in a different way. In ADSL2, the shared overhead channel consists of octets associated with each of the latency path functions. The data rates and number of octets used for a message-based overhead channel are negotiated during the initialization sequence of the line. The overhead channel within a latency path can carry the following parts of the overhead, over a repeating sequence of sync octets transported over the path:

- The CRC portion used to protect overhead data
- The bit-oriented portion used to optionally transport the 8 kHz network timing reference to ATU-R and indicate defect conditions such as loss of signal or loss of timing reference
- The message-oriented portion to carry management messages, with a message structure derived from the HDLC protocol described in Chapter 3, Section 3.1.3.3

2.2.6 ADSL Bearers and Transport Classes

ADSL transport class as specified in ANSI T1.413 [10] and ITU-T G.992.1 [11] is a concept to determine how many and what kind of bearer channels are used on the line and the bitrates provided by those channels. A transport class can be chosen to provide either lower or higher bit rates, as fit for the characteristics of the subscriber line. The shorter the subscriber line is, the higher bitrate transport class can be used.

Usage of the AS0 bearer channel is mandatory for every transport class, but other bearer channels may be used within the constraints of the chosen transport class. The bitrates of AS bearers may follow one of the following regional options shown in Table 2.3:

Tables 2.4 and 2.5 provide a summary for the ADSL transport classes as defined for different markets.

Note: In the tables of this chapter, the term ANSI means the markets where the transported trunk signals are typically based on the 1.536 Mbps T1 structure, and ETSI means markets where the trunk signals use the 2.048 Mbps E1 structure.

The bitrate of each bidirectional LS channel can be 160, 384, or 576 kbps.

According to the standards, in each transport class it is possible to configure the bearer channels in a few different ways. For instance, when using transport class 2M-2 the device may use the AS channels in the following two ways:

- AS0 with 4.096 Mbps bitrate
- AS0 and AS1 with 2.048 Mbps each

Table 2.3 ADSL AS bearers.

Bearer	ANSI bitrate Mbps	ETSI bitrate Mbps
AS0	1.536, 3.072, 4.608, or 6.144	2.048, 4.096, or 6.144
AS1	1.536, 3.072, or 4.608	2.048 or 4.096
AS2	1.536 or 3.072	2.048
AS3	1.536	–

Table 2.4 ADLS ANSI market transport classes.

Class (ANSI)	Maximum bitrate downstream Mbps	Available downstream bearer channels	Maximum bitrate of bidirectional data bearers	Available bidirectional bearer channels
1	6.144	AS0–AS3	640	LS0–LS2
2	4.608	AS0–AS2	608	LS0 and (LS1 or LS2)
3	3.072	AS0–AS1	608	LS0 and (LS1 or LS2)
4	1.536	AS0	176	LS0–LS1

Table 2.5 ADLS ETSI market transport classes.

Class (ETSI)	Maximum bitrate downstream Mbps	Available downstream bearer channels	Maximum bitrate of bidirectional data bearers	Available bidirectional bearer channels
2M-1	6.144	AS0–AS2	640	LS0–LS2
2M-2	4.096	AS0–AS1	608	LS0 and (LS1 or LS2)
2M-3	2.048	AS0	176	LS0–LS1

Table 2.6 ADSL ATM transport classes.

Class (ATM)	Maximum bitrate downstream Mbps	Available downstream bearer channels	Maximum bitrate of bidirectional data bearers	Available bidirectional bearer channels
1	6.928	AS0	64	LS0
2	5.196	AS0	64	LS0
3	3.464	AS0	64	LS0
4	1.732	AS0	64	LS0

The standard defines transport classes also for the case where the full capacity of ADSL signal is allocated for ATM cell transport, without dividing it to multiple bearer channels. In this case, the transport classes are specified as shown in Table 2.6:

ADSL2 specification ITU-T G.992.3 [12] does not use the concept of transport class, as its bearer structure is more flexible than that of ADSL. ADSL2 line has one to four frame bearers, which can be flexibly mapped to 1–4 latency paths. The achievable total transport bi rate over the line can be freely allocated between those frame bearer channels that are taken into use during the ADSL2 line initialization procedure.

2.2.7 ADSL Line Initialization

This chapter describes the initialization sequence of an ADSL line as per T.413 [10] and G.992.1 [11] specifications and an ADSL2 line as per G.992.3 [12]. The line initialization sequences of these two ADSL versions follow similar high-level phasing. The details inside of phases are, however, quite different as the frame structures of the protocol versions differ and since new improved initialization methods have been developed for the newer ADSL versions. The following description is generalized from both "basic" ADSL and ADSL2 procedures and omits various details, but a few major differences between the two protocol versions are highlighted.

Initialization of an ADSL line is done when the ATU modems in both ends of the line have been switched on. In the line initialization, the modems exchange signals of predefined lengths in a tightly synchronized sequence. They adjust their transmit and reception parameters for the line based on measurements done for the received signals. The initialization is done over four or five phases as follows:

1) Activation and acknowledgment
   - In the handshake phase, both the ATUs detect the start of the initialization sequence and reach clock synchronization. The ATU-C decides which of the modems shall provide the clock frequency. Typically, ATU-C provides the clock to which ATU-R synchronizes itself. In ADSL2, ATUs also exchange messages within the handshake steps to tell each other the supported capabilities and to select specific operation modes. The handshake procedures are defined in ITU-T recommendation G.994.1 [14].

2) Channel discovery (ADSL2 only)
- In this phase, the ATUs may perform a coarse timing recovery, channel probing, and power cutback. The cutback lowers the power spectral density (PSD) of the signals, which means the level of power distributed over different ADSL2 subcarriers. Its purpose is to avoid causing excessive interference to other lines in adjacent wires.

3) Transceiver training
- In the training phase ATUs transmit training signals in turns. The known power spectral density properties of a training signal provide a reference against which the device on the other end of the line is able to:
  - Measure the power of the training signal received over each subcarrier and adjust the power of transmitted tones according to the attenuation determined separately for each subchannel.
  - Adjust operation parameters of receiver amplifier, equalizer, and echo canceller circuits. An example of such parameters is the automatic gain control used to maintain a suitable signal amplitude at amplifier output.

4) Channel analysis
- At the analysis phase, the exchange of training signals is continued to estimate the signal-to-noise ratio on every subcarrier being used.
- In this phase, ATUs provide proposals for the transport configuration (in ADSL) or framing configuration (in ADSL2) to be used. This information is protected with CRC checksum.
  - With "basic" ADSL, the ATU devices exchange an initial proposal of four configuration alternatives for the transport formats and data rates (QAM levels) as well as the number of parity bits used per QAM symbol for both interleaved and fast data paths.
  - With ADSL2, the ATU devices exchange the following proposals: how to divide the full bitrate of the line up to frame bearer channels and how to configure basic physical layer parameters of the subcarriers, such as QAM levels and related noise margins.
- With "basic" ADSL, in the analysis phase the ATUs also exchange description of modem properties covering items such as vendor names, abilities of using trellis coding, echo cancellation, and different optional frame structures. Note that the data structure defined in ITU-T G.992.1 [11] lacks some pieces of information specified in T1.413 [10]. With ADSL2, the exchange of properties is part of the initial handshake, rather than the channel analysis phase.

5) Exchange (of transport parameters and configuration)
- In this phase, the ATUs exchange information collected and derived during the channel analysis phase, such as subcarrier characteristics, estimated attenuation, maximum depth of interleaving used, coding gains, or performance margins.
- ATUs provide further proposals or information related to transport configuration (in ADSL) or framing configuration (in ADSL2) to be used on the channel. Exchange of this information is protected with CRC checksum.
  - With "basic" ADSL, the ATU devices exchange the second proposal of four configuration alternatives for the transport formats, data rates, and QAM levels used. Compared to the first proposal given in the analysis phase, the second proposal has been optimized for the subscriber line, based on its properties measured during the analysis. In the end of the negotiation, the ATUs will agree on the single configuration chosen from the four options provided.
  - With ADSL2, the ATU devices exchange the following pieces of configuration: mapping between frame bearers and latency paths, parameters for each latency path, such as interleaving depth and overhead channel configuration, as well as the physical layer parameters for subcarriers.

After completing the initialization sequence, ATU devices start their normal operation. This phase is called showtime, referring to the start of user data transport.

The initialization process of ADSL2+ as described in ITU-T G.992.5 [13] is essentially similar to ADSL2 initialization at the high level. Once again, differences are in the details.

2.3 VDSL

2.3.1 Architecture and Bands of VDSL System

Very high-speed digital subscriber line (VDSL) supports highest bit rates of all the DSL variants. VDSL systems have been standardized by ITU-T as follows:

- VDSL specification G.993.1 [15] was published in 2001.
- VDSL2 specification G.993.2 [16] was published in 2006.

The VDSL system bandwidth is 12 MHz. VDSL uses frequency division duplex (FDD) to separate upstream and downstream transmissions. The VDSL frequency band is divided into four subbands: DS1, US1, DS2, and US2, as shown in Figure 2.12. The two DS bands are used for the downstream subcarriers and the two US bands are used for upstream subcarriers.

The band plan of VDSL2 is identical to VDSL up to 12 MHz. Usage of a VDSL2 extra downstream band between 12 and 30 MHz is subject to the regional profiles defined within the annexes A-C of ITU-T recommendation G.993.2.

Since VDSL standards have been created from the earlier ADSL standards, their specifications share many common features, such as the following:

- The basic architecture of the VDSL system is similar to the ADSL system. The modem in customer premises is called VTU-R and the corresponding modem unit at the network premises is called VTU-O. The VTU-O multiplexes and adapts the VDSL signal from the customer line to an optical trunk. The functional block model of VDSL is identical to ADSL2. Both the models have PMD, PMS-TC, TPS-TC, and I/F blocks with similar responsibilities per block.
- Like ADSL, VDSL is an asymmetric technology with downstream bitrates higher than upstream rates. VDSL uses the same DMT modulation method and similar subcarrier structure as ADSL. VDSL subcarrier spacing is the same 4.3125 kHz as used for ADSL
- Frequencies used for VDSL are those above the analog voice channel. The VDSL frequency band starts from 138 kHz. Splitters are used at both ends of the subscriber line to separate VDSL frequencies from the analog voice frequencies. However, to avoid interference to any ADSL or analog voice signals carried in other cables of the same cable bunch, VDSL uses low power for its subcarriers below 1.1 MHz.
- Both VDSL and ADSL use interleaving and Reed-Solomon forward error correction codes to protect signals against transmission errors. The VDSL specifications define convergence functionality for ATM, STM, or packet-based payloads, as ADSL specifications do.

The original VDSL standard was based on the ADSL G.992.1 [11] standard with which VDSL G.993.1 shares certain structural properties. The VDSL2 standard was based on the work done for ADSL2, so VDSL2 also inherits some of its features from the G.992.3 [12] specification of ADSL2. Examples of the similarities between the first and second versions of ADSL and VDSL standards are as follows:

- The phases of ADSL and VDSL line initialization are very similar to each other. For the second versions of both standards, ADSL2 and VDSL2, the structure was changed by introducing a new phase called channel discovery to be executed right after the initial handshake process.

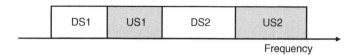

Figure 2.12 VDSL band allocation.

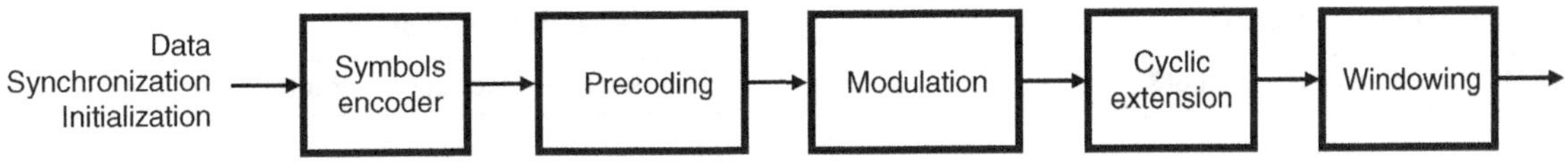

Figure 2.13 VDSL2 PMD transmitter block diagram.

- Both the original ADSL and VDSL versions have only two latency paths: fast and interleaved data. Only one single interleave depth can be used on a VDSL line, as set by the network management system.
- The second versions of both standards ADSL2 and VDSL2 introduced flexible mapping between bearer channels and latency paths with interleaving depths adjustable at line initialization. Both the standards also support dynamic adjustment of the interleaving depth while the user data transmission is already enabled.

The most significant differences between VDSL and ADSL are as follows:

- The bitrates supported by VDSL depend on the VDSL version and properties of the subscriber line. VDSL downstream rates range between 13 and 300 Mbps and upstream bitrates between 1.5 and 100 Mbps, when both VDSL versions are taken into account. With subscriber lines shorter than those typically used for ADSL, the bitrates of VDSL are significantly higher than the maximum ADSL bitrates. To keep the promise of VDSL high bitrates, the span of twisted pair wires should be kept short and an optical cable should be available at either the customer premises or a utility cabinet which is not too far away. The VTU-O unit connects the optical cable with the subscriber lines, so it is installed in the room in which these cables are terminated. In terraced houses or apartment buildings, this typically is a telecom room from where the apartment-specific twisted pairs are built to individual apartments. In such an environment, one single optical cable may serve a large base of dwellers.
- To achieve high bitrates, VDSL uses a wider system band and larger number of subcarriers than ADSL. The total bandwidth of the VDSL channel is up to 12 MHz but is lower for the longer spans of subscriber lines. The specific set of subcarriers used over a subscriber line is agreed on line initialization and depends on the device configuration and measured values of signal-to-noise ratio on different subchannels. The number of VDSL subcarriers taken into use for a subscriber line varies between 256 and 4096 as powers of two. Higher-frequency subcarriers can be used on short enough subscriber lines, since high frequencies attenuate quicker than the lower frequencies. To increase downstream bitrates even further, VDSL2 may use one additional frequency band between 12 and 30 MHz over very short subscriber lines. With the first version of VDSL, the relationship between downstream bitrates and length of the subscriber line was approximately as follows:
  - 52–55 Mbps over 0.3 km
  - 26–28 Mbps over 1 km
  - 13–14 Mbps over 1.5 km
- Vectoring is a feature that may be used with VDLS2 to reduce far end crosstalk between different lines connected to a single VTU-O. ITU-T recommendation G.993.5 [17] specifies a downstream vectoring scheme where VTU-O PMD precoding blocks (see Figure 2.13) coordinate precoding for a group of VDSL2 lines combined into a single cable bunch. When precoding a signal to one of the lines, the precoder takes into account symbols sent over other lines and modifies the generated signal to compensate crosstalk in the far end.

2.3.2 VDSL Frame Structure

The VDSL frame consists of the information and overhead bytes carried over all the used VDSL subcarriers within one DMT symbol slot. The frame has either both the fast and slow (interleaved) data latency paths or just the latter if no fast data is transported over the line. VDSL frames are generated with the following process as shown in Figure 2.14:

The stream of information bytes to be transported is split to packets to which VDSL overhead bytes are added. Dummy padding bytes may be added to each packet to align their sizes. The data packets are scrambled and

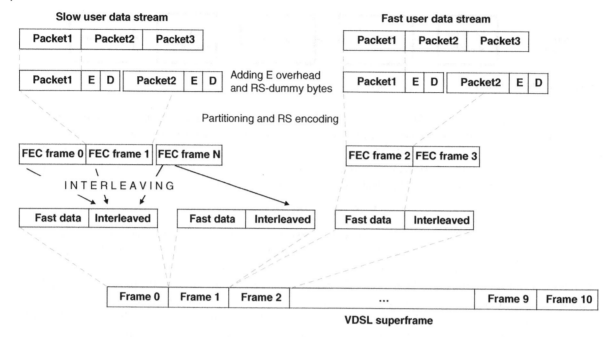

Figure 2.14 VDSL frame structure for slow and fast latency paths.

protected with Reed-Solomon codes. The packets are partitioned into constant size forward error correction (FEC) frames. Up to this point, the processing is similar to both fast and interleaved data paths but diverge thereafter. Packets of the slow data path are processed by interleaver, while fast packets are not interleaved. Thereafter, the data from the interleaved and fast latency paths are merged to VDSL frames so that in each frame there is a segment of fast data followed by a segment of interleaved data. The VDSL frame is carried by DMT symbols transmitted within a single slot on the parallel subcarriers. The number of bits per symbol is determined separately for each latency path based on the system configuration.

As mentioned earlier, VDSL2 supports a flexible mapping between bearer channels and latency paths, which makes VDSL2 frames fundamentally different from the first version of VDSL. The process of generating VDSL2 frames is shown in Figure 2.15:

- A mux data frame (MDF) is built from VDSL2 overhead octets followed by a number of octets for each bearer channel mapped to one single latency path. The number of octets per bearer channel within the MDF is agreed on the line initialization.
- A number of consecutive MDFs are scrambled and protected with Reed-Solomon coding. The resulting structure is called RS code word. The number of MDFs and RS redundancy octets per RS code word is agreed on during the line initialization.
- Interleaved RS code words from different latency paths are multiplexed to the VDSL2 data frames. The VDSL2 data frames are carried by DMT symbols sent within a single slot on parallel subcarriers. The number of bits per symbol for each latency path are determined dynamically during the line initialization.

2.3.3 VDSL Overhead

VDSL overhead consists of the following channels:

- The VDSL overhead control channel (VOC) transfers VDSL link activation and configuration messages between the VTUs during the initialization. VOC is transported over the overhead bytes within the slow data packets.

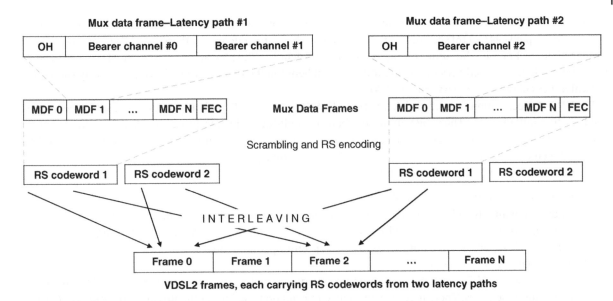

Figure 2.15 VDSL2 frame structure for two latency paths.

- The embedded operations channel (EOC) is used for network management messages transported between VTUs with HDLC link protocol. EOC messages are carried within the overhead bytes of both fast and slow data packets. EOC is used by the network managements to reach VTU-R over the subscriber line via VTU-O.
- Indicator bits (IB) are used to transport state indications (such as loss of signal) of lines and paths, potentially requiring immediate actions at the other end.

VDSL2 overhead carries only EOC and IB channels. Instead of the VOC channel, VDSL2 uses a special operations channel (SOC) to transport line initialization and diagnostics messages with HDLC. The SOC channel is carried over the DMT symbols transmitted during the initialization and loop diagnostics phases. As SOC messages are used only for initialization and diagnostics, the SOC channel is inactive most of the time in normal operation. Due to that, the SOC channel does not need any capacity reservation after the initialization, when the user data is already transported over the line. After the initialization has been completed, the SOC channel is used only when entering the loop diagnostics mode.

VDSL2 specification defines a mapping of diagnostics SOC bits into the DMT symbols of different subchannels. These bits are used for diagnostics when the SOC channel is reactivated in the diagnostics mode. Diagnostics may be used to sort out any problems on the line. As the problems may impact the transmission, all SOC messages are sent with high redundancy. Every symbol has a single SOC information bit, the value of which is repeated in five consecutive DMT symbols for increased robustness.

2.3.4 VDSL Line Initialization

The details of the VDSL line initialization process depend on the case for which the line activation is done. There are different variants of the process to satisfy the startup time requirements specified for the following four cases:

- Cold start is the first activation of the line or activation after significant changes in the line conditions. A cold start must be completed within 10 seconds.
- Warm start is an activation after power down so that line conditions have not significantly changed from the latest activation. A warm start must be completed within 5 seconds
- Resume-on-error is an activation after losing synchronization during transmission and must be completed within 300 ms.
- Warm resume is an activation from the dynamic power savings state and it must be completed within 100 ms.

This chapter describes the cold start initialization process of a VDSL line as per ITU-T G.993.1 [15] and VDSL2 line as per G.993.2 [16] specifications. The line initialization sequences of these two VDSL versions follow similar high-level phasing but differ on the phase-specific details. While VDSL was based on ADSL and VDSL2 based on ADSL2, the differences in the line initialization between the two versions of VDSL are partly similar to the differences between the two versions of ADSL.

The VDSL2 specification G.993.2 [16] defines specific SOC bits of DMT symbols over different subcarriers to be used for line initialization. Those bits are used when transmitting various training signals during the channel discovery and training phases. For channel analysis and exchange phase the line constantly transports MEDLEY signals, which contain those SOC bits as agreed in the line training phase.

The cold start of a VDSL line is done as follows:

1) Activation – ITU-T G.994.1 [14] handshake
   - Handshake with which both the VTUs detect the start of the initialization sequence and reach the clock synchronization. VTU-R derives its clock from the signals sent by VTU-O. VTUs also exchange messages to describe their supported capabilities and to select specific operation modes.
2) Channel discovery (VDSL2 only)
   - In this phase, the VTUs may perform timing recovery, set pilot tones, and establish the SOC to exchange initialization messages. Thereafter, they exchange information used to set the initial power spectral density for the line and the values of other parameters needed for the training phase.
3) Training
   - During the training phase, VTUs exchange training signals in turns. The known power spectral density properties of a training signal provide a reference against which the device on the other end of the line is able to make adjustments:
     - VDSL: Adjust timing advance; exchange capability information for bands used and PSD limits; update gains used for receiver amplifiers; set PSD levels for subcarriers and optionally train the echo canceller.
     - VDSL2: Adjust the operation parameters of equalizer and echo canceller as well as the timing advance used.
4) Channel analysis and exchange
   - In the analysis phase, VTUs exchange further training signals to estimate the signal-to-noise ratio on each subchannel.
   - The following steps are used in the analysis and exchange phase of VDSL to exchange messages over the VOC channel:
     - At first, the VTUs exchange information about the supported PMS-TC settings for interleaving and Reed-Solomon coding as well as the supported PMD values, such as power and QAM levels. They also describe the number of bytes per frame that can be used for EOC and VOC channels.
     - Thereafter, the VTUs agree on the contracts for PMS-TC configuration used on the line. The configuration covers the data rates used for both fast and interleaved paths as well as the final settings used for both the interleaver and Reed-Solomon coding. The contract is agreed so that at first VTU-R sends its contract specification after which VTU-O returns its own proposed specification. VTU-R calculates the SNR margins based on the contract as proposed by VTU-O and sends them back to VTU-O. If VTU-O is happy for the margins, the negotiation is completed, but otherwise VTU-O will send another contract proposal to VTU-R and the process continues.
     - Finally, the VTUs exchange configuration of the physical PMD layer such as the number of bits and gains to be used for each of the subcarriers.
   - The following steps are used in the analysis and exchange phase of VDSL2 to exchange messages over the SOC channel:
     - Both VTUs exchange their supported capabilities such as data rates for different latency paths, interleaver parameters, and transmission requirements such as SNR margins.

- Thereafter, VTUs exchange the final PMS-TC block configuration for both upstream and downstream, describing the bearer channels used and their mapping to latency paths. Various properties of latency paths such as interleaving depth and number of overhead channel octets are agreed on. TPS-TC convergence functions for ATM, STM, or packet-based payloads are also taken into use.
- Finally, the VTUs exchange configuration of the physical PMD layer, such as the possible use of trellis coding, the QAM levels, and gains to be used per each subcarrier.
- After completing the initialization sequence, the VTU devices start the normal operation mode known as showtime, in which the user data transport can take place.

2.4 Questions

1 Why were analog modems used for remote data connectivity?

2 Please list the components of an ADSL system.

3 What is the difference between ADSL fast and interleaved data?

4 How does ADSL2 latency path design differ from the ADSL design?

5 Which transport classes does ADSL have for ANSI markets?

6 How are ADSL lines initialized?

7 What is the main difference between VDSL and ADSL?

References

1 ITU-T Recommendation V.21, 300 bits per second duplex modem standardized for use in the general switched telephone network.

2 ITU-T Recommendation V.22, 1200 bits per second duplex modem standardized for use in the general switched telephone network and on point-to-point 2-wire leased telephone-type circuits.

3 ITU-T Recommendation V.22bis, 2400 bits per second duplex modem using the frequency division technique standardized for use on the general switched telephone network and on point-to-point 2-wire leased telephone-type circuits.

4 ITU-T Recommendation V.32, A family of 2-wire, duplex modems operating at data signalling rates of up to 9600 bit/s for use on the general switched telephone network and on leased telephone-type circuits.

5 ITU-T Recommendation V.32bis, A duplex modem operating at data signalling rates of up to 14 400 bit/s for use on the general switched telephone network and on leased point-to-point 2-wire telephone-type circuits.

6 ITU-T Recommendation V.34, A modem operating at data signalling rates of up to 33 600 bit/s for use on the general switched telephone network and on leased point-to-point 2-wire telephone-type circuits.

7 ITU-T Recommendation V.90, A digital modem and analogue modem pair for use on the Public Switched Telephone Network (PSTN) at data signalling rates of up to 56 000 bit/s downstream and up to 33 600 bit/s upstream.

8 Anttalainen, T. and Jääskeläinen, V. (2015). *Introduction to Communications Networks*. Norwood: Artech House.

9 Goralski, W. (2001). *ADSL & DSL Technologies*. New York: McGraw-Hill.

10 ANSI T1.413, Network and Customer Installation Interfaces — Asymmetric Digital Subscriber Line (ADSL) Metallic Interface.

11 ITU-T Recommendation G.992.1, Asymmetric digital subscriber line (ADSL) transceivers.

12 ITU-T Recommendation G.992.3, Asymmetric digital subscriber line transceivers 2 (ADSL2).

13 ITU-T Recommendation G.992.5, Asymmetric digital subscriber line 2 transceivers (ADSL2)- Extended bandwidth ADSL2 (ADSL2plus).

14 ITU-T Recommendation G.994.1, Handshake procedures for digital subscriber line transceivers.

15 ITU-T Recommendation G.993.1, Very high speed digital subscriber line transceivers (VDSL).

16 ITU-T Recommendation G.993.2, Very high speed digital subscriber line transceivers 2 (VDSL2).

17 ITU-T Recommendation G.993.5, Self-FEXT cancellation (vectoring) for use with VDSL2 transceivers.

3

Data Network Technologies

The digital modem technologies described in the Chapter 2 provide the physical layer and a part of the link layer functionality to be used for fixed Internet access over a subscriber line. This chapter introduces key protocols used on the link, network, and transport layers of OSI (open systems interconnection) model. Some of these protocols are used over the DSL (digital subscriber line) connection and others within a local network, the Internet service provider network, or Internet core network. The topics elaborated in Chapter 3 are related to three different areas of data protocols:

- Data link protocols, according to the OSI model, are used to transport data between two adjacent devices connected by a single link. A number of different link protocols are briefly described. Ethernet, wireless local area network (WLAN), and point-to-point protocols (PPP) are used to carry transmission control protocol/Internet protocol (TCP/IP) packets over various types of physical links such as local Ethernet cables or WLAN radio connections, DSL access lines, as well as wide area networks (WAN) with optical synchronous digital hierarchy (SDH) or WDM (wavelength division multiplexing) structures. Another high-link data level control (HDLC) protocol is introduced, mainly because of its role as an ancestor of a few link protocols used for ISDN (integrated services digital network), local area networks (LANs), and second-generation mobile networks. Additionally, HDLC is used to transport DSL management messages, as described in Chapter 2.
- Switching protocols extend the link layer functionality from a single physical link between two adjacent devices to a virtual link over a whole network. Such an extension is done with a concept of virtual connection. From the user point of view, the virtual connection appears as a single contiguous link between its endpoints. But actually, the virtual connection is a chain of separate links connected by switching devices between them. The switching devices forward data link layer frames from one link to another, based on a locally relevant link identifier within the frame. This makes switching quick and efficient, but the price is paid with the effort needed to maintain the virtual connections between the physical links. Frame relay and asynchronous transfer mode (ATM) were early examples of such switching protocols, which were used in some limited contexts, such as third-generation mobile networks. Multiprotocol label switching (MPLS) was a step forward as it provided a good solution for virtual connection management. MPLS is currently in wide use within the Internet and enterprise core networks.
- Network and transport protocols are used to route and carry user data flows between terminals and servers of different kinds. TCP/IP protocol suite is the de facto data networking and transport protocol used on OSI layers 3 and 4. The Internet traffic consists of IP protocol packets, which carry various application-specific protocols and messages. Both the fixed and mobile networks have evolved from voice only networks to data centric networks where most of the network traffic is transported with TCP/IP protocols. In 4G and 5G mobile networks, voice is just an additional data service provided on top of TCP/IP stack as Voice over IP. Protocols of the TCP/IP suite are also used to transport both end user data flows and the signaling messages used for network control.

Converged Communications: Evolution from Telephony to 5G Mobile Internet, First Edition. Erkki Koivusalo.
© 2023 The Institute of Electrical and Electronics Engineers, Inc. Published 2023 by John Wiley & Sons, Inc.
Companion website: www.wiley.com/go/koivusalo/convergedcommunications

The universal usage of TCP/IP for various purposes is the reason why knowledge of TCP/IP is essential for understanding modern communications networks. This book provides insight for both Internet protocol (IP) versions 4 and 6, as well as the key protocols of TCP/IP suite making the Internet fully functional.

In the end of the chapter, an end-to-end network scenario is provided to describe how protocols of different OSI model layers work together to enable a Web browsing session from a home computer connected to the Internet.

3.1 Data Link Protocols

3.1.1 Ethernet

3.1.1.1 Ethernet Standardization

Ethernet was born as a local area network (LAN) technology in the 1970s. The very first Ethernet standard was published in 1980 by an ad hoc industry consortium. LAN is a local network which connects end user **host devices** (such as desktop or laptop computers) with the information technology (IT) infrastructure, such as local **servers** of different kinds. LAN can also provide connectivity for the hosts toward external services via routers or gateways also connected to the LAN. During the 1980s, Ethernet was not the only LAN solution, but competing systems such as token ring and token bus took a share of the wired LAN market. Since its early days, Ethernet has become the dominant LAN standard, and the protocol has experienced many functional changes. The Ethernet family of protocols is currently used both in the contexts of LAN and WAN.

The Institute of Electrical and Electronic Engineers (IEEE) took over the maintenance of Ethernet standards from 1985 onwards. As of now, every individual standard of the Ethernet family has a name which refers to the IEEE Ethernet working group 802.3. The various amendments are identified by suffixes as extensions to the family name 802.3. Major Ethernet protocol variants are also referred with a short name consisting of the word "Base" and other identifiers for supported data rate and type of the transmission media. The steps of Ethernet evolution toward ever higher bitrates were the following:

- Original 802.3 [1] Ethernet was used to interconnect servers and personal computers within a building or a small campus. The topology of the network was a bus and maximum data rate provided was 10 Mbps. An Ethernet LAN consisted of one shared coaxial cable to which the host computers were attached with special connectors. Thicker 10Base-5 (1983) cables supported maximum segment lengths of half a kilometer, and thinner 10Base-2 (1985) cable could be used up to 200 m. Even if the design was innovative, in retrospective it had many drawbacks: Only one device at a time could successfully communicate over the shared bus, which limited the capacity of the LAN. The capacity was limited even further by collisions of Ethernet frames caused by two devices trying to transmit frames to the bus at the same time. Such a collision corrupted both the frames and wasted the otherwise useful transmission time. Further on, one single bad connector or missing terminal resistor in the end of the coaxial cable segment could stop the whole segment. To nail it, coaxial cable was a rather expensive type of media, at least when compared to cheaper twisted-pair wires.
- Ethernet 802.3i [2] 10Base-T (1990) and 802.3u [3] 100Base-T (1995) solutions transformed Ethernet architecture to solve the issues caused by using coaxial cable as a shared bus. These newer Ethernet variants used twisted pairs with a star topology, as shown in Figure 3.1. Every device connected to the LAN had its own dedicated pair of wires toward a central Ethernet hub. The first types of hubs reproduced the original bus functionality by relaying the frames transmitted by one of the devices to all the other devices connected to the hub. Such Ethernet hubs were soon replaced with Ethernet multiport bridges, also commonly known as Ethernet switches. The multiport bridge is an intelligent device, which is able to forward the frame to only

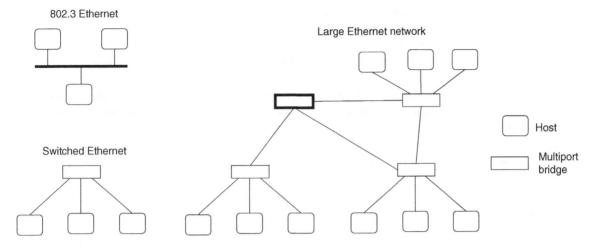

Figure 3.1 Examples of Ethernet LAN networks.

that link, which is connected to the specific Ethernet destination address of the transmitted frame. With such bridges, a network can support multiple simultaneous data flows provided that there is only one single flow per link (or a direction of a full duplex link). 100Base-T was able to provide full duplex data rates up to 100 Mbps over links up to 100 m.

- Ethernet 802.3j [4] 10Base-F (1993) and 802.3u [3] 100Base-F (1995) used optical multimode cables as transmission media between devices and the switch. With optical cables, 10 Mbps data rates could be supported for distances up to 2 km. When increasing the data rate to 100 Mbps, the maximum distance was limited to 200 m. In addition to using Ethernet as a LAN solution, it became possible to use Ethernet as a point-to-point link for interconnecting different networks or individual network devices.
- 802.3z [5] 1000Base (1998) and 802.3ae [6] 10GBase (2002) were the first members of gigabit Ethernet family supporting data rates over 1 Gbps. These variants used optical multimode, single mode, or coaxial cables as media. With multimode cables, distances were limited to a few hundred meters and with coaxial cables to tens of meters. The 1000Base-X supported data rates up to 1 Gbps up to 70 km with a single-mode cable, and 10GBase supported data rates up to 10 Gbps. In shorter range, up to 100 m, twisted-pair cable could be used with 1000Base-T specified within 802.3ab [7] (1999) and 10GBase-T specified within 802.3ae [6] (2006) amendments, offering full duplex data rates up to 1 Gb/s and 10 Gb/s, respectively.
- Later amendments of the Ethernet standard have increased the data rates even further. At the time of this writing, the latest versions such as 802.3cm [8] (2020) and 802.3cu [9] (2021) can support up to 400 Gbps data rates over multimode and single-mode optical fibers.

3.1.1.2 Ethernet Media Access Control and Networking

Methods used for Ethernet media access control (MAC) have evolved over the time. As mentioned above, the original 802.3 Ethernet design relied on a shared coaxial cable bus to connect hosts and servers attached to the LAN. In such a LAN, only one device at a time could communicate over the LAN while the other devices must wait for their turn. For shared MAC, 802.3 Ethernet used an algorithm known as **carrier sense multiple access with collision detection (CSMA/CD)**. With CSMA/CD the shared media (physical or logical bus) is in idle state when none of the connected devices has anything to transmit. When a need for communication arises, a device shall check if any transmission is going on over the bus. This check is known as the carrier sense. If no transmission is detected, the bus is considered to be idle and the device may transmit a frame to it. In case multiple host devices have started transmitting frames simultaneously, there is a collision and the frames are corrupted.

When the devices which sent the frames detect such a collision, they must wait for a random time and thereafter retry sending the frame. The one which has the shortest waiting time will win and gets the access to the link. The winning host may continue transmitting frames to the link as long as needed until a maximum threshold time is exceeded. The threshold time is there to ensure that none of the devices permanently block other devices to use the link.

The same CSMA/CD mechanism was also used in Ethernet networks where the coaxial cable was replaced with twisted-pair wires and a hub. In bus-type Ethernet design, all the devices connected to the bus will receive the transmitted frames. The frames contain the MAC address of the intended destination, based on which the other devices of the bus can filter out any frames not sent to them. The MAC address is a unique string of 48 bits assigned to the Ethernet interface unit hardware during its manufacturing process. The way to make MAC addresses unique is to construct them from a manufacturer ID (allocated by IEEE) and a serial number [10].

Ethernet networks which use switches instead of hubs use neither shared media nor CSMA/CD access control mechanisms. Instead, point-to-point links between host devices and Ethernet switches are used, working bidirectionally in the full duplex manner. With full duplex configuration, traffic may flow at the same time toward both directions of transmission over separate upstream and downstream wires or cables. As one single transmitter owns the whole wire, transmission can happen any time and there is no need for any access control mechanism. When the devices have no data packets to send, an idle pattern is generated to the link to maintain the link synchronization.

Earliest Ethernet networks were standalone LANs. It did not take long, before there was a need to interconnect different LANs within an office block or a whole campus. The technology gradually evolved toward supporting large, switched Ethernet networks. At the early days of Ethernet, the different LAN segments could be interconnected over repeaters, which just passed through any traffic from one segment to another. To make sure that there are not too frequent collisions or that the collisions would not be detected too late, the number of networks interconnected with repeaters had to be limited. The major drawback of the original Ethernet bus topology and the CSMA/CD mechanism was that with higher than 40–50% of the maximum theoretical load, the growing frequency of collisions started to eat up capacity. When LAN interconnection was done with repeaters, the large number of hosts connected to a single piece of shared media just increased the probability of collisions. Introducing multiport bridges and the full duplex mode essentially solved that problem. Multiple Ethernet LAN segments can now be interconnected with Ethernet switches (using link level technology) or with routers of a network layer protocol, such as IP. In this way, a whole campus can be covered with a large contiguous Ethernet network, like that shown in Figure 3.1.

As mentioned earlier, Ethernet switches are able to forward Ethernet frames to the links toward the destination host, whether the host is directly connected to that link or if it would be multiple hops away and the frame must pass over other switches to reach its destination. Mappings between links and destination MAC addresses are not configured to the switches. Instead, switches learn mappings via dynamic MAC learning. Initially, the switch does not know where destination MAC addresses are located, so the switch can only flood the frames to all other links except the one from which the frame arrived. Whenever a MAC frame arrives from a link, the switch checks its Ethernet source MAC address. If that MAC address is not yet known, the mapping between the link and the MAC address is recorded to an address table. To save space in the table, switches use aging mechanism to drop mappings from the table in a few minutes, if there is no traffic to/from that MAC address.

The biggest problems in large switched Ethernet networks are risks for network loops and oversized broadcast domains. If there is a loop in the network, the Ethernet frames may start circulating in the loop, which in the worst case may collapse the whole network. To prevent that, IEEE 802.1w [11] specifies a spanning tree protocol to prevent such loops and to provide an option to implement fault tolerant networks with redundant links. As mentioned above, when an Ethernet switch receives a frame with an unknown destination address, the switch broadcasts the frame to all other ports, except the one it was received from. In a large network, this can cause broadcast storm. To mitigate such a risk, IEEE 802.1q [12] specifies virtual LAN (VLAN) mechanism. The whole Ethernet network can be divided into multiple VLAN networks. A tag field of 12 bits is added into the Ethernet frame to carry a virtual LAN ID (VID). The frame is forwarded only to those ports that are configured to belong to that VLAN.

Even though Ethernet emerged in the context of LAN, it is currently the de facto link protocol for WAN. Transport of TCP/IP traffic over Ethernet frames is a well-known and proven technology. Ethernet evolution has created variants supporting ever higher data rates over different types of physical media for both short and long distances. All that makes the full duplex point-to-point Ethernet ideal for trunk and core networks connecting routers or other network nodes. In WANs, members of the gigabit Ethernet family are used for various purposes. High-speed Ethernet is used to transport IP packets over optical WDM trunk networks connecting core Internet routers, and other slower Ethernet variants are used to transport IP packets over DSL access links to customer premises.

3.1.1.3 Ethernet Layers and Frames

Ethernet 802.3 standards rely on the common IEEE 802 architecture, which separates protocol functions to two layers, as follows [13]:

- Physical layer defines the electrical or optical characteristics of the media and the line coding (such as Manchester; see *Online Appendix A.5.3*) to be used to carry binary data over the media.
- MAC layer is responsible for device addressing, data encapsulation into Ethernet frames and the MAC. The classic 802.3 Ethernet MAC uses CSMA/CD algorithm for access control.

An Ethernet frame covers both physical and MAC layers.

As shown in Figure 3.2, the frame has the following structure used on both bus and point-to-point topology Ethernet variants:

- Preamble: seven octets with constant bit string, used for bit synchronization and collision detection
- Start of frame delimiter (SFD): fixed bit string 10101011
- Destination MAC address
- Source MAC address
- EtherType field to indicate the upper layer protocol carried within the payload
- Data is the payload of the frame to carry upper layer protocol data.
- Padding octets are possibly used to adjust the frame length so that it is long enough for collision detection.
- Frame checksum: CRC-32

The size of the Ethernet frame is between 64 and 1518 bytes. To reach the minimum size, padding is used whenever needed. To avoid oversized frames, the payload data is fragmented to multiple frames.

Within the payload, the Ethernet frame carries upper layer protocol messages. In the context of Ethernet LANs, the transported protocol is often logical link control (LLC) (described in Section 3.1.3.4). With LLC, the EtherType field may be used to indicate the length of the data field, since LLC itself provided information about the upper layer protocol. In addition to LLC, other protocols, such as IP, can also be directly transported over Ethernet. For high- performance links, this is the typical configuration, as usage of LLC would introduce some extra overhead and waste capacity.

Preamble	Start of Frame	Dest. address	Source address	Ether Type or Length	Data	CRC

Figure 3.2 Ethernet frame.

3.1.2 WLAN Systems

Wireless local area network (WLAN) systems were developed as a wireless enhancement to LAN solutions, such as Ethernet relying on fixed cabling. As long as personal computers were mainly desktops used at office spaces, it was feasible to build LAN cabling to attach those desktops to it. But with the introduction of portable laptop computers and other form factors of mobile equipment, such as tablets and mobile phones with WLAN support, wireless networks have enabled many attractive use cases such as these:

- Using a laptop in different working environments, such as at an office desk, in a meeting room, at a hotel, or during a customer visit. Wireless LANs support corporate user mobility without any need to connect or disconnect cables when changing the working space.
- Accessing the Internet over complimentary WLAN networks provided by restaurants, cafes, or airports for their customers. With WLAN, it is easy to provide network connectivity to a complete restaurant room or waiting space. Many hotels also opted for wireless LANs rather than building LAN cabling to their hotel rooms.
- Accessing the Internet at home without the need for LAN cabling to cover the whole house or apartment. Only the most modern houses and buildings were equipped with LAN-compatible cabling. For the rest, wireless LAN provided an inexpensive and easy way to build connectivity to every room.
- Accessing the Internet with mobile devices, such as tablets or mobile phones. These devices' form factors are too small to support standard Ethernet cable connectors. To offload traffic from cellular networks and avoid any related data costs, WLAN support was introduced to many smartphone models well before 2010 and has been a standard feature of smartphones and tablets since then.

As WLAN is a wireless local area network, its purpose is to connect end user devices locally – either with each other or to the wired network infrastructure used for wider connectivity. The coverage of a WLAN network may be one apartment or restaurant premises, a floor of an office building, or even all the premises of an enterprise. WLAN networks are typically built with WLAN base stations connected to wired LAN. The base station is also called **WLAN access point (AP),** providing access to the LAN network. The WLAN network supported by the access point is identified by its **service set identifier (SSID),** which is a human readable piece of text. SSID can be shown at the user interface of the host device when used for WLAN network selection.

WLAN network may be limited to the coverage area of a wireless signal of one single base station (for instance, at home) or the coverage area of multiple base stations connected to a shared LAN. Indoor coverage of a single base station is approximately between 20 and 70 m, depending on the room layout and physical layer specification of the WLAN variant deployed. Enterprise-wide WLAN networks are built by assigning the same WLAN SSID identifier to multiple interconnected WLAN access points and even to separate WLAN networks in different geographical locations. To enable location independent WLAN authentication, enterprise networks typically rely on a centralized authentication server, which supports all related WLANs. After connecting to a WLAN access point of the office, the enterprise user can access enterprise internal IT resources. The cable and DSL modems used at homes for Internet access typically provide both LAN ports and a WLAN access point so that the network users can connect their devices to the modem via wired or wireless LAN. For Internet connectivity, the modem has an embedded IP router function and a DSL or TV cable port toward the Internet service provider network.

The first WLAN implementations used either radio or infrared waves as their transmission media. Two frequency areas around 5–6 GHz and 2.4–2.5 GHz are commonly used by WLAN radio systems. As infrared requires line-of-sight connectivity, such systems were not deployed widely and soon become obsolete. WLAN radio systems use unlicensed frequencies with strict limitations to the used transmission power. The choice of the frequency bands and the used low power levels make the effective signal range small, typically in the order of few tens of meters before the attenuation renders the signal unusable. Such an approach enables independent actors setting up their own WLAN networks without licensing the frequencies or making it necessary to coordinate the frequency allocations with other WLAN network providers nearby. In the enterprise environments, coordination of used frequencies (or WLAN channels) is still useful to optimize the performance of different WLAN networks in close proximity.

WLAN systems are standardized in the IEEE 802.11 standard series. The first standard of the series was the baseline IEEE 802.11 [14] itself, which was published in 1997. The original specification supported three types of physical layers: one using infrared light medium and two others using radio medium in the 2.4 GHz frequency range. For radio medium, FHSS and DSSS (see *Online Appendix A.9.1*) multiplexing options were specified. Bitrates between 1–2 Mbps were reached, depending on the type of the physical layer. Architecture of the system was based on IEEE 802.3 Ethernet LAN specification, as the aim was to specify a wireless version of Ethernet LAN.

Since then, a number of WLAN variants have been specified as amendments to the IEEE 802.3 core specification. Early variants defined limited additions and amendments to the baseline standard. Later on, newer variants made more extensive changes, especially to the WLAN physical layer, while trying to keep the topmost layers of WLAN protocol stack compatible with the older versions. WLAN physical layer variants differ from each other in respect to the applied frequencies, multiplexing, and modulation methods as well as for the achievable bitrates. Nowadays, WLAN is the primary local network access method for various types of terminals, including laptops, tablets, smartphones, and gaming consoles. Wired Ethernet is still used for terminal access when convenient, but Ethernet is the de facto solution as a local backbone network within a building or campus. The backbone network is there to connect the local network with various fixed infrastructure elements, such as servers, printers, and WLAN base stations.

WLAN systems are also called Wi-Fi (wireless fidelity) systems. The Wireless Fidelity Alliance Consortium (known as Wi-Fi Alliance) is an organization that takes care of standard conformance and mutual compatibility testing of WLAN vendor implementations. "Wi-Fi" is a trademark of the Wi-Fi Alliance and the brand name for products using WFA programs based on the IEEE 802.11 family of standards. Wi-Fi Alliance has also labeled certain major variants of the 802.11 standard as Wi-Fi generations 4, 5, and 6. Earlier generations 1 to 3 were not officially announced, even if they might be commonly referred to. The six WLAN system major variants are as shown in Table 3.1 [15]:

- 802.11b [16] uses unlicensed frequencies in the 2.4 GHz band. The precise frequency ranges and bandwidths were country specific. Since the lower frequencies penetrate obstacles better than higher frequencies, the 802.11b signal has better reach than the 802.11a signal.
- 802.11a [17] uses unlicensed frequencies in the 5 GHz band. The drawback of the higher band is shorter transmission range due to higher attenuation of high frequencies.
- 802.11g [18] uses the 2.4 GHz band. This variant supports the same bandwidth, orthogonal frequency division multiple access (OFDM), and quadrature amplitude modulation (QAM) modulation methods as the 802.11a variant. Also, direct sequence spread spectrum (DSSS) signals are supported to make variant g compatible with both earlier variants a and b.

Table 3.1 WLAN variants.

Variant	Released	Wi-Fi Gen.	Freq. Band	System Bandwidth	Multiplexing and MIMO	Modulation	Maximum Bitrate
802.11b	1999	(1)	2.4 GHz	22 MHz	DSSS	64-QAM	11 Mbps
802.11a	1999	(2)	5 GHz	5–20 MHz	OFDM	64-QAM	54 Mbps
802.11g	2003	(3)	2.4 GHz	5–20 MHz	OFDM, DSSS	64-QAM	54 Mbps
802.11n	2009	4	2.4 or 5 GHz	40 MHz	OFDM4 x MIMO	64-QAM	600 Mbps
802.11ac	2013	5	5 GHz	160 MHz	OFDM8 x MIMO	256-QAM	6.9 Gbps
802.11ax	2021	6	2.4, 5, or 6 GHz	160 MHz	OFDMA8 x MIMO	1024-QAM	9.5 Gbps

- 802.11n [19] uses both 2.4 GHz and 5 GHz bands. This variant was specified for the WLAN to support high data rates provided by Internet access technologies, such as ADSL2+, VDSL, or DOCSIS 3.x. Additionally, this specification version introduced QoS mechanisms for WLAN.
- 802.11 ac [20] uses 5 GHz band. Limited multiuser use cases are supported by the specification with a downlink multiuser multiple input, multiple output (MIMO) technique, where different users relying on the same frequencies are separated by different base station transmission antennas.
- 802.11ax [21] uses 2.4, 5, and 6 GHz bands. Instead of ever-increasing bitrates to a single host, the focus of this variant is to support hotspots with a large number of hosts in a dense area, such as a sports arena or a crowded office. To support such multiuser use cases, 11ax uses OFDMA multiplexing. OFDMA enables multiple terminals to communicate simultaneously with a single WLAN access point, as each terminal is allocated with its own OFDM subcarriers.

The maximum bitrates mentioned in the table are the theoretical maximum speeds in optimal conditions. The sustainable bitrates are typically much lower as the bitrate decreases with longer distances and obstacles of different kinds between the WLAN-equipped host and base station.

In 2007, IEEE published standard 802.11-2007, which combined variants a, b, and g into one single specification. The standard version 802.11-2012 published five years later added variant n to the specification and version 802.11-2016 added variant ac. In addition to these major variants, both of these specifications covered a number of other amendments of the 802.11 standard family, which are not covered by this book. These amendments do not redefine the WLAN physical interface but provide new methods for aspects like network operation, quality of service, and security. At the time of this writing, the latest combined version of the specification is 802.11-2020, into which recent amendments 11ah, 11ai, 11aj, 11ak, and 11aq were added.

For further details of WLAN network architecture and operation, please study *Online Appendix G.1*.

3.1.3 HDLC and LLC

3.1.3.1 Architecture of the HDLC System

High-level data link (HDLC) protocol is a link layer protocol specified in ISO/IEC 13239 standard [22]. A significant number of other protocols have been derived from HDLC for different contexts, such as LANs, data modems, ISDN, and GSM (global system for mobile) mobile networks. As described in *Online Appendix C*, the LAPD protocol used on the ISDN UNI channel D is a variant of HDLC. In addition to ISDN, the LAPD protocol is used in GSM networks to transport short messages between network elements. Radio link protocol (RLP) is an HDLC-like protocol used for non-transparent GSM data transfer. An HDLC variant called LAPDm is used in the air interface of GSM to transport signaling messages between the mobile phone and the network. The LLC protocol used in LAN networks is also derived from HDLC. Frame relay switching protocol also used HDLC-like frame structure. The HDLC protocol itself is used to transport management messages between DSL modems.

HDLC protocol uses a model where multiple devices are connected over a shared link [23]. One of the devices is the primary device, which controls the link. The other, secondary devices are only able to communicate with the primary device. The primary device does not need an address since all messages from secondary devices are sent to the single primary device on the link. Every secondary device has its own address so that the primary device is able to send a message to a specific secondary device.

HDLC supports flow control with sliding window and message retransmissions with both types of continuous RQ mechanism introduced in *Online Appendix A.6.3*.

There are only two types of HDLC messages, as shown in Figure 3.3:

- Commands (select, information, poll) sent by the primary device
- Responses (acknowledgment, information) sent by a secondary device

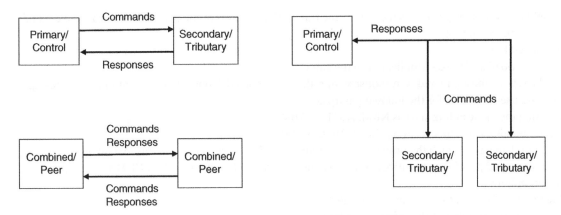

Figure 3.3 Different topologies of HDLC link. *Source:* Adapted from ISO/IEC 13239 [22].

The primary device may send commands to secondary devices whenever it wishes. There are three different modes defined for HDLC, which determine when a secondary device may send responses to the primary device:

- Normal response mode (NRM), where the primary explicitly requests a secondary device to return a response.
- Asynchronous response mode (ARM), where the secondary device may send a message to the primary device without receiving an explicit permission for it.
- Asynchronous balanced mode (ABM), where two devices in the end of a point-to-point link have both the roles of the primary and secondary device.

3.1.3.2 HDLC Frame Structures

HDLC has three different types of frames:

- Unnumbered frames used for link startup, recovery from error situations, or for acknowledgments. The unnumbered frames themselves are not acknowledged.
- Information frames, which carry upper layer data in the message payload but may also provide acknowledgments of earlier HDLC frames successfully received.
- Supervisory frames used for flow control and retransmission requests.

HDLC is a bit-oriented protocol. As shown in Figure 3.4, the frame consists of the following fields:

- Frame start flag: One octet carrying a fixed bit sequence 011111110 to indicate the start of the HDLC frame.
- Address identifying the secondary device to which the command was sent or which has sent the response. The length of the address is one or two octets. The address within the HDLC command may also mean all the secondary devices on the link or a predefined subset of them.

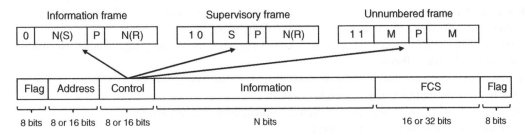

Figure 3.4 The structure of control field within HDLC frame.

- Control field whose first bit(s) defines the type of the frame. The contents of the rest of the control field depend on the frame type as follows:
  - Unnumbered frame:
    - The first two bits of the control field have the value 11.
    - M: The type of the command or response within the unnumbered frame. Different unnumbered message types are used in HDLC for the following purposes:
      - Setting the mode at link start as NRM, ARM, or ABM.
      - Commands used at link recovery: RSET, DISC, FRMR.
      - Unnumbered poll which the primary uses to grant secondary devices permissions to send messages: UP.
      - Unnumbered positive or negative acknowledgment for a numbered frame: UA, CDMR, FRMR, DM.
  - Information frame:
    - The first bit of the control field has the value 0.
    - N(S): Sequence number of the information frame.
    - N(R): The sequence number of the latest successfully received HDLC information frame.
  - Supervisory frame:
    - The first two bits of the control field have the value 10.
    - S: Type of the frame used for flow control or retransmission request.
      - RR: Secondary is ready for receiving information frame.
      - RNR: Secondary is not ready for receiving information frame.
      - SREJ: Receiver requests retransmission of the frame with sequence number as given in N(R).
      - REJ: Receiver requests retransmission of the frames starting from the sequence number given in N(R).
    - N(R): The sequence number related to SREJ and REJ acknowledgments.
  - The control field always has a Poll/Final bit (P). The primary sets the value of P/F bit as 1 when it requests an acknowledgment from the secondary. The P/F bit shall also be set as 1 in the acknowledgment returned.
- Information field to carry the data from upper layers. All information frames have the information field, and some of the unnumbered message types also have it. Supervisory frames never contain an information field.
- Frame check sequence (FCS), containing a CRC of two or four octets to detect bit errors in frames.
- Frame end flag: One octet carrying a fixed bit sequence to indicate the end of the HDLC frame.

3.1.3.3 Operation of HDLC

The basic operation of HDLC protocol is as follows:

- The primary device starts the link by sending an unnumbered frame to all the secondary devices. The type of the command message tells which link mode (NRM, ARM, or ABM) shall be used. The message also initializes the sequence numbers. After accepting the received command, the secondary devices respond to it with an unnumbered UA response. If a device does not accept the link start command, it may reject it with a DM response.
- The primary device may send information frames whenever it wishes or it may grant a turn for a secondary device to send information fames by sending an UP command to it. HDLC protocol uses the sliding window method for flow control. If receiving a message fails, a retransmission request may be sent for the single failed frame or all frames starting from a sequence number given in the request.
  - If the information exchange is bidirectional, positive acknowledgments are usually piggybacked to the information frames sent to the other direction.
  - It is possible to send acknowledgments in separate unnumbered or supervisory frames in the following cases:
    - The receiver has no information frames to send.
    - The receiver wants to send a negative acknowledgment or request a retransmission.
    - The receiver wants to send a flow control request.
- To shut down the link, the primary device sends an unnumbered DISC frame, to which the secondary devices shall respond with a UA acknowledgment.

3.1.3.4 LLC Protocol

Logical link control (LLC) is a protocol specified for LAN networks. In LAN environment, LLC is run on top of the LAN specific MAC layer. LLC protocol frames are carried as a payload of MAC frames. LLC is a derivative of HDLC protocol (described in Section 3.1.2). In addition to LANs, it is possible to run LLC protocol also on top of other MAC layers, like the ones provided by DSL or cable modems. LLC is specified in IEEE specification 802.2.

The purpose of the LLC protocol is to take care of flow and error control over the link connection as well as associating the link level frames to network layer protocols. Compared to HDLC frame, LLC frame is a bit simpler due to the split of link layer responsibilities between LLC and MAC layers. LLC frames do not have endpoint addresses, frame checksums, frame start, or end flags because those belong to the underlying MAC frame.

LLC frames consist of the following fields:

- Destination service access point (DSAP), which identifies the upper layer service or protocol that shall process the payload of the LLC frame.
- Source service access point (SSAP) identifying the source service or protocol. Value of SSAP is typically identical to DSAP.
- Control field as defined for HDLC.
- Subnetwork access protocol (SNAP) header optionally is present for certain service access points, consisting of two subfields:
  - Organizationally unique identifier (OUI) of the organization that has a defined value of PID.
  - Protocol identifier (PID) specifying the protocol carried within the LLC frame as payload. If present, PID in practice overrides SSAP. PID was defined so that its values could match the values of EtherType field of the original Ethernet protocol.
- Payload of the LLC frame.

Like HDLC, LLC has three modes:

1) Connectionless without acknowledgments.
2) Connectionless with acknowledgments.
3) Connection-oriented, asynchronous balanced mode.

The first of these modes is typically used in LANs where reliability is provided by upper protocol layers. The connection-oriented asynchronous balanced mode can be used when the upper layer service needs reliable service from a link. Connectionless service with acknowledgments provides additional reliability for time-critical services, such as factory automation.

3.1.4 PPP

Point-to-point protocol (PPP) was designed to provide link level services for network level protocols, such as the Internet protocol (IP) [24]. Such services were needed for links, which provided just a raw bit stream, like analog modem connection or 64 kbps ISDN connection over telephone network. PPP provides the following services:

- Setup of a PPP link and negotiation of connection parameters between link endpoints.
- Selecting the authentication mechanism for the PPP link and completing the authentication with the chosen authentication protocol.
- Multiplexing and forwarding upper layer protocols over one physical link.
- Compressing certain upper layer protocol header fields to reduce the protocol overhead.
- Detection and correction of errors and optional message numbering with help of HDLC-type frames encapsulating the core PPP frames.

Due to its support for connection parameter negotiation and authentication, PPP is used also on such types of links that provide basic link level transport services, such as DSL. The basic PPP functionality is specified within the following IETF RFC documents:

- RFC 1332 [25]: Usage of IPCP protocol to set up IP connectivity over PPP links
- RFC 1661 [26]: Functionality and frame structures of PPP and LCP protocols
- RFC 1662 [27], 1663 [28]: Usage of HDLC-type frame structure on bit and byte synchronous links
- RFC 1994 [29]: PPP Challenge handshake authentication protocol (CHAP)
- RFC 3748 [30]: Extensible authentication protocol (EAP)

In addition to these RFCs, IETF has delivered many other RFCs related to PPP for topics such as

- Adapting various upper layer protocols for transport over PPP
- Adapting PPP itself to the underlying protocols such as SDH, ATM, or Ethernet
- Methods for negotiating various configuration parameters and options, such as data encryption or compression, over PPP links

To encapsulate an upper layer protocol, PPP uses a very simple message structure with three parts:

- Protocol field of 1 to 2 bytes identifies the carried protocol
- Information field contains the upper layer packet
- Padding may be used to adjust the PPP message length

For link management, PPP uses two control protocols: link control protocol (LCP) and network control protocol (NCP). Both of these protocols are used at the link setup phase to negotiate parameter values for the link. LCP is used for parameters not related to upper layer protocols, while different protocol specific NCP implementations are used for parameters specific to one upper layer networking protocol [25]. IPCP is the NCP protocol used for IPv4 protocol, and IPV6CP is used with IPv6. In addition to link setup, LCP supports framing of IP packets and error control.

LCP message structure is as follows [26]:

- Code field identifies the LCP packet type.
- Identifier is used to match replies with the requests.
- Length indicates the length of LCP packet.
- Data contains the data related to the LCP packet type.

Setting up PPP link for IP protocol is done as follows:

1) The network terminal sends an LCP configure request message to the other end of the link. This message contains proposed parameter values, such as the maximum size of PPP frame, authentication protocol, compression, and error control methods to be used.
2) The other end either accepts the proposed values and returns LCP Configure Ack message or rejects them with LCP Configure Nack message. Thereafter it sends an LCP configure request message with new proposals and/or additional parameters to continue negotiation.
3) After the parameter values have been mutually accepted, the PPP link is open. In case authentication was agreed, the end user can be authenticated with CHAP or EAP protocols.
4) Finally, the terminal sends an IPCP configure request to acquire the IP address and agree on compression method, such as RoHC, to be used to compress IP headers. The other end responds with an IPCP Configure Ack message.

3.2 Switching Protocols for Virtual Connections

3.2.1 Frame Relay

Frame relay (FR) protocol was designed as an inexpensive WAN communication protocol supporting data rates between 9.6 kbps and 2 Mbps [24]. Frame relay uses HDLC-like frames to switch those between network links. As network level routing (see Section 3.5) was deemed as a complex and slow piece of functionality, switching was proposed as a lightweight alternative. HDLC frames are transported over a single physical link, and FR frames are transported over an end-to-end path, which consists of physical links and **virtual connections** in the network nodes between the links. Frame relay is specified in CCITT/ITU-T Recommendations I.233.1 [31] and Q.922 [32].

The frames of FR use HDLC frame structure but with a major exception. HDLC frame contains a destination address field. In FR the address field has been redefined so that it contains a data link connection identifier (DLCI) and a few bits related to network congestion management. DLCI value is local to every network node along with the end-to-end virtual connection path. DLCI tells the node to which link the frame should be forwarded and what is the new DLCI value to be written to the forwarded frame. This mechanism enables the node to have a small DLCI lookup table used for frame switching. The FR node reads DLCI from any FR frame received and finds the corresponding entry from its local lookup table. This entry tells the new DLCI value and next link to be used for the frame. Statistical multiplexing (see *Online Appendix A.9.2*) makes it possible to share the capacity of one physical link over multiple frame relay virtual connections. Every frame sent to a link may have its own DLCI and belong to a different virtual connection.

Two ways could be used to create the needed virtual connections and populate the lookup tables:

- Configure permanent virtual connections (PVC) to the FR nodes with a management system.
- Create switched virtual connections (SVC) to the nodes with signaling messages exchanged during connection opening phase. ITU-T recommendation Q.933 [33] defines signaling procedures for setting up FR SVCs.

3.2.2 ATM

Asynchronous transfer mode (ATM) was designed to be a general-purpose protocol to be used for transporting either circuit or packet switched data. ATM is a switched link level protocol, using PVC and SVC virtual connections like frame relay [34]. Both ATM and frame relay use statistical multiplexing to share the capacity of a physical link for multiple virtual connections. The fundamental difference between ATM and FR is the frame structure used. ATM was designed to use short fixed-length frames of 53 bytes, known as ATM cells. Short cell size is an enabler for efficient statistical multiplexing and QoS management, as they introduce minimal delays for processing, transmission, or buffering. The fixed data paths of virtual connections minimize variations to end-to-end latency. Latency variations are only caused by queuing in data buffers, but they never emerge from cells of a single stream traversing different routes from source to destination.

During the 1980s, ATM was designed to be used for broadband ISDN (B-ISDN) service. In the 1990s, only a few telecommunications service providers decided to build ATM-based standalone networks. In the beginning of the 2000s, ATM was deployed in 3G cellular networks to interconnect various elements of the base station subsystem and the core network. ATM has also been used as a connectivity solution over DSL links toward a router or ATM multiplexer of an Internet service provider. ATM is specified by ITU-T in recommendations I.361 [35], I.363 [36], and I.365 [37]. ATM signaling for the B-ISDN connection setup protocol B-ISUP is specified in ITU-T Q.2761 [38], Q.2764 [39], and Q.2931 [40].

ATM protocol uses the following layering concept:

- Physical layer to adapt ATM traffic to the underlying physical link. Depending on the functionality of the underlying layer, this layer may be either very simple or more complex. When used on top of complete link protocols such as SDH or DSL, the only task of the ATM physical layer is to map cells to/from underlying frames. When using ATM over a raw transport media, such as a cable, the ATM physical layer needs to implement complete physical layer functionality.

- ATM layer is responsible of cell creation, statistical multiplexing, switching, buffering, flow, and congestion control.
- ATM adaptation layer (AAL) responsible of segmenting and mapping upper-level protocol packets to ATM cells. The adaptation layer may also take care of detecting and correcting errors as well as providing end-to-end clock synchronization signals. Different adaptation layer implementations have been defined to support different Quality of Service (QoS) types. AAL1 supports constant bit rate connections, AAL2 short and variable length packets with latency limitations, and AAL5 is used for "best effort" data connections without strict data rate or latency restrictions. The AAL5 adaptation layer is the one used to transport Internet protocol. It takes care of IP packet segmentation, reassembly, and error control.

An ATM cell has five header bytes and 48 bytes of payload [35]. The most important part of the header is the routing field, which consists of VPI/VCI identifiers of the virtual path and virtual circuit used for ATM cell switching. In ATM, one virtual path may transport multiple virtual circuits, one circuit for every data flow which requires specific QoS treatment. The VPI/VCI values are local to a node and used to access lookup tables of virtual connections. When forwarding a cell to the next link, the node searches for an entry in the virtual connection lookup table with the VPI/VCI values that the cell had at its arrival. After finding the correct entry, the node overwrites the VPI/VCI values of the cell according to the new values retrieved from the found lookup table record. The cell is thereafter buffered for transmission or transmitted to the next link as given within the lookup table record. A few VCIs are permanently reserved for signaling connections, used, for instance, for managing SVCs.

Like any for statistically multiplexed packet switched protocol, congestion may take place at ATM links and nodes, filling up cell buffers. ATM uses a few mechanisms for congestion control. The generic flow control header field of the cell can be used for flow control at the cell source. The cell payload type header field carries an explicit forward congestion control bit, informing the congestion to upstream nodes, which can use, for instance, TCP flow and congestion control mechanisms accordingly. The cell loss priority header field tells whether the cell can be dropped in congestion. Real-time media packets which do not tolerate delays can often be dropped to alleviate congestion. ATM connections may also use resource management (RM) cells with which the ATM nodes could inform the data source about congestion and/or the explicit cell rate which the node is able to provide for the virtual connection. The RM cells can be used to adjust data rates dynamically. Further details about ATM can be found from *Online Appendix H*.

After the early 2000s, ATM was not widely used, since ATM was found to cause extra overhead and complexity to networks, compared to its low value added. ATM headers take 10% of the cell size, which is a considerably large amount of overhead. ATM design for virtual connection management also had a fundamental mismatch with the IP payloads. Broadband ISDN (B-ISDN) terminals were designed to set up ATM virtual connections with a signaling protocol called B-ISDN user part (B-ISUP) [38] [41]. Unfortunately, such a connection management model, copied from the circuit switched telephone networks, does not fit too well with the routed IP packet concept. Typical IP protocol use case is Web browsing. In a typical session, a Web browser may generate a large number of IP packets toward many different destinations in a relatively short time. It would not be feasible for any terminals or the network to dynamically manage virtual circuits toward those destinations, as the signaling used for creating and releasing the circuits would consume both time and network capacity. Instead, terminals should just have link level connections toward a local router, which will forward IP packets toward their destination networks with means of IP routing.

3.2.3 MPLS

While packet switching is good technology for high-capacity, mesh topology core networks, it is not without drawbacks. The end user traffic transported in those networks uses typically IP network protocol. As will be explained in Section 3.3, IP protocol is routed rather than switched. With frame relay or ATM, significant effort or extra complexity would be needed to maintain the virtual connections over the core network to transport IP flows between all connected IP networks. Since neither FR nor ATM provide good native solutions for this problem, the datacom industry tried out a few different solutions to align switching with the needs of IP protocol [34]:

- Multi-protocol over ATM (MPOA), where ATM was used to emulate LAN network functionality in a much bigger scale [42].
- IP switching, where ATM virtual connections were created on-demand based on the network elements tracking the packet flows between specific IP addresses. IP switching was a proprietary technology by Ipsilon Inc., but its specifications were published by IETF as RFCs 1953 [43] and 1954 [44].
- **Multi-protocol label switching (MPLS),** where virtual connections are created between MPLS **label switched routers (LSR)** for **forwarding equivalence classes (FEC)** of transported packets. For the IP protocol, a FEC consists of IP packets that share the same values of certain header fields such as DiffServ QoS class, UDP/TCP port numbers, and network or subnetwork prefixes of source and destination IP addresses. When routing IP packets, LSR at first partitions them to FECs and then maps FECs to the next hops and related links. Each FEC is assigned a label based on which the switching will be done in the downstream LSRs, without consulting IP headers anymore. See Figure 3.5 for a simple example.

From these solutions, MPLS was the one that gained popularity due to its automated mechanisms for maintaining virtual connections and related **label switched paths (LSP)** for FECs identified by the LSRs. While MPLS deployments have grown, other switching protocols like frame relay and ATM have lost their ground. MPLS resembles ATM with its approach of use switching based on locally relevant labels within the packets to access local label switching mapping tables. The label may be an MPLS specific label, FR DLCI label, or ATM VPI/VCI label when the MPLS node has been developed from an older FR or ATM implementation.

MPLS is specified in the following IETF RFC documents:

- RFC 3031 [45]: MPLS architecture
- RFC 3032 [46]: MPLS frame structures and label stack processing
- RFC 3034 [47]: MPLS implementation based on frame relay
- RFC 3035 [48]: MPLS implementation based on ATM

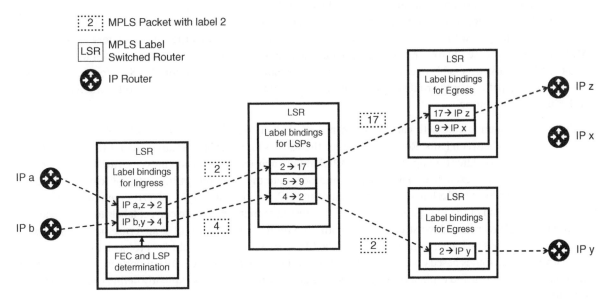

Figure 3.5 MPLS packets and label switching with forwarding equivalence classes.

- RFC 3036 [49]: LDP specification
- RFC 3212 [50]: CR-LDP specification

All the switching protocols use the common approach of packet switching based on locally relevant labels, whereas the following features are specific to MPLS [10]:

- MPLS packet encapsulates a complete upper layer protocol frame rather than segmenting it to multiple switched frames or cells.
- MPLS virtual connections are dynamically created based on upper layer protocol addressing. IP routing protocols may be used in the connection setup phase to route the virtual connections over MPLS network. When an MPLS router forwards a packet without FEC classification and related LSP path, it may work like an IP router.
- MPLS supports nested MPLS connections as MPLS tunneling, which is a useful feature when an MPLS core network is used to interconnect regional MPLS networks or when building enterprise virtual private networks on top of a shared MPLS backbone. In MPLS specifications, such nesting is referred as MPLS stacking.

MPLS packet structure is as follows [46]:

- MPLS label used to access local lookup table for packet switching.
- S bit to tell if the MPLS packet encapsulates another nested MPLS packet.
- TTL is the time-to-live value, which is copied to the MPLS packet from the encapsulated IP packet when the IP packet enters the MPLS network. TTL is decremented by every MPLS LSR on the path and copied back to the IP packet in the end of the LSP path.
- Payload of the MPLS packet, which is typically the IP packet transported over LSP path.

MPLS networks may use a few different mechanisms to derive FECs, LSP paths, and related node-local virtual connections for them:

- FEC and/or LSP path can be determined by a network management system.
- MPLS router may determine FEC independently of other routers. The router will communicate the locally established FEC and the related MPLS label to upstream MPLS routers from where packets arrive for the FEC. This mechanism is known as independent LSP control.
- MPLS router at the network edge may derive an FEC class from the IP subnetwork addresses between which the router regularly forwards traffic. The router may set up the LSP path via other MPLS routers with help of a MPLS routing database the router has built. The database stores records about the other known MPLS routers and the IP networks accessible over those. The edge router may then request other MPLS routers to set up FEC classes and LSP paths over the route determined by the edge router. This mechanism is MPLS ordered LSP control, with explicit routing.

MPLS routers communicate with each other with the Label Distribution Protocol (LDP) [49]. LDP is used to distribute the labels between LSRs. Routers announce themselves for other MPLS routers with LDP Hello packets used for LSR discovery. After finding each other, MPLS routers may use LDP session messages to set up LDP management connections to exchange LDP messages and negotiate the deployed MPLS optional features. When a router has created a local MPLS label and attached it to an FEC, it will send an LDP advertisement message to the MPLS router upstream of the FEC path. In this way, the upstream router can write correct labels to MPLS packets sent to the next MPLS router downstream of the path. The router may also use LDP advertisement messages to release a label which is no longer used for any FEC. CR-LDP [50] is an extension of LDP protocol, used for managing constraint-based routes based on QoS requirements or other constraints. CR-LDP supports traffic engineering with ordered LSP control and explicit routing, where the label request messages explicitly define routes as a list of nodes for a route.

At the time of this writing, MPLS is widely used in the core IP network used to interconnect IP networks. In the context of enterprise networks, MPLS is commonly used as a backbone to interconnect local sites and their IP subnetworks. MPLS can be used to build a virtual private network within a physical network of a network service provider when the physical network is shared between multiple client enterprises.

3.3 Internet Protocol Version 4

3.3.1 History of IPv4 Protocol Suite

Development of the Internet and its protocols was started in 1966 within the ARPANET project funded by the U.S. Department of Defense (DoD). The goal of the project was at first to enable access to remote computers and later on to create a computer network which would be resilient against losing a part of the network nodes in warfare. The term Internet refers to inter-networking, where network-connected computers could communicate with each other within a large distributed network, regardless of the individual link technologies between any two nodes in the network. Internet protocols were designed to be used on top of different types of data links. The TCP/IP protocol suite, being the core of the Internet protocols, was originally specified within this project. TCP/IP was deployed both in the ARPANET and MILNET networks, the latter of which was split from ARPANET during the 1980s to interconnect U.S. military bases. The ARPANET network itself continued to be used by researchers. The ARPANET project cooperated with Berkeley University, where a TCP/IP stack was developed as part of the BSD Unix operating system. When the BSD Unix was published, the source code of its TCP/IP stack became freely available. Soon, within the 1980s, the TCP/IP protocol suite was commonly used by universities and research communities around the world.

The Internet itself emerged when separate network islands were connected together with TCP/IP. In the early days of the Internet, the ARPANET network served as its interconnecting core. Later on, when the focus of the Internet gradually shifted from the military to the research community, a new NSFNET core was created. The NSFNET network was maintained by the National Science Foundation, an organization based in the United States. The capacity of the NSFNET had to be increased many times during the 1980s. In 1974, the links of ARPANET were leased lines with modest 56 kbps bitrates, but after many modifications in 1990, the network backbone had been rebuilt with T1 links with 1.5 Mbps bit rate [51]. Essentially, the Internet was, at that stage, a network run over telephone infrastructure to connect relatively small numbers of computers.

When LAN became common in the 1980s, a number of vendor-specific application protocols were run in them. Those protocols allowed users of LAN-connected personal computers to access some centralized resources, such as printers or databases. Internet protocols were soon proven to be suitable also for this environment, and in a few years the TCP/IP protocol suite replaced various vendor-specific protocols used earlier on LANs. When the Internet was adopted for commercial purposes during the 1990s, the capacity needs for the Internet core exploded. Corporate LANs were connected to the Internet with routers. Home users could connect to the Internet via a modem access service provided by an Internet service provider. The responsibility of Internet core maintenance was moved to commercial enterprises in the middle of the 1990s. By the end of the century, the bit rates of the core network links were increased to gigabits per second.

The early TCP/IP networks supported only a few use cases, such as remote use of computers over modem connections, sending email, and moving files between computers connected over the network. The HTTP protocol and World Wide Web introduced in the 1990s provided totally new possibilities for using the Internet. It did not take long until Internet services made a breakthrough, at first to workplaces and soon thereafter for consumers. Various public and commercial organizations published their Internet pages, but a bit later, technically oriented individuals were able to publish their own content with homepages and blogs. Eventually, in the 2000s, the social

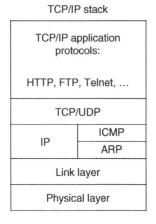

OSI layer model

| Application layer |
| Presentation layer |
| Session layer |
| Transport layer |
| Network layer |
| Link layer |
| Physical layer |

TCP/IP stack

TCP/IP application protocols:

HTTP, FTP, Telnet, …

TCP/UDP

| IP | ICMP |
| | ARP |

| Link layer |
| Physical layer |

Figure 3.6 OSI model versus the traditional TCP/IP protocol stack.

media applications become available over the Internet, making virtually everyone a potential content provider. In the 1990s the Internet was a data network relying on telephone infrastructure, and only 30 years later voice services are provided over the Internet, and the independent telephone network has largely ceased to exist. In the background, the proven TCP/IP protocol suite (shown in Figure 3.6) is still the workhorse taking care of moving all the data in the Internet.

The original Internet protocol design philosophy can be summarized with a motto, "IP over everything and everything over IP." This means an approach where the Internet protocol (IP) would be the one and only protocol used at the network layer, providing ubiquitous worldwide packet switched data connectivity. IP itself can be transported over many different link layer protocols, such as Ethernet, WLAN, frame relay, ATM, SDH, PPP over ISDN, cellular data link protocols, and so on. On the other hand, IP can carry any packet switched application protocol messages, some of those standard protocols (like HTTP, SIP, RTP) and others proprietary.

Since the year 1985, the specification and maintenance of protocols belonging to the TCP/IP family is done by the Internet Engineering Task Force (IETF). The specifications published by the IETF are called request for comments (RFC), and specifications under development are called Internet drafts. Typically, an Internet draft is evolved over tens of versions, subject to wide debate in the IETF, before the work converges into a formally agreed RFC. Certain RFCs have been promoted as Internet Standards, which expresses the status and wide deployment of the specified mechanisms in the Internet. An Internet standard may cover only a single RFC or consists of multiple related RFCs.

3.3.2 IPv4

Internet protocol (IP) is a network layer protocol which provides a connectionless packet switched communication service between devices of an IP network. The first globally deployed version of the protocol was **IP protocol version 4 (IPv4)** published in the beginning of the 1980s and deployed in ARPANET in 1983. Earlier versions of IP were experimental protocols never promoted as Internet standards. Despite of its age and limitations, IPv4 is still the most widely used core protocol used to access services available in the Internet or local networks. IPv4 is defined in the following IETF specifications:

- RFC 791 [52]: Internet Protocol - IP
- RFC 919 [53] ja 922 [54]: Broadcasting of IP packets
- RFC 950 [55]: Internet subnets

3.3.2.1 Architecture and Services of IPv4

The endpoints of IP communication which own IP addresses are called **hosts**. A large **IP network** consists of a number of interconnected **subnetworks**. The devices which interconnect the subnetworks are called IP **routers**. The task of a router is to know the IP addresses of devices directly connected to the router over a link and the related link layer addresses. Additionally, the router must understand to which other routers such IP packets should be forwarded, whose destination is not within any subnetwork directly connected to the router. **The Internet** itself is a global network consisting of interconnected IP networks [24]. Figure 3.7 shows a highly simplified model of the Internet structure. Note that it is also possible to establish local closed IP networks not connected to the global Internet.

The IP protocol provides the following services:

- Forwarding of IP packets between two or multiple endpoints of an IP network. An endpoint is identified with its IP address, as used in the header of the IP packet.
- Routing of IP packets between interconnected IP subnetworks. Routing is done with help of the hierarchically composed IP address and routing mechanisms within IP router nodes. The routers maintain their routing databases by exchanging routing protocol messages. Routing protocols are used to spread information about the IP network structure among those routers, which use the same routing protocol.
- Fragmentation and reassembly of IP packets forwarded over a link when the size of the packet exceeds the maximum payload size of the link layer protocol frames. Fragmentation splits the packet to shorter fragments, each carrying a part of the payload of the original IP packet. Every fragment has an IP header similar to the fragmented packet, except that the header of the fragment has a fragment sequence number and indication of the last fragment.

Every host that communicates with IP protocol may have either one or many IP **network interfaces**. Such an interface is a connection point toward an IP subnetwork. The network interface is identified with its IP address

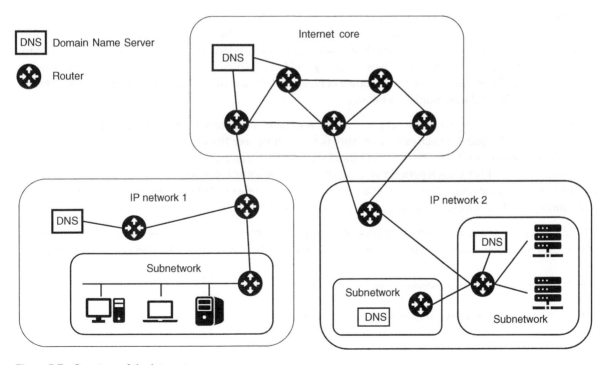

Figure 3.7 Structure of the Internet.

and it provides the network connectivity to some physical network with a link, such as LAN, GPRS mobile radio, or DSL connection over telephone subscriber line. If the host has multiple network interfaces to different IP sub-networks, it also has separate IP addresses for each of those interfaces. This is called **IP multihoming**.

3.3.2.2 IPv4 Addressing

IP address of the IPv4 protocol is a bit string of 32 bits. IPv4 addresses are typically expressed in a form where each set of 8 consecutive bits from the address is represented as a decimal number. The whole 32-bit address is divided into four such decimal numbers, separated as dots. An example of such an IPv4 address is 173.22.10.246. The IP address of a host serves two purposes: it is an identifier of the host and its network interface. But like telephone numbers used to be, an IP address identifies also the specific network to which the host currently belongs. IP addresses can be assigned to hosts either permanently or dynamically. Even when a host would have a permanent IP address assigned, if the host is moved to a new network, it will get a new IP address from the address range of the new network.

The internal structure of an IPv4 address was originally defined to consist of two parts, as shown in Figure 3.8, where the undefined individual bit values are denoted with "x":

- Subnetwork prefix, consisting of the network class identifier and a network number (netID) within the class as follows:
  - Class A networks have the first bit of subnetwork prefix as 0, after which the next 7 bits are used for NetID.
  - Class B networks have the first 2 bits of subnetwork prefix as 10, after which the next 14 bits are used for NetID.
  - Class C networks have the first 3 bits of subnetwork prefix as 110, after which the next 21 bits are used for NetID.
- Host number (hostID): The identifier of a single host/interface within the subnetwork, following the NetID and covering the rest of the least significant bits of the address.

Internet Corporation for Assigned Names and Numbers (ICANN) is a nonprofit organization responsible for IPv4 address allocation. The top-level allocation work is carried out by the Internet Assigned Numbers Authority (IANA) and five regional Internet registries (RIRs), which allocate IP address blocks for different Internet service providers. The IP network address space was originally divided into three parts based on the size of the subnet-works, to simplify the allocation process. As shown in Figure 3.8, the bits of an IPv4 address were divided between the network number and host number as follows:

- Class A networks are very large. The length of a class A network number is 7 bits, so there can be no more than 127 such networks globally. Each class A network is able to host over 10 million devices.
- Class B networks are mid-size. The length of the class B network number is 14 bits, so there can be 16 384 such networks globally. Each class B network is able to host roughly 650 000 devices.
- Class C networks are small. The length of a class C network number is 21 bits, so there can be 2 million such networks globally. Each class C network is able to host only 254 devices.

In addition to these classes, a part of the IPv4 address space has been reserved as class D to represent multicast-addresses, which allow sending packets to a group of endpoints. The class of an IPv4 address can be detected from

Class A: `0xxxxxxx.xxxxxxxx.xxxxxxxx.xxxxxxxx`   Subnetwork

Class B: `10xxxxxx.xxxxxxxx.xxxxxxxx.xxxxxxxx`   Host

Class C: `110xxxxx.xxxxxxxx.xxxxxxxx.xxxxxxxx`

Figure 3.8 IPv4 address structure, original design for different network classes.

the 1–3 bits in the beginning of the address, as shown in Figure 3.8. When looking into the decimal value of the first byte of the IPv4 address, the values are divided between classes as follows:

- Class A: 1–126
- Class B: 128–191
- Class C: 192–223
- Class D: 224–239

Addresses with value 127 in their first byte indicate the localhost or loopback address. The destination of a packet sent to the localhost address is the host itself. Address consisting of only 1 bits (255.255.255.255) are the broadcast address to reach all the IP endpoints of the local subnetwork [53].

Traditional IP routers make their routing decisions based on the network number of the IP address. With it, the routers check to which subnetwork (and the next hop router) the packet should be forwarded. If the subnetwork is the local one, then the host number of the address is used to figure out the link level address of the host to which the packet should be sent.

The division of the address space to such fixed classes in a mechanical way has been proven problematic for a number of reasons:

- If the subnetwork grows so that the number of its hosts exceeds the maximum number of hosts of its current class, either the subnetwork must be split or a new network number of higher class must be allocated to the subnetwork. In the latter case, the IP address of all hosts of the subnetwork must be changed.
- If the number of hosts in a subnetwork is not close to the maximum number of hosts in the subnetwork, a range of the address space is wasted as unused. This is troublesome because the number of IP-enabled hosts has grown so large that the available public IP addresses have become scarce. As the Internet continues to grow, the number of hosts and IP network interfaces in the Internet has already exceeded the number of available IPv4 addresses. This issue, known as IPv4 address depletion, is addressed with various approaches such as IPv4 subnet masking and IPv6 discussed in Sections 3.3 and 3.4.
- The fixed granularity of different address classes is not optimal in many cases. If an organization manages a number of consecutive network numbers of class C, the routers must keep in their routing tables one record for each 21-bit network number even if those networks could all be identified with a single identifier with a different size, like 18 bits. This causes the routing tables to become long, which decreases the performance of the routers.

Because of these problems, the original IPv4 address design with classes A, B, and C has been abandoned. Instead, the hosts and routers must learn how the bits of an IPv4 address have dynamically been divided between the network prefix and host ID. This is achieved with new designs known as **classless inter-domain routing (CIDR)** and **subnet masks**. As shown in Figure 3.8, classes A, B, and C addresses had a fixed number of bits for the network number. Classless inter-domain routing, specified in IETF RFC 4632 [56], introduces a new approach by dividing the address to **network prefix** and host ID. Length of the CIDR network prefix can be any number of bits between 1 and 32. The boundary between the network prefix and host ID parts of the IPv4 address can be represented as a **subnet mask**, which is a string of 32 bits [55]. The subnet mask has value 1 for those bit positions which are reserved for subnet number (or network prefix) and value 0 for those bit positions used for host ID. Routers maintain their own network prefixes for each known IP network in the routing table. Routers learn the subnet masks to be used for different networks from the routing protocol messages, as described in Section 3.5. Routing decisions are done based on the network prefix part of the destination IPv4 address.

CIDR introduced a new notation for representing IP addresses. The length of the network prefix is given right after the address itself, separated by a slash, like this: 172.16.0.0/12. This means that the 12 most significant bits of the IP address are used for the network prefix. In this case, the binary network ID is 101011000001, as represented by the decimal numbers 172.16. Flexible length of the network prefix allows organizations to maintain a good balance for the allocated subnetwork sizes, to avoid too big subnetworks with many unused IPv4 addresses. When the network evolves, the length of the network prefix can be adjusted.

Even with CIDR, the Internet has run out of IPv4 addresses, as the number of IP endpoints globally exceeded the number of different values of 32 bits used for IPv4 addresses. When the IPv4 address was designed the number of available addresses appeared as abundant. Nobody was able to predict how large a network the Internet would become in less than 50 years. According to ICANN, the regional RIRs had already run out of unallocated top level IPv4 address pools in 2011 [57]. The IPv4 address exhaustion problem [51] has been addressed with the following approaches:

- Instead of allocating IP addresses to IP network interfaces permanently, a dynamic address allocation scheme DHCP is used. With DHCP, if a device like a home computer is shut down, its IP address can be released and reallocated for another device. When the computer is restarted, it will get another IP address from the DHCP server. This model works well for such client hosts that do not provide any public service interfaces but just want to use services provided by others.
- The usage of so-called **private IP addresses** as defined in IETF RFC 1918 [58]. Certain ranges of IP address space have been reserved for private use, so that anyone can use (and even reuse) these addresses in their internal IP networks, provided that the private addresses are not visible to the public Internet. A private address is not unique over all the IP networks in the world, as multiple different networks may use the same private address. However, the private address must stay private and unique within a single IP network, otherwise its routers would be confused. The private address spaces of RFC 1918 are as follows:
  - 10.0.0.0/8 IP addresses: 10.0.0.0–10.255.255.255
  - 172.16.0.0/12 IP addresses: 172.16.0.0–172.31.255.255
  - 192.168.0.0/16 IP addresses: 192.168.0.0–192.168.255.255
- To deploy IP protocol version 6 (IPv6), in which the length of the address is 128 bits and the address space is much larger than that of IPv4. IPv6 was specified years ago, but it was deployed slowly due to the difficulties of coexistence of IPv4 and IPv6 infra as well as the extra effort needed to support and use IPv6.

If a host with a private IP address wants to send messages to an endpoint in another IP network, **network address translation** has to be performed for the address [10]. When a packet with a private IP source address is routed between subnetworks, a NAT device must be used to change the used private source address and the UDP or TCP source port number of the packet to another pair of public IP address and port number. The NAT takes the new pair of IP address and port number from its pool of unallocated pairs. NATs typically have only one or a few public addresses in its pool, but they can use many different **ephemeral port numbers** with those addresses. An ephemeral port number is one which has not been fixedly reserved for any well-known service. The mapping between the original IP address and port versus the new ones must be stored by the NAT as long as communication is going on over the NAT. The new IP address used must be a valid one in the destination subnetwork and the combination of the IP address and port unique, so that a reverse mapping could be performed by the NAT when forwarding responses back to the source. When the NAT receives a response to the packet sent earlier, it shall make a reverse address and port translation for the destination address of the response, to set those as the original private IP address and port which identified the source of the original packet to which is now being responded. An example of this process is shown in Figure 3.9.

NATs typically allocate the public IP address and port pairs against outgoing IP packets and keep their mapping to the private addresses (and ports) valid only for a short period to receive responses. Unless new outgoing packets are sent from the private address, the mapping is dropped so that the public address and port can be reallocated for another host. This scheme has the following consequences:

- Hosts with private IP addresses cannot be reached from external IP networks unless the host has initiated communication and has a valid NAT mapping to a public address and port pair.

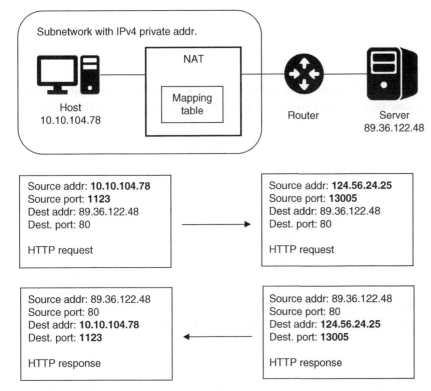

Figure 3.9 Network address translation.

- To keep its NAT address mapping valid for a longer time, the host shall periodically send some sort of keepalive messages to prevent the NAT from dropping the mapping.
- Some types of NATs allow incoming packets only from sources toward which the host has already sent a packet. This means that the NAT restricts the communication only for such external nodes of which the host has already reached out. For any other external hosts, the host behind the NAT would stay unreachable.

NAT technology is widely used to support reuse of the IPv4 private addresses. In many countries, most of the IP-capable end user hosts (such as laptops and smartphones) only get private IPv4 addresses from DHCP. Many popular Internet applications and application protocols working on top of IP are designed to cope with NAT traversal. The public IPv4 addresses are used for two main purposes: for servers that must be visible in the public Internet and for NATs using pools of public IP addresses to be mapped to the private IP addresses of hosts, to support communication over the Internet. Network address translation has been the key solution to extend the lifetime of IPv4 in the Internet even when the public IPv4 addresses are in short supply.

3.3.2.3 IPv4 Packet Structure
IPv4 protocol message shown in Figure 3.10 has the following fields:

- IP version: IPv4.
- The length of IP packet header, depending on the optional header fields used in the packet.
- Type of Service (TOS) is a field which describes the quality of service (QoS) needs of the packet. In the original TOS specification IETF RFC 1349 [59] this field was used to tell which QoS aspect to optimize: delay, throughput, or reliability. IETF RFC 2474 [60] later redefined the usage of field so that the field describes the QoS class

ver	hdl	TOS	len	flags	offset	TTL	Proto	Checksum	Src addr	Dest addr	Information

Figure 3.10 Structure of the IPv4 packet.

to be used for the packet. In the recent specifications, 6 bits of the TOS field were reallocated to differentiated services code point (DSCP) field and the other 2 bits as explicit congestion notification (ECN). QoS mechanisms are not widely used in the Internet, so DSCP value is used only within certain IP networks supporting QoS for QoS aware services such as Voice over IP.

- The total length of IP packet which can be 65 535 octets in maximum.
- The Identification field that has the same value for successive fragments of the original IP packet.
- Flags to control and identify fragments. These bits may be used to request that the packet would not be fragmented or tell which packet is the last fragment of the original packet so that the packet can be reassembled at the destination.
- Fragment offset, which is used to reassemble the fragments of an IP packet in the correct order.
- Time to live (TTL), which tells when to drop a packet that has gone to a routing loop. Each router decrements the TTL value of the packet with one, and when the value reaches zero the packet is discarded.
- Protocol identifier of the upper-level protocol (such as UPD, TCP, ICMP, IGMP, or OSPF) carried within the IP packet payload.
- Header checksum used for error detection. If the checksum does not match with the header field values as received, the router will discard the packet. The checksum must be recalculated by a router after decrementing the value of TTL.
- Source IPv4 address.
- Destination IPv4 address.
- Optional header fields, which may contain, for instance, a timestamp indicating when the packet was sent or a route selected by the sender for the packet (when using IP source routing).
- The payload of the IP packet.

3.3.3   ICMP and IGMP

Internet control message protocol (ICMP) provides error control and diagnostics services for IP network users. To provide those services, the following ICMP message types have been defined:

- ICMP errors, which a router or an IP endpoint may send back to the source of an IP packet when the packet could not be delivered to its destination. The ICMP error packet contains a subset of the headers of the original IP packet so that the recipient of the ICMP error could correlate the error with the corresponding IP packet.
- ICMP echo packets, to which the recipient must immediately respond. Based on any ICMP replies received, the sender of the echo packets can check the reachability of the destination IP address and the latency of the route used for the ICMP packets.
- ICMP source quench packets, which can be used in congestion cases to ask a source to slow down sending IP packets.
- ICMP router advertisement packets sent by IP routers, to allow other hosts in the subnetwork to find out the IP addresses of the routers.

The IP protocol itself does not have any response message types. If any endpoint or router processing IP packets would like to send feedback to the source of the IP packets concerning the IP protocol layer, ICMP protocol shall be used. IETF has published the following RFCs to specify the structure of ICMP packets and usage of them:

- RFC 792 [61] specifies ICMP protocol and its different message types.
- RFC 1256 [62] defines how a host can use ICMP protocol to find a local router.

The structure of ICMP packet is as follows:

- ICMP packet type.
- Error code, which describes the problem related to the received IP packet. Typical errors are routing failures so that a destination is unreachable or routing loops detected by IP packet TTL field reaching value zero.
- ICMP packet checksum.
- Other information within the ICMP packet, which most typically is the copy of headers of the received IP packet to correlate it with the ICMP packet.

Internet Group Management Protocol (IGMP) is an independent and optional protocol defined in IETF RFC 1122 [63]. It can be seen as an extension of ICMP, used by hosts to join or leave IPv4 multicast groups. As mentioned in Section 3.3.2.2, multicast groups are identified by IPv4 addresses of class D. To receive messages from a multicast group, the host joins the group by sending an IGMP report identifying the multicast group address in it. IGMP reports are processed by a multicast router of the subnet. When the router finds that the subnet has at least one host which has joined to the multicast group, the router will deliver the multicast messages to the subnet to be available for interested hosts in the subnetwork.

3.3.4 UDP

User datagram protocol (UDP) is a minimal transport layer protocol, which provides connectionless transport service of data between two IP hosts. UDP is specified in IETF RFC 768 [64]. UDP provides the following functions:

- Multiplexing flows of upper layer protocol messages for transport between two IP endpoints. Multiplexing is done with the help of UDP port numbers. UDP protocol uses the port numbers to map a flow of messages to an upper layer service of some sort. The mechanism is similar to the concept of TCP port described in the Section 3.3.5.
- Transport of data as UDP packets from the source to a destination. The source and destination addresses are not within the UDP packet but in the underlying IP packet. UDP packets contain only the source and destination ports mentioned above. UDP packets are sent one by one without opening any connections between endpoints. Any flow control, error correction, and retransmission mechanisms are left to the upper layer protocols.

An UDP packet has the following fields:

- Source port, which is the port number of the sender of the UDP packet.
- Destination port, which is the port number of the recipient of the UDP packet.
- Length of the UDP packet payload.
- Checksum, which covers UDP headers and a certain part of the headers of the IP packet, which carries the UDP packet.
- The payload of the UDP packet.

3.3.5 TCP

Transmission control protocol (TCP) is a transport layer protocol which provides connection-oriented and reliable transport of data between two IP endpoints. TCP as a protocol has evolved over time by extending the basic protocol with a number of additional mechanisms. Some of these mechanisms are mandatory, others recommended, and a few are experimental. The most important IETF specifications for TCP are as follows:

- RFC 793 [65]: Original transmission control protocol (TCP) specification
- RFC 1122 [63]: Additions and corrections to RFC 793

- RFC 2018 [66], 2883 [67], 6675 [68]: Selective TCP acknowledgments
- RFC 3042 [69], 3390 [70], 5681 [71]: Management of TCP congestion and slow start
- RFC 6298 [72]: Method for computing the retransmission timer for TCP
- RFC 7323 [73]: TCP extensions for paths with high bandwidth and delay

In addition to these baseline documents, RFC 7414 [74] provides an overview of all IETF documentation related to TCP. The list of the document is long because multiple optional mechanisms have been specified for TCP, to be used in various contexts. This book briefly describes the most common basic TCP mechanisms.

TCP has the following features:

- TCP multiplexes different data flows and upper layer protocols between two IP endpoints with the help of **TCP port numbers**. The servers within the Internet providing services over TCP use either a well-known, service-specific TCP port number or a port number which the clients can find out from a DNS server. For instance, when a user connects to any WWW server with HTTP protocol, an underlying connection is created with TCP to the well-known server port 80, which is reserved for HTTP protocol. The Web browser (being the client of the TCP connection) reserves dynamically a TCP client port from the TCP/IP stack of the host. The client port is reserved from the range of ports not allocated for any servers. The client port identifies the specific TCP connection among a set of TCP connections which the host or the browser may have opened toward different destinations.
- The data transported over TCP is a byte stream whose internal structure is unknown to the TCP protocol. The byte stream is split to parts, called TCP segments, for the transport. Every segment carries the sequence number of the first data byte in the segment. Additionally, every segment contains a checksum calculated over the segment. All the TCP segments are carried as payload of IP protocol. While IP makes its routing decisions totally unaware of the carried payload, the TCP segments sent between endpoints may be dropped or delivered via different IP routes, so that they may arrive at their destination in a different order than how they were sent. TCP uses the sequence numbers and checksums of segments to rearrange the segments at the destination and request retransmission of any corrupted or lost packets.
- TCP uses connections to manage delivery of TCP segments. Endpoints of a TCP data flow are identified by their IP addresses and TCP ports. A bidirectional TCP connection is opened between the endpoints before starting data transfer and closed after data transfer has been completed. A connection is created by exchanging TCP SYN packets between the endpoints and mutually acknowledging them with a three-way TCP handshake sequence. The maximum segment size is agreed for the connection within this initial sequence. A connection is closed by sending FIN packets over the connection.
- TCP uses the sliding window mechanism (see *Online Appendix A.10.2*) for flow control. TCP adjusts the window size based on the acknowledgments received from the remote end of the connection. The destination will return an acknowledgment (ACK) response for every other received TCP segment either as a separate TCP packet or piggybacked with TCP data sent back to the remote end.
- TCP uses multiple interrelated mechanisms for congestion control. The aim of TCP congestion control is to provide equal fair share of the bottleneck link capacity for each active connection over that link, while making sure that the link is fully utilized. TCP uses the following combination of congestion control mechanisms [71]:
  - Slow start of the connection
  - Congestion avoidance
  - Fast retransmit
  - Fast recovery

Let's start with a look into the TCP segment retransmission process. Typically, a received ACK message acknowledges successfully transported segments cumulatively up to the last segment indicated in ACK. If the TCP client does not receive an acknowledgment for a segment within the retransmission time calculated from measured round-trip time (RTT) [72], the client retransmits the segment. When using fast retransmit, if three consecutive received acknowledgments refer to the same sequence number, the client considers transmission of that segment failed and sends the segment again. This logic is based on the fact that a missing segment prevents the server increasing the sequence number while segments arriving later still generate acknowledgments. With the TCP selective acknowledgment option SACK, the endpoint may indicate specific segments received correctly so that only the missing segments must be retransmitted.

The TCP client maintains two different windows for the congestion control of a connection: receiver window and congestion window. The sizes of these windows are adjusted dynamically. The receiver window size reflects the capability of the remote TCP endpoint for processing the segments and the congestion window the capability of the network to forward the segments. At any moment, the smaller of the windows determines how many additional segments the client is allowed to send while waiting for an acknowledgment to a segment sent earlier. The client must not send segments with sequence numbers higher than the highest acknowledged sequence number plus the minimum of receiver or congestion window sizes.

The TCP ACK responses received over the TCP connection tell the client what is the size of free receive buffers at the destination. This value is set as the size of the receiver window. Management of the congestion window size is a more complicated process because the client does not get any direct indication about what is the capability of the network for segment forwarding. Further on, that capability is not static but varies as a function of network load. TCP uses heuristics to learn and adjust the feasible value of the congestion window with two complementary processes, known as slow start and congestion avoidance. The client determines which of those two processes to apply with a dynamically managed state variable called slow start threshold (ssthresh). When the congestion window size is below the slow start threshold value, slow start is used, and when it is above the threshold, congestion avoidance is used. The idea is to start sending segments slowly and increase the congestion window size (and data rate) exponentially per received acknowledgment until the threshold value is met or acknowledgments are no longer received (slow start). After meeting the threshold value, the window size is increased in a linear way once per round-trip time to maintain the optimal data rate (congestion avoidance). This process means that the client will gradually increase the bitrate of its TCP connection, eventually saturating the bottleneck link. The details of the process are as follows:

- When opening the TCP connection, the size of the congestion window is set to 2–10 maximum size segments, depending on what the sender maximum segment size (SMSS) is for the connection. The maximum size can be detected with a path MTU discovery process [75]. Whenever the client receives TCP ACK, it may increase the size of the congestion window by SMSS bytes until the window size exceeds the initial value of slow start threshold value or until no further acknowledgments arrive. This procedure is called slow start, used to find an initial sustainable bitrate for the new connection. Slow start process leads to exponential growth of the window size since when the window grows, there will be an increasing number of segments and acknowledgments on their way per RTT. If no ACKs arrive, that is probably caused by a router that has discarded some segments and/or ACKs due to congestion. In such a case, the client shall reset the value of slow start threshold to minimum of either twice the SMSS or half the amount of outstanding TCP data. Further on, the congestion window size is reset to one SMSS, and the slow start process is restarted. This mechanism helps in clearing the congestion quickly and prevents network collapse due to excessive load.
- The congestion avoidance method is used to maintain the size of the congestion window after the slow start has finished and congestion window size has exceeded current ssthresh value. As the load on the routers along with the connection path may change while the connection is used, the client shall adjust its congestion window size. With congestion avoidance, the congestion window is increased linearly by SMSS bytes (or a lower number of bytes acknowledged by latest ACK) once per RTT of the connection. If a segment is lost and no acknowledgment received, the segment is retransmitted and the ssthresh value is set to half of the current congestion window value. Thereafter, the congestion window size is reset and the client restarts the slow start process to rapidly increase the congestion window size to half of its earlier value. After that point, congestion avoidance is

Src port	Dest port	Seq # of first byte of the segment	Seq # of first byte of first next segment	Length of TCP headers	Control bits	Size of the receiver window	Checksum of the segement	Urgent pointer	Optional headers, padding	Data

Figure 3.11 Structure of TCP segment.

continued with a slightly less aggressive value of the congestion window. This method aims at dynamically maintaining the maximum sustainable bitrate of the connection but at the same time temporarily cutting the data rate when detecting a congestion.

- If TCP client receives three duplicated ACKs, the segment is considered lost and retransmitted with fast retransmit. However, the arriving ACKs indicate segments passing through the network; thus, in this case data rate will not be dropped as dramatically as when ACKs stop arriving due to bad congestion. Instead of returning to slow start, the fast recovery process simply cuts the congestion window to half and maintains the number of segments in transit until the transmission recovers and the congestion avoidance process is restarted. In fast recovery, the value of slow start threshold is reset to half of the congestion window, as when ACKs time out. The congestion window is set to the new ssthresh value + three times SMSS. This "inflates" the congestion window size by the number of segments which already have left the network and were acknowledged.

The TCP packet shown in Figure 3.11 contains the following fields:

- Source port, which is the port number of the client.
- Destination port to which the segments are sent.
- The sequence number of the first byte of the data in the segment.
- The sequence number of the first byte of data, which is expected to be received in the next TCP segment from the other end.
- Length of TCP headers, indicating the start of data in the TCP packet.
- Control bits used to indicate congestion or tell the type of TCP command carried by the TCP packet, such as SYN, FIN, ACK, or RST.
- The size of the receive window, which in the ACK segment tells the size of the free receive buffers.
- Checksum, which covers TCP segment and a certain part of the headers of the underlying IP packet.
- Urgent pointer, which tells the starting point of any urgent data sent within the segment.
- Optional headers of TCP.
- The payload of the segment.

3.3.6 SCTP

Stream Control Transmission Protocol (SCTP) is defined in IETF RFC 4960 [76]. Like TCP, SCTP is a connection-oriented protocol used on top of the IP. SCTP was designed to support telephone network signaling use cases, which require reliable messaging sessions between network elements. UDP is message-oriented but it lacks the needed reliability. TCP, on the other hand, supports reliable transport of byte streams, rather than messages.

SCTP has the following features:

- SCTP supports reliable transmission of messages.
- SCTP supports data rate adaptation.
- SCTP uses multistreaming where data can be partitioned to multiple streams in which messages are sequenced independently. If one of the streams is temporarily blocked, the other streams are able to continue.
- SCTP supports multihoming for redundancy and increased reliability. An SCTP connection may have multiple source and destination addresses, even if they always represent exactly two communication endpoints.

Src port	Dest port	Verification tag	Checksum	Data chunk				Data chunk			
				Type	Flags	Length	Value	Type	Flags	Length	Value

Figure 3.12 Structure of SCTP message.

- SCTP flow and congestion control algorithms work like those of TCP. SCTP supports selective ACK procedures and data segmentation.
- SCTP uses a four-way handshake to protect elements against flooding attacks.

The SCTP message consists of a common header followed by one or more data chunks, which follow the TLV structure as shown in Figure 3.12. Additionally, chunks contain flags which also are specific to chunk type. Flags are used, for instance, to control data segmentation. The chunk value field is a data structure, the format of which is defined for every chunk type. Chunk type zero carries payload data, and other types of chunks are used for SCTP acknowledgments and protocol control functions. Data chunks have an SCTP stream identifier and stream sequence number within the value data structure; thus, an SCTP message may contain chunks from one or multiple streams.

SCTP header has the following fields:

- Source and destination port numbers
- Verification tag to protect the message against man-in-the-middle attacks
- CRC-32 checksum

3.3.7 QUIC

Quick UDP Internet connections (QUIC) was a transport protocol developed at Google. It was deployed in 2012 with the Google Chrome browser to replace TCP as the transport protocol of HTTP traffic. The aim of QUIC was to speed up the connection setup process and support multiple parallel independent streams within a single connection. In 2015, QUIC was introduced into IETF as Internet-draft. For IETF, QUIC became the actual name of the protocol rather than an acronym. After a few years of further specification work, QUIC was published in 2021 within the following IETF RFC documents:

- RFC 8999 [77]: Version-Independent Properties of QUIC
- RFC 9000 [78]: QUIC: A UDP-Based Multiplexed and Secure Transport —the main QUIC specification
- RFC 9001 [79]: Using TLS to Secure QUIC
- RFC 9002 [80]: QUIC Loss Detection and Congestion Control

QUIC (as specified by IETF) is a secure and connection-oriented, general-purpose transport protocol, which runs over the UDP protocol. QUIC has the following features:

- QUIC uses multi-streaming, where data can be partitioned to multiple parallel streams in which QUIC packets are sequenced independently. If there is any problem, such as packet loss, only the affected streams are blocked for retransmission. The other QUIC streams can continue operating normally. Information within QUIC streams is transported as QUIC packets, containing one or multiple QUIC frames. QUIC has different frame types, used for connection and stream management, transport of user data, retransmission, flow, and congestion control.
- QUIC provides reliable transport of data as it establishes both flow control and packet retransmission processes for the streams. QUIC uses limit-based flow control where the receiver announces the limit of bytes it is prepared to receive on a given stream or for the whole connection. Lost QUIC packets are not retransmitted as such. Instead, the information carried in the lost QUIC frames is sent again within new frames, as necessary.

- QUIC supports congestion control with generic signals, such as ECN in the IPv4 packet header. The sender may use an implementation-specific algorithm, such as CUBIC or the default QUIC congestion controller introduced in RFC 9002. The default algorithm uses slow start and congestion avoidance processes similar to TCP. QUIC establishes rules for declaring a packet as lost instead of being delayed or reordered. QUIC sender enters a recovery period when loss of packet is detected or ECN is reported by the peer. At this point, both slow start threshold and congestion window are set to half of the earlier value of congestion window. After packet loss, the QUIC client waits for a probe timeout before sending probe packets to find out if the congestion was only a temporary one. The recovery period ends and congestion avoidance is started when a packet sent during the recovery period is acknowledged. In case of more persistent congestion detected by failing probes, the congestion window is set to its minimum value and slow start is triggered to help the congestion to be cleared.
- QUIC supports confidentiality and integrity protection of the streams with the methods specified for TLS and essentially brings the best properties of TCP and TLS together. Each QUIC packet is individually encrypted to allow the destinations to decrypt them without delay. QUIC reduces the overhead for connection setup when securing the data with TLS as QUIC combines the negotiation of the security keys and exchange of other transport parameters into one single integrated connection setup process. This is different from TCP, where the TCP connection has to be established at first and only thereafter the TLS handshake can take place, as described in Section 3.3.10.2.
- QUIC introduces connection identifiers to QUIC packets, which allows seamless migration of QUIC streams over a new access link. The connection ID reference of a stream would stay as is, even if the IP address of the host is changed. With this feature, QUIC supports a mobile client to switch between networks (like WiFi and cellular) causing the used IP address to be changed. In such a case, a TCP client would simply drop and reestablish the connections, which takes time and disrupts the session. The QUIC client is able to avoid such connection drops by relying on immutable connection identifiers.

QUIC packets have multiple different header formats, depending on the packet type. The following are examples of QUIC packet header fields:

- Header form: Indicates the type of QUIC packet header format.
- Packet type: Identifies the type of the packet.
- Version: Identifies the QUIC protocol version.
- Destination and source connection ID are the identifiers of the connection at both of its endpoints.
- Packet number is the sequence number of the QUIC packet.

3.3.8 DNS

Domain name system (DNS) is one of the key mechanisms used for endpoint identification in IP networks. With the help of DNS, the users do not have to use or remember numeric IP addresses of the endpoints but can instead use human readable **domain names**, such as http://wiley.com. The domain name space has a hierarchical tree structure, and a domain name represents a single branch of the tree [81]. The different parts of the name separated with dots are used for the different levels of the naming hierarchy. The last part of the name is the highest hierarchy level. Domain names owned by organizations or private citizens can be used in many contexts. Domain names can be found within email addresses like john.smith@myorg.com or embedded into Web site names such as www.wikipedia.org or www.gov.uk.

The administration of domain names is also organized in a hierarchical manner. Within each level of naming hierarchy, any domain has an owner responsible for allocating names for its subdomains. The top-level domains are managed by ICANN, which also allocates IP address blocks. The most widely used top-level domains are as follows:

- com: Commercial enterprises
- gov: Public authorities of the United States

- edu: Universities
- org: Nonprofit organizations
- Two-letter country codes such as us, uk, de, or fi as defined in ISO 3611 standard [82]

The complete list of top-level domains maintained by the Internet Assigned Numbers Authority (IANA, a body of ICANN) can be found from http://www.iana.org/domains/root/db.

Subdomains are administered by registrars which ICANN accredits to do the job. Each top-level domain has a single owner which is responsible for allocating any subdomain names under the top-level name. For instance, at the time of this writing, VeriSign Global Registry Services administers the subdomains of the .com domain. When an organization has been given a top-level domain or a subdomain, it may thereafter create the needed subdomains under its assigned root domain. With the help of subdomains, the organization may direct the users to the right part of its Web portal. For instance, a multinational commercial enterprise might like to have subdomains for its country-specific Web pages or for the various subsidiaries of the corporation.

Mapping numerical IP addresses to domain names has been done since the dawn of the Internet. In the early days of the Internet, the Network Information Center (NIC) in the United States maintained a single file which mapped the domain names with the corresponding numeric IP addresses. The users of the Internet loaded this file regularly to their computers with the help of a file transfer protocol known with its abbreviated name FTP. When the Internet gradually expanded, it soon become impossible to manually administer a centralized database of all the domain names and their IP addresses. A decision was made to create a distributed domain name service (DNS) to support automatic domain name queries. Each organization which administrates a given domain and its subdomains must maintain a domain name server with a database to record the mappings of the domain names and corresponding IP addresses within the managed subdomain. For big subdomains, one domain name server may not be enough, but instead a hierarchy of those would be needed. Instead of managing its own DNS servers, the organization may outsource this task from a specialized DNS service provider. When the users of the Internet know the domain name of a server, they can fetch the IP address of the server from the authoritative DNS server of the same domain. Resolving domain names to IP addresses is done with the help of automated domain name service (DNS) protocol queries behind the scenes, for instance, when a Web browser tries to access some pages from a certain domain.

IETF has defined the operation of the DNS system in a number of specifications, including the following RFCs:

- RFC 1034 [81] and 1035 [83]: Original concept and implementation specifications of DNS.
- RFC 1101 [84]: Method for mapping network name to a number in a DNS system.
- RFC 2535 [85], 4033 [86] – 4035 [87]: Security extensions for DNS.
- RFC 2181 [88], 4343 [89], 4592 [90]: Clarifications to earlier DNS standards.
- RFC 2308 [91]: Improvements to DNS caching.
- RFC 2782 [92]: Ways to search a domain name of a server via DNS based on the domain name of the organization and type of protocol (such as SIP Voice over IP) used for the service.
- RFC 5966 [93]: Requirements for supporting TCP as a transport protocol for DNS.

The domain name system is composed of a hierarchy of domain name servers. When a computer wants to resolve a domain name to the corresponding IP address, it sends a **DNS lookup request** to its local DNS. If the local DNS server knows the answer, it returns the mapping in a DNS lookup response message [81]. In many cases, the local DNS server does not know the domain name and a recursive DNS lookup has to be used. In that case, the local name server sends a new DNS lookup request to another DNS server, which may be in the DNS server hierarchy at the same, higher, or lower level than the local server. The address of the second DNS server should at least partly match with the domain name being resolved. When the lookup query eventually reaches a DNS that is able to complete the resolution, the lookup response will be returned back via the chain of DNS servers that participated to the recursive lookups. On the top of the global Internet DNS server hierarchy, there are a few DNS root servers to store records about the top-level domains administered by IANA.

Each DNS server in the chain of recursive lookups stores the returned answer to its own temporary storage, called **DNS cache** [83]. The purpose of caching is to improve performance of answering further lookups of the same domain name. By storing the answers temporarily closer to users who were interested of the domain, further requests could be completed sooner. Thus, a domain name server is able to provide answers for lookup requests based on data within its own permanent database or the data temporary stored into its cache. Most of the end user computers also store answers received for recent DNS lookups to their local DNS cache so that when using a server multiple times a day or loading many different Web pages from the one server in a session, no further DNS messages need to be sent to the network for resolving a domain after the first lookup, for it has recently been answered. When data in the cache becomes old enough, it will be purged.

The database of a domain name server contains DNS resource records. The most common type of resource record is A, which stores the IPv4 address corresponding to a domain name. DNS is designed to be a generic name service so that the DNS server may possess many different types of resource records. Examples of other record types are the following:

- AAAA record storing the IPv6 address corresponding to a domain name [94].
- CNAME record storing another alias name corresponding to a domain name.
- NS record, which stores the name of an authoritative name server responsible for managing domain names of a given domain name space.
- SRV record, which contains a more specific domain name against the domain of an organization and protocol used for the service [92].

The domain name lookup queries can be sent over either UDP or TCP transport protocols. Typically, the DNS queries are sent over UDP, but if the answer does not fit the maximum size of UDP packet, TCP can be used to avoid problems caused by IP packet fragmentation [93].

DNS lookup and response messages have a common structure shown in Figure 3.13:

- ID: Identification number given for the lookup query, used to map the response to the original query.
- Parameters which tell, among other things, if the message is a query or response, the specific type of the query, and whether recursive query is requested or supported.
- QDcount: Number of separate queries in the message.
- ANcount: Number of separate answers in the response.
- NScount: Number of name server records in the response, used to describe the authoritative name servers responsible for the name space referred by the original lookup.
- ARcount: Number of additional info records in the response.
- The individual queries in the lookup message, each of which gives the type of the record being checked and the corresponding domain name.
- Returned domain name server records in the response.
- Additional information in the response. If the returned records contain references to more specific domain names, the additional information may give records which contain IP addresses of those additional domain names returned in the same response. In this way, DNS is able to return all the needed pieces of information in one response, rather than forcing the DNS clients to send a number of queries after finding out that more information is still needed.

ID	Params	QDcount	ANcount	NScount	ARcount	Queries	Returned records	Additional information

Figure 3.13 Structure of a DNS message.

3.3.9 DHCP

Dynamic host configuration protocol (DHCP) is a protocol with which a host that wants to join an IP network can automatically fetch the most important network parameters. Such parameters are, for instance, the local subnet mask or addresses of other key servers vital for the correct operation of Internet protocol, such as the address of a DNS server [95]. If the host does not have an IP address permanently assigned for the host, it can use DHCP to get an IP address temporarily assigned for itself [96]. The importance of DHCP has grown as the number of IP-enabled hosts has exploded via the proliferation of workstations, home computers, IP-enabled mobile phones, and other network-connected appliances.

DHCP servers allocate IP addresses dynamically for hosts connected to the network, based on requests from the hosts. This method has the following benefits:

- There is no need to manually configure IP addresses to the hosts as they will automatically fetch addresses from a DHCP server.
- There is no need for all the hosts to have IP addresses all the time. When a host is disconnected from a network or switched off, it does not need an IP address. When assigning IP addresses dynamically, only when necessary, the total number of IP addresses needed for the network can be reduced.

Every time when a host requests a fresh IP address for itself, it may receive a different IP address from the DHCP server than the one used at the earlier time. This is not a major drawback because most IP-based services such as Web browsing or emails are based on the famous client–server model. When a host has a role of a client, it sends its own IP address (as allocated by the DHCP server) as the source address of any IP packets sent out to the Web or email server. The server can return its response to that source address. It does not cause any harm if a host uses one IP address today and another one tomorrow. For the cases in which a host would for some reason need to have the same IP address all the time, DHCP supports also such fixed IP address allocation. A host may request a known IP address by adding that address to its DHCP discovery or request message.

Figure 3.14 shows the DHCP message exchange with which a host can get an IP address for itself. When the host is powered on and proceeds with setting up its network interfaces for Internet protocol, it sends a DHCPDISCOVER request message to reach out to any local DHCP server. The host does not know the IP address of the DHCP server, so it sends the request to all the computers of the network by using the IP broadcast address 255.255.255.255 as the destination address of the IP packet. The source address of DHCPDISCOVER message is set as 0.0.0.0 because the computer does not yet have an IP address that it could use. Despite the lack of the source IP address, the

Figure 3.14 IP address allocation process over DHCP.

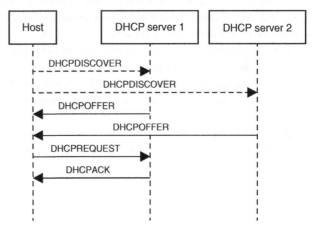

IP packet is carried within a link layer frame which has the valid link layer source MAC address of the host. This MAC address is also written into the DHCPDISCOVER message so that the DHCP server could return its response to the correct host that owns the MAC address.

When a DHCP server receives a DHCPDISCOVER message, it responds with a DHCPOFFER message. This message contains an IP address allocated to the host and any other parameters that the host may have requested in the DHCPDISCOVER message, such as the address of a local DNS server. The DHCPOFFER message also contains the IP address of the DHCP server itself, so that the host could send any further DHCP messages directly to that address rather than to the broadcast address.

The host may receive either a single or multiple DHCPOFFER messages from different DHCP servers. The host picks the first of them and takes the IP address from it into use. The host thereafter responds to the chosen DHCPOFFER message with a DHCPREQUEST message. When the DHCP server receives the response from the host, it records the allocated IP address to have been reserved and returns a DHCPACK message to the host. If a DHCP server does not receive a DHCPREQUEST message as a response to a DHCPOFFER message sent, it deems that the IP address sent within the DHCPOFFER message is still available for some other computer.

Allocation of the IP address to the host with a DHCPOFFER is a temporary one. It will expire after a period, which is told in the DHCPACK message in the end of the DHCP message sequence. If the host does not request extension of the period before it expires, the DHCP server will release the address automatically and may offer the same address later to another computer. This may happen, for instance, if the host is switched off. When the host wants to continue using its IP address and extend the allocation period, it shall send a new DHCPREQUEST message to the DHCP server before expiration. In that message, the host tells the IP address that it wants to keep. After extending the lease period, the DHCP server returns to the host a DHCPACK message with a renewed period.

The DHCP protocol is based on an earlier BOOTP protocol, which was used by diskless workstations for getting IP addresses and fetching copies of an operating system from a BOOTP server [97].

DHCP and BOOTP protocols share a common message structure shown in Figure 3.15:

- Op code tells if the message is a request or response.
- HWtype defines the type of the underlying link layer protocol, such as Ethernet LAN.
- Length is the length of the link layer address in the message.
- Hops is a counter that is increased every time a router forwards the message to another network. When the counter reaches a predefined threshold value, the routers stop forwarding the message any further.
- Transaction ID is used to match the response with the original request.
- Seconds records the time elapsed from starting the DHCP lookup process.
- Flags carry additional DHCP protocol information.
- Client IP address is either 0.0.0.0 or the IP address the host has requested for itself.
- Your IP address is the address that the DHCP server offers to the host.
- Server IP address is the address of the DHCP server, as given in DHCPOFFER sent by the server.
- Router IP address is the address of a router connected to the network, capable of routing messages between networks.
- Client HW address is the link layer address of the host.
- Server host name is the domain name of the DHCP server.

Op code	HW type	Len	Hops	Trasaction ID	Seconds	Flags	Client IP addr	Your IP addr	Server IP addr	Router IP addr	Client HW addr	Server host name	Boot file name	Options

Figure 3.15 Structure of DHCP message.

- Boot file name is a field used within BOOTP protocol to tell the name of the operating system file, which a workstation could download with the TFTP protocol.
- Options are used for providing hosts with additional network parameters. In a request from a host, options contain the identifiers of requested parameters, and in a response from the DHCP server options contain the requested information [98].

The specifications of DHCP protocol refer to the BOOTP specifications. The following RFCs specify the basic structures and operations of these protocols:

- RFC 951 [97]: BOOTP protocol
- RFC 2131 [96]: DHCP protocol
- RFC 2132 [98], 3942 [99]: DHCP options

Various DHCP options have been defined in a number of other IETF RFC documents such as 2937 [95], 3011 [100], 3046 [101], and 3361 [102].

3.3.10 Security of IPv4 Data Flows

Due to the open nature of IP networks, information security is a crucial aspect of IP-based communications. Many applications are publicly available for capturing and analyzing IP network traffic. IETF requires any new protocol proposals to contain analysis of related security threats, which must then be addressed in the protocol design. Security must be a feature built into every new standard protocol.

IETF has specified two major protocols focused on securing the data transported over IP:

- IP security (IPSec)
- Transport layer security (TLS)

These connection-oriented protocols support reliable identification of connection endpoints as well as encryption and integrity protection of the data transported between them.

3.3.10.1 IPSec

IP security (IPSec) is a generic protocol for securing IP packet data flows. IPSec uses the Internet key exchange (IKE) protocol used for managing the security keys. IETF RFC 4301 [103] provides an overview of IPSec.

IPSec provides the following services for its clients:

- Confidentiality: Encryption of transported data to prevent eavesdropping.
- Integrity protection: Preventing unauthorized modification of data in transit.
- Authentication: Identifying the sender of the data in a reliable way.

IPSec has been used for two major purposes:

- Virtual private networking (VPN) to provide a way to create secure connections over the public Internet. Messages sent over a VPN are encrypted and can be interpreted only by the endpoints of the VPN tunnel. VPN tunnels may be used to interconnect different sites of a corporation and protect the confidential corporate internal data flows. Another typical VPN use case is secure access to corporate networks by employees remotely from their homes or while traveling. The endpoints of such VPN tunnels are the VPN client in a user's laptop and the VPN concentrator at the corporate data center. VPN has also become a consumer service via multiple VPN service providers, who offer Internet access over VPN. Such VPN service providers claim VPN to provide various sorts of benefits, ranging from additional security to enabling access for services that might otherwise be unavailable or blocked. VPNs typically use either IPSec or TLS protocols for building secure tunnels between the VPN endpoints.

- 3GPP IP multimedia subsystem (IMS), which uses IPSec to protect IMS signaling between the terminals and call control servers. IMS is used for multimedia services, such as voice or video calls, provided by network operators over their packet-switched mobile networks.

IPSec consists of two separate protocols, which may be used either separately or together to protect a connection:

- Authentication header (AH) specified in IETF RFC 4302 [104] is an integrity protection protocol with which a receiver can reliably identify the sender of IP packets and ensure that the packets have not been modified in transit.
- Encapsulation security payload (ESP) specified in IETF RFC 4303 [105] supports encryption enabling the sender to ensure that no one else other than the legitimate recipient of IP packets is able to interpret the data within them. ESP also supports integrity protection and authentication, like AH.

Both AH and ESP add their own headers around the IP packet to be protected. The recipient of the packet is able to decrypt the packet and/or ensure the authenticity of the data based on these headers and the secret keys shared between IPSec tunnel endpoints. Each bidirectional IPSec connection is described as a pair of security association (SA) data records in the security association database of the IPSec protocol module.

IPSec connection can be used in two different modes, shown in Figure 3.16:

- **Transport mode,** where the AH or ESP headers are added between the IP headers of protected packets and the headers of upper layer protocol such as UDP or TCP. Transport mode can be used between the endpoints of the protected IP data flow.

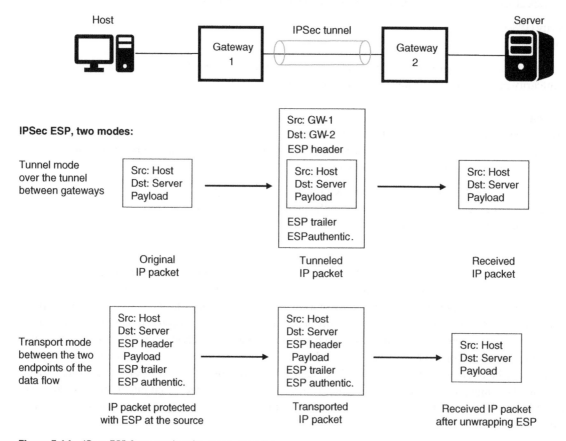

Figure 3.16 IPsec ESP for tunnel and transport modes.

- **Tunnel mode,** where the protected IP packet is encapsulated within an outer IP packet. The tunnel mode is typically used when one of the IPSec connection endpoints is not the endpoint of the protected IP data flow but instead a router or gateway forwarding the inner IP packets toward their final endpoint. In this mode, the IP packets belonging to the inner data flow are transported within the outer IPSec tunnel up to the router, such as the VPN concentrator in the corporate network, used to terminate the tunnel. The IPSec AH or ESP headers are added between the IP headers of inner and outer packets.

There are two databases related to IPSec in both endpoints of the IPSec connection:

- Security policy database (SPD), which contains such pieces of information with which IPSec layer can decide whether a certain IP packet shall be sent over a specific IPSec connection (or security association). Such pieces of information are, for instance,
  - The address or domain name of packet source or destination.
  - The transport protocol used (UDP or TCP) and the related port numbers.
- Security association database (SAD), which describes security associations (SA) created for each IPSec connection. Each SA record contains the following kind of data:
  - The identifier of the SA, consisting of the destination IP address, IPSec protocol used (AH or ESP), and the security parameter index (SPI) used to separate the security associations toward one single destination.
  - The values of counter fields within the AH and ESP headers for each SA.
  - The algorithms, integrity, and encryption keys used for AH and ESP.
  - Whether transport or tunnel mode is used for the SA.
  - The remaining lifetime of the IPSec SA.
  - The maximum transfer unit (MTU) defining the maximum size of IP packet to be transported over the IPSec SA without fragmentation.

The algorithms and master keys can be either configured to the endpoints manually or agreed automatically when opening the SA. The first option can be used, for instance, when creating a fixed VPN connection between sites of an enterprise. Master keys can also be delivered to an endpoint within a smart card such as universal integrated circuit card (UICC), also known as a SIM card used within mobile phones. When using the IKE protocol during the IPSec connection setup phase, the applied security algorithms and used keys will be agreed in two phases. The first phase is done only once, but the second phase is repeated periodically as long as the IPSec connection is used. In the first phase, the endpoints derive a common master key that is used later to derive session keys for encryption. In the second phase, the endpoints agree on the session keys to be used for data encryption and integrity protection for a certain period. The session keys are changed periodically, and the master keys are permanent. This arrangement limits the impact of a third party learning a session key in some way. IKE protocol works as follows:

1) At first, the connection endpoints exchange plaintext messages to agree on the authentication and encryption algorithms used for creating the master key. They also agree on the method for creating random numbers needed in the key exchange process.
2) Next, the endpoints exchange cleartext messages, which contain data used for the Diffie-Hellman algorithm (see *Online Appendix A.13*) used to derive the secret master key.
3) After the same master key has been derived separately by both the endpoints, the endpoints exchange the first encrypted and digitally signed messages to authenticate themselves. The authentication is based on the public key certificates of both the endpoints. This step completes the first phase of IKE.
4) In the second phase of the IKE, the endpoints exchange encrypted messages to authenticate earlier messages, exchange their identities and certificates, create IPSec security associations, and agree on the security algorithms and session keys to be used on those. The keys may either be based on the secret master key that was created at the first phase or they can be separately created during the second phase, again using the Diffie-Hellman algorithm. In the latter case, the session keys will be independent of the master key and any encryption keys used on the first phase.

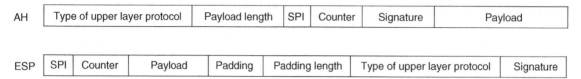

Figure 3.17 Structure of IPSec AH and ESP messages.

5) The second phase is completed so that the endpoint that initiated the phase sends a new message encrypted with the session keys. By decrypting that message, the other endpoint can ensure that session keys were taken successfully into use.

The current version [103] of the IPSec protocol relies on version two of the IKE key exchange protocol [106]. IKEv2 supports certain advanced SA negotiation methods, for instance, for a range of ports used instead of only individual ports.

After opening the IPSec SAs, AH and/or ESP headers shown in Figure 3.17 will be added to all the IP packets sent over the IPSec connection. When using ESP headers, the upper layer data in the packets is encrypted.

The AH message consists of the following fields:

- The type of upper layer protocol carried by the IP packet.
- The length of the packet payload.
- Value of SPI.
- Counter, which is incremented for each IPSec packet sent. This counter is used to prevent replay attack, where a packet captured from a network would be maliciously sent once again.
- Digital signature calculated over the packet payload and part of the headers. There are many different algorithms (such as HMAC-MD5-96 and HMAC-SHA-1-96) that can be used for creating the signature.
- Payload, which is the signed IP packet.

The ESP message consists of the following fields:

- Value of SPI.
- Counter, which is incremented for each IPSec packet sent. This counter is used to prevent replay attack, where a packet captured from a network would be maliciously sent again.
- Payload encrypted with an encryption algorithm.
- Empty padding bytes added to the end of the payload to ensure that the length of the payload and padding can be divided by four.
- The number of padding bytes added.
- The type of upper layer protocol carried by the IP packet.
- Digital signature calculated over the packet payload and the ESP headers.

3.3.10.2 TLS

Secure socket layer (SSL) was a security solution used to protect HTTP messages used for Web browsing. SSL was developed by Netscape Communications during the 1990s. Initially, SSL was deployed wider than any competing solutions as SSL was available without charge as part of the Netscape Web browser. The protocol was improved several times and SSL version 3.0 was used in 1996. Later, SSL was largely replaced by **transport layer security (TLS),** which was derived from SSL with minor changes to it. TLS is a standard protocol used to protect any connection-oriented IP data flows. Version 1.0 of TLS was defined in IETF RFC 2246 [107], version 1.1 in RFC 4346 [108], version 1.2 in RFC 5246 [109], and version 1.3 in RFC 8446 [110]. The TLS protocol standardized by IETF differs from SSL only in a few respects. The TLS functionality for digital signatures is improved from SSL, for instance, digital signature standard (DSS) is supported in the TLS handshake.

TLS provides the following services:

- Encryption of data.
- Integrity protection of the data against alterations during transit.
- Authenticating the sender of the data.

Within the OSI layer reference model, the TLS protocol can be considered to belong to the session layer. TLS supports many different encryption algorithms. One of them is used for a TLS protected connection. TLS versions earlier than 1.3 supported creating encryption keys in two different ways. They could be created either with the Diffie-Hellman algorithm or by the client selecting the master key and sending it to the server, encrypted with the public key retrieved from the server certificate. In version 1.3 of TLS, the usage of the Diffie-Hellman algorithm and the key exchange messages were deprecated; instead, usage of the public key certificates stayed as the only approach for the key exchange.

It is possible to manage the encryption mechanisms used for complete HTTP sessions with the help of TLS. TLS is able to reuse encryption keys created for one TCP connection for another TCP connection opened later between the same endpoints. When a Web browser fetches multiple Web pages from a server, the browser may open and close multiple TCP connections for transporting the data, all protected by a single TLS session with the server. TLS is also commonly used to protect VPN connections of remote users to corporate networks.

TLS creates a protected session with the help of a TLS handshake protocol as follows:

1) The client (such as a Web browser) sends a TLS client a Hello message in which it tells the TLS version being used, the identifier of the session being created, and algorithms that can be used for key exchange, encryption, digital signatures, and data compression.
2) The server responds with a TLS Server Hello message in which the server tells the algorithms it supports TLS. If the server has a public key certificate, the server can send a certificate message or server key exchange message to the client. The server may also request the client to send its own certificate by sending a client request message. In practice, this is not done too often because the clients very seldom are in possession of client certificates. In the end, the server allows the client to send its responses by sending a Server Hello Done message to the client.
3) The client answers either by sending its own certificate (if any) within certificate message or, alternatively, continuing the key exchange process by sending a client key exchange message. Thereafter, the client sends a Finished message to tell it is ready.
4) Finally, the server completes the handshake with the Finished message.

At the end of the initial handshake, the endpoints may either start the encryption or agree with the session keys and algorithms by exchanging change cipher spec messages before sending encrypted Finished messages.

If the client wants to map a new TCP connection to an existing TLS session, the client shall send the identifier of the existing session in the Client Hello message. In this case, the server will send change cipher spec and Finished messages right after its Server Hello message. The client acknowledges those with same types of messages. Reusing an existing TLS session is much simpler and quicker than creating a new session. All the TLS control messages and encrypted HTTP messages are transported as payload of TLS record protocol messages. Record protocol has header fields to tell the type of TLS control protocol used, the TLS version, and length of the payload.

3.4 Internet Protocol Version 6

3.4.1 Standardization of IPv6 and the Initial Challenges

In the early 1990s, it was found that the length of the IPv4 address was too short to support the projected number of hosts to be connected to the Internet. The length of the IPv4 address is 32 bits, and in theory it is possible to create roughly 4 000 000 000 different addresses. In practice, due to the hierarchical structure of IPv4 addresses, not all the addresses can be used as the IP networks simply do not contain exactly as many hosts as would be available for the network prefix chosen, even when using the CIDR mechanism. Because of that, the number of potentially available IPv4 addresses is lower than the theoretical maximum number. It became clear that the length of the IPv4 address would eventually limit the expansion of the Internet. After all the available IPv4 addresses had been taken into use, the existing Internet would still work but no new networks or servers could be connected.

To solve such an alarming problem, in 1994 IETF started the work to design a new version of IP protocol. The outcome was that the specifications for **Internet protocol version 6 (IPv6)** provided a new structure for IP addresses with 128 bits. The length of the IPv6 address is sufficient for creating more than 1 000 addresses per every square meter of earth, even with quite inefficient address allocation schemes [10]. The format of IPv6 packets differs quite much from IPv4 packets. The most important new features of IPv6 in comparison to IPv4 are as follows:

- IPv6 has a much larger address space than IPv4. There are no limitations for the length of network and subnetwork prefixes in IPv6.
- IPv6 supports automatic generation of unique IPv6 addresses for any IPv6-enabled host connected to the network.
- IPv6 natively supports embedded security mechanisms based on approaches that were earlier available with a separate IPSec protocol.
- IPv6 has an improved multicast mechanism to support sending a message to a group of endpoints.
- IPv6 supports more efficient routing as IPv6 routers do not fragment packets, and they rely on link level error detection and correction, rather than error detection within the IP header.

Deployment of IPv6 has significant impacts to other basic protocols widely used in IP networks, such as ICMP, DHCP, DNS, and IPSec, which have been respecified to support IPv6. At the time of this writing, the following RFCs are the most important IETF specifications related to IPv6:

- RFC 8200 [111]: Message structure and operation of IPv6.
- RFC 2675 [112]: Very large IPv6 packets called jumbograms.
- RFC 3056 [113], 4213 [114], 4241 [115], 4380 [116], 4798 [117], 6052 [118], 6146 [119], 6147 [120], 6180 [121], 7949 [122]: Mechanisms and practices to support parallel use of IPv4 and IPv6.
- RFC 4704 [123], 8415 [124]: DHCPv6.
- RFC 3363 [125], 3596 [94]: IPv6 and DNS.
- RFC 6724 [126]: Default selection of IPv6 address.
- RFC 3587 [127], 4007 [128], 4193 [129], 4291 [130]: IPv6 addressing architecture.
- RFC 4443 [131]: ICMPv6
- RFC 4861 [132], 5942 [133]: Discovery of the neighboring IPv6 hosts on a shared link.
- RFC 4862 [134]: IPv6 stateless address autoconfiguration without DHCPv6.

When the specification of IPv6 was started during the 1990s, it was foreseen that IPv4 address space would be depleted in only a few years, causing a hard stop for growth of the Internet. However, with the introduction of methods like CIDR, DHCP, and network address translation, the need for public IPv4 addresses slowed down. Even when there are no new public IPv4 address available, it has been possible to grow the IPv4 network thanks to these methods. Facing both technical difficulties and economic challenges, the IPv6 deployment has been sluggish, which has meant a longer than expected lifetime for IPv4 and slower than expected takeoff for IPv6. That said, the Internet is making its way toward IPv6 step by step.

Although IPv6 specifications were largely completed well before 2010, the progress of its deployment has been slow for the following reasons. At first, it took a long time for the most important device and software vendors to add IPv6 support to their protocol stacks and operating systems used in routers, servers, and end user computers. However, adding IPv6 support was simply far from enough. The most important obstacle for IPv6 was the ubiquitous use of IPv4 protocol, which just makes replacing IPv4 very difficult. At the end of the day, nobody wanted to break the existing interconnectivity between the Internet-enabled hosts by having some hosts communicating with the old proven IPv4 while the others used newer and more capable IPv6, which were not compatible between each other.

Since 2000, IETF has tried to address the **IPv6 transition** problem by specifying a number of different transition mechanisms that support parallel use and interoperability of IPv4 and IPv6. Despite of all the engineering effort, the transition has not been without technical problems, economic issues, and risks for service continuity. In 2007, over 400 million IPv4 addresses were already allocated. As the Internet was still growing, at that time it was predicted that the Internet would run out of all the available IPv4 addresses before 2012. This prediction eventually came true in 2011. Still, the usage of DHCP and network address translators (NAT) together with private IPv4 address space (see Section 3.3.2.2) has been able to keep IPv4 networking as the mainstream technology.

Development of NAT technology took place in parallel with IPv6 development. Both can be seen as competing approaches to solve the same IPv4 address shortage problem. The benefit of network address translation is that it does not break compatibility of different hosts and it supports well the main use cases of the World Wide Web. Its drawback is that NAT does not support well use cases like VoIP or public servers where the connection endpoints should have IP addresses unique and publicly known within the Internet. The deployment of NATs has been significantly easier than deployment of IPv6, but the wide acceptance of network address translation has slowed down the adoption of IPv6. Nevertheless, IPv4 address shortage is still an issue as the new networks, servers, and NATs still need some public IP addresses for their connections toward the Internet. Service providers are moving toward IPv6 step by step. According to the measurements by Google, the number of hosts accessing their services over IPv6 has grown from 0% in 2010 to 30% in 2020. According to http://worldipv6launch.org the percentage of top 1 000 websites reachable over IPv6 was slightly over 30% in the end of 2021. The share of IPv6 is expected to continue growing in the coming years.

3.4.2 IPv6

3.4.2.1 IPv6 Addressing

As mentioned earlier, the length of IPv6 address is 128 bits. Like IPv4 addresses, neither IPv6 addresses are represented in writing using their long binary format. Written format of an IPv6 address consists of eight hexadecimal numbers separated by colons [130]. Each of the hexadecimal numbers represents 16 consecutive bits of the address. An example of such an IPv6 address is FE80:0000:0000:0000:0001:07A2:5AAC:D281.

Even such a format is long, so there are two conventions to make it shorter:

- Any leading zeros of the hexadecimal numbers can be omitted.
- A single group of hexadecimal zeros in the IPv6 address can be represented by two colons without any number between them.

Based on the two preceding rules, the IPv6 address given in the previous example can be expressed in a brief format as FE80::1:7A2:5AAC:D281.

Unlike the IPv4 address space, the IPv6 address space is not divided into any classes based on network sizes. However, IPv6 uses a division based on the purpose of the address. The purpose of an IPv6 address can be deduced from its leading bits, known as the format prefix. Every prefix has its own length, which comes from the size of the address subspace allocated for the purpose, which is identified by its prefix. The prefix is expressed with the following notation: a slash followed by the prefix length as a number of bits, e.g., /7.

The most important IPv6 prefixes are specified as follows using the binary format for clarity:

- 1111 110: Site-local unicast addresses, which are unique within an organization [129]. Addresses belonging to this block are private and can be reused in the same way as the private IPv4 addresses. The corresponding address block is FC00::/7. Site-local addresses have later been deprecated by IETF.
- 1111 1110 10: Link-local unicast addresses. The corresponding address block is FE80::/10. Link-local addresses can be used for IP address autoconfiguration and communication over a link such as a LAN. Messages with link-local destination address are discarded by IPv6 routers.
- 1111 1111: Multicast addresses. Corresponding address block is FF00::/8.

The address given in our earlier example is a link-local unicast address where the length of the prefixes is 10 bits. Consequently, the address can be expressed as follows to highlight the length of its prefix: FE80::1:7A2:5AAC: D281/10.

An IPv6 address hides the internal topology of IP network. The structure of an IPv6 unicast address [127], mapped to one single host, is as follows:

- Format prefix, if any, indicating the address type.
- IPv6 global routing prefix is the identifier of the IPv6 network. The length of this prefix depends on the network and is defined with a network-specific network mask, using the CIDR method earlier described for IPv4.
- IPv6 subnet ID, the length of which is also defined by a network-specific mask. The subnet ID shall be unique within the IPv6 network identified by its global routing prefix.
- IPv6 interface ID, which defines the link level interface of a certain IPv6- enabled host. The Interface ID must be unique within the IPv6 subnet.

IPv6 global unicast addresses (GUA) are unique within the whole Internet. Originally, IETF defined the GUA address space so that there was a prefix 001, after which the GUA would have a global routing prefix of 45 bits. In the IETF RFC 3587 [127], the 3-bit prefix 001 was made obsolete and the routing prefix was defined in another way. The first 64 bits of a GUA are divided in some way between the global routing prefix and subnet ID, based on the subnetwork configuration. The major requirement is that the prefix used for GUA shall not match with a prefix assigned for other types of IPv6 addresses.

The allocation strategy for the global routing prefixes is based on hierarchical address space management. Chunks of global address spaces have been given by IANA for regional Internet registries. The registries will allocate subsets of their spaces to various Internet service providers (ISP), which allocate address ranges for individual organizations and companies [10].

The structure of an IPv6 multicast address, mapped to multiple hosts, is as follows:

- Format prefix: Eight consecutive 1-bits, being the prefix for IPv6 multicast addresses.
- Flags are bits that tell if the address is a permanent one or a temporary, application specific one.
- Scope tells the scope of the multicast address, such as being the local link, a whole organization, or a global multicast address.
- Group ID is the identifier of the multicast group.

Some of the multicast groups are fixedly defined, such as the multicast groups identifying all routers or DHCP servers on a link or site. These groups have fixed IPv6 multicast addresses in global use. Other groups may be managed with ICMPv6 protocol, which supports creating and deleting groups as well as members joining or leaving existing groups.

A host may get an IPv6 address for itself via the DHCP service or use **IPv6 address autoconfiguration** to generate a unique IPv6 address for itself [134]. The IPv6 autoconfiguration process works as follows:

1) The host creates for itself a link-local IPv6 address by combining the prefix FE80:: to its own interface identifier. The identifier must be unique on the link, so typically the link level MAC address is used for the purpose of generating a 64-bit interface ID of type EUI-64, as described in IETF RFC 2373 [135]. In the derivation process,

the MAC address is split into two parts, then 16-bit hexadecimal value 0xFFFE is inserted between the parts, and finally the seventh bit of the resulting interface ID is inverted.

2) The host tries to ensure the uniqueness of the created IPv6 address by sending an ICMPv6 neighbor solicitation message on the link to that address. A response received indicates another host already using the address.

3) As its next step, the host might send ICMPv6 router solicitation message to the multicast address of link-local routers. Any router receiving such a message will respond to it with ICMPv6 router advertisement message, which tells the address of the router and the IPv6 network prefix of the link. Alternatively, the host might just wait for a periodically sent ICMPv6 router advertisement message.

4) Finally, the host acquires a globally unique IPv6 address (GUA) for itself by either fetching it with DHCPv6 or replacing the prefix of the automatically generated link-local IPv6 address with the correct network prefix as received from the router.

Even when the host has a GUA, it continues listening to the periodically sent ICMPv6 router advertisement messages. These messages allow the host to find out whether the routers are alive or if there are any changes in critical network parameters.

3.4.2.2 IPv6 Packet Structure

The structure of an IPv6 packet and its headers is defined in IETF RFC 8200 [111]. An IPv6 packet has the following fields, shown in Figure 3.18:

- IP version: IPv6.
- Traffic class used to indicate the diffserv QoS category and Explicit Congestion Notification.
- IPv6 flow label.
- Payload length of the packet.
- Next header, which identifies the type of extension header right after the mandatory IPv6 headers.
- Hop limit defining how many times the packet can still be routed between networks. The value is decremented by any router forwarding the packet, just as done for the IPv4 TTL field.
- Source address (IPv6).
- Destination address (IPv6).

Figure 3.18 IPv6 message structure and the extension headers.

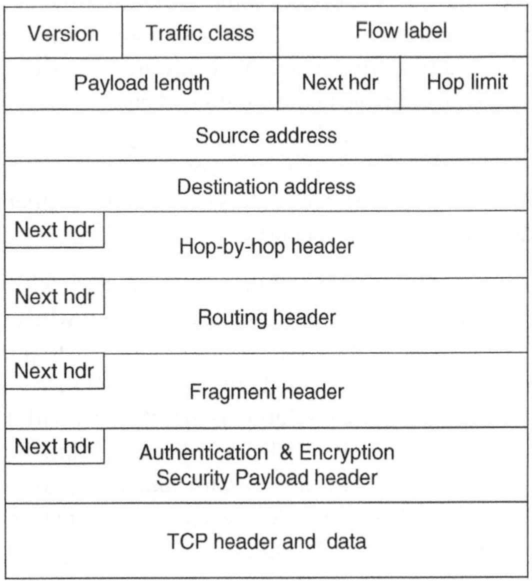

Version	Traffic class	Flow label	
Payload length		Next hdr	Hop limit
Source address			
Destination address			
Next hdr	Hop-by-hop header		
Next hdr	Routing header		
Next hdr	Fragment header		
Next hdr	Authentication & Encryption Security Payload header		
TCP header and data			

- The optional extension headers.
- Packet payload carrying the user data.

The IPv6 packet may contain a set of optional extension headers. There are two fields in the beginning of an extension header record, the first of which tells the length of the record and the second the type of the record, which can be found right after the type field. The structure of the record and the purposes of its fields depend on the record type. Most important extension headers are, for instance,

- Hop-by-hop header containing information for routers, which forward the packet. An example is a field describing the length of IPv6 jumbogram. This field is needed because the payload of the jumbogram is larger than what can be described within the Payload Length field of the mandatory IPv6 header.
- Routing header to list the IPv6 addresses of those routers via which the packet shall travel. The sender of the packet may add this kind of source routing record if it has some special reason for forcing the packet to be routed via a predefined path.
- Fragment header contains the pieces of information needed to reassemble a fragmented packet. Routers are not allowed to fragment IPv6 packets, but the sender of the packet shall check the maximum size of the packet that can be sent over all the links along with the route. Any bigger packets must be fragmented by the sender.
- Authentication and Encapsulating Security Payload header containing the header information used for protecting the packet with IPSec. While IPSec is a protocol separate from the IPv4, similar functionality is an integrated part of the IPv6 protocol.

3.4.3 Methods to Support the Parallel Use of IPv4 and IPv6

Since IPv4 is used globally by millions of hosts, it was not possible for them to switch for IPv6 and stop using IPv4 instantaneously. Instead, the transition will take decades, or perhaps both the protocols will stay in operation forever in the Internet. IPv6 transition is a very complex topic, but some guidelines for it can be found from IETF RFCs 4213 [114] and 6180 [121].

The basic approach for IPv6 transition is to allow host devices to choose between either IPv4 or IPv6 when communicating with each other. The hosts would need to have dual IPv4 and IPv6 stacks, capable of processing IP packets in both of those formats. Whenever the network infrastructure between the devices would support both versions, then the devices could use either version of the IP protocol without any problem. The aim is to have all the routers to have both dual stacks and routing mechanisms for both protocol versions, to equally support both IPv4 and IPv6. To drive the world toward IPv6 over the transition, preference could be given to IPv6 whenever that would be available in the end-to-end path. Eventually, when IPv6 is supported by every single IP-enabled device, support for IPv4 could be dropped altogether.

The complexity of transition comes from scenarios where IPv6 or dual protocol capability is not yet there either in some of the hosts or routers between the endpoints of the communication. The following cases can be identified as problematic scenarios:

- A host that uses IPv6 wants to communicate with a server that only supports IPv4. The source host may find out the lack of IPv6 support in the destination when it uses DNS to convert the domain name of the server into the IP address. In this case, the DNS is not able to provide any IPv6 address but only the IPv4 address of the server.
  - In the simplest case, both the host and server are connected to one single shared link such as LAN. In this case, the typical solution is that the host with IPv6 has an IPv4/v6 dual stack supporting both the protocols, as defined in IETF RFC 4241 [115]. The link is mapped to both IPv4 and IPv6 subnetworks so that every host connected to the link may use and belong to both of those IP subnetworks. The host selects one of the protocols and subnetworks, depending on what the destination supports. In our example case, the host would revert to using IPv4 toward the IPv4-only capable server.

- In a more complex case, the host and server do not share a common link or subnetwork but communicate over a router. The communication relies on an intermediate node, which converts the used protocol and addresses. Such mechanisms are defined in IETF RFC 6052 [118] and RFC 6146 [119]. A NAT64 node works as such a translator between IPv6 and IPv4 protocols. It has both IPv4/v6 dual stack and NAT software capable of converting IPv6 packets and addresses to IPv4 and vice versa. When connecting the IPv6 host with the IPv4-only server, NAT64 node maps the used IPv6 and IPv4 address pairs together, like IPv4 NAT does for private and public pairs of IPv4 address and port, when connecting private IPv4 hosts to the public Internet. The IPv6 host can encode the IPv4 destination address into its IPv6 packet as an IPv4 mapped IPv6 address. The IPv6 address block::ffff:0:0/96 has been reserved for this purpose [130]. Since this block is not bigger than the IPv4 address space, any IPv6 host should use an address of this block only when they must communicate with an IPv4-only endpoint. In practice, the IPv6 host typically retrieves the destination IP address with a DNS query for the destination hostname. As defined in IETF RFC 6147 [120], the DNS64 server is able to return the IPv4 mapped IPv6 address as a response to AAAA query for hostname, which only has A records for IPv4 addresses. In addition to the embedded IPv4 address, the IPv6 address returned by DNS64 uses a prefix assigned to a NAT64 translator node. When the IPv6 host uses this address as the destination address of an IPv6 packet, that packet will be routed toward the IPv4 destination via the NAT64 translator.
- Two IPv6 networks or subnetworks are connected over IPv4-only trunk network. In this case, the IPv6 packet has to be carried over an IPv4 network segment as a payload of IPv4 packet. This method is called tunneling IPv6 over IPv4. The IPv4 tunnel within the trunk network can be created either manually or automatically.
  - If the tunnel is created manually, the routers in both the ends of the tunnel must be configured with the IPv4 address of the other endpoint as well as the IPv6 prefix of the network behind the tunnel. When a router receives from its local interface an IPv6 packet, whose destination belongs to the IPv6 network behind the tunnel, the router encapsulates the IPv6 packet into an IPv4 packet sent to the tunnel. The destination of the IPv4 packet is the router in the other end of the tunnel. When the IPv4 packet arrives to its destination, the router unwraps the IPv4 headers and forwards the encapsulated IPv6 packet to the correct network. IETF RFC 4213 [114] describes such tunneling setups between routers.
  - Automatic tunneling can be used when the source and destination IPv6 addresses both belong to the IPv4 mapped address space [51]. Devices use such addresses if they want to communicate between hosts in two separate IPv6 islands within the IPv4 network. The automatically created tunnel may be started either in the source host or router but it is always terminated at the final destination of IPv6 packet rather than at the edge of the destination IPv6 network. IPv6 islands may use 6to4 border routers, which automatically enclose IPv6 packets into IPv4 packets, when routing the packets over an IPv4 cloud to another network with 6to4 routing. This scheme is defined in IETF RFC 3056 [113], but it suffers from the presence of NATs in the networks. Creating tunnels requires help of external servers and relays when the endpoints are behind NATs and do not have globally routable addresses. Teredo, defined in IETF RFC 4380 [116], is an IPv6 transition mechanism that supports such external server setup. IETF RFC 4798 [117] suggest yet another approach to interconnect IPv6 islands over an MPLS tunnel.
- Two IPv4 networks or subnetworks are connected over an IPv6 trunk network. In this case, the IPv4 packets have to be carried over the IPv6 network segment either by tunneling them over IPv6 or converting the IPv4 packets into IPv6 packets. In both cases, the IPv6 addresses in the trunk network may be IPv4 mapped addresses.

Since the number of IPv4 mapped IPv6 addresses is limited, deploying IPv6 has initially been challenging while IPv6 support was gradually ramped up in the Internet. Regardless of whether a host supports or does not support IPv6, it very often has a need to communicate with hosts, such as servers supporting only IPv4. In this case, there is no other way than using IPv4 mapped addresses, the scarcity of which has limited the growth of IPv6 networks. The way to overcome this is to use DHCP to temporarily allocate IPv4 mapped addresses to IPv6-enabled client hosts. While the transition is going on, servers typically need both permanent IPv4 and IPv6 addresses, and some of their IPv6 addresses shall be IPv4 mapped.

3.4.4 ICMPv6

ICMPv6 is an evolution of ICMP protocol as used for IPv4. ICMPv6 covers the message types as specified for IPv4 but additionally has the following IPv6 specific types:

- ICMP router and neighbor solicitation and advertisement messages to detect other IPv6-enabled routers and hosts on the link. This functionality [132] replaces the address resolution protocol (ARP) used in the context of IPv4 to find out the link level addresses mapped to the host IP addresses.
- IPv6 multicast groups are managed with the multicast listener discovery (MLD) protocol [136], which in IPv6 architecture replaces the IGMP protocol used for IPv4. MLD uses ICMPv6 message types for group management rather than defining its own message structure.

IETF RFC 4443 [131] defines ICMPv6 message format as well as ICMPv6 informational and error message types. ICMPv6 has the same fields as ICMP message.

3.4.5 DHCPv6

DHCPv6 is developed further from the DHCP protocol of IPv4. Like DHCP also, DHCPv6 is used by hosts to acquire IP addresses and other necessary parameters for themselves. DHCPv6 is defined in IETF RFC 8415 [124]. DHCPv6 supports the following functions:

- Host retrieving IPv6 address for itself from DHCPv6 server. This mode of operation, known as stateful DHCPv6, is rather similar to how DHCP provides hosts with IPv4 addresses.
- Host which has retrieved an IPv6 address with IPv6 stateless address autoconfiguration may still use DHCPv6 stateless service to retrieve configuration parameters, such as IPv6 addresses of DNS or SIP servers, as DHCP options.
- DHCPv6 server can support delegating network prefixes to different routers of the managed IP network. A DHCPv6 server, also known as a delegating router, is configured with prefixes to be used for the IPv6 subnets of the IP network. The other routers of the network request network prefixes for themselves from the delegating router. In this way, the network operation staff can avoid configuring prefixes separately to every router of the network, some of which might be owned by clients of the Internet service provider who manages the network.

If a host wants to use services of a DHCPv6 server, it at first sends a DHCPv6 SOLICIT message to the multicast address of DHCPv6 servers to find the address of such a server. In that message, the source address is a link-local IPv6 address that the host has created by itself. The DHCPv6 server responds with ADVERTIZE message returned to the link-local address of the host. This message contains the address of the DHCPv6 server and some other parameters, such as the address of a local DNS server. These DHCPv6 messages may be sent between the client and server directly when they share a common link, but otherwise a DHCPv6 relay will forward these messages between links.

After finding out the address of the DHCPv6 server, the host may send further DHCPv6 requests directly to the found server, without using the DCHPv6 server multicast address. The same process applies also to routers, which use DHCPv6 for acquiring the network prefixes. A DHCPv6 REQUEST message is used to request a new IPv6 address or delegated prefix. A DHCPv6 INFORMATION-REQUEST message is used to retrieve some configuration parameters only. The server returns the allocated IPv6 address and/or the other requested parameters within a DHCPv6 REPLY message.

Stateful DHCPv6 grants IPv6 addresses only for a period, the length of which is told in the REPLY message. The host may request extending the period with a DHCPv6 RENEW message. The DHCPv6 server releases the granted access either when it receives a DHCPv6 RELEASE message from the host or the period expires.

The DHCPv6 server may also inform the clients about any changes for parameters delivered earlier by sending a DHCPv6 RECONFIGURE message to them. This message contains the new values for the changed parameters.

A DHCPv6 message has the following fields:

- DHCPv6 message type
- Transaction identifier
- A number of DHCPv6 options, encoded as TLV structures

3.5 IP Routing

IP routers are devices that forward IP packets between IP subnetworks. A router is a multihomed device, having multiple IP network interfaces. Routing effectively means the decision making of the router to select the network interface toward which the IP packet is sent, toward its final destination.

Since there are millions of IP-enabled hosts connected to the Internet, very efficient routing mechanisms are needed. IP routing mechanisms are based on the following basic approaches [51]:

- Interconnected IP networks form hierarchical structures. Routers operating on the top level of the hierarchy within the IP core network must know the addresses of routers that are responsible of known subnetworks. Any IP packet having a specific subnetwork as its destination is forwarded to the designated router of that subnetwork. Routers operating at the subnetwork level must know addresses of the IP endpoints (hosts) within the subnetwork directly connected to the router as well as addresses of some core routers and routers of closely related or peer subnetworks. As described earlier, an IPv4 address consists of the subnetwork prefix and host ID. If the destination subnetwork is not known to the router, the packet is forwarded to a core router, which is expected to be able to forward the packet toward the right subnetwork. An IP network under its own administration and with a specific routing technology is called an **autonomous system (AS)**. Routers maintaining routes within an autonomous system use interior gateway protocols (IGP), such as OSPF. Top level routers, which maintain routes between autonomous systems, use exterior gateway protocols (EGP), such as BGP.
- The routers maintain internal **routing tables**, which are sort of maps over the structure of a network which the router knows. Routers on the same hierarchy level of the IP network use automated **routing protocols** to exchange information about the network structure and links between the routers in the network. The routing protocol messages are used to maintain the contents of the routing tables. When new hosts or links are added to the network or old ones removed, routing protocol messages are sent to distribute relevant filtered knowledge of those changes between routers. In addition to any permanent changes to the network structure, routing protocol messages inform the routers for any temporary changes caused by faults that persist long enough.

In its routing table, a router maintains records that describe the known IP subnets versus routes to reach them. A route is expressed as the IP address of the closest IP router on the route and the network interface to be used to reach that router. The routing table may have three types of records:

- Direct routes to those IP networks or subnets to which the router is connected with a single link. The routing database contains information about those subnet IDs. For any packets sent to those subnets, the router is able to dynamically find out the link-level address of the host, with protocols such as ARP defined in IETF RFC 826 [137] or ICMPv6 neighbor solicitation process, as described in Section 3.4.4.
- Indirect routes to those IP networks, which are reachable over other known routers.
- The address of the default router to which all such IP packets will be forwarded, which contain an unrecognized network or subnetwork number.

Compared to the routing used in telephone networks, the IP routing has both similarities and differences. Both the networks have hierarchical structures, partially represented in the addresses used in the networks. While the telephone network relies on manual maintenance of the routing tables, IP routers maintain their routing tables automatically with the routing protocols. Because of this, the maintenance of an IP network is consequently much easier than that of a traditional telephone network. Further on, IP networks are able to automatically provide service continuity in many typical fault situations, which was one of the design goals of IP networking.

Because the size of an IP network can range between a small local network to the global Internet, it is not easy to provide a single generic method for routing table maintenance. Therefore, many different types of IP routing protocols have been defined, to be used in different contexts. As technology has evolved, some of the newer protocols have replaced older, less efficient protocols. The routing protocols can be divided into two main categories based on the type of algorithm they rely on [13]:

- Distance-vector algorithm (DVA)
- Link state algorithm (LSA)

Distance-vector algorithm maintains knowledge in the routing table about the shortest route to reach a destination network. The route is described with two pieces of data stored to the table: the length of the route and the first router on it. In its startup, the router checks to which networks it is directly connected. In the routing database, the length of the routes to any directly connected subnetworks is marked as one. The neighboring, directly connected routers within the same network use routing protocol to regularly pass each other the contents of their routing table entries. Each router updates its routing tables based on the received routing protocol messages. In the first phase, the router quickly learns new routes to subnetworks directly connected to any of the neighboring routers. As these subnetworks are reachable over one hop of routers, the lengths of those routes are marked as two. At the next step of the process, the neighboring routers will additionally tell routes they have learned from their neighbors. Those subnetworks are then reachable over two hops of routers, so lengths of these routes are marked as three. When the process is continued, step by step the router learns routes to subnetworks even farther away from the router. After the process goes on long enough, the router knows all the subnetworks and corresponding routes of the whole IP network.

Routing protocols based on DVA are simple, but the drawback is the amount of routing information passed between routers. As routers announce the contents of their complete routing tables over the protocol, the routing protocol messages are large even when there have been no changes in the network. Lots of network capacity is wasted for sending routing protocol messages that do not provide any new information to anyone, unless sending of those messages is somehow suppressed. Another drawback is that information of any changes in the network is propagated between the routers slowly, hop by hop between the directly connected routers. Because of such properties, routing protocols based on distance-vector algorithms are only used for small IP networks.

Link state algorithms rely on the routers to store information about individual network links rather than complete routes to their routing tables. Each router internally creates a database of the network topology in terms of known subnetworks, routers, and links between them. Each link may have a cost related to it. The router is able to calculate the shortest route to any known subnetwork with the help of the shortest path first (SPF) algorithm and the contents of the topology database. The algorithm finds a route that has a minimum total cost of all available routes. Routers connected to the same network recognize each other by broadcasting hello-packets, which other routers in the network are listening to. When a router detects a change on any of its directly connected links or stops getting hello packets from directly connected routers, the router changes the state of the link in its database and sends a link state packet (LSP) to other neighboring routers. When receiving an LSP packet, a router updates its topology database accordingly and forwards the LSP packet to other links, except to the one from which the packet was received. With this kind of LSP packet flooding, it is ensured that all the routers of the network will quickly get the update of the changed state of the link. Eventually, forwarding of the LSP packet is finished and every router in the network has its topology map updated accordingly.

Compared to the DVA, the LSA approach has two benefits: routing packets are sent only for any network changes and the packets propagate quickly to all the routers of the network. Even if the link state algorithms are more complex than distance vector algorithms, they are much better for bigger networks, where the benefits easily outweigh the additional complexity.

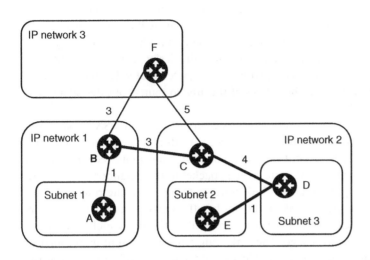

Router B: Distance-Vector		
Subnet	Next hop	Len
1	A	1
2	C	3
3	C	2

Any router: Link State		
EP1	EP2, subn	Cost
B	A, 1	1
B	C	3
B	F	3
C	D, 3	4
C	F	5
D	E, 2	1

Figure 3.19 Routing tables with DVA and link state algorithms for the same network.

Figure 3.19 illustrates the different operations of DVA and LSA algorithms. The diagram on the left shows a simple network configuration with three networks and subnetworks, six routers, and six links between them. The two tables on the right show the contents of DVA and LSA router tables of router B. When calculating the shortest route to Subnet 2, router B operates as follows:

- DVA: The routing table has one single entry for Subnet 2, indicating that the shortest route to it is via router C, over three links.
- LSA: Using shortest-path-first algorithm, router B finds the shortest route to Subnet 2 to be over routers C, D, and E. The total cost of that route is 8.
- If link B–C fails, router B with DVA does not have any clue about the route to Subnet 2 until contents of its DVA routing table are updated. But if the router B uses link state algorithm, it can immediately find an alternative route to Subnet 2 via router F using the current entries in its LSA routing table. Total cost of the new route is 13.

Routing information protocol (RIP) is a well-known routing protocol based on DVA. RIP is defined in IETF RFC 1058 [138], and its improved version RIP-2 is defined in RFC 2453 [139]. A router that uses RIP sends the contents of its routing database within a routing packet in three cases: to respond to a request from another router, to respond to a change of route metric change (e.g., due to link failure), and periodically every half a minute. The RIP protocol message contains an IP address identifying a network or subnetwork and the route distance metric. A commonly used metric is the round-trip delay of a message over the route toward the address, as known by the router. RIP-2 message has some additional elements, like the subnetwork mask related to the IP address, address of the next hop router of the route, and a route tag used for route identification. The routing protocol message is sent as payload of a UDP packet sent to all hosts in the subnetwork.

Open shortest path first (OSPF) protocol is the most common interior gateway routing protocol based on LSA. OSPF is specified in various IETF RFCs such as 1245 [140], 2328 [141], and 5340 [142]. The IP network where OSPF is used is an OSPF routing domain and also an autonomous system (AS). An AS is divided into OSPF areas [140]. The routers within an OSPF area maintain a mutually similar topology database over the area. The area border routers maintain additional information about the links between the area and other connected areas. AS boundary routers located to the edge of the whole IP network maintain additional information about the connections toward external networks [141]. Routers using OSPF detect the neighboring routers in the area with help of hello-packets exchanged between routers. If a link has only two routers, they will become adjacent and exchange link state advertisements (LSA) within link state update packets to synchronize their topology databases. If there are multiple routers on the link, one of them will be chosen as the designated router of the link. Other routers on the link will exchange the link state packets only with the designated router. The border routers provide summarized state information in summary LSAs about links of the area to routers of external areas or networks.

OSPF attaches an IP subnet mask to every advertised route. Routers exchanging OSPF LSA packets synchronize their routing tables in two phases. In the first phase, the adjacent routers will exchange database description packets. Those packets let the routers check if the neighbor router has more recent LSA link status information. In such a case, the router sends a link state request packet to the neighbor router to get the complete status of that link. After checking the received link state update packet, the router decides whether to flood it to other routers so that changes to link states can be efficiently propagated to routers within the OSPF area. The OSPF link state packets provide metrics related to the links such as capacity constraints, latencies, or other costs related to link usage. The link state packets are sent as payload of high-priority IP packets. Those IP packets are multicast to an address mapped to all OSPF routers of the network.

Border gateway protocol (BGP) is an exterior gateway routing protocol commonly used in IP core networks interconnecting individual IP networks. BGP uses DVA and is specified in IETF RFC 4271 [143]. BGP routers are located on the edge of an IP network. In its routing tables, BGP routers maintain paths toward various destinations as chains of autonomous systems (AS). Every AS has its own unique AS ID, and a route is a list of AS IDs used to reach the destination. The destinations are described as CIDR address ranges or network prefixes. Fueled by the growth of the Internet, one problem with BGP has been the ever-increasing number of network prefixes in BGP routing tables with limited length. To limit the number of network prefix entries in the tables, the prefixes can be aggregated. Instead of storing multiple long network prefixes for separate IP networks, a BGP router may decide to aggregate those as a single shorter prefix, consisting of the most significant bits of the address, which are the same for all those individual aggregated prefixes. Aggregation is, however, not efficient for routing decisions, if the aggregated prefixes belong to physically and geographically separate networks, not sharing a common router. The BGP router may learn multiple routes toward one destination, but it gives them an order of preference based on policies configured to the router. Only the most preferred route is used as far as it is working, and the other routes are standby. Neighbor BGB routers exchange routing information about preferred paths over connections established with TCP.

While IP routers use their own routing protocols to detect each other, also the hosts must find local routers to be able to connect to the Internet. For this purpose, IPv4 routers periodically send ICMP Router Advertisement messages [62]. After its startup, the host starts listening to such ICMP messages and will learn the IP addresses of routers from them. For IPv6, the router discovery is done with IPv6 neighbor discovery protocol defined in IETF RFC 4861 [132]. IPv6 routers send Router Advertisement messages both periodically and to respond to router solicitation messages from hosts.

3.6 Web Browsing with HTTP Protocol

User data flow is a term that means different types of end-to-end data sessions, such as Web browsing, streaming music, video, instant messaging, or a VoIP call. The network infrastructure transports user data flows from the data source to the destination. Probably the most typical user data flow is the one related to browsing Web pages, with the help of **hypertext transfer protocol (HTTP)**. HTTP is a protocol used to retrieve Web pages and any elements of them, such as text, pictures, or video clips embedded to the pages. At the time of this writing, two HTTP protocol versions are in common use: HTTP 1.1 and 2.0. Those versions of HTTP have been specified in the following IETF RFCs:

- HTTP version 1.1 was originally specified in IETF RFC 2616 [144] but was later revised by IETF RFCs 7230 [145] – 7235 [146] and is still a mainstream version of HTTP.
- HTTP version 2 is specified in IETF RFC 7540, which was published in 2015. This HTTP version is also in wide use. RFC 8740 defines how TLS version 1.3 is used together with HTTP version 2.

HTTP 1.1 is a text-based protocol without a rigid frame structure. Instead, HTTP 1.1 messages consist of the following parts:

1) The first line of an HTTP request message is known as the **request line** and of responses the **response line**. The request line contains the HTTP request type and an HTTP uniform resource locator (URL) address. The response line has an HTTP response code.
2) The first line is followed by **header lines,** which start with the header name and a colon. The value of the header is given after the colon. HTTP headers are used for various purposes, such as routing or authenticating the message, negotiating acceptable media types, or defining the properties of returned content. The order of different header fields is not strictly defined in the specification.
3) In the end of the message, there is a **message body**, which carries the payload data retrieved by HTTP GET requests or new data to be inserted to the remote server by HTTP PUT or HTTP POST requests.

HTTP messages are transported between an HTTP client and an HTTP server over TCP protocol. Instead of TCP, the newer QUIC protocol may also be used. The logical structure of HTTP messages – requests and responses with headers and a body – is essentially the same in both HTTP versions 1.1 and 2.0. HTTP 2.0, however, enhances the structure of the messages and the way the HTTP client and server manage HTTP message exchange or parallel HTTP transactions. HTTP 1.1 client can send a new request over a TCP connection after getting the response to the previous request. If the client needs to retrieve multiple resources from the same server, it can open multiple parallel TCP connections toward the server to send many HTTP requests to it. HTTP 2.0 improves handling of parallel requests by introducing multiple parallel independent streams within one single TCP connection. A stream carries an HTTP request and response pair as frames. HTTP 2.0 frames have a fixed binary message structure allowing them to express HTTP headers in a compressed format. Different types of HTTP 2.0 frames are defined for opening and closing streams, managing stream-level flow control and stream priorities, as well as sending header blocks and data over a stream.

Within Web browsing sessions, HTTP works as follows [24] over an end-to-end network connection:

- An end user has a Web browser in his/her computer. The browser is a client that is able to generate HTTP requests, such as HTTP GET to retrieve Web pages from remote Web servers [147]. The HTTP GET request is transported over a network connection to the correct server, which returns the page (or a part of it) within HTTP 200 OK response. The returned data is formatted with HTML language, which describes the layout of the page to the Web browser. Typically, to render one page, a sequence of HTTP requests and corresponding responses is needed to retrieve all the elements of the page (such as text, images, formatting instructions, and active content such as Javascript code) from one or multiple servers. To render a large Web page the browser might need dozens of HTTP requests toward different servers to collect all the contents of the page to be shown.
- HTTP requests use HTTP uniform resource identifiers (URI) as destination addresses to identify resources (such as Web sites or individual pages) in the network. HTTP uniform resource locators (URL) start with a prefix http:// or https://, the latter of which indicates a secured connection. You can type an HTTP URL to the browser address bar or find suitable URLs by making a search with your preferred Web search service. After the prefix, the URL has a string that describes the domain name of a Web server (or proxy) and a specific resource

within the server being accessed. To provide an example, let's break down the URL https://www.wiley.com/ en-gb/aboutus. After the https:// prefix, the string www.wiley.com is the domain of Wiley. The final part of the URL 'en-gb/aboutus' points to the "About Us" pages of Wiley in the United Kingdom. When retrieving the page, the Wiley server redirects the browser to make further HTTP requests to a number of more specific URLs, pointing to different parts of the page, such as pictures, text, and a video clip, which the user can play if so desired.

- The browser extracts the hostname of the Web server from the beginning of the HTTP URL, which identifies the resource being accessed. The hostname is entered to the Host header field of the HTTP request and is used to determine the IP address of the destination server. The IP address can be found with the domain name service so that the client sends a DNS query for the hostname to the local DNS server. The domain name system checks the type of the query and returns an IPv4 or IPv6 address to which the HTTP request should be sent. The client may optionally use a local HTTP proxy configured to browser settings. In that case, the IP address of the destination proxy is either retrieved from the local configuration or derived with DNS from the proxy hostname stored locally. The proxy itself is able to route the HTTP request onwards based on its HTTP URL address arriving within the HTTP request.

- The HTTP client, such as a Web browser, may get a response to the HTTP request either from the server identified in the URL or any HTTP proxy between the client and the server. HTTP proxies are intermediate server nodes that route and forward HTTP requests toward their ultimate destinations and responses back. Proxies are able to temporarily store or cache resources, such as Web pages recently fetched, just in case those are soon tried to be retrieved again [148]. In that case, it may be the proxy, which provides the response to the client. Even the Web browser itself has a local cache to help with retrieving recently accessed information. If the user moves back and forth between Web pages, the Web browser needs to fetch those pages only once over the network connection and can later on return to the same pages locally. The HTTP protocol defines extensive logic in how the freshness and validity of cached pages can be verified so that the proxy would not return any stale data. Data considered too old shall be purged from the caches.

- HTTP uses TCP protocol as its transport layer. The HTTP client opens a TCP connection between the end user host and the server that owns the resources identified by the HTTP URL. Typically, the TCP connection is opened to an HTTP proxy, which processes the HTTP request and may forward it to the server over another TCP connection. The remote endpoint of the TCP connection is the IP address resolved earlier with DNS. For any IP network nodes between these two communication endpoints, TCP is handled transparently. This means that the intermediate nodes do not process TCP segments or headers at all, but just pass them through within the IP packets routed toward its destination. All processing of TCP takes place only at the TCP connection endpoints, which run TCP protocol software. For HTTP traffic, the TCP connections are kept open as long as they are explicitly closed, so that multiple HTTP messages toward a single server can be sent over a single TCP connection. If contents shall be retrieved from multiple servers (each having their own HTTP URLs), separate TCP connections are needed for each of them.

- When accessing an https URL, the connection shall be secured with TLS protocol. Right after the TCP connection is opened, a TLS session is established between TCP connection endpoints, as described in Section 3.3.10.2. TLS protocol encapsulates and encrypts HTTP messages before handing them over TCP for end-to-end transport. The TCP headers carry the TCP port number, which at the destination indicates the specific process to which the TCP payload belongs to. For unprotected HTTP messages, the TCP port numbers 80 or 8080 are commonly used, and TCP port 443 is used for HTTP traffic secured with TLS.

- TCP segments (which carry the HTTP messages as payload) are written into IP packets, which are routed toward the endpoint of the TCP connection with the destination address written to the IP packet header. The routers along with the path check the network prefixes of the destination IP addresses to find corresponding entries in their routing tables. The routing table record tells to which network interface the router shall forward the packet to reach the next hop router toward the destination. The network interface is connected to a link of some kind, such as a LAN, an ADSL line, SDH link, or a virtual link such as an ATM virtual connection or MPLS label switched path. The next hop router is in the other end of the link.

- On the links, the IP packets are transported between the network nodes with different kinds of link protocols specific to the type of the link. A link protocol is typically used over a single physical link between two adjacent devices, connected over a cable or radio connection. Virtual link switching protocols, such as MPLS or ATM, are an exception. They support virtual connections over multi-hop network segments. MPLS establishes label switched paths across a chain of MPLS label switching routers. For the user of MPLS protocol, the label switched path appears as a single link, even if it spans over multiple MPLS-enabled devices, interconnected with lower layer link protocols. Another example of a technology providing multi-hop links is wavelength division multiplexing (WDM), which provides paths over a network on physical layer only. A light path provided over the WDM network is able to carry link protocols (such as Ethernet) over the WDM nodes, which only process transmission at the physical optical layer.

Figure 3.20 shows an example of how an HTTP request is transported within IP packets from a laptop to an ADSL modem. The protocol layer structure of the figure follows the description above for unprotected HTTP requests; thus, the TLS layer is not shown. The laptop uses a WLAN connection to the modem. The HTTP GET request is embedded to the payload field of a TCP segment. Only the request line and first header of the HTTP request are shown, and any other headers are omitted for simplicity. In practice, the HTTP request is long enough to be fragmented into multiple TCP segments rather than a single one. The request is carried to the remote destination within multiple TCP segments, the last of which indicates it contains the final bytes of the message to be now reassembled. Below the TCP layer, the protocol stack has the network layer IP protocol and the link layer LLC protocol. The WLAN protocol on the bottom of the stack has both link and physical layer functions, as described in *Online Appendix G.1*.

The laptop composes the nested structure of protocol frames as shown on the right side of the figure and sends the nested frame over the WLAN radio to the modem. After receiving the WLAN frame, the modem extracts the WLAN and LLC frame structures because the physical and link protocol layers are terminated at the modem. The modem has IP routing functionality, which checks the next hop link for the IP packet from the destination IPv4 address of the packet. The modem finds that the IP packet with its TCP segment payload should be forwarded to the ADSL link of the modem toward the Internet service provider network. The ADSL link is not shown in Figure 3.20 but it can be found in Figure 3.21.

Figure 3.21 is a simplified diagram about an end-to-end Web browsing session. The figure shows certain network elements on the end-to-end communication path as well as the protocol stacks used between them. The

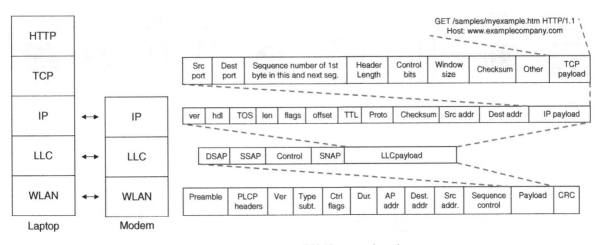

Figure 3.20 Transporting HTTP message with an underlying TCP/IP protocol stack.

stacked boxes under each link depict the message structures on each of the links, how HTTP messages are encapsulated into messages of underlying protocol layers. The vertical bolded lines on both sides of the stack indicate which protocol layers are locally processed at the network element. Where the bolded line is missing and the protocol name is written in gray, it means that those protocol messages are passed transparently through the node, without processing the message headers or payload at that layer of the stack. As we can see, only the endpoints of the Web browsing session, the laptop and the Web server, process protocol layers HTTP, TLS, and TCP. Any intermediate IP router nodes (including the ADSL modem) process IP packets while the other nodes process protocols on link and physical layers. The leftmost link of the figure is the WLAN link between the laptop and the ADSL modem, as depicted in detail within Figure 3.20, except that TLS is omitted.

Figure 3.21 depicts the nested message structures on each of the links, and Figure 3.22 focuses on the network nodes. The stacked boxes below each node represent the protocol stacks running in the node. As each of the nodes

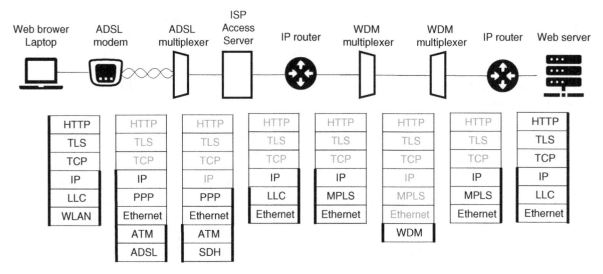

Figure 3.21 Web browsing session over Internet.

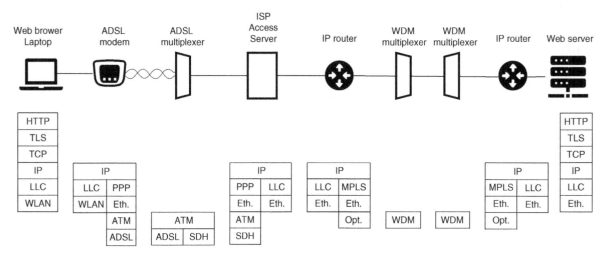

Figure 3.22 Protocol stacks in the network nodes.

along with the end-to-end path has two network interfaces between which the traffic is forwarded, both of those interfaces have their own protocol stacks. In most of the cases, an instance of a protocol in the diagram shows an endpoint of that protocol. In the source endpoint, the upper layer message is wrapped within the headers of the protocol, and in the destination the headers of the protocol are removed, like peeling an onion. The span of the protocol is shown as the empty space between the two adjacent instances of the protocol in the same layer. In the following cases of Figure 3.22, the nodes do not terminate the protocol but apply a protocol-specific routing or switching process to find out to which network interface the processed protocol message or frame should be forwarded:

- IP packets are routed between different interfaces of ADSL modem, ISP access server, and the IP routers. The IP packets are generated and consumed at the two terminating points of the IP protocol path, which are the laptop and the Web server.
- The ADSL multiplexer forwards ATM cells over an ATM permanent virtual connection (PVC) created between the ADSL modem and ISP access server. Within the ADSL multiplexer, ATM switching is used to forward cells over the PVC between the subscriber line and the trunk interface.
- The two IP routers on the right use MPLS to carry the IP packets between them. This is because there is an MPLS network between the routers, which is not drawn in the figure. The MPLS label switched routers (LSR) have been omitted only due to the lack of space in the figure. If they were drawn, they would have a three-layer protocol stack from optical to MPLS layer. The LSR switches the MPLS packets between its interfaces without looking into or processing the IP packets, when an MPLS label switched path has been set up through the LSR between the two IP routers shown in the figure.
- The WDM multiplexers process and multiplex physical layer optical signals in the WDM network. For this end-to-end path, the optical signal is the optical layer of the Ethernet signal passed between the two IP routers in the diagram. The network might also have a WDM add-drop multiplexer or cross connect between the two WDM multiplexers, to switch a light wavelength between its WDM network interfaces. As shown in the figure, the WDM equipment is not aware of any protocols above the optical physical layer.

In summary, Figures 3.21 and 3.22 demonstrate the following aspects of the end-to-end data flow, highlighting the basic "everything over IP and IP over everything" principle of Internet communications:

- IP protocol is the lowest level end-to-end protocol in the network layer. HTTP data within the TCP segments is routed as IP packets between the endpoints of the browsing session.
- On top of IP there are three protocol layers, which are terminated in the endpoints of end-to-end data flow, namely the laptop and Web server. All the other network elements on the path only pass through these protocol data units without processing them:
  - TCP is used on top of IP to form a connection between Web browser and Web server. TCP transports HTTP data reliably over the IP protocol.
  - TLS protocols are used on top of TCP to secure the HTTP messages between endpoints.
  - HTTP is the application protocol used to access the Web site resources and deliver the contents of Web pages back to the client.
- The link and physical layers below the IP layer are specific to each link section. In this diagram, the following major sections are shown:
  - WLAN link between the laptop and the ADSL modem in customer premises. On this link, IP packets are carried over WLAN and LLC link frames.
  - ADSL link from the modem to the ADSL multiplexer over the telephone subscriber line. On this link, IP packets are encapsulated into PPP protocol frames, which are carried in Ethernet frames mapped to ATM cells. The PPPoEoATM virtual link is used to reach the access server of the Internet service provider (ISP).

– Link from ADSL multiplexer to the access server of ISP. On this link, SDH is used instead of ADSL, but otherwise the PPPoEoATM structure is the same. The PPP protocol is terminated at the access server and used to authenticate the subscriber.
– Wired Ethernet link from the access server to IP router of ISP. Here, the protocol layer structure is similar to the one within the customer premises, except that wired rather than wireless LAN is used.
– Connections between core IP routers over WDM optical network. The IP routers support MPLS, so they are also edge LSR routers. MPLS packets are encapsulated into gigabit Ethernet frames, which are carried in the core network over optical WDM links. The optical nodes of the WDM infrastructure only see and process the optical wavelengths; they are not aware of the upper layer digital protocols carried within the optical signals and do not modify the digital content in any way.
– Final Ethernet connection between the router on the Web service provider premises and the Web server. Since MPLS connection has been terminated by the router, the protocol layering is similar to that used at ISP.

Depending on the specific HTTP proxy setup, to access a Web server, a chain of TCP connections may be needed. Those connections are between the client and HTTP proxy, between the proxies on the path and the Web server itself [145]. TCP connections are terminated at any node that needs to access and process the HTTP payload carried by TCP. Other nodes between the TCP connection endpoints will just forward TCP segments and cannot access HTTP messages inside of them. The ISP may supply an outbound HTTP proxy for the client so that the domain name of the proxy is configured to the settings of the Web browser. Web service providers typically have an inbound proxy which receives all the HTTP requests for the domain of the service provider and distributes them to various Web servers, based on the resources being accessed. To route the HTTP message onwards, each proxy has its way to find the next hop HTTP proxy or server. This is typically done based on HTTP URI of the request or some local configuration. The hostname of the next hop is then recorded by the proxy to the Host header field of the forwarded HTTP request. This hostname is resolved with DNS to the IP address, which will be used as the destination IP address for the TCP connection toward the next hop HTTP proxy or server.

3.7 Questions

1 How does Ethernet media access control work?

2 What is common between switching protocols such as frame relay, ATM and MPLS?

3 What are the main advantages of MPLS compared to ATM?

4 Please describe the high-level structure of the Internet

5 What is the IPv4 address shortage problem and how is it managed?

6 What is network address translator (NAT) and how does it work?

7 What is TCP slow start and how is it related to TCP congestion avoidance?

8 What is the domain name system?

9 Why would an IP host use DHCP protocol?

10 Why is the transition from IP4 to IPv6 so difficult?

11 How can a host get an IPv6 address for itself?

12 How does a link state routing algorithm work?

References

1 IEEE 802.3 Carrier Sense Multiple Access with Collision Detection (CSMA/CD) Access Method and Physical Layer Specifications.

2 IEEE 802.3i System Considerations for Multi-segment 10 Mb/S Baseband Networks and Twisted-Pair Medium Attachment Unit (MAU) and Baseband Medium, Type 10BASE-T.

3 IEEE 802.3u Supplement - Media Access Control (MAC) Parameters, Physical Layer, Medium Attachment Units, and Repeater for 100Mb/s Operation, Type 100BASE-T.

4 IEEE 802.3j Supplement to 802.3 — Fiber Optic Active and Passive Star-Based Segments, Type 10BASE-F.

5 IEEE 802.3z Media Access Control Parameters, Physical Layers, Repeater and Management Parameters for 1,000 Mb/s Operation.

6 IEEE 802.3ae CSMA/CD Access Method and Physical Layer Specifications — Media Access Control (MAC) Parameters, Physical Layer, and Management Parameters for 10 Gb/s Operation.

7 IEEE 802.3ab CSMA/CD Access Method and Physical Layer Specifications — Physical Layer Parameters and Specifications for 1000 Mb/s Operation over 4 Pair of Category 5 Balanced Copper Cabling, Type 1000BASE-T.

8 IEEE 802.3cm Ethernet — Amendment 7: Physical Layer and Management Parameters for 400 Gb/s over Multimode Fiber.

9 IEEE 802.3cu Ethernet — Amendment 11: Physical Layers and Management Parameters for 100 Gb/s and 400 Gb/s Operation over Single-Mode Fiber at 100 Gb/s per Wavelength.

10 Anttalainen, T. and Jääskeläinen, T.A. (2015). *Introduction to Communications Networks*. Norwood: Artech House.

11 IEEE 802.1w Media Access Control (MAC) Bridges: Amendment 2 — Rapid Reconfiguration.

12 IEEE 802.1Q Bridges and Bridged Networks.

13 Held, G. (2003). *Ethernet Networks: Design, Implementation, Operation, Management*. West Sussex: Wiley.

14 IEEE 802.11 Wireless LAN Medium Access Control (MAC) and Physical Layer (PHY) Specifications.

15 Sauter, M. (2021). *From GSM to LTE-Advanced pro and 5G: An Introduction to Mobile Networks and Mobile Broadband*. West Sussex: Wiley.

16 IEEE 802.11b Wireless LAN Medium Access Control (MAC) and Physical Layer (PHY) specifications: Higher Speed Physical Layer (PHY) Extension in the 2.4 GHz Band.

17 IEEE 802.11a Wireless Medium Access Control (MAC) and Physical Layer (PHY) Specifications: High Speed Physical Layer in the 5 GHz Band.

18 IEEE 802.11g Wireless LAN Medium Access Control (MAC) and Physical Layer (PHY) Specifications: Further Higher Data Rate Extension in the 2.4 GHz Band.

19 IEEE 802.11n Wireless LAN Medium Access Control (MAC)and Physical Layer (PHY) Specifications Amendment 5: Enhancements for Higher Throughput.

20 IEEE 802.11ac Wireless LAN Medium Access Control (MAC) and Physical Layer (PHY) Specifications— Amendment 4: Enhancements for Very High Throughput for Operation in Bands Below 6 GHz.

21 IEEE 802.11ax Wireless LAN Medium Access Control (MAC) and Physical Layer (PHY) Specifications Amendment 1: Enhancements for High-Efficiency WLAN.

22 ISO 13239, Information technology — Telecommunications and information exchange between systems — High-level data link control (HDLC) procedures.

23 Halsall, F. (1996). *Data Communications, Computer Networks and Open Systems*. Boston: Addison-Wesley.

24 Clark, M.P. (2003). *Data Networks, IP and the Internet: Protocols, Design and Operation*. West Sussex: Wiley.

25 IETF RFC 1332 The PPP Internet Protocol Control Protocol (IPCP).

26 IETF RFC 1661 The Point-to-Point Protocol (PPP).

27 IETF RFC 1662 PPP in HDLC-like Framing.

28 IETF RFC 1663 PPP Reliable Transmission.

29 IETF RFC 1994 PPP Challenge Handshake Authentication Protocol (CHAP).

30 IETF RFC 3748 Extensible Authentication Protocol (EAP).

31 CCITT Recommendation I.233.1, Frame mode bearer services: ISDN frame relaying bearer service.

32 ITU-T recommendation Q.922 ISDN data link layer specification for frame mode bearer services.

33 ITU-T recommendation Q.933 ISDN Digital Subscriber Signalling System No. 1 (DSS1) — Signalling specifications for frame mode switched and permanent virtual connection control and status monitoring.

34 K. Ahmad, Sourcebook of ATM and IP Internetworking, IEEE, 2002.

35 ITU-T Recommendation I.361, B-ISDN ATM layer specification.

36 ITU-T Recommendation I.363, B-ISDN ATM adaptation layer specification.

37 ITU-T Recommendation I.365, B-ISDN ATM adaptation layer sublayers.

38 ITU-T Recommendation Q.2761, Functional description of the B-ISDN user part (B-ISUP) of signalling system No. 7.

39 ITU-T Recommendation Q.2764, Signalling System No. 7 B-ISDN User Part (B-ISUP) — Basic call procedures.

40 ITU-T Recommendation Q.2931, Digital Subscriber Signalling System No. 2 — User-Network Interface (UNI) layer 3 specification for basic call/connection control.

41 ITU-T Recommendation Q.2762, General functions of messages and signals of the B-ISDN User Part (B-ISUP) of Signalling System No. 7.

42 IETF RFC 2684 Multiprotocol Encapsulation over ATM Adaptation Layer 5.

43 IETF RFC 1953 Ipsilon Flow Management Protocol Specification for IPv4 Version 1.0.

44 IETF RFC 1954 Transmission of Flow Labelled IPv4 on ATM Data Links Ipsilon Version 1.0.

45 IETF RFC 3031 Multiprotocol Label Switching Architecture.

46 IETF RFC 3032 MPLS Label Stack Encoding.

47 IETF RFC 3034 Use of Label Switching on Frame Relay Networks Specification.

48 IETF RFC 3035 MPLS Using LDP and ATM VC Switching.

49 IETF RFC 3036 LDP Specification.

50 IETF RFC 3212 Constraint-Based LSP Setup Using LDP.

51 M. Murhammer, O. Atakan, S. Bretz, L. Pugh, K. Suzuki, D. Wood, TCP/IP Tutorial and Technical Overview, IBM, 1998.

52 IETF RFC 791 Internet Protocol.

53 IETF RFC 919 Broadcasting Internet Datagrams.

54 IETF RFC 922 Broadcasting Internet Datagrams in the Presence of Subnets.

55 IETF RFC 950 Internet Standard Subnetting Procedure.

56 IETF RFC 4632 Classless Inter-domain Routing (CIDR): The Internet Address Assignment and Aggregation Plan.

57 ICANN, Available Pool of Unallocated IPv4 Internet Addresses Now Completely Emptied, 2011.

58 IETF RFC 1918 Address Allocation for Private Internets.

59 IETF RFC 1349 Type of Service in the Internet Protocol Suite.

60 IETF RFC 2474 Definition of the Differentiated Services Field (DS Field) in the IPv4 and IPv6 Headers.

61 IETF RFC 792 Internet Control Message Protocol.

62 IETF RFC 1256 ICMP Router Discovery Messages.

63 IETF RFC 1122 Requirements for Internet Hosts — Communication Layers.

64 IETF RFC 768 User Datagram Protocol.

65 IETF RFC 793 Transmission Control Protocol.

66 IETF RFC 2018 TCP Selective Acknowledgment Options.

67 IETF RFC 2883 An Extension to the Selective Acknowledgement (SACK) Option for TCP.

68 IETF RFC 6675 A Conservative Loss Recovery Algorithm Based on Selective Acknowledgment (SACK) for TCP.

69 IETF RFC 3042 Enhancing TCP's Loss Recovery Using Limited Transmit.

70 IETF RFC 3390 Increasing TCP's Initial Window.

71 IETF RFC 5681 TCP Congestion Control.

72 IETF RFC 6298 Computing TCP's Retransmission Timer.

73 IETF RFC 7323 TCP Extensions for High Performance.

74 IETF RFC 7414 A Roadmap for Transmission Control Protocol (TCP) Specification Documents.

75 IETF RFC 1191 Path MTU Discovery.

76 IETF RFC 4960 Stream Control Transmission Protocol.

77 IETF RFC 8999 Version-Independent Properties of QUIC.

78 IETF RFC 9000 QUIC: A UDP-Based Multiplexed and Secure Transport.

79 IETF RFC 9001 Using TLS to Secure QUIC.

80 IETF RFC 9002 QUIC Loss Detection and Congestion Control.

81 IETF RFC 1034 Domain Names — Concepts and Facilities.

82 ISO 3166 Codes for the Representation of Names of Countries and Their Subdivisions.

83 IETF RFC 1035 Domain Names — Implementation and Specification.

84 IETF RFC 1101 DNS Encoding of Network Names and Other Types.

85 IETF RFC 2535 Domain Name System Security Extensions.

86 IETF RFC 4033 DNS Security Introduction and Requirements.

87 IETF RFC 4035 Protocol Modifications for the DNS Security Extensions.

88 IETF RFC 2181 Clarifications to the DNS Specification.

89 IETF RFC 4343 Domain Name System (DNS) Case Insensitivity Clarification.

90 IETF RFC 4592 The Role of Wildcards in the Domain Name System.

91 IETF RFC 2308 Negative Caching of DNS Queries (DNS NCACHE).

92 IETF RFC 2782 A DNS RR for Specifying the Location of Services (DNS SRV).

93 IETF RFC 5966 DNS Transport over TCP — Implementation Requirements.

94 IETF RFC 3596 DNS Extensions to Support IP Version 6.

95 IETF RFC 2937 The Name Service Search Option for DHCP.

96 IETF RFC 2131 Dynamic Host Configuration Protocol.

97 IETF RFC 951 Bootstrap Protocol.

98 IETF RFC 2132 DHCP Options and BOOTP Vendor Extensions.

99 IETF RFC 3942 Reclassifying Dynamic Host Configuration Protocol version 4 (DHCPv4) Options.

100 IETF RFC 3011 The IPv4 Subnet Selection Option for DHCP.

101 IETF RFC 3046 DHCP Relay Agent Information Option.

102 IETF RFC 3361 Dynamic Host Configuration Protocol (DHCP-for-IPv4) Option for Session Initiation Protocol (SIP) Servers.

103 IETF RFC 4301 Security Architecture for the Internet Protocol.

104 IETF RFC 4302 IP Authentication Header.

105 IETF RFC 4303 IP Encapsulating Security Payload (ESP).

106 IETF RFC 4306 Internet Key Exchange (IKEv2) Protocol.

107 IETF RFC 2246 The TLS Protocol Version 1.0.

108 IETF RFC 4346 The Transport Layer Security (TLS) Protocol Version 1.1.

109 IETF RFC 5246 The Transport Layer Security (TLS) Protocol Version 1.2.

110 IETF RFC 8446 The Transport Layer Security (TLS) Protocol Version 1.3.

111 IETF RFC 8200 Internet Protocol, Version 6 (IPv6) Specification.

112 IETF RFC 2675 IPv6 Jumbograms.

113 IETF RFC 3056 Connection of IPv6 Domains via IPv4 Clouds.

114 IETF RFC 4213 Basic Transition Mechanisms for IPv6 Hosts and Routers.

115 IETF RFC 4241 A Model of IPv6/IPv4 Dual Stack Internet Access Service.

116 IETF RFC 4380 Teredo: Tunneling IPv6 over UDP through Network Address Translations (NATs).

117 IETF RFC 4798 Connecting IPv6 Islands over IPv4 MPLS Using IPv6 Provider Edge Routers (6PE).

118 IETF RFC 6052 IPv6 Addressing of IPv4/IPv6 Translators.

119 IETF RFC 6146 Stateful NAT64: Network Address and Protocol Translation from IPv6 Clients to IPv4 Servers.

120 IETF RFC 6147 DNS64: DNS Extensions for Network Address Translation from IPv6 Clients to IPv4 Servers.

121 IETF RFC 6180 Guidelines for Using IPv6 Transition Mechanisms During IPv6 Deployment.

122 IETF RFC 7949 OSPFv3 over IPv4 for IPv6 Transition.

123 IETF RFC 4704 The Dynamic Host Configuration Protocol for IPv6 (DHCPv6) Client Fully Qualified Domain Name (FQDN) Option.

124 IETF RFC 8415 Dynamic Host Configuration Protocol for IPv6 (DHCPv6).

125 IETF RFC 3363 Representing Internet Protocol Version 6 (IPv6) Addresses in the Domain Name System (DNS).

126 IETF RFC 6724 Default Address Selection for Internet Protocol Version 6 (IPv6).

127 IETF RFC 3587 IPv6 Global Unicast Address Format.

128 IETF RFC 4007 IPv6 Scoped Address Architecture.

129 IETF RFC 4193 Unique Local IPv6 Unicast Addresses.

130 IETF RFC 4291 IP Version 6 Addressing Architecture.

131 IETF RFC 4443 Internet Control Message Protocol (ICMPv6) for the Internet Protocol Version 6 (IPv6) Specification.

132 IETF RFC 4861 Neighbor Discovery for IP version 6 (IPv6).

133 IETF RFC 5942 IPv6 Subnet Model: The Relationship between Links and Subnet Prefixes.

134 IETF RFC 4862 IPv6 Stateless Address Autoconfiguration.

135 IETF RFC 2373 IP Version 6 Addressing Architecture.

136 IETF RFC 2710 Multicast Listener Discovery (MLD) for IPv6.

137 IETF RFC 826 An Ethernet Address Resolution Protocol.

138 IETF RFC 1058 Routing Information Protocol.

139 IETF RFC 2453 RIP Version 2.

140 IETF RFC 1245 OSPF Protocol Analysis.

141 IETF RFC 2328 OSPF Version 2.

142 IETF RFC 5340 OSPF for IPv6.

143 IETF RFC 4271 A Border Gateway Protocol 4 (BGP-4).

144 IETF RFC 2616 Hypertext Transfer Protocol — HTTP/1.1.

145 IETF RFC 7230 Hypertext Transfer Protocol (HTTP/1.1): Message Syntax and Routing.

146 IETF RFC 7235 Hypertext Transfer Protocol (HTTP/1.1): Authentication.

147 IETF RFC 7231 Hypertext Transfer Protocol (HTTP/1.1): Semantics and Content.

148 IETF RFC 7234 Hypertext Transfer Protocol (HTTP/1.1): Caching.

Part III

Mobile Cellular Systems

4

Cellular Networks

Cellular networks support mobility of their users. When relying on a fixed network, the user is reachable and can access a network only when sitting next to the terminal connected to the network with a cable. Mobile phones were originally designed to make it possible to communicate and be reachable on the road. When mobile phones became smaller so that they could be held in pockets, the users became reachable regardless of time or place. The mobile devices started ultimately to support Internet access and use cases equaling those available with personal computers. All this was enabled by the cellular network technology.

4.1 Cellular Networking Concepts

Cellular network is a radio network architecture capable of supporting large numbers of mobile radio terminals. Cellular network technologies have been used for both commercial mobile phone and data networks as well as private mobile radio networks used by the military, police, and firefighters. Cellular network design must solve many problems, which simply do not exist for fixed networks, to support terminal mobility. This introductory chapter about cellular systems provides the reader with an overview of such problems and related solutions.

4.1.1 Structure of a Cellular Network

Cellular communication systems have two fundamental device types: **mobile radio terminals,** such as mobile phones, and **base stations**, which connect radio signals to a fixed telecommunications network. The **coverage area** of the network is the geographical area within which terminals can access base stations of the network over radio. A **cell** is a smaller area within the network where a terminal is able to communicate with a single base station in control of the cell.

The earliest mobile phone networks were deployed in the Unites States during the 1940s, and they relied on a single base station in a fixed central location. The radio signals transmitted by the base station covered a very wide geographical area. The base station consisted of an omnidirectional antenna, a powerful radio transmitter, and a sensitive receiver. The base station had a cable link to a local telephone center. Calls were connected manually between the mobile phone network and the fixed telephone network by an operator working at the telephone center. Because the base station could use only a limited radio frequency band, only a small number of simultaneous calls could be supported. Calls were multiplexed with frequency division multiplexing (FDM) so that each call required its own radio channel with a certain bandwidth and a center frequency. In the next few decades, switching of the calls was automated but the mobile network capacity was not increased until cellular network technologies were introduced during the 1970s and 1980s.

Converged Communications: Evolution from Telephony to 5G Mobile Internet, First Edition. Erkki Koivusalo.
© 2023 The Institute of Electrical and Electronics Engineers, Inc. Published 2023 by John Wiley & Sons, Inc.
Companion website: www.wiley.com/go/koivusalo/convergedcommunications

To support larger number of mobile network users, a new method was needed to allocate the scarce radio frequencies between different devices. Bell Labs invented the cellular radio network approach during the 1960s [1]. Cellular network design became a breakthrough innovation for increasing capacity of mobile telephone networks. Cellular networks introduced a way to reuse the same frequency for multiple devices within a geographical area that is large enough. The basic idea is very simple: the same frequencies can be used for radio connections that are far enough from each other not to disturb each other. Cellular networks use multiple base stations, each having its own dedicated coverage area. As radio signals attenuate proportionally to the distance, the same frequency can be used by base stations that are not adjacent. When the transmission power of both the base stations and terminals is limited, the interference can be limited by not using the same frequencies for multiple terminals close to each other.

To provide mobile network coverage over a large geographical area, the cellular network uses multiple base stations, each of which covers only some fraction of the area. The base stations are connected by a **core network**, which takes care of connecting calls and/or routing of user data flows. The area served by a single base station is called a **cell** and the coverage area of the whole network is consequently known as a **cellular network**. The size of a cell depends on the used transmission power, any obstacles such as hills nearby the base station, and the used frequency areas. As lower frequencies attenuate slower than higher frequencies, lower frequency bands can be used for building larger cells. The earliest cellular networks used relatively low- frequency bands to be able to provide wide network coverage with a small number of base stations. Higher frequencies have been taken into use for newer cellular technologies to provide additional capacity with densely located base stations.

When planning and building a cellular network, the locations and transmission power budgets of the base stations are planned so that the size of the cells are optimized for the deployment:

- At rural areas where the density of terminals is low, high transmission power levels and low frequencies are used. Antennas of base stations are installed on tall masts or towers to avoid any obstacles blocking signal propagation. The radius of a cell may range from a few kilometers to tens of kilometers when the lowest available frequencies are used. A single base station can cover an area of several square kilometers. This kind of cell is called **macrocell**. Interference is not a problem in a large cell when the number of supported terminals is low.
- At urban areas the density of terminals is high. When the number of users per square kilometer grows, smaller cells and lower transmission power levels are used. Small indoor cells and base stations can be used in addition to the larger outdoor cells. Cells may have small coverage areas, but the frequencies used by one cell can be reused by other cells not too far away. The radius of a cell might range from a few tens of meters to 1 km. Small cells are called **microcell, picocell,** or **femtocell** – from the largest to smallest sizes. Pico- and femtocells are typically used as indoor base stations to cover one floor of open office space or a corridor of a shopping mall. Femtocells are special types of functionally limited cells without full handover and power management support. They might be used as home cells to improve network coverage when the signal to the closest macrocell is too weak.

Small cells provide various benefits. Small transmission power levels save battery of the terminals. With short distances, the signal quality is good and more advanced modulation methods can be used with higher signal-to-noise ratio, so that bitrates can be increased. Small cell radius also means fewer users sharing the capacity of the cell. The drawback of small cells is the additional investment needed to build and maintain a large number of base stations and links from them to the core network.

Even though Figure 4.1 shows base stations located in the center of the cell, that is not always the case. The base station may also be located at the edge of two or three cells, when directional antennas are used. In such a case, the base station has antennas pointing to different directions and forming sectors with potentially different frequencies. Each such **sector** may be configured as an independent cell.

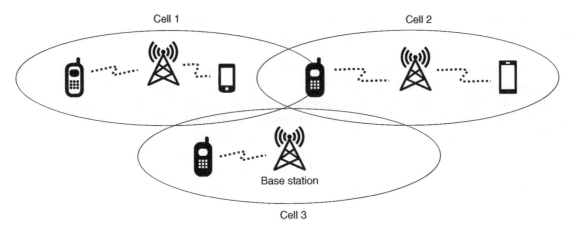

Figure 4.1 Structure of a cellular network.

4.1.2 Operation of Cellular Network

When a mobile phone communicates with a cellular network, in most of the cases it has a radio connection with one base station at the time. The phone is said to **camp** in the cell, which the mobile phone uses to communicate with the base station. When the phone moves, the quality of its radio connection to the currently used cell may decline. When approaching the border area of two adjacent cells, the phone may start receiving stronger and better radio signals of another cell. The process of switching the serving cell is called **cell reselection** when the phone is idle but a **handover** when performed while a call or data session is active. Handovers are more complicated processes than cell reselections. In a handover, the network tries to ensure that the cell change will not disturb the ongoing communications.

If the region has multiple cellular systems of various operators and the phone uses one of them, the phone is said to **roam** in that network. In order to use the services of certain network, the user of the phone must either have a **subscription** with the network operator or the operator must have a **roaming contract** with another network which the subscriber may use. The network to which the user has a subscription is known as the **home network** and the other networks are known as **visited networks**. Subscribers are allowed to use visited networks only when they are outside of their home network coverage area, such as when traveling abroad.

When the phone is switched on, it performs a **cell search**. After finding cells for any supported radio access technologies, the phone selects the network to use, either the home network service or another network when the home network is unavailable. The phone tries to register to the selected network. After the phone has synchronized to the cell and opened the radio connection with it, the network **authenticates** the subscriber to verify the subscriber identity and **authorizes** the user for services subscribed. Authentication is typically done with help of a smartcard (subscriber identity module [SIM]) equipped on the phone. The card is supplied by the home network operator, and it contains security algorithms and subscriber-specific credentials for performing authentication. The network may also check the identity of the phone to lock stolen phones out of service or to deliver some settings to new phones just taken into use. After authenticating the user, the network associates the telephone number of the subscriber with the phone registered to the service. During the registration procedure, the network records the location of the phone so that calls can thereafter be connected to it.

Mobile phones communicate with the network for two distinct purposes. Phones and the network exchange **signaling** messages, which are used to control phone functionality, such as tracking the phone location or managing calls. But the ultimate purpose of the cellular system is to support **user data** flows, whether those are audio streams for voice calls, instant messages, Internet pages downloaded, or any other application-specific communications. Phones receive **mobile terminated** calls and initiate **mobile originated** calls. The phone is in an **idle** state when no communication is going on. Idle phones listen to messages broadcast by the serving base station. When a need for communication emerges, the phone makes an access and/or service request to get a radio channel allocated for the communications. The network assigns some radio resources to the phone and moves the phone to a **connected** state so that the requested communication can take place. In the connected state, the phone sends **uplink** traffic to the base station and receives **downlink** from the base. After finishing with the communications, the radio channel is eventually released and the phones returns to an idle state.

It is worth noting that while this chapter used a mobile phone as an example of a mobile terminal, also other types of terminals exist. The terminal may be a tablet or laptop equipped with a SIM card and supporting some cellular technology. Besides these common types of end-user devices, also cars and other appliances or devices may use cellular network connections for various data communication purposes.

4.1.3 Antenna Technologies

Base stations of early cellular networks used standalone omnidirectional antennas. An omnidirectional antenna transmits radio signals to all the directions with equal strength. In an ideal case, an omnidirectional antenna can support a cell, which has a perfectly circular form. In practice, any physical obstacles within the circle radius weaken the signal and shorten its range, which makes cell coverage areas irregular. The locations of other base stations also impact cell coverage areas when the cells are partially overlapping, as shown in Figure 4.1.

A more recent approach is to build base stations with multiple antennas or antenna arrays rather than just one antenna. In such a case, the antenna mast may have a few or even tens of antennas arranged as a matrix. Multiple antennas can be used to form directional signal beams. This is done by carefully selecting the distances between antennas and phase differences of signals transmitted from different antennas. Those signals are then combined as stronger in the desired direction and weaker in the others. The form of the **beam** depends on the number of antennas used. Narrower beams can be formed by using a higher number of antennas, as shown in Figure 4.2. Beams can be used either as independent cells, sectors, or beams within one cell, depending on the radio technology used.

It is also possible to build the network so that the same area is served by multiple cells, which might even have different sizes. The overlapping cells would use different frequency bands not to disturb each other. Small cells may support a large number of users at a hot spot area such as sports arena. The large cell would then be used to support mobile users to avoid handovers between many small cells while the user is moving.

Spectral efficiency is a term that refers to the information data rate that can be transmitted over a given frequency bandwidth. Improving the spectral efficiency has been a major goal for cellular system design over all the system generations. The recent 4G and 5G radio systems use advanced antenna technologies to improve the spectral efficiency and network capacity with the following methods:

Figure 4.2 Using a higher number of antennas for **beamforming** results narrower beams.

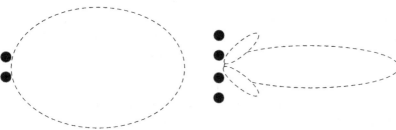

Beam formed with two antennas   Beam formed with four antennas

- **Adaptive antenna systems** use multiple antennas to produce beams of radio signals. The transmitted power of a beam is focused to a certain direction by adjusting the phases of the signal as sent from different antennas within an antenna array. The transmitted signals amplify each other in some direction and cancel out in other directions, depending on the distances between the antennas, used frequencies, and phase adjustments. The mobile terminal can find the strongest beam toward itself by measuring the strengths of different beams. After finding the best beam, the terminal informs the network about the beam to be used. The network consequently uses that specific beam when communicating with the terminal. This kind of beamforming technology enables the base station to increase the coverage range of its signals, as the transmitted power can be focused to a narrow beam toward the destination rather than distributed to directions where the destination is not located. Beamforming can also decrease interference toward adjacent cells. The base station can use weaker beams toward directions where the cell borders to the neighboring cells are closer to the base station.
- **Multiple input, multiple output (MIMO)**, where both the base station and terminal have multiple antennas [2]. 2×2 MIMO means a system in which both of them have two antennas. The transmitting device sends either the same or different signals from its antennas. The same frequency is used for all the transmissions, and the receiver must use advanced signal processing techniques to separate the received signals from each other. To support that, each antenna typically sends its own dedicated reference signal along with the rest of its transmission. The receiving device can interpret the received signals from both its antennas, taking advantage of the fact that different antennas have different probabilities to catch a specific copy of the transmitted signal as the strongest one. The number of antennas in a MIMO system can be selected as powers of two, reaching even as high as 256 antennas for base stations. Due to their small form factors, mobile terminals are typically limited to two or four antennas. Depending on the specific way of using MIMO, the system can either decrease the bit error ratio and increase the transmission reliability or keep the bit error ratio in acceptable limits but increase the bitrate. The MIMO system can transmit one single symbol via different antennas to increase the redundancy and reliability of the communications channel. The other option is to split the data flow to multiple branches so that a group of adjacent symbols would be sent simultaneously by different antennas, one symbol per antenna to increase the bitrate. By selecting the approach optimally, MIMO is able to improve the quality of the connection, especially in such environments where multiple copies of the transmitted signal are received via different paths, due to reflections and refractions. Dense urban areas are typically such environments, and MIMO can thus be used to increase the network total capacity in cities.

4.1.4 Multiplexing Methods in Cellular Networks

All the cellular networks have a radio band that the network is licensed to use. This band is known as the system band and its bandwidth the system bandwidth. The system band resources can be allocated to the cells and terminals of the network with the following three approaches.

4.1.4.1 Frequency Division Multiple Access (FDMA)

FDMA-based cellular systems, such as Nordic Mobile Telephone (NMT) and Global System for Mobile (GSM), divide the whole frequency range of the network to a number of narrow radio channels to be allocated for cells. Allocation of the system subbands to cells is done so that no adjacent cells use the same subband. In this way, the neighboring cells do not interfere with each other. A few standard frequency allocation patterns have been defined for this purpose, like the pattern with seven subbands shown in Figure 4.3.

The frequency allocation pattern can be repeated to groups of seven cells as many times as needed to cover a geographical area. Other similar models have been developed, which use, for instance, 4, 12, or 21 different subbands or carriers. How often a carrier is used determines the frequency reuse factor of the patterns. The larger number of subbands the system uses, the farther away from each other those cells can be located which use the same subband. But a large number of subbands also means splitting the system bandwidth to narrow subbands,

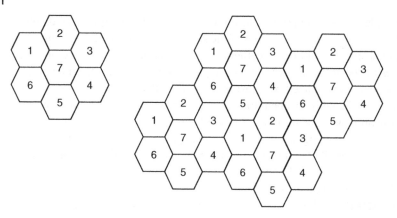

Figure 4.3 Cellular network band allocation pattern using seven subbands (European Telecommunications Standards Institute [ETSI]).

which do not support too many active connections or simultaneous network uses per cell. When only a few subbands are used, the same frequencies are reused in cells closer to each other, which increases their mutual interference. To strike a balance, a typical solution is to use patterns with a few subbands for macro cells and patterns with many narrower subbands for smaller cells. In this way, a macro cell can host a larger number of users compared to a small microcell with a limited range.

The mobile terminal can maintain a connection toward one base station at a time in the frequency division multiple access (FDMA) type of network. The base station allocates a fraction of its own radio band for the terminal as a single modulated carrier for some period of time. The first generation analog systems allocated a single carrier for a complete call. GSM, on the other hand, uses frequency hopping on traffic channels, so that the GSM frame is divided into timeslots, which can be allocated to different GSM phones by changing the allocated frequency between every timeslot. This means that the GSM multiplexing method is a combination of FDMA and time division multiple access (TDMA), as further described in Chapter 5, Section 5.1.4.1 of the book. When the terminal moves from one cell to another, any actively used data connection must be moved from the old to the new cell at a precisely defined moment, as controlled by the network. This kind of instantaneous switch between base stations is called a **hard handover**.

4.1.4.2 Code Division Multiple Access (CDMA)

In code division multiple access (CDMA) networks, the cells can use the same frequency bands, as different transmissions are distinguished with code words rather than frequencies. In 3G wideband code division multiple access (WCDMA) networks, all the cells of the network actually use the whole system bandwidth allocated to the network. With CDMA DSSS (see *Online Appendix A.9.1*), each of the terminals are given one or multiple code words for communicating with the base stations. 3G WCDMA mobile terminals may communicate either with one or multiple base stations at any time, as all the base stations of the network use the same shared wideband frequency range. When the terminal has a radio connection to multiple base stations, all the connected base stations know the code words allocated to the terminal. Those base stations may simultaneously send the same signals with the same code words. The terminal is able to listen to all those transmissions and combine the information received from multiple sources, even if some of the signals would be slightly distorted. As the terminal is able to be connected with multiple base stations, it is possible to do the handovers gradually so that the connection to the old base station is not cut immediately when a new connection is created with another base station. These kinds of gradual switches between base stations are called **soft handovers**.

4.1.4.3 Orthogonal Frequency Division Multiple Access (OFDMA)

Orthogonal frequency division multiple access (OFDMA)–based cellular systems, such as long-term evolution (LTE) and 5G, divide the system band to a number of narrow subcarriers. Every cell uses the full system bandwidth and the same subcarriers. The mobile terminal is allocated with a set of subcarriers for a short period. Each subcarrier transports its own quadrature amplitude modulation (QAM) modulated symbol stream. Adjacent cells use the same subcarriers for different data flows as mobile terminals are connected to only one cell at a time. This causes a problem for mobile terminals located close to the cell edge, as they receive strong signals on the same frequencies from at least two cells. The transmission from one of the cells is the desired one while transmission on same subcarriers from other cells is just noise for the terminal. OFDMA systems use **inter-cell interference coordination (ICIC)** methods to resolve this problem. With ICIC, the base stations of the network coordinate their scheduling decisions so that adjacent cells would use different subcarriers with high transmission power when trying to reach terminals located near the edge of the cell. OFDMA networks support only hard handovers between cells because the terminals can only be connected with one cell any time, as in FDMA networks.

4.1.5 Mobility Management

In the fixed telephone network, the location of the phone can be derived either directly or indirectly from the telephone number. Before the introduction of intelligent networks, the location was encoded into the number itself. With an intelligent network, the mapping between the phone location and the telephone number is stored into a subscriber database, but the relation is static until the subscriber either changes the telephone number or moves to another residence and transfers the old number there. In cellular networks, the telephone number itself does not reveal the location of the mobile phone, but instead the network must dynamically track the phone and store its current location to some kind of location database.

When a network tries to reach a mobile terminal for incoming call or message, it sends a **paging** message to the terminal. The purpose of the paging message is to wake up the terminal and trigger it to open radio channels toward the network to receive the transmission. The network may use two basic approaches for selecting the cells where the paging message is sent to reach the mobile terminal:

1) The terminal may announce its current location to the network whenever it moves from one cell to another. In case of an incoming call, the network has to page the terminal only in that cell where the terminal has most recently announced its location. This approach causes a high amount of signaling load for frequent location announcements.
2) The terminal does not inform the network at all about any changes of its location. In case of an incoming call, the network has to page the terminal in all the cells of the network. This approach causes a high amount of signaling load for paging messages.

Cellular networks use a combination of these two approaches. The cells of the whole network are divided into subsets where a number of cells belong to a single area called a **location area, routing area,** or **tracking area.** The network learns the initial location of the terminal when the terminal registers to the network service. Thereafter, the terminal sends a location update message when it moves between such areas but remains silent when it changes a cell within its current area. To reach the terminal, paging messages are sent only within the cells of the currently registered area. This method is used to balance the amount of location update messages and paging messages. For the circuit switched mobile telephone network, one location area typically (but not necessarily) consists of cells that are managed by a base station controller or a radio network controller. A mobile switching center serves multiple location areas. The concepts of routing and tracking areas are related to packet switched data networking, in which location tracking is done either by packet gateways or nodes dedicated to terminal mobility tracking. The sizes of areas in the circuit and packet switched sides of the network are often different, as optimized for the nature of those services. The current area of the terminal is recorded to a single central register of subscriber's home operator so that the terminal can be reached for any incoming calls or packet connections toward the subscriber.

4.2 History of Cellular Technologies

The era of cellular networks started in the 1980s. At that time, the first analog cellular networks with proper mobility (handover) support were taken into commercial use in many countries. The aim was to open a mass market of automated telephony services for mobile phones. The first cellular systems were country-specific so that the mobile phone could not be used abroad. When traveling to another country, the user would need another phone for the local mobile network.

During the 1990s, the GSM mobile networks were deployed almost globally, except in North America and Japan, which still mostly relied on their own market-specific technologies. GSM was the first digital cellular technology widely used across the countries. GSM networks were introduced in Europe from 1991 onwards. Australia was the first non-European country to deploy GSM in 1993. In North America, the first GSM network was opened in 1995, but the competing CDMA IS-95 was still the mainstream cellular technology in the United States until the early 2000s. GSM supported an unprecedented set of mobile services, such as telephony, short messages, telefax, and narrowband circuit switched data connections. Roaming agreements between the operators made it possible to use a mobile phone and a subscription also from abroad. In the end of the century, GSM networks were extended to support packet switched data with general radio packet service (GPRS) technology.

In the 2000s, the next step was the adoption of WCDMA networks, which provided faster data connections and more compact network equipment than GSM. The new mobile phone models supported both WCDMA and GSM technologies. In this way, the phone could use the best radio access technology available at its current location. This was very important since initially the coverage of mature GSM networks was much wider than that of WDCMA networks being built. The specification of the WCDMA network relied on a common network core shared with GSM. This made it possible to roll out WCDMA access networks consisting of base stations and radio network controllers, while the GSM mobile switching infrastructure could be reused for the new radio network. The WCDMA standard was created as Eurasian cooperation, which paved its quick adoption also in Japan and South Korea.

In North America, a new set of CDMA2000 standards were created to support 3G telephony and data services comparable to WCDMA. Despite of the efforts of many major operators to rely on the American CDMA technology with CDMA2000 networks, the deployment of GSM and WCDMA networks gradually grew also in those markets, initially thanks to certain challenger operators. Eventually, GSM and WCDMA reached the status of global standard of cellular telephony and networking. The owner of a WCDMA/GSM phone could use the phone virtually in any country visited.

In the 2010s, the fourth-generation LTE cellular technology was rolled out. The new global 4G LTE technology supported only packet switched data connections, but not the traditional circuit switched telephony. As optimized for packet data, LTE was able to provide very high data rates and low latencies compared to earlier cellular systems. The first devices supporting LTE were data modems, which could be connected to a personal computer over the universal serial bus (USB) connector interface. The first LTE phones were deployed in global markets during 2012. To support telephony, the early LTE phones used a circuit switched fallback method where the phone moved to a WCDMA or GSM network for the duration of a phone call. The phone used LTE access only for data connectivity such as Internet browsing or email download. Native operator voice support for LTE was added a bit later with voice over IP technology known as Voice over LTE (VoLTE). A single American operator, Verizon, had launched VoLTE service already in 2014, it was in 2015 that was the breakthrough of VoLTE, when many other operators launched VoLTE in both their networks and VoLTE-compatible LTE phones. VoLTE uses 3rd Generation Partnership Project (3GPP) IP multimedia subsystem architecture designed to support various real-time IP communication services, such as voice or video calls over packet switched mobile networks.

The 5G networks were deployed from 2020 onwards, providing even higher data rates and lower latency data connections compared with LTE. The radio technology and architecture of a 5G network is based on and evolved

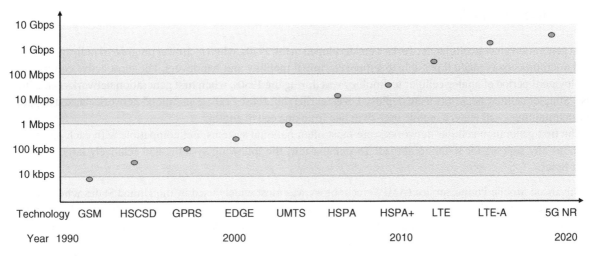

Figure 4.4 Downstream data rates of GSM and 3GPP cellular technologies.

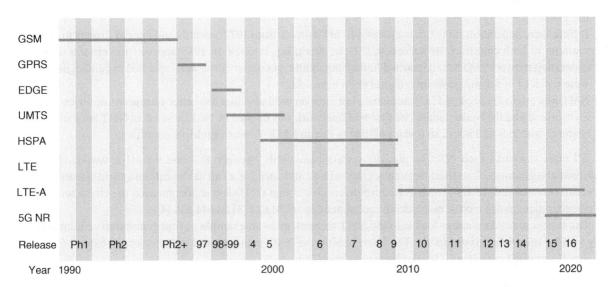

Figure 4.5 GSM and 3GPP specification releases.

from the LTE design. The increase of downstream data rates from GSM to 5G is approximately depicted in Figure 4.4. The logarithmic scale on the left is worth noting.

The major cellular network technologies described in this book were specified by 3GPP and its predecessors. As described in *Online Appendix A.2.3*, 3GPP develops technologies incrementally, within 3GPP specification releases. Each of the releases has its own scope and schedules. Figure 4.5 depicts how different technologies were developed in 3GPP over various specification releases, shown in the bottom of the diagram:

Figure 4.5 is only an approximation, as earlier studies of specific technologies were been started and maintenance of them has continued later than what the horizontal bars indicate. The bars are mapped to those specification releases in which the majority of the new content for these technologies were published.

4.3 First Generation

First generation analog cellular networks were developed in the 1970s, when technical progress of radio technology and microprocessors made it possible to support terminal mobility and handovers. The most active commercial deployment period of analog cellular technology was during the 1980s, when first generation networks came into use. Simpler cellular networks were deployed even earlier, but those "zero generation" networks lacked support for maintaining a call when a mobile user switched from one cell to another.

The first generation cellular networks were most often national systems, not compatible with each other and without any support for roaming abroad. To name a few, the following systems had relatively large national user bases:

- Advanced Mobile Phone Service (AMPS) technology was most widely used in the United States where AMPS networks had nearly a country-wide coverage. The first AMPS network was opened in 1979. AMPS services were also opened in other countries, including Israel, Australia, and Pakistan. AMPS services were discontinued in the United States in 2008 and elsewhere by 2010.
- Total Access Communication System (TACS) technology was specified in Great Britain. TACS was based on the earlier AMPS specifications. In addition to Britain, TACS networks were deployed also in Italy, Austria, and Spain. The first TACS network was opened in 1983, and its coverage in Great Britain was extended throughout the 1980s.
- B-Netz "zero generation" network was deployed in Germany from 1972 onwards and later in Austria to support roaming in both countries. Its successor, C-Netz, was taken into use in 1985 and met the criteria of a first generation network as C-Netz supported handovers. Both types of networks were operated in parallel until they were replaced by second generation digital GSM networks.
- NMT was originally deployed in Scandinavia but later on used also in a few other European countries, such as Switzerland and the Netherlands. The first NMT network was opened in 1981 in Sweden. NMT later supported roaming within Scandinavia, so it was possible to use an NMT phone in multiple countries.

All these networks used FDM so that the system bandwidth, between 800 and 900 MHz, was divided into narrowband channels. One fixed-frequency channel was allocated for a call at the call setup and released when the call was finished. International Telecommunications Union (ITU)-R has documented the characteristics of first generation analog mobile networks in recommendations M.622 [3] and M.624 [4].

The reader may take a closer look into the NMT technology as an example of first generation cellular system by visiting the *Online Appendix I.1*. The ideas behind the NMT standard and its deployment experiences had a significant role later when GSM standards were specified.

4.4 Questions

1 In which way does cellular network technology increase capacity of a radio network?

2 In the context of cellular network, what is the difference of camping and roaming?

3 What is beamforming?

4 What are the three major ways used in cellular networks to multiplex terminals to the shared radio band?

5 What is paging, and how is it related to mobility management?

References

1 Anttalainen, T. and Jääskeläinen, V. (2015). *Introduction to Communications Networks*. Norwood: Artech House.

2 Holma, H. and Toskala, A. (2009). *LTE for UMTS: OFDMA and SC-FDMA Based Radio Access*. West Sussex: Wiley.

3 ITU-R Recommendation M.622 Technical and operational characteristics of analogue cellular systems for public land mobile telephone use.

4 ITU-R Recommendation M.624 Public land mobile communication systems location registration.

5

Second Generation

The second generation (2G) cellular networks were designed as digital instead of analog. The main purpose of the digitalization was to minimize the bandwidth needed for a voice channel to support wide deployment of mobile phones. The emerging digital technologies provided a means to improve the quality of the voice connection by making the voice traffic resilient against any disturbances on the radio channel. Security could also be enhanced by encrypting the digital radio signals. An additional benefit of digital transmission was that it also enabled a straightforward way of transporting data over the cellular network without digital-analog conversions.

5.1 GSM

5.1.1 Standardization of Second Generation Cellular Systems

The first generation (1G) analog mobile systems were deployed nationally. By the mid-1980s, the concept of mobile telephony had proven to be a viable one. The first generation mobile telephony market covered advanced business users and staff who worked far from their office environments with fixed phones. Mobile telephony supported superior reachability of their users on the road. There were still a few major problems to solve. First of all, being national systems, the mobile telephone networks did not support international travelers, except Nordic Mobile Telephone (NMT) in Scandinavia. Second, the analog networks could not be scaled up for larger numbers of users, as their spectral efficiency was just too low. Further, their terminals were expensive and heavy. When electronic components become smaller, the first hand-portable mobile phones were introduced, which indicated that mobile telephony might also have potential for the consumer market in addition to the business market.

Global System for Mobile Communications (GSM) was born as a pan-European cellular system standard [1]. GSM became the leading 2G cellular system. The original aim of GSM standardization was to create a single mobile phone system to cover the whole of Europe. This would be achieved by getting telephone operators in different European countries to deploy compatible GSM systems and creating international **roaming** contracts between the operators. The following goals were defined for the standard:

- Provide network elements and mobile stations compliant to the GSM standard for a large market area. Due to the economies of scale, this was expected to decrease the cost of equipment and network deployment as the initial research and development cost would be shared by a very large number of customers.
- Enable a single GSM user to use the mobile station and GSM subscription in different countries while traveling within Europe.

Converged Communications: Evolution from Telephony to 5G Mobile Internet, First Edition. Erkki Koivusalo.
© 2023 The Institute of Electrical and Electronics Engineers, Inc. Published 2023 by John Wiley & Sons, Inc.
Companion website: www.wiley.com/go/koivusalo/convergedcommunications

- Enable connecting a voice call to the GSM user regardless of which country the user is located in and which operator network the phone is currently using. Such a roaming service is provided by the visited operator, which has made a roaming agreement with the home network operator of the user.
- Optimize the usage of radio frequencies reserved for GSM channels and provide the user with high-quality and secure voice and data services.
- Provide services that would be compatible with the fixed public switched telephone network (PSTN) network, especially with the Integrated Services Digital Network (ISDN) standards, which emerged in the 1980s parallel to GSM standardization.
- Enable development of different types of mobile stations such as portable phones, vehicle-based stations, and modems attached to computers.
- Support maritime usage of mobile stations.

GSM standardization was started in 1982, and the service was deployed commercially the first time a decade later, in 1991. The standardization work was started by a new committee of European Conference of Postal and Telecommunications Administrations (CEPT), called GSM (Groupe Spécial Mobile). In 1989, the GSM standardization was moved to a new standardization forum, European Telecommunications Standards Institute (ETSI), which was established in 1988. Currently, the GSM standards are maintained by the 3rd Generation Partnership Project (3GPP) standardization forum. In 3GPP, the GSM radio access network standards are referred to with a new acronym, GERAN (GSM EDGE Radio Access Network), which covers both GSM and Enhanced Data rates for Global Evolution (EDGE) technologies, part of the GSM evolution.

In the beginning of the 1990s, networks compliant to GSM specifications were opened first within Europe, but soon such networks were deployed also on other continents, especially Asia and Australia. In 2008, there were GSM networks in over 200 countries, and the total number of GSM subscribers exceeded 3 billion. At the time of this writing in 2022, GSM networks are still widely supported for low-price entry level phones, building network coverage to sparsely populated areas and supporting telephony for users roaming abroad.

GSM standards had three phases of evolution:

- GSM phase 1 standard was frozen in 1991. The standard specified GSM voice calls, **short messaging service (SMS),** and basic supplementary services, such as call forwarding and network roaming. GSM air interface and SMS were novelties for GSM, but the GSM core network specifications for voice call support were largely based on existing circuit switched technologies such as digital exchanges and the SS7 protocol stack. GSM extended SS7 protocols to cover areas relevant for cellular networks, such as mobility and radio resource management. This made it possible for the vendors to use their existing digital exchange products to support GSM just with additional software packages [2].
- GSM phase 2 standard was frozen in 1995. This version of the standard provided an extended set of supplementary services, such as conference calls, call hold, call waiting, and originating identification presentation (rendering the number of the caller on the GSM phone of callee). Additionally, the specification defined how GSM could be used for transporting data or telefaxes.
- GSM phase 2+ standards were completed by 1997. The release 96 version of the standard specified the **high-speed circuit switched data (HSCSD)** service, and the release 97 version introduced packet switched data support on GSM networks with the **general radio packet service (GPRS).**

The structure and functions of the GSM system are described in the GSM specifications of ETSI, later adopted by 3GPP. There are more than a hundred technical specifications in the original GSM specification library. The original ETSI GSM specifications are divided into different standard series according to their topics, as follows:

- TS GSM 01: General, GSM terminology and abbreviations
- TS GSM 02: Services provided by GSM system
- TS GSM 03: The functions of GSM network and general descriptions of those

- TS GSM 04: GSM radio interface and protocols
- TS GSM 05: The physical layer of GSM radio interface
- TS GSM 06: Voice coding algorithms used for GSM
- TS GSM 07: Terminal adaptors for mobile station
- TS GSM 08: Interfaces and protocols between GSM base station and mobile switching center (MSC)
- TS GSM 09: Interconnection between GSM network and fixed PSTN network
- TS GSM 11: GSM SIM card, equipment, and type approval of GSM devices
- TS GSM 12: Operation, maintenance, and charging in GSM networks

The listed standardization series covers all GSM standards up to 3GPP standard release 99. This release was the last of the 3GPP releases with a name that referred to the target year for completing the release. During the 1990s, 3GPP tried to produce a set of standards every year, but soon it turned out that completing the standards release within the target year was too challenging. After completing releases 97, 98, and 99, 3GPP published its fourth standards release, named Rel-4. From 3GPP Rel-4 onwards, the GSM standardization series was renumbered. The number of the new standardization series was its original number plus 40; thus, numbers in the range 41–52 were allocated for GSM specifications. Two additional standardization series, 33 and 55, were created for information security aspects.

In addition to these technical standards, the GSM operators had their own agreements, which aim at harmonizing the commercial aspects of joint usage of GSM networks. This agreement framework, known as GSM Memorandum of Understanding (MoU), covers the following topics:

- Deployment of GSM networks
- Telephone numbers used in GSM networks
- Tariffs and pricing of GSM services

5.1.2 Frequency Bands Used for GSM

The following frequency areas were reserved globally for GSM systems:

- GSM 900
  - The original GSM 900 standard specified two 25 MHz bands for GSM:
    - o Frequency band 890–915 MHz for uplink channel from the mobile station to the base station
    - o Frequency band 935–960 MHz for downlink channel from the base station to the mobile station
  - Each of those bands is divided into 125 subbands. GSM dedicates a subband for one single mobile station for only a short period of time known as a GSM timeslot (or burst) and uses a frequency hopping scheme to renew subband allocations for every timeslot. The subbands are numbered as 0 . . . 124.
  - In the later version of the GSM standard, both the uplink and downlink bands were extended with 10 MHz additional bandwidth to the lower end of the original band. These additional bands were divided into 50 subbands, which were numbered as 974 . . . 1023.
- GSM/DCS 1800:
  - DCS 1800 standard was created in the years 1990–1991 and allocated two new 75 MHz bands for GSM:
    - o Frequency band 1710–1785 MHz for uplink channel from the mobile station to the base station
    - o Frequency band 1805–1880 MHz for downlink channel from the base station to the mobile station
  - Each of those bands is divided into 375 subbands.

Additional GSM 1900 and 850 bands were used in North America, where the global GSM bands were already used for other systems. After the 4G LTE (long-term evolution) technology required more bandwidth than was initially available, the higher GSM frequency bands have in many cases been reallocated from GSM to LTE networks [2].

5.1.3 Architecture and Services of GSM Systems

5.1.3.1 GSM Services
The GSM Phase 2 system provides the following services:

- Voice calls within the GSM network or with the fixed PSTN/ISDN network
  - Ordinary voice calls between two subscribers
  - Emergency calls to a local emergency center
- **Short message service (SMS)** between GSM mobile stations. The short message was specified as a message of a maximum 160 characters, written with the keypad of the mobile phone. The short message is forwarded to its recipient via the short message center within the GSM network. In the later versions of the specification, the upper limit of the message length was relaxed, and more advanced ways were specified to deliver different types of content than only plain text.
- Transport of Telefax messages according to International Telecommunications Union (ITU)-T standard T.30 [3]
- Data transport using GSM connection with a maximum speed of 9600 bps toward other data networks:
  - Digital connection from a GSM mobile station to a modem in the edge of the GSM network. The modem converts the data to analog format compatible with PSTN modems, to provide connectivity from a GSM mobile station to a PSTN modem service.
  - Digital connection from a GSM mobile station to the digital ISDN network and an ISDN terminal.
  - Digital connection from a GSM mobile station to a packet switched public data network (PSPDN) and its terminals. The connection to the packet data network can be set up in different ways, either directly or via the fixed telephone network (PSTN or ISDN) using an equipment called packet assembler and disassembler (PAD). The PAD takes care of needed conversions of the packet and circuit switched data (CSD).
  - Digital connection between two GSM mobile stations that use the same GSM network.
- **Supplementary services** related to the voice call. GSM users may make use of the supplementary services by defining how the network shall process incoming or outgoing calls. GSM Phase 2 has the following supplementary services:
  - Presentation of the calling number or connected number [4]
  - Transferring an ongoing call to another party [5]
  - Forwarding the call to another number in different cases such as no answer or the called phone is busy, powered off, or out of the network coverage [6]
  - Queuing of the incoming call and call waiting indication [7]
  - Putting the call on hold, for instance, if the user wants to answer another call or discuss with others before continuing the call [7]
  - Barring of outgoing or incoming calls based on conditions defined for the user's subscription, such as barring of all outgoing calls or barring of international calls [8]
  - Charging indication [9]
  - Multiparty conference calls [10]
  - Closed user groups, which allow any calls between the members of the group but none outside of the group [11]
  - Unstructured supplementary service data (USSD) with which operators can build their own supplementary service messages for operator proprietary services [12]

The availability and pricing of these services for the user depend on the cellular operator as well as user's subscription type. Also, the features supported by the mobile station may limit the supplementary services available. GSM teleservices are briefly listed in 3GPP TS 02.03 [13].

5.1.3.2 GSM System Architecture

GSM system architecture and interfaces were defined in 3GPP TS 03.02 [14]. GSM architecture description was later incorporated to the 3GPP multi-RAT (radio access technology) architecture specification TS 23.002 [15].

The GSM system consists of the following parts, shown in Figures 5.1 and 5.2:

- **Mobile equipment (ME),** which may be a mobile phone or a GSM modem attached to a computer. A GSM mobile phone consists of a radio modem (transmitter and receiver); antenna, microphone, and loudspeaker; display and keypad; and a battery power source. GSM modems typically consisted of a low-level digital signal processing (DSP) unit and an application-specific integrated circuit (ASIC) for higher-level L1 signal processing. Compared to the earlier first-generation phones, the major difference was the smaller size of hand-portable phones, which was achieved with the great progress of electronics and battery technology throughout the 1990s. Such progress also made it possible to introduce separate application microprocessors to the phones, to run software for more advanced applications and user interface functions. When equipped with a SIM card, the mobile equipment is called a **mobile station (MS).**

- **Subscriber identity module (SIM),** which is a smart card used to identify the service subscription of a user. A GSM phone has a SIM card holder. When the card is put into the holder, the contact points on the surface of the card touch the SIM connectors of the phone, allowing the phone to exchange data with the card. The card stores secret data related to subscriber **authentication**. There is also some processing power on the SIM card to run the authentication algorithms and some small applications created with a SIM application toolkit (SAT). SAT was used to create some operator-specific UI applications, but more recently its main purpose has been to support remote update of configuration data within the phone or on the SIM card. While a GSM phone is switched off, the SIM card stores certain pieces of state information (such as the location area and frequencies used in the cell before switch-off) needed when the phone is powered on once again. The SIM card has also limited amount of memory capacity to store phonebook entries or short messages. The SIM card is separate from the mobile station to allow the user to switch the phone while keeping the subscription or change the subscription while keeping the phone. The SIM card can be protected against unauthorized use with a personal identification number (PIN) code, which the user is asked to enter when powering on the phone. The SIM card specification can be found from 3GPP TS 42.017 [16].

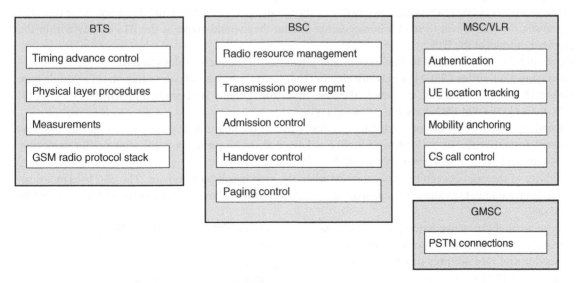

Figure 5.1 Functional split between GSM network elements.

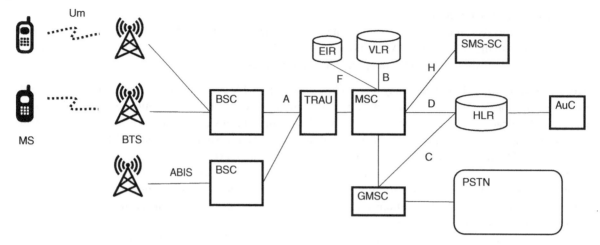

Figure 5.2 Architecture and interfaces of the GSM system.

- **Base transceiver station (BTS)** is a GSM base station that has radio transmitters, receivers, and the antennas serving the GSM cell. GSM BTS may have 1–16 different **transceivers (TRX)**. A transceiver has a transmitter and receiver working together as a pair. Additionally, BTS has a computer system controlling the BTS functions and a telecommunications link toward the base station controller (BSC). BTS functionality is described in 3GPP TS 48.052 [17]. GSM base stations have the following tasks:
  - Modulation, multiplexing, binary coding, and encryption of transmitted GSM radio signals.
  - Demodulation, demultiplexing, decoding, and decryption of received GSM radio signals in order to retrieve the digital data sent by the mobile stations.
  - Managing GSM radio frequencies, frame structures, and synchronization.
  - Recognizing **random access** requests from mobile stations.
  - Managing **timing advance** of the mobile stations camping in a GSM cell. The correct time of a GSM mobile station for sending GSM bursts depends on its distance from the BTS. When a mobile station is far from the cell tower, the BTS tells it to send it bursts earlier so that they would arrive at the BTS exactly within the expected timeslot at the BTS.
  - Measuring the signals sent by mobile stations and providing the measured values to BSC.
- **Base Station Controller (BSC)** is a network element that controls and coordinates functions of multiple base stations and connects those base stations to the MSC. BSC functionality is elaborated in 3GPP TS 48.002 [18] and TS 48.052 [17]. GSM BSCs have the following tasks:
  - Allocation of radio channels for voice calls. The BSC controls the channel allocation for all the base stations connected to the BSC. The radio channel is allocated as a defined frequency hopping sequence for successive GSM bursts, starting from a given timeslot of the GSM frame.
  - Handover of the call between two base stations when the mobile station is moving. The BSC uses measurements from a number of candidate base stations to select the best one to serve the mobile station for an active GSM call.
  - Management of the transmission power of base stations and the mobile stations served by the base stations.
  - Management of the telecommunications links toward the MSC.
- **Transcoder and rate adapter unit (TRAU)** is a piece of equipment used to convert the 16 bps (or less) voice coding used in GSM systems to the 64 kbps PCM signal used in the fixed PSTN network and vice versa. For downstream, TRAU multiplexes four GSM speech channels into a single 64 kbps timeslot toward the base

station. The GSM standards assume that TRAU would be a part of the BSC, but in practical implementations TRAU is located as close to the MSC as possible, in order to minimize the transmission capacity needed between the BSC and the MSC.

- **Mobile switching center (MSC or** mobile telephone exchange [**MTX**]): GSM MSC takes care of routing and switching calls between mobile phones. GSM MSCs perform the following tasks:
  - Routing and connecting mobile originated and mobile terminated calls, and calling the phone (paging) located under certain BSC when an incoming mobile terminated call arrives for the GSM phone.
  - Control of the call handover if the handover is done between two base stations controlled by two different BSCs.
  - Participating in the location management and user authentication in GSM network when the mobile station registers to the GSM network service.
- **Interworking function (IWF)** is a piece of equipment used to adapt the GSM data connection with an external data network.
- **Gateway mobile MTX (GMSC)** is a special MSC that has to find out which MSC is currently serving a mobile station, which would receive a mobile terminating call attempt. The GMSC connects the mobile terminating call to the right MSC. GMSCs connect the calls to PSTN and control echo cancelers used between PSTN and PLMN due to the long speech coding delay [19].
- **Home location register (HLR)** is a database used in GSM network to permanently maintain information about subscribers, services available for them, related settings, and other GSM networks, which the subscriber is entitled to use. The HLR also knows which MSC/VLR (visitor location register) currently serves the subscriber. HLR ensures that the subscriber information is copied to only one single VLR database at any moment.
- **Visitor Location Register (VLR)** is a database used by a single MSC to store subscriber information about those users who are currently camping on any base station connected to that MSC. The VLR database knows the current location area of the user and has a copy of HLR subscriber data as needed to connect calls to the subscriber. Copying the data from HLR to VLR reduces traffic between MSC and HLR while the MS is located under the area managed by MSC. In cases where the user is roaming abroad, the serving MSC and VLR do not belong to the home network of the user but to the visited network. MSCs in both home and visited network must interact to connect calls to or from the roaming users.
- **Authentication center (AuC)** is a server that permanently stores the secret data used to authenticate users of GSM network. The AuC exchanges authentication data with HLR and VLR databases when a GSM phone with SIM card is powered on and it registers to the GSM network. The AuC also stores the encryption keys used to secure the data sent over the radio interface.
- **Equipment identity register (EIR)** is a database that stores information about the GSM mobile stations. Each mobile station is identified by its unique International Mobile Equipment Identity (IMEI) code. The EIR knows IMEI codes of stolen phones so that they can be blocked for GSM network access.
- **Short message service serving center (SMS-SC)** is a server that forwards GSM short messages between mobile stations. SMS-SC is able to temporarily store the message if the destination mobile station is not reachable or is switched off. The message will be sent when the MS becomes reachable again.
- **SMS-gateway** refers to two types of gateways that contribute to forwarding of short messages:
  - SMS-GMSC has the task to find out under which MSC the target mobile station camps and thereafter forwards the short messages to that MSC.
  - SMS-IWMSC has the task of forwarding the short messages to the SMSC of the short message destination, when the SMS endpoints have different home networks.
- **Billing center (BC)** is a database used for collecting the billing data.
- **Operations and maintenance center (OAMC)** is the GSM network management system.

GSM network has the following subsystems:

- **Mobile station (MS)** is the GSM phone and the SIM card inside of it.
- **Base station subsystem (BSS)** covers BSCs and the base stations controlled by them.
- **Network and switching subsystem (NSS)** has MSCs and the related databases such as HLR, VLR, AuC, EIR, and SMS-SC.
- **Operations and support subsystem (OSS)** is the GSM network management system supporting the following functions:
  - Subscriber management
  - Configuring the network and its parameters
  - Measuring and following up network performance
  - Supervision of different error situations in the network

GSM standards define named **interfaces** between different network elements. Detailed standardization of those interfaces is necessary to make it possible to build networks with elements from different vendors so that those elements are able to interoperate. The most important interfaces of the GSM system, shown in Figure 5.2, are as follows:

- Um: Radio interface between mobile station and base station
- Abis: Telecommunications interface between GSM base station and BSC
- A: Telecommunications interface between BSC and the MSC
- B: Interface between MSC and the VLR database
- C: Interface between GMSC and HLR database
- D: Interface between MSC/VLR and HLR databases
- E: Interface between two MSCs that contribute setting up connections to a certain mobile station, for instance, in an inter-MSC handover scenario
- F: Interface between MSC and EIR database
- G: Interface between two VLR databases
- H: Interface between MSC and SMS-SC
- I: Interface between MSC and mobile station

A standard **protocol stack** is defined for most of these interfaces. The stack is used to pass signaling messages or user data between network elements over the interface. The radio interface relies on protocols specified only for the GSM network while the protocol stacks used on other interfaces are extensions of the stacks defined in ISDN and SS7 protocol specifications. GSM standards also specify how the protocols interoperate when signaling messages are forwarded over a chain of interfaces that use different protocol stacks.

It is worth noting that this interface list corresponds to the traditional GSM architecture. After 3G UMTS was deployed, another option was introduced to connect GSM BSS to Serving GPRS Support Node (SGSN) and MSC via the Iu interface that was used also by 3G RNC (Radio network controller). Both of those architectures are depicted in 3GPP TS 23.002 [15] and the related two protocol stack architectures in TS 23.060 [20].

The following different types of identifiers have been defined in 3GPP TS 23.003 [21] for the GSM subscriber and mobile station:

- **International mobile subscriber identity (IMSI)** is a unique and permanent identifier of a GSM subscriber and the home network. Every SIM card has its own IMSI code, which is different from IMSI codes of other SIM cards.
- **Temporary mobile subscriber identity (TMSI)** is a temporary identifier given for a GSM user by the MSC, which currently serves the subscriber. TMSI is used in the clear text messages sent over the radio interface between the mobile station and the GSM network. TMSI is used to hide the real identity of the user for any third parties who might listen to those messages.

- **International Mobile Equipment Identity (IMEI)** is a unique permanent identifier of the mobile equipment. IMEI is used to block a stolen GSM mobile station from accessing GSM networks and also to ensure that every mobile station model connecting to the network has received a type approval. IMEI code consists of the following parts: the type approval code, the factory which has manufactured the mobile station, and the serial number of the device.
- **Mobile station ISDN number (MSISDN)** is the telephone number of the mobile station (or rather its SIM card) used for creating circuit switched voice or data connections. The international number format has a country code, area or network code, and the subscriber number within the network. The IMSI is a permanent property of the SIM card, but the MSISDN telephone number can be changed if the subscriber wants to have a new telephone number for any reason.
- **Mobile station roaming number (MSRN)** is an ISDN number granted by the VLR to a mobile station when there is a mobile terminated call attempt for the station. The MSRN identifies the mobile station and the MSC currently serving it. The MSC/VLR that has reserved the MSRN number is able to associate the MSRN to the IMSI code of the subscriber. Other MSCs use the MSRN number to route the MT call to the current serving MSC.

Additionally, GSM specifications define the following two identifiers related to the location of the mobile station in the network:

- **Cell global identity (CGI):** Globally unique identifier of a GSM cell and the corresponding base station.
- **Location area identity (LAI):** Identifier for a set of cells under a single MSC. The mobile station shall send a location update message to the network when it moves from one location area to another.

5.1.3.3 GSM Functions and Procedures

GSM system specifications cover three functional areas of interaction between the GSM mobile station and the network:

- **Radio resource management (RR)** for managing the radio connections between a mobile station and the MSC. The RR procedures are not focused on GSM air interface alone. Instead, they also cover interfaces between and among the base station, BSC, and MSC irrespective of the specific transmission technology used on those links. The main tasks of radio resource management are the following:
  - Allocating radio channels between a base station and those mobile stations that have circuit switched GSM connections and releasing the allocated radio channels after disconnecting the circuits.
  - Handover of a circuit switched call between two base stations when the mobile station moves from a cell to another.
  - **Power control** to minimize the transmission power used by mobile and base stations while keeping the quality of received radio signals within predefined limits. The power control mechanism has two purposes: to minimize the power consumption and minimize interference between mobile stations in different cells. Power saving is important for both the operator and the subscriber. Operators want to minimize the base station electricity bills and subscribers want to maximize the idle and talk time for a single charge of the battery.
  - Adjusting timing of **GSM bursts** sent by mobile stations to synchronize them to the frame structure of the base station. **Timing advance** commands align the reception time of bursts from mobile stations to the common timeslot structure from the base station perspective, regardless of the distance of the mobile stations from the base station.
- **Mobility management (MM),** which covers the following tasks:
  - Tracking the location of a mobile station. The main task of the mobility management is to keep the network aware of the location area (and related cells) within which the mobile station should be paged for any incoming mobile terminated calls.

Table 5.1 Functional areas of GSM.

	RR	MM	CM
Mobile station	Request radio resources, adjust the transmission power and timing, handovers	Camp to a GSM cell, send location updates	Initiate MO calls, answer MT calls, send and receive short messages
Base station	Give commands for timing advance, generate GSM radio frames	–	–
Base station controller	Select the base station and radio channel for the mobile station, power control, handover between base stations under a single BSC	–	–
Network and switching subsystem(MSC, VLR, HLR, SMSC, AuC)	Handovers between two BSCs	Location information management, user authentication	Connect calls and forward SMS to and from mobile stations

- Actions related to cell selection (and reselection when an idle mobile station is moving) by both mobile station and the network.
- **Security management (SM),** which covers authentication of GSM subscriber and encryption of the traffic sent over radio interface. The information security of the GSM system covers both traffic encryption over the GSM radio interface and hiding the location of the user within the network.
• **Communication management (CM)** takes care of call control, forwarding of short messages, and management of supplementary services. Call control means routing, switching, setup, and release of circuit switched calls.

The devices defined for GSM architecture have the responsibilities described in Table 5.1 related to these three functional areas.

The GSM system elements communicate with each other with signaling protocols to exchange needed pieces of information to perform these tasks. GSM specifications define a set of signaling protocols to be used over the interfaces between the network elements. 3GPP TS 29.010 [22] describes GSM communication procedures and mapping of parameters between messages of different protocols used in the communication path over multiple interfaces.

5.1.3.4 GSM Protocol Stack Architecture

GSM was designed to use layered protocol stacks following the open systems interconnection (OSI) model, according to 3GPP TS 44.001 [23].

On the link layer of Um radio interface, the signaling messages are transported over LAPDm protocol, which is a member of the high-level data link control (HDLC) protocol family. Within the BSS, the link layer uses LAPD protocol, which is another variant of HDLC protocol. Connections between and within NSS up to BSC use the message transfer part (MTP), SSCP, and TCAP protocols of SS7 protocol family [24]. Links between BTS, BSC, and NSS elements are run either over cables or fixed microwave radio links.

GSM radio resource control (RRC) protocol is used over the radio interface between BSC and MS for radio resource management. The BSC uses TS 48.058 [25] protocol to control BTS radio resources. NSS uses base station subsystem management part (BSSMAP) and MAP/E protocols for radio resource management.

Mobile station and MSC are the endpoints for mobility and communication management protocols MM and CM, which are members of the direct transfer application part (DTAP) protocol family. BSS only forwards

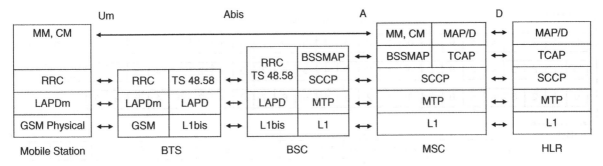

Figure 5.3 GSM control plane signaling protocols.

messages belonging to these protocols but does not participate in the actual communication. In NSS, the VRL uses SS7 MAP/D protocol to discuss with the HLR database.

The GSM control plane (Figure 5.3) and user plane protocol stacks are further described in 3GPP TS 43.051 [26] and TS 48.008 [27]. For the description of radio resource management, location management, and communication management, please refer to Section 5.1.3.3.

5.1.4 GSM Radio Interface

The interface between the GSM mobile station and base station is called GSM **radio interface** or **air interface**. On the lowest physical layer of the radio interface, the GSM protocol stack has GSM radio channels.

GSM radio interface is specified in the 45-series of 3GPP specifications, such as these:

- TS 45.001 [28]: General description covering, for instance, GSM frame structures
- TS 45.002 [29]: Multiplexing and mapping of logical to physical channels, frequency hopping parameters
- TS 45.003 [30]: Channel coding
- TS 45.004 [31]: Modulation
- TS 45.005 [32]: Transceiver requirements
- TS 45.008 [33]: Power control, measurements, cell reselection
- TS 45.009 [34]: Link adaptation

5.1.4.1 Modulation and Multiplexing

GSM uses two different approaches combined for multiplexing radio channels of multiple users over the shared radio media, as seen in Figure 5.4.

1) **Time division multiple access (TDMA):** The GSM frame of 4.62 ms is divided into eight timeslots, 577 µs each. The full-speed voice channel of a mobile station uses only one single timeslot for each GSM frame. In other words, a frame supports eight mobile stations at a time. The uplink timeslots are delayed by three slots compared to the downlink, allowing the devices some time to switch between reception and transmission.
2) **Frequency division multiple access (FDMA):** Each base station divides its available GSM radio band to subbands or radio frequency (RF) physical channels of 200 kHz. The subband is used by one single mobile station for the duration of one timeslot. The GSM cell uses one frequency band for transmission and another for reception. The distance between adjacent uplink and downlink RF channels allocated for one mobile station are always 45 MHz. The specific RF channel pair is identified with its absolute radio frequency channel number (ARFCN). The adjacent cells do not use the same subbands to avoid any intercell interference.

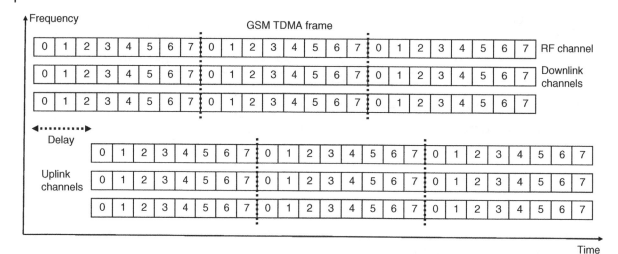

Figure 5.4 Multiplexing on GSM radio interface. *Source:* Adapted from 3GPP TS 05.02 [35].

Transmission done within one timeslot and subband of the GSM radio interface is called a **burst**. The modulation method for the GSM radio signal is called **gaussian minimum shift keying (GMSK)**, defined in 3GPP TS 45.004 [31]. GMSK is a mathematically complex modulation method where the frequency of transmitted signal is chosen as a function of multiple bits sent after each other. The value of 1 bit in theory impacts a GMSK modulated signal until its end, but in practice the impact is noticeable in the signal generated for transmitting three sequential bits. The benefit of GMSK modulation is that the transmission power is focused to a rather narrow band, which makes it possible to divide the GSM bandwidth to many 200 kHz RF physical channels.

One GSM cell can in theory use 31 subbands, but in practice a cell typically has up to 16 RF channels. The RF channels are reallocated between served mobile stations for every timeslot. Thus, GSM uses frequency hopping radio technology where the mobile station changes its frequency for every burst it sends and receives, once for a GSM TDMA frame. The mobile station knows the RF channels it shall use for its timeslots based on its frequency hopping sequence according to the HSN and Mobile allocation index offset (MAIO) parameters that the MS gets from BSC, as described in Section 5.1.4.4. The hopping sequences are defined in such a way that two mobile stations will never use same RF channel of a cell simultaneously.

As can be seen in Figure 5.5, the TDMA frames of downlink and uplink are not aligned, but there is a difference of three timeslots between them. When a mobile station is given with a channel assigned to timeslot TN3, the transceiver of the station has 1154 ms (two timeslots) to switch between the transmit and receive modes and to tune itself to the correct RF channel.

The number of simultaneous calls within GSM cell is approximately eight times to the number of 200 kHz RF channels allocated to the cell. In practice, the number of simultaneously served mobile stations within the cell is slightly less since certain radio channels are used as shared channels on which common information related to the cell and GSM network, such as **system information** messages, are broadcast to all the mobile stations within the cell. System information messages provide the mobile stations with common network and cell specific parameter values.

5.1.4.2 Frame Structure and Logical Channels

The GSM mobile station has multiple information flows toward the network. To support those flows, a number of logical channels have been defined for the GSM air interface in 3GPP TS 44.003 [36].

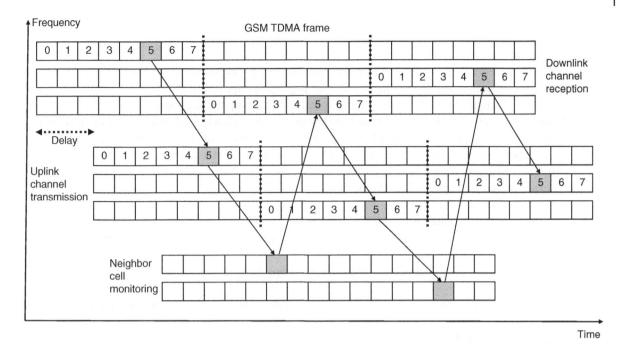

Figure 5.5 GSM frequency hopping scheme used by mobile station. *Source:* Adapted from 3GPP TS 05.02 [35].

GSM logical channels are as follows:

1) Shared channels used by all mobile stations camping on the cell
   - Broadcast channels (BCH) used by the base station to send commonly used information to all mobile stations within the cell.
     - Frequency correction channel (FCCH): Channel used to transport a fixed sine wave within its every burst. When searching a GSM cell, the mobile station tries to find such a burst pattern in the time-frequency space as used for GSM. After detecting an FCCH burst, the mobile station is able to recover the clock frequency, timeslot boundaries, and the radio frequency used for BCH channels. This enables the mobile station to recognize the synchronization channel (SCH) of the cell.
     - Synchronization channel (SCH): Bursts sent on the SCH help mobile stations to detect the frame structure as transmitted by the base station. After the mobile station has successfully received SCH bursts, it is synchronized to both the timeslot and frame structure of the cell and is thereafter able to decode other channels.
     - Broadcast control channel (BCCH): Channel used to broadcast GSM system information messages to all mobile stations within the cell, to provide them with information about the network, and various control parameter values such as these:
       o The name of the network operator, which is used for the operator selection algorithm of the mobile station
       o The location area to which the cell belongs
       o The frequencies used by the cell
       o The frequencies used by the BCH channels of adjacent cells
       o Information of the common control channel (CCCH) and cell broadcast channel (CBCH) timeslot configuration

- Common control channels (CCCH) are in shared use of all mobile stations within the cell, but without a broadcast mechanism. Only one mobile station may use an uplink channel at a time, and the messages on the downlink channel have the destination mobile station address.
  - Paging channel (PCH): Channel on which the network sends paging messages to a mobile station to notify it about an incoming mobile terminated call. The paging message is sent in all the cells of that location area where the mobile station has most recently sent a location update message.
  - Random access channel (RACH): Channel used by the mobile station to request a dedicated radio connection to initiate a mobile originated call, send a short message, or a location update message.
  - Access grant channel (AGCH): Channel on which the network tells the mobile station about a dedicated radio channel granted to the MS.
- Cell broadcast channel (CBCH): Channel on which special short messages can be broadcasted to all mobile stations synchronized to the GSM cell.

2) Voice or data channels dedicated to one single mobile station
- Traffic channel (TCH): Voice channel dedicated to one single mobile station at a time. TCH can also be used for CSD transport, depending on the application.
  - TCH/F – full-speed voice channel, which provides 13 kbps speed for voice and 12, 6, or 3.6 kbps speed for data connections.
  - TCH/H – half-speed voice channel, which provides 7 kbps speed for voice and 6 or 3.6 kbps speed for data connections.

3) Signaling channels dedicated for one single mobile station
- Dedicated control channels (DCCH): Channels used to transport control information between GSM network and one mobile station.
  - Slow associated control channel (SACCH): Bidirectional channel allocated to a mobile station capable of slow exchange of signaling messages. The transmission speed of this channel is on the average sufficient for two messages per second. Mobile stations measure signals received from neighboring GSM cells and use SACCH to inform the network about the measurement results. The network sends timing advance and power control commands to mobile stations over SACCH. It is possible to use SACCH also for transporting short messages.
  - Fast associated control channel (FACCH): Channel allocated to a mobile station capable of fast exchange of signaling messages. FACCH is not a separate channel in the GSM frame structure but it uses the capacity of a TCH channel when no voice or data is sent over that TCH. FACCH is used at the opening and closing phases of TCH when transport of user data over the TCH is not yet started or is already finished. Every burst of a TCH has 2 bits, which tell if the timeslot belongs to the FACCH or whether the TCH carries user data or voice.
  - Standalone dedicated control channel (SDCCH): Channel which the network may allocate to a mobile station when no other user data than short messages have to be transported or for call setup signaling prior to the traffic channel has been allocated for the call. SDCCH is typically used when the mobile station only needs to send signaling messages, such as location updates or messages related to call transfer. The transmission speed of SDCCH is one-eighth of the TCH/F.

Each of the above-mentioned radio channels are implemented as a set of bursts repeating in defined intervals in the GSM TDMA frame structure cycle of eight timeslots, as shown in Figure 5.6.

The bursts sent by mobile stations located at different distances from the base station shall still be received at the base station as aligned to the frame structure of the base station and its timeslots. As propagation of radio signal over distance takes a small amount of time, the base station asks those mobile stations that are farther away to send their bursts a bit earlier compared to the stations closer to the base station, in order to compensate the transmission delay. This timing advance command is sent on SACCH.

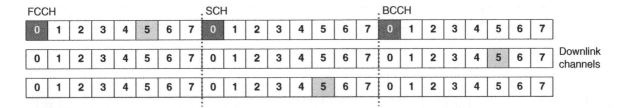

FCCH · SCH · BCCH

<table>
<tr><td>0</td><td>1</td><td>2</td><td>3</td><td>4</td><td>5</td><td>6</td><td>7</td><td>0</td><td>1</td><td>2</td><td>3</td><td>4</td><td>5</td><td>6</td><td>7</td><td>0</td><td>1</td><td>2</td><td>3</td><td>4</td><td>5</td><td>6</td><td>7</td></tr>
</table>

0 Timeslots that belong to shared channels FCCH, SCH, BCCH, PCH and AGCH

5 Timeslots that belong to a single frequency hopping TCH channel

Figure 5.6 Allocation of GSM physical channels over GSM frame structure. *Source:* Adapted from 3GPP TS 05.02 [35].

The logical channels are mapped to physical channels of GSM frames and RF channels with mechanisms defined in TS 44.004 [37] and 45.002 [29]. The mapping of GSM logical radio channels to GSM timeslots is defined using a concept of GSM **multiframe** and rules about which RF channels the logical channel may use.

- **Shared channels** BCH and CCCH: These channels are transported over one single GSM RF channel of the cell, using the BCCH beacon frequency. No frequency hopping is used. The physical channel mapping of this C0 RF channel is defined within a multiframe consisting of 51 successive TDMA frames, eight timeslot each. The shared channels are transported in the first timeslot TN0 of the TDMA frame. The RACH channel consists of uplink TN0 bursts from the mobile station to the base station. Mapping of the shared downlink channels to TN0 bursts from the base station to the mobile station is as follows:
  - FCCH and SCH channels are composed of one burst per 10 frames in timeslot TN0. Within the multiframe, the first TN0 burst of TDMA frame #0 belongs to FCCH and second TN0 burst of frame #1 to SCH.
  - BCCH is transported in TN0 bursts of frames #2–#5.
  - PCH and AGCH use the rest of the TN0 bursts of the multiframe (except those carrying FCCH or SCH bursts), either up to burst #19 or to the end of the multiframe, depending on the capacity reserved for these channels in the cell. These channels can be divided into multiple subchannels so that the mobile station must listen to only one single subchannel and save its battery while other subchannels are being used. This mechanism specified in 3GPP TS 43.013 [38] is called GSM discontinuous reception (DRX). The mobile station can deduce the subchannel it shall listen to from its own IMSI code.
  - In a cell with large capacity, the structure of the shared channels can be mapped also to other timeslots TN2, TN4, and TN6. Exceptions to this are the FCCH and SCH used for cell synchronization, which may appear only at TN0.
- **Dedicated channels** TCH and DCCH: These channels may be mapped to any RF channel of GSM cell. Frequency hopping is used. The channel structure is defined within a multiframe consisting of 26 successive TDMA frames, eight timeslots each:
  - TCH/F use the bursts #0–#11 and #13–#24 for the timeslot allocated to mobile station. The related SACCH uses bursts #12 and #25. The second burst is not used for GSM transmission, and the mobile station may use it to listen to frequencies used by adjacent cells for their shared channels.
  - TCH/H uses every other timeslot of bursts #0–#11 and #13–#24. The related SACCH uses either burst #12 or #25 of the timeslots used for TCH/H.
- The dedicated SDCCH signaling channel: SDCCH and the associated SACCH may be mapped to any RF channel of the cell. The channel structure is defined within a multiframe consisting of 102 successive frames: SDCCH uses two groups of four bursts within the multiframe of 102 frames (see Figure 5.7). The SACCH related to this channel uses one group of four bursts within the same multiframe. Within the multiframe a total of 12 bursts are needed to carry these two channels for a mobile station. These bursts may be mapped to a timeslot

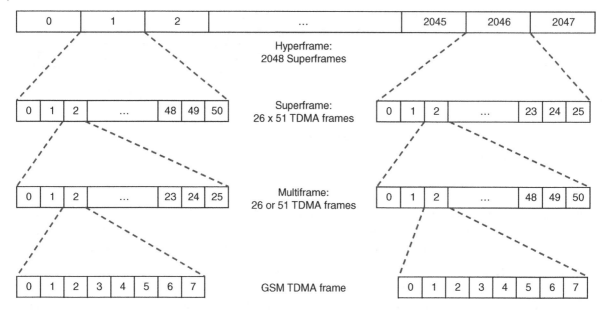

Figure 5.7 The multiframe structure of GSM radio interface.

that does not carry any other channels or unused TN0 timeslots when PCH and AGCH do not use any timeslots after TN0 #19 of the multiframe.

The lengths of multiframes defined for shared and dedicated channels are different to ensure that a mobile station actively engaged for voice or data connection could measure and detect the synchronization channels of adjacent cells. The fact that the multiframes have different sizes prevents the case in which the burst belonging to the synchronization channel would always overlap with the timeslot the mobile station is using for its traffic.

5.1.4.3 GSM Bursts and Channel Coding

As mentioned in Section 5.1.4, the nominal length of a GSM timeslot is 577 µs. In the beginning and end of the burst, there is a period of 30 ms to ramp the transmission power up and down. Taking that into account, it is possible to send a maximum 148 bits of information within a single burst. The internal structure of the burst depends on the type of the radio channel to which the burst belongs to [28] (see Figure 5.8):

1) On the FCH channel, F-bursts are sent, which contain 148 zero bits. After GMSK modulation, the signal is pure sine wave. The purpose of such a burst is to be very easily recognizable so that a mobile station can easily detect the FCH channel and can thereafter calibrate its own clock frequency to the frequency of sine wave of the F-burst as sent by the base station. The leading and trailing edges of the sine wave burst indicate the boundaries of timeslots within the GSM frame.

2) On the SCH channel, S-bursts are sent with the following structure:
   - In the middle of the burst, there is a predefined constant string of 64 bits called a training sequence. This sequence enables the mobile station to recognize the S-burst and adjust its receiver to compensate for any signal distortion on the radio channel.
   - Both in the front of and right after the training sequence, the burst has 39 bits that describe the BSIC code of the cell and the TDMA frame number.
   - In both ends of the burst, there are three zero bits.

3) On the RACH channel, the length of the burst is only 87 bits. This burst is short because the mobile station that sends a random access burst has not yet received a timing advance command. As the station does not know

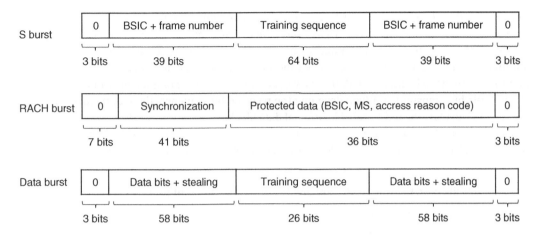

Figure 5.8 GSM transmission burst types.

how far it is from the base station, usage of a short burst aims at ensuring that the whole burst can be received by the base station within one timeslot. The structure of a random access burst is as follows:

- There are 7 zero bits in the beginning and 3 zero bits in the end of the burst.
- Right after the leading zero bits, the burst has 41 bits with a constant predefined synchronization bit sequence.
- Thereafter, the burst carries 36 bits of protected data. This bit sequence is derived from 8 bits of information, protected with 6 parity bits added bitwise with 6 bits of BSIC code. The result is encoded with 1/2 convolutional code to produce the transmitted data bit sequence. The 8 information bits contain a random bit sequence for identifying the mobile station and a reason code for its random access attempt.

4) Bursts sent on other channels have the following structure:
   - In the middle of the burst there are 26 bits with constant predefined training sequence, helping the receiver to adjust itself to compensate for any distortion over the radio channel.
   - On both sides of the training sequence, there is 1 stealing bit and 57 bits of user data, such as encoded voice. The total number of data bits per burst is 114. Stealing bits are used to distinguish control and data bursts. When the stealing bit is set, the burst belongs to the FACCH rather than the TCH channel.
   - There are 3 zero tail bits in both ends of the burst.

GSM **channel coding** process is used to make transmission of voice and data tolerant against transmission errors. The channel coding process specified in 3GPP TS 45.003 [30] consists of the following steps:

- Block coding
- Convolutional coding
- Interleaving
- Burst generation

In the channel coding process, the user data bitstream to be transmitted is divided into data bursts in the following manner:

1) The bit stream is divided into fixed-size blocks for further processing.
2) Each block is encoded to protect the data against transmission bit errors. The encoding increases the redundancy of data so that the size of the block can be doubled from its original size. Encoding increases the number of data bits in two ways:
   - A cyclic redundancy check (CRC) may be calculated over and added to the original block. The CRC checksum enables the receiver to detect and correct single bit errors.

- The bit sequence of the block can be used to generate a convolutional forward error correction code. Convolutional code may be created from the original bit sequence by combining it with exclusive-or operation with copies of the same bit sequence shifted with one or multiple bit positions. It is also possible to calculate multiple different convolutional codes from the same bit sequence and combine selected bits of each code for the bit stream to be transmitted. In this way, any single bit in the original bit sequence impacts the value of multiple bits transmitted. When decoding the received bit sequence, the redundancy within it helps the receiver to calculate the most probable original data block.
3) The encoded data block is split into shorter blocks, each of which can be transmitted within a burst. A burst may carry either one such block or parts of multiple interleaved blocks for further robustness against transmission errors.

In block coding, the 260 bits from the voice codec are divided into three different priority classes:

- Class I-A of 50 bits protected by 3-bit CRC
- Class I-B of 132 bits not protected by CRC
- Class II of 78 bits

The class I bits are then provided to a convolutional coder while class II bits are provided directly to the interleaver. In a convolutional coder the class I bits are protected with a forward error correction method, which expands the number of transmitted bits to two or three times the number of information bits. After GSM convolutional encoding, 1 information bit affects 4 transmitted bits. The strongly protected class I-A bits carry the most important parameters for voice reproduction, such as higher-order bits of the filter parameters [2].

When the bits generated by convolutional encoder are combined with class II bits, a bit sequence of 456 bits is provided to the interleaver. Those 456 bits can be carried by four GSM bursts, each containing 114 information bits. The basic **interleaving** process is performed as follows: At first the interleaver divides the 456 bits to eight subblocks of 57 bits, so that bits from every group of 8 bits are put to its own subblock. For instance, the first subblock gets bits 1, 9, 17, . . ., 449 of the original sequence. GSM burst is able to carry two of these subblocks. Further on, the bits carried in the GSM burst may be also interleaved to achieve a second level of interleaving.

This kind of channel coding method makes the transmission robust against typical short disturbances on the radio path affecting some narrow areas of frequency. Even if a single burst is corrupted, the convolutional coding can be used to reconstruct the original data bit stream without any retransmissions. The drawback of channel coding is the additional delay caused by both convolutional coding and interleaving processes. The receiver must wait for all the bursts over which the original stream of 260 bits has been spread, before giving the decoded bits to the voice decoder.

GSM standards specify the details of convolutional coding and interleaving processes separately for each type of radio channel and bit rate used. The ways differ from each other in the following respects: the length of the original and encoded blocks, the exact way of calculating the convolutional code, usage of a separate checksum, how the bit stream is divided into bursts, and how multiple blocks of encoded data are interleaved to a single burst.

5.1.4.4 GSM Frequency Hopping

GSM base stations use one single fixed RF channel for the shared BCH and CCCH logical channels. As frequency hopping is not used for the shared channels, the base station has no need to inform mobile stations about its hopping patterns. Also, the cell search algorithm of the mobile station is kept simple as both FCCH and SCH channels stay on a single frequency.

GSM uses **frequency hopping** for TCH, DCCH, and SDCCH dedicated to one mobile station. In frequency hopping, the mobile station changes its operating frequency (subband) always for a new burst, like that shown in Figure 5.5. As the TDMA frame of GSM carries eight different timeslots, the frequency hopping is done per frame so that the frequency is changed for every complete frame. The frequency hopping method has many goals. At first it decreases the effect of any disturbance source impacting a specific narrow frequency area. It also minimizes interference between two GSM calls running simultaneously. Finally, frequency hopping decreases fading of a channel caused by any obstacles on the radio path, since the impact of an obstacle depends on the radio frequency used.

The frequency hopping sequences given to different mobile stations within a cell must satisfy the following two properties:

- No two mobile stations may use the same RF channel at the same time. Otherwise, their bursts would clash with and corrupt each other.
- To support a maximum number of voice calls per cell, the hopping sequences used in parallel must allocate every available timeslot of every RF channel of a cell to the dedicated channels. There should be no unused available timeslots on any RF channel that a mobile station is unable to use.

The mobile station and base station derive the frequency hopping sequences with a function defined in GSM specifications. This function uses the following three parameters as its input:

- The hopping sequence number (HSN) identifies 1 of the 64 different hopping patterns to be used for picking subchannels for every new burst of the multiframe.
- MAIO defines the RF subchannel, which is used for the first burst of the hopping sequence. The hopping sequence starting times are synchronized within the GSM superframe structure of the cell.

The parameter values are allocated so that every cell has its own value for HSN. Every transceiver of the cell has its own MAIO value, so the MAIO defines one single physical channel per timeslot of the GSM frame. The impact of these parameters to the frequency hopping sequence is as follows:

- Two sequences with different HSN will use the same frequency in 1/n timeslots where n is the number of frequencies used for the timeslot. In practice, this means that two GSM calls in different cells will interfere with each other only occasionally for a single burst.
- Two sequences with the same HSN but different MAIO never use the same frequency for the same timeslot. This means that two calls within a GSM cell will never interfere with each other.

The mobile station learns the value of the HSN parameter and frequencies used by the cell from the system information messages that the BTS sends on the BCCH channel. When a new dedicated channel is given to the mobile station, the BTS tells the MS the values of MAIO and the timeslot number to be used for the channel. Equipped with those pieces of information, the MS knows how to apply the frequency hopping sequence to the given subchannels.

5.1.5 Signaling Protocols between MS and GSM Network

5.1.5.1 LAPDm Protocol

The link layer protocol used on DCCH signaling channels is called LAPDm. The LAPDm protocol transports upper layer signaling messages between the mobile station and base station. LAPDm is specified in 3GPP TS 44.005 [39] and TS 44.006 [40].

LAPDm is a variant of HDLC, which was described in Chapter 3, Section 3.1.33.1. The frame structure of LAPDm has been optimized to take advantage of the synchronization and error correction mechanisms of the GSM physical layer. The biggest differences between LAPDm and HDLC protocol are as follows:

- The LAPDm protocol frame is not delimited by start and end markers. Instead, it has been defined that the LAPDm frame is always 23 octets, which is the length of blocks transported on DCCH channels. Depending on the specific channel (SACCH, FACHH, or SDCCH), such a block is transported either within four or eight GSM bursts. In the latter case, the block is interleaved with either the previous or next block. If the size of the upper

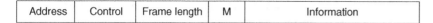

Address	Control	Frame length	M	Information

Figure 5.9 Structure of LAPDm frame.

layer message is longer than what fits in one LAPDm frame, the LAPDm protocol segments the upper layer message before transmission and reassembles it after receiving the related LAPDm frames.

- Unlike the HDLC frame, the LAPDm frame does not contain a cyclical redundancy check since the data block sent over the radio interface is protected by mechanisms of the GSM physical layer. The LAPDm protocol uses the information provided by the GSM physical layer to decide if the LAPDm frame needs to be retransmitted. Retransmissions are not used at all for time critical data, such as radio signal measurements, which only have value over a very short lifetime.
- Size of the LAPDm window for unacknowledged message is one.
- LAPDm protocol does not use flow control commands.

The LAPDm frame shown in Figure 5.9 consists of the following fields:

- Address of one octet tells if the frame is used for signaling messages (service access point identifier [SAPI] 0) or short messages (SAPI 3).
- Control field, which tells the type of the frame that has the same structure as defined for HDLC.
- The length of LAPDm frame together with the more-bit used for segmentation and reassembly function.
- Information field, which contains the upper layer data to be transported over the link.

5.1.5.2 RIL3 Protocols

Three different GSM radio interface layer 3 (RIL3) signaling protocols are used over the radio interface between the mobile station and GSM network. These protocols are members of the DTAP protocol family of SS7:

- RIL3-RR or RRC: radio resource control
- RIL3-MM: mobility management
- RIL3-CC: call control

The RIL3 protocols for MM and CC and the related procedures are specified in 3GPP TS 24.008 [41], and the protocols and procedures for RRC are specified in TS 44.018 [42].

Most of the messages of any RIL3 protocols use the same common frame structure shown in Figure 5.10. This structure consists of the following fields:

- Protocol discriminator as either RR, MM, CC, or some other value defined for other services.
- Transaction identifier, which is used in RIL3-CC protocol to match the response message to the corresponding request message. In the other two protocols, a skip indicator with value 0 is used to indicate that the transaction identifier is not used.
- Message type, which tells the purpose of the message and determines how the rest of the message shall be interpreted.
- The mandatory parameters of the message, which depend on the message type. Since the message type defines also the order and lengths of those mandatory parameters, the message itself contains only the values of those parameters in the predefined order.

Protocol discriminator	Transaction identifier	Message type	Mandatory parameters	Optional parameters

Figure 5.10 Generic structure of RIL3 protocol frame.

- The optional parameters of the message. For these, the message contains at least the parameter name and value if the length of the parameter is fixed in the specification. For variable length parameters, the message uses the type-length-value (TLV) pattern where the parameter is encoded to the message with three fields: the name of the parameter, the length of its value field, and the value itself.

RIL3 protocol messages are transported over the GSM air interface as payload of the LAPDm link layer protocol.

5.1.6 Signaling Protocols of GSM Network

5.1.6.1 Layer 1
Connections between BTS, BSC, and MSC are provided as 64 kbit DS0 channels over E1 or T1 data links. See 3GPP TS 48.004 [43] and TS 48.054 [44]. In addition to signaling, the main purpose of these channels is to carry voice circuits. A single BTS may not need all the channels of an E1 link, but just a subset of them to support the needed voice connections.

5.1.6.2 Layer 2
Signaling messages are transported between the GSM base station and BSC over Abis interface within a dedicated 64 kbit channel. The LAPD link layer protocol is used to carry the signaling messages, as specified in ISDN standards.

RIL3 protocol messages for mobile station are transported as payload of LAPD protocol frames over the Abis interface. Additionally, the LAPD protocol is used to transport base station control messages, which are specified in 3GPP TS 48.058 [25]. The functionality of LAPD protocol is described in TS 48.056 [45].

5.1.6.3 Layer 3
Layer 3 commands between BTS and BSC are defined in 3GPP TS 48.058 [25]. These commands are used to manage GSM channels, radio links, and BTS TRX units.

5.1.6.4 SS7 Protocols
GSM MSC specifications were based on the existing specifications for fixed network switches. Because of that, MSC uses the SS7 protocol stack when communicating with BSC or other switching centers, whether mobile or fixed.

The original GSM design for protocol layers 1–3 was based on delivering MTP protocols of SS7 stack over the E0 channels of the E1 link. In the A interface, the signaling messages between BSC and MSC were transported within a dedicated 64 kbit E0 channel. The link layer protocol is the MTP2 as specified for SS7. In the network layer, the signaling messages are carried with MTP3 and SCCP protocols. The MTP3 protocol is used to manage the chain of links between MSC and BSC and carry the SCCP protocol messages. For detailed descriptions of these SS7 protocols, please refer to *Online Appendix B.1*.

While GSM networks evolved, 3GPP introduced another SS7 over IP (or SIGTRAN) option to replace the traditional MTP protocols of SS7 with an IP-based stack. In this option, point-to-point Ethernet over fiber links (see Chapter 3, Section 3.1.1) are used at the physical and link layer to replace E1 links and MTP1. The functionalities of MTP layers 2 and 3 are replaced in a SIGTRAN solution with the following stack:

- IP protocol is used on the network layer to support SS7 message routing.
- SCTP is used on the transport layer to support reliable transport of signaling messages between endpoints. See Chapter 3, Section 3.3.6 for further details.
- MTP3 User Adaptation Layer (M3UA) protocol was defined to emulate the MTP3 service interface toward the upper layer SS7 protocols, such as SCCP. M3UA is an adaptation layer between the SCTP transport protocol and the original SS7 protocols as specified on top of MTP3.

At the time of this writing, GSM networks being deployed are already relying on the IP option so that the operators can maintain just a single IP-based transport network for all types of their cellular networks.

The SCCP protocol is used as follows:

- The control commands from the MSC to the BSC are transported with SCCP class 0 connectionless service.
- The signaling messages for a specific mobile station are transported to the BSC with SCCP class 2 connected service. The MSC sets up a new SCCP connection for the mobile station when it has to send messages related to a handover or receive location update messages from the mobile station. The lifetime of the connection is limited to the execution of the related procedures.

As its payload the SCCP protocol carries messages of one the following two upper layer protocols:

- BSSMAP: signaling messages between MSC and BSC as specified in 3GPP TS 48.008 [27].
- DTAP: RIL3 signaling messages between MSC and mobile station, forwarded over the BSC.

The SS7 protocol stack is used also for the interfaces B to H within the NSS subsystem. The difference to interface A to the BSS is that within the NSS subsystem the MAP protocol is used as the topmost protocol of the stack. There are many variants of the MAP protocol specified for GSM, called MAP/B to MAP/I, where the letter after the slash means the GSM interface over which the MAP subprotocol is used. MAP protocol is specified in 3GPP TS 29.002 [46].

Within the NSS, the protocols underlying the MAP are as follows:

- SCCP protocol is used to route messages to the correct GSM network element, using SCCP class 0 connectionless service.
- TCAP protocol messages are carried as the payload of SCCP.
- MAP/x protocol messages are carried as the payload of the TCAP protocol. These messages may contain either control information between GSM network elements or signaling messages for the mobile station.

There is no frame structure defined for the MAP protocol. Instead, the frame structure defined for the TCAP protocol is used for MAP. The MAP messages are defined as GSM specific operations (with operation-specific information elements) to be transported within the TCAP protocol frame. The names of MAP messages used later in this book refer to the operation codes used within the corresponding TCAP message.

5.1.7 Radio Resource Management

Management of radio resources aims at maximizing the number of GSM users that can be served at the network coverage area with the frequency band given. A dedicated GSM radio channel is reserved for a mobile station only for the duration of the voice or data call. To minimize power consumption and interference, GSM mobile stations should use cells with which they have the best radio connectivity. All this is achieved by the GSM radio resource management.

Radio resource management is the main task of the BSC. The MSC participates in radio resource management only at a handover between two different BSCs or MSCs.

5.1.7.1 GSM Radio Channel Assignment

The GSM mobile station has two different modes related to its state of the network connectivity and usage of radio channels:

1) **Idle mode:** The mobile station in the idle mode has not yet been allocated with any dedicated channels.
2) **Dedicated mode:** The mobile station in the dedicated mode has been allocated with a dedicated channel, such as TCH or SDCCH. The station uses the dedicated channel for signaling, voice, or data call. The station

granted with TCH also uses either slow or fast DCCH for signaling. In addition to using these channels, the mobile station may periodically listen to the FCCH and SCH of any nearby GSM cells to prepare itself for a possible handover between cells.

After being switched on, a GSM mobile station gets into the idle mode as follows: The mobile station searches a GSM network. The station detects the frequencies and frame structures of GSM cell by finding its FCCH and SCH bursts. After **synchronizing** itself to the cell, the station starts to listen to SYSTEM INFORMATION messages sent on the BCCH and CBCH. Eventually, the mobile station starts listening to the paging requests sent on the PCH. The mobile station in idle mode can also send a **random access request** on the RACH channel to request a dedicated channel. The network grants the dedicated channel over AGCH, causing the mobile station to move from idle to dedicated mode.

If the mobile station enters the dedicated mode for a voice or data call, it is granted with a TCH. The TCH is connected to the anchor MSC of the GSM core network, to connect the call to the remote endpoint. In the dedicated mode, the mobile station always has a dedicated signaling channel. If a TCH has been granted, the signaling channel is either SACCH or FACCH. Without TCH, the standalone SDCCH signaling channel is used. In the dedicated mode the signaling channel is connected to the BSC, but the controller is able to forward signaling messages between the mobile station and MSC. As the protocol stacks are different in the interfaces between BSC – MSC versus BSC – MS, the BSC does all the necessary conversions between protocols used on different layers of the stacks.

The mobile station transfers from the idle mode to the dedicated mode over the access procedure, which can be performed due to the following reasons:

1) To initiate a call or send a short message.
2) To receive a call or short message. In this case, the mobile station performs the access procedure after the network has paged the mobile station.
3) To send a location update message.

The following two procedures are depicted in Figure 5.11. The GSM network pages a mobile station for a mobile terminated call as follows:

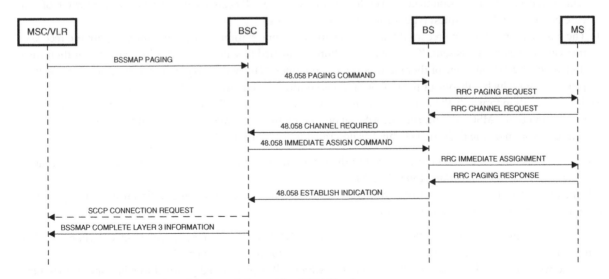

Figure 5.11 Paging and opening of dedicated channel for GSM MT call.

1) The GMSC, which routes the call attempt to the user, asks the location of the user's mobile station from the HLR database. The HLR tells the GMSC which MSC/VLR currently serves the user.
2) The GMSC sends a connection request to the MSC/VLR switching center, which sends a BSSMAP PAGING message to the BSC serving the user's location area. The message tells the BSC the IMSI identifier of the called user and the user's location area as known by the HLR.
3) The BSC sends a 48.058 PAGING COMMAND message to all the base stations of the location area. This message tells the TMSI identifier of the user and the PCH subchannel used by the mobile station when using DRX.
4) The base station sends an RRC PAGING REQUEST message to the given PCH subchannel to reach the MS. This message contains the TMSI identifier of the called user.

When the mobile station has received a paging request or when the user initiates a call, a random access procedure for acquiring the dedicated channel is performed as shown in Figure 5.11:

1) The mobile station sends a single RACH burst (also known as RRC CHANNEL REQUEST) to the base station. The burst contains only 8 bits of protected information about
   - The reason of the channel request: answering to a mobile terminated call, mobile originated call, emergency call, or location update. Based on the reason, the network can decide whether to accept or reject the request and choose the type of dedicated channel to be provided when accepting the request.
   - A random bit sequence with which the mobile station may recognize the response from the network. When multiple mobile stations try a random access procedure incidentally over the very same RACH burst, there is a collision and the network does not necessarily receive either of the requests. When the mobile station does not receive any response, it shall repeat its access request after a random period.
2) The base station forwards the received access request to the BSC, accompanied with an initial estimate of the transmission delay between the mobile station and base station. The BSC reserves a dedicated channel for the mobile station and informs both the base station and the mobile station about it. With the RRC IMMEDIATE ASSIGNMENT message that the BSC sends to the MS, the BSC tells the MAIO and timeslot number parameters used for hopping sequence control as well as the transmission power and timing advance to be used on the dedicated channel.
3) The mobile station opens a LAPDm signaling link to the base station and sends a request message, the type of which is specific to the reason to go into dedicated mode: MT or MO call, location update, or MS switch off. In case of an MT call, the request type is RRC PAGING RESPONSE.
4) The base station echoes the received request back to the mobile station, which can now compare the received message to the one it has sent. If two mobile stations happened to send the same 8-bit sequence in their random access bursts, only one of them now gets its own request returned. This completes the content resolution process, after which only one of the mobile stations continues using the dedicated channel.
5) The base station forwards the request from the mobile station to the BSC, which sets up a new SCCP protocol connection to the MSC and informs the MSC about the request. The MSC can now authenticate the user and start encryption on the dedicated channel as described in Section 5.1.8.2.

For further details about the messages used in these procedures and various options for the UE to open radio connection, please refer to *Online Appendix I.2.1*.

If the network or the cell suffers from overload, the network may stop the mobile station from sending further random access requests in the following two ways:

- The base station may send an RRC IMMEDIATE ASSIGNMENT REJECT message on AGCH channel. This message denies the mobile station to try out random access during a given period.
- The network may send a request on BCCH channel, which tells a predefined group of mobile stations to not try out random access for any other reason than an emergency call.

5.1.7.2 Changing Channel Type or Data Rate

After the MS has been assigned with a dedicated GSM channel, the network can later change the parameters or the type of the channel in one of the following ways:

1) Change the type of the dedicated channel (SDCCH, TCH/H, or TCH/F). A typical case is that the mobile station was initially given only with a standalone signaling channel but later the mobile station requests a TCH for a voice or data call.
2) Change the bit rate or other parameters of the circuit switched TCH data channel.

Changing the channel type or data rate is done as follows:

- The MSC initiates the change with the BSSMAP assignment request procedure toward the BSC to describe the new characteristics of the dedicated connection.
- Procedures within the BSS depend on whether the channel type is changed or not.
  - To change data rate or other parameters of the channel, BSC uses 48.058 mode modify procedures toward the base station and the RRC channel mode modify toward the mobile station.
  - To change the type of the channel and simultaneously other necessary connection parameters, BSC uses 48.058 channel activation procedures toward the base station and the RRC assignment toward the mobile station.

The process for changing the type of the dedicated channel is a bit more complex than the process for changing other parameters. When the channel type is changed, the signaling is also moved from the old to the new channel. The BSC does not release the old channel until it has received a message from mobile station over the new channel. That message confirms that the new channel has been successfully taken into use. In any error case, the mobile station can stay on the old channel and avoid loss of the connection.

For further details about the messages used in these procedures, please refer to *Online Appendix I.2.2*.

5.1.7.3 Releasing GSM Radio Channel

When a call has been finished or location update has been performed, the anchor MSC will tear down the connection toward the mobile station and release any resources reserved for it. The MS goes back to idle mode. The release procedure is performed as shown in Figure 5.12:

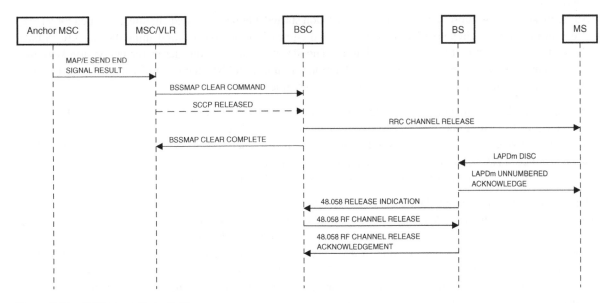

Figure 5.12 GSM connection release.

1) If the anchor MSC does not directly control the serving BSC, the anchor MSC sends a MAP/E SEND END SIGNAL RESULT message to the MSC connected to the BSC. The anchor MSC releases its circuit switched connections to the mobile station.
2) The MSC which is currently serving the mobile station contacts the BSC and triggers teardown of the SCCP connection. The BSC sends an RRC CHANNEL RELEASE message to the mobile station and releases the SCCP connection toward the MSC after responding to the MSC.
3) The mobile station disconnects the LAPDm signaling link toward the base station. Thereafter, the base station and the BSC release the channels used for the mobile station.

For further details about the messages used in these procedures, please refer to *Online Appendix I.2.3* (Figure 5.12).

5.1.8 Security Management

Security mechanisms have been defined in GSM specifications for two different purposes:

- Authentication of the user to ensure that the calls are routed to correct recipients and that the right user is charged for the GSM services used.
- Encryption of the data over the radio interface so that no third parties could eavesdrop on the communication or learn the location of a GSM subscriber.

GSM security architecture and functions are defined in 3GPP TS 03.20 [14].

5.1.8.1 Security Algorithms

GSM specifications define the framework for GSM security but not the detailed algorithms used for security key derivation or encryption. The algorithm specifications were left as a task for the industry, including the network operators and security software vendors.

Both user authentication and data encryption are based on a secret user-specific **master key** Ki, which is stored to two places: the SIM card of the subscriber and the AuC of the network. The Ki key is never exposed outside of these two entities.

Authentication of the subscriber is done and encryption started when a dedicated channel is opened for a location update or a circuit switched call. Consequently, the lifetime of encryption keys and authentication credentials is the same as the lifetime of the dedicated channel. The SIM card and the AuC derive the encryption keys and credentials based on the secret master key Ki and a random number RAND, the latter of which is sent from the network to the mobile station when opening a dedicated channel.

- During the user authentication process, the mobile station uses values of Ki and RAND for algorithm A3 to calculate the SRES authentication code to be returned to the network. The A3 algorithm was not specified in the original GSM specifications, but it was left to be defined by each service provider (or SIM card vendor). The same instance of A3 algorithm shall be used on both SIM card and at AuC.
- The encryption key Kc is calculated with algorithm A8, which is not specified in GSM standards.

The encryption of transported data is done with the A5/1 algorithm, which is defined in the GSM MoU agreement between GSM operators. This algorithm uses the encryption key Kc and the number of the frame being encrypted to produce bit sequences S1 and S2, each 114 bits. The S1 and S2 sequences are used to encrypt and decrypt transported frames. A later version of A5/3 was added in 2002 to support longer frames of GPRS and EDGE networks, described in Sections 5.2 and 5.3. Initially, the security level of the initial GSM security algorithms used was deemed, sufficient but while the processing power of devices increased, it eventually became possible to crack the encryption with reverse engineering and raw brute force trial-and-error approaches.

5.1.8.2 Security Procedures

The MSC/VLR gets the authentication credentials – RAND and corresponding SRES – from the HLR within a MAP/D INSERT SUBSCRIBER DATA message. The MSC/VLR authenticates the subscriber by sending a RIL3-MM AUTHENTICATION REQUEST message to the mobile station. The request contains one of the RAND numbers received from the HLR. The mobile station responds with an authentication response message that contains the SRES code calculated by the SIM against the given RAND. If the SRES number received from the HLR and stored to the VLR database for the used RAND matches with the SRES number received from the mobile station, the authentication is completed successfully.

Encryption is started on a dedicated channel following carefully designed steps to ensure that corruption of any message at a critical moment would not break the connection. This process is as shown in Figure 5.13:

1) The MSC activates encryption by sending a BSSMAP CIPHER MODE COMMAND to the BSC. This message tells which encryption mode to use.
2) The BSC sends a 48.058 ENCRYPTION COMMAND to the base station. This message contains an RRC CIPHERING MODE COMMAND to be forwarded to the mobile station.
3) The base station starts decrypting the information sent by the mobile station on the dedicated channel. The base station then sends an RRC CIPHERING MODE COMMAND message in clear text to the mobile station. If reception of this message fails, the base station continues sending the message until it finds the mobile station to have started traffic encryption on the dedicated channel.
4) After receiving the ciphering mode command, the mobile station starts encryption of the transmitted traffic and decryption of the received traffic from the base station. As the base station does not yet encrypt its traffic, the signaling connection from the network to the mobile station is temporarily lost. The mobile station sends an encrypted RRC CIPHERING MODE COMPLETE to the base station.
5) When the base station receives the first encrypted message and is able to decrypt it, the base station starts to use encryption also for transmitted messages. The base station forwards the RRC CIPHERING MODE COMPLETE message from the MS to the BSC.
6) In the end, the BSC sends a BSSMAP CIPHER MODE COMPLETE message to the MSC. This message tells the MSC that the encryption was successfully taken into use (Figure 5.13).

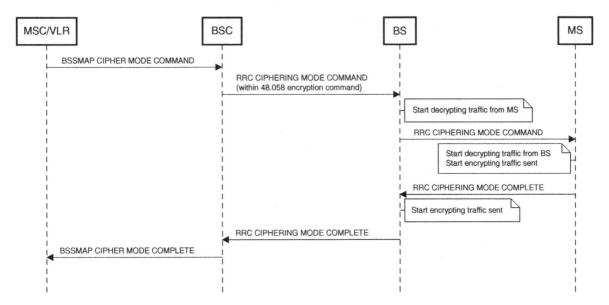

Figure 5.13 Starting GSM encryption.

5.1.8.3   Hiding The Identity of the User

To prevent external parties from finding out the location of a certain GSM subscriber within the network coverage area, any signaling messages sent without encryption should not contain the IMSI identifier of the subscriber. As GSM encryption is user-specific, the user identity shall be told in the signaling messages before encryption can be started. To solve this dilemma, GSM specifications use a temporary user identity TMSI in cleartext signaling messages. The TMSI is allocated to the mobile station by the MSC after performing subscriber authentication once. The TMSI identifier contains a temporary identifier of the mobile station and the identifier of the location area under the MSC. If the mobile station changes its location area, the new MSC can find it from the old TMSI of the location update message from which the MSC got the IMSI code behind the TMSI. The new MSC sends a MAP SEND PARAMETERS message to the old MSC. After being returned with the user's IMSI, the new MSC can allocate a new TMSI code for the mobile station, to be used under its new location area.

5.1.9   Communication Management

5.1.9.1   Mobile Originated Call

To initiate a voice call, the GSM user selects the called number and presses the "send" button of the phone. The mobile station performs random access to get a dedicated TCH channel for the call. After a dedicated signaling channel has been opened, the voice call is set up as shown in Figure 5.14:

1) The mobile station sends a RIL3-MM CM SERVICE REQUEST message to the MSC with random access procedures described in Section 5.1.7.1. This request contains the TMSI identifier of the subscriber and the reason code for requesting a dedicated channel.

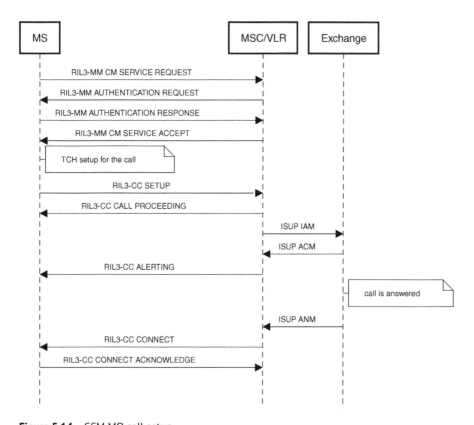

Figure 5.14   GSM MO call setup.

2) The MSC may thereafter authenticate the user as described in Section 5.1.8.2. After performing the authentication, the MSC may change the channel type as TCH and start encryption on it. Finally, the MSC sends a RIL3-MM CM SERVICE ACCEPT message to the mobile station.

3) The mobile station sends a RIL3-CC SETUP message that contains the telephone number of the callee and some information about the call type (voice or data call). The MSC checks if the request is acceptable and figures out the switching center via which the call must be connected toward the callee. If the MSC accepts the setup request, it sends a RIL3-CC CALL PROCEEDING message to the mobile station and an ISDN user part (ISUP) IAM message to the other switching center to get the call connected.

4) When the MSC eventually receives an ISUP ACM message from the other switching center indicating the target phone to be ringing, the MSC sends a RIL3-CC ALERTING message to the mobile station.

5) When the callee eventually answers the call, the MSC is informed about it with an ISUP ANM message. At this point, the MSC sends a RIL3-CC CONNECT message to the calling mobile station, which responds the MSC with a RIL3-CC CONNECT ACKNOWLEDGE message. Now transport of digitally encoded voice can be started over the dedicated TCH channel and the circuit switched connection established (Figure 5.14).

The MSC may also reject the call due to a number of reasons while the call is being set up:

- The MSC may send a RIL3-MM CM SERVICE REJECT message as a response to the RIL3-MM CM SERVICE REQUEST message, for instance, if subscriber authentication fails.
- The MSC may send a RIL3-CC RELEASE COMPLETE message as a response to the RIL3-CC SETUP message, for instance, if there are no available timeslots for the TCH or if the user is not entitled to call the specific number. The latter case may be due to an agreement with the operator to block calls to expensive service numbers or due to the subscriber activating a call barring supplementary service.
- The MSC may send a RIL3-CC DISCONNECT message after the RIL3-CC CALL PROCEEDING message, for instance, if the called phone cannot be reached, the phone is busy, or if the callee does not simply answer soon enough. After receiving the RIL3-CC DISCONNECT message, the mobile station shall send a RIL3-CC RELEASE message to get a RIL3-CC RELEASE COMPLETE back from the MSC.

5.1.9.2 Mobile Terminated Call

A mobile terminated call is connected to a GSM mobile station as shown in Figure 5.15:

1) The MSC which serves the caller sends an ISUP IAM message, which is routed to the GMSC center of the callee's home network, based on the called MSISDN telephone number. The GMSC has the task to figure out the location of the called phone and continue routing the call setup request toward the right MSC/VLR. The GMSC contacts the HLR to get the MSRN for the subscriber. The HLR knows which MSC/VLR currently serves the subscriber and contacts the MSC to get an MSRN allocated to the subscriber. The HLR forwards the received MSRN number to the GMSC, which uses the information from it to route the ISUP IAM message to the correct MSC/VLR. In the forwarded message, the subscriber is identified with the MSRN number rather than the originally used MSISDN number.

2) When the MSC/VLR receives the IAM message, it sends a BSSMAP PAGING message to the BSC serving the mobile station. With that message, the MSC/VLR requests the BSC to page the mobile station. After being paged and granted with a dedicated channel, the mobile station sends an RRC PAGING RESPONSE to the BSC, which now correlates the RRC message with the paging request. The BSC opens an SCCP connection toward the MSC/VLR, which then sends a RIL3-CC SETUP message toward the mobile station. The mobile station confirms the call and requests a specific type of TCH channel for it. The MSC can thereafter change the type of the existing dedicated channel as described in Section 5.1.7.2.

3) The mobile station starts ringing and sends an alerting message to the MSC/VLR, which informs the calling switching center about the callee being alerted. When the subscriber answers the call, the mobile station sends a RIL3-CC CONNECT message to the MSC/VLR, which informs the calling exchange over ISUP. The digitally

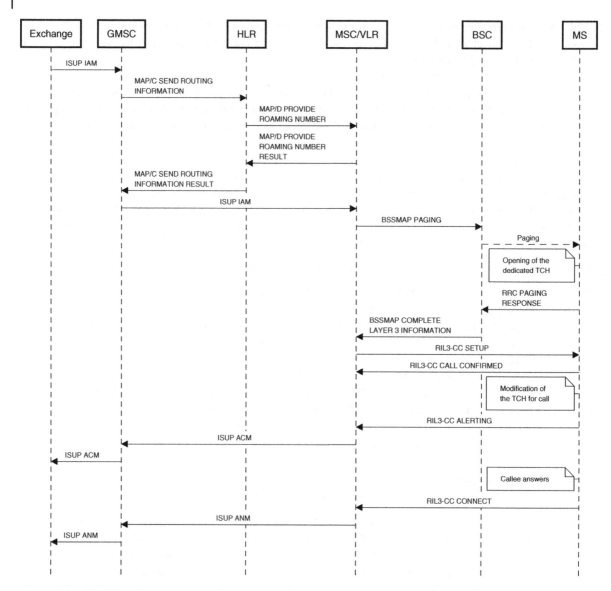

Figure 5.15 GSM MT call setup.

encoded voice can now be transported over the established circuit switched connection and the dedicated TCH channel assigned to the mobile station.

For further details about the messages used in these procedures and MSRN identifier, please refer to *Online Appendix I.2.4* (Figure 5.15).

5.1.9.3 Call Release
An ongoing call is terminated as follows:

1) When the user of the mobile station ends the call, the mobile station sends a RIL3-CC DISCONNECT message to its MSC. The MSC then sends an ISUP RELEASE message to the other switching center. The call ends and the circuit switched connection is torn down.

2) At the remote end, the mobile station receives a RIL3-CC DISCONNECT message from its MSC.
3) The DISCONNECT message causes its receiver to send a RIL3-CC RELEASE message, which is acknowledged with a RIL3-CC RELEASE COMPLETE message. This message concludes the three-way call termination handshake process.

5.1.9.4 Other Communication Management Functions

The following functions of GSM system are also part of the communications management:

- Changing the channel type between a voice and data call, when necessary. The mobile station initiates the change by sending a RIL3-CC MODIFY message to the MSC, which will change the channel type as described in Section 5.1.7.2.
- Putting the call on hold and removing it from hold with the help of a RIL3-CC HOLD and a RIL3-CC RETRIEVE message.
- Supporting dual-tone multifrequency (DTMF) tones from the keypad of the mobile station. Dual frequency tones would be a rather exceptional use case for GSM voice codecs, causing extra complexity for them. Because of this, DTMF tones are not sent over GSM as audio but instead with signaling messages RIL3-CC START DTMF and RIL3-CC STOP DTMF.
- Managing supplementary services. The mobile station uses MAP/I protocol messages, such as ACTIVATE, REGISTER, or INVOKE, to manage supplementary service settings in the HLR database. These messages are sent within a RIL3-CC protocol messages, such as a RIL3-CC FACILITY message. Other RIL3-CC message types may also be used if the mobile station has also some other needs for using RIL-CC protocols at the same time that it is modifying supplementary service states. GSM supplementary service procedures are defined in TS 23.081 [4] – 23.096 [47], and TS 24.010 [48] provides a generic view about supplementary service control procedures.

5.1.10 Voice and Message Communications

5.1.10.1 Voice Encoding for GSM Circuit Switched Call

Voice is transported in digital form over GSM radio connections. GSM systems use several types of **voice codecs** to convert analog voice waveforms to digital formats. The basic process of voice digitalization is the same as used for PCM: taking frequent samples from the voice acoustic waveform captured with a microphone, encoding the values of those samples to digital numbers, sending the numbers to the remote end where they are used to reproduce the sound with a loudspeaker. While the basic PCM is a simple approach to run this process, it is not an optimal way to do the job. A PCM stream requires relatively high data rate of 64 kbps. Further on, the PCM quantization process filters out the frequencies above 4 kHz, so its sound reproduction properties are far from perfect. Better outcome and lower bitrates can be achieved with vocoding techniques, which utilize the specific characteristics of human voice. Vocoding works by dividing a continuous voice waveform to segments of a few tens of milliseconds, matching the segments against predefined voice models, trying to adjust the chosen model with the segment using parameters of the model, and eventually transmitting the resulting parameters to other end where the voice can be reproduced by decoder with the help of models locally stored as codebooks.

A number of standard algorithms have been specified for digital encoding of voice to meet the following goals:

- The audio shall be reproduced well and close enough to the original sound. What is "well enough" is typically measured with **mean opinion score (MOS),** where a number of listeners would rate the quality of audio samples played. Each sample would then get an average score indicating its perceived quality. To reach high MOS score, the voice encoding and decoding process should retain most of the information from the form of the original analog electrical signal from the microphone, regardless of the frequencies within the sound wave.
- The encoded voice stream should need a minimal bit rate to maximize the number of voice connections for a given radio bandwidth.

- Voice encoding and decoding processes must not cause significant latency for interactive end-to-end voice transport.
- The voice codec implementation should consume minimal resources of the GSM phone. Compared to the modern smartphones, GSM phones of the 1990s were really minimal computing platforms. The voice codec software had to work only with a very small amount memory and processing power.

Voice is transported between the GSM mobile station and the TRAU transcoding element in one of the digital formats defined in the GSM standards. The TRAU takes care of conversions between the GSM digital voice encoding and the 64 kbps PCM voice encoding used in fixed PSTN networks. Since the TRAU is typically located at an MSC site, the compact GSM voice encoding methods save transmission capacity within the whole GSM BSS rather than air interface only. Even if the links between GSM base stations and MSC/TRAU would be standard E0 channels used for PCM voice, the GSM network can multiplex four 16 kbps voice connections into one E0 channel. This multiplexing structure is removed at the TRAU so that a single 64 kbps E0 channel from the MSC toward the fixed PSTN or another PLMN mobile network is occupied by a single voice call only.

The GSM phase 1 standard TS 06.10 [49] defined the GSM Full Rate voice codec, which uses regular pulse excitation-long-term prediction (RPE-LTP) encoding algorithm, depicted diagrammatically in Figure 5.16. This algorithm generates a 13 kbits bitstream that can be transported over a GSM full-speed dedicated channel TCH/F. The latest version of its specification can be found from TS 46.010 [50].

The RPE-LTP algorithm groups the taken voice samples to segments of 20 ms. At first, PCM encoding is applied to the samples to encode them into digital form. Thereafter, the digitalized samples are analyzed and encoded

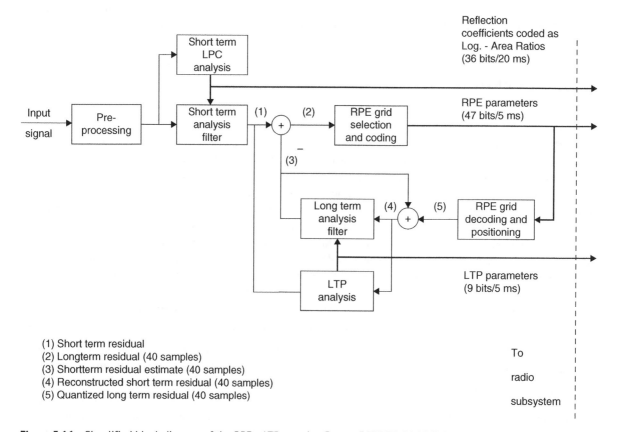

(1) Short term residual
(2) Longterm residual (40 samples)
(3) Shortterm residual estimate (40 samples)
(4) Reconstructed short term residual (40 samples)
(5) Quantized long term residual (40 samples)

Figure 5.16 Simplified block diagram of the RPE – LTP encoder. *Source:* 3GPP TS 06.10. Fair use.

with code excited linear prediction (CELP) filtering technology for transmission. For samples taken in 20 ms, the algorithm generates data blocks of 260 bits, to carry dynamic parameters and values created during the encoding process for CELP filters to reconstruct the voice. The CELP technology uses the following approach for voice modeling:

- The vibration produced by human vocal cords can be described by two components: frequency and amplitude. Because the frequency and amplitude of human voice do not typically change very quickly in some unexpected way, it is possible to predict the frequency and amplitude values of the next sample by extrapolating values of a few earlier samples. This approach is called linear prediction coding (LPC).
- The tongue and lips of the speaker modify this basic vibration produced by vocal cords and cause certain non-linear changes to it. CELP technology uses a codebook to encode such changes typical to human voice. Each code of the codebook means a certain waveform pattern. When encoding a voice sample, the codec uses such code of the codebook which makes the voice synthesizer to reproduce voice waveform very close to the original one.

The data blocks encoded with RPE-LTP then carry data about differences of amplitude and phase between voice samples and codebook codes for generating specific voice synthetization waveforms. The amount of encoded data bits is optimized with the long-term prediction (LTP) technique. The stream produced by an LPC filter contains a periodically repeating component, which can be removed from transmitted blocks by the long-term prediction approach. In the receiving end, the removed component can be algorithmically reproduced, rather than using additional data bits to transport the information.

The TCH/F data rate of 16 kbps covers the 13 kbps RPE-LTP bitstream and the following types of additional data items:

- Indicator of the voice codec type used
- Timing and synchronization data of the voice frames
- Currently used DTX mode (ON/OFF)

The **discontinuous transmission (DTX)** method can be used to optimize GSM voice transport. The phone uses **voice activity detection (VAD)** technique to enable DTX. On a typical voice conversation, only one of the parties speaks at a time, so at every moment of the call the voice channel to one of the directions is idle. When the user does not speak, instead of sending voice blocks every 20 ms, the phone sends samples encoded from the background noise more infrequently. Only 0.5 kbps bitrate is needed to send such comfort noise samples to avoid the remote user believing that the voice channel has been dropped. Since GSM uses bidirectional TCH, the DTX method does not save any air interface or transport capacity but it has the following other benefits:

- Longer effective voice call time with one battery charging cycle
- Minimization of the interference level of the GSM system

The GSM phase 2 standard TS 06.20 [51] defined the GSM Half Rate voice codec using vector sum excited linear prediction (VSELP) algorithm. VSELP generates a 6.5 kbps bitstream that can be transported over the half-speed dedicated channel TCH/H. VSELP is a variant of the CELP algorithm with an improved way of using codebooks. The GSM 06.20 VSELP codec generates a block of 112 bits every 20 ms. The latest version of the specification can be found from TS 46.020 [52].

GSM Enhanced Full-Rate (EFR) is a codec evolved from the original GSM Full Rate coded. GSM EFR uses the algebraic code excited linear prediction (ACELP) technique, which is able to use a very large codebook. GSM FER produces a 12.2 kbps bitstream. As further enhancement, the AMR encoding defined for 3G networks as described in Chapter 6, Section 6.1.11.1 has been used also in GSM networks with GSM Phase 2+ release 99.

The full and half rate speech processing functions are currently described in 3GPP technical specifications TS 46.001 [53] and TS 46.002 [54], respectively.

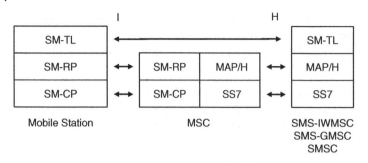

Figure 5.17 GSM short message protocols.

5.1.10.2 Short Messages

As described earlier, short messages are transmitted between GSM mobile stations and base stations on SACCH or SDCCH, over the SAPI3 link of the LAPDm protocol. The following nested protocols and layers specified in 3GPP TS 23.040 [55] and TS 24.011 [56] are used to transport short messages between a mobile station and the MSC:

- The Short Message Control Protocol (SM-CP) supports sending one SM-RP message between the MS and the MSC. The SM-CP also supports message acknowledgments and retransmissions.
- The Short Message Relay Protocol (SM-RP) provides the MSC with the SMSC address and reference number to identify a specific short message among multiple ones in transit. The SM-RP implements the protocol service for the SM-RL layer.

There are two layers related to SMS service, as shown in Figure 5.17:

- The short message relay layer (SM-RL) protocol provides support for transporting short messages within SM-TL packets between mobile stations and an MSC. The SM-RL RP message contains the addresses of the SMSC short message center and the mobile station between which the SM-TL message is relayed. The MSC uses the SMSC address to communicate with the correct SMSC with the MAP/H protocol.
- The short message transfer layer (SM-TL) protocol provides support for sending and receiving one short message or SMS receiver report at the time. The SM-TL protocol is run between the mobile station and the SMS center or gateway. The SM-TL uses SMS reference numbers to identify short messages within parallel SM-TL transactions.

Between the mobile station and the base station, the above-mentioned protocol messages are carried by the LAPDm protocol while the LAPD protocol is used between the base station, BSC, and MSC on the link layer. Between the MSC and SMS-gateway, the SS7 protocol stack is used. The MSC exchanges SM-TL messages with the gateway or SMSC over MAP/H protocol.

Sending a short message from a mobile station to another is done as shown in Figure 5.18:

1) The mobile station sends the short message within a SM-TL SMS-SUBMIT message. This message is encapsulated into a SM-RL RP-MO-DATA message sent to the MSC.
2) The MSC forwards the SM-TL short message to an SMS-IWMSC, which routes the message to the correct SMSC based on the MSISDN number of the short message recipient. The SMSC sends the short message to the SMS-GMSC that serves the SMS recipient.
3) The SMS-GMSC sends a MAP/C SEND ROUTING INFO FOR SHORT MESSAGE request to the HLR of SMS recipient. The HLR returns the address of the MSC/VLR that currently serves the SMS recipient. The SMS-GSMC sends the short message onwards to the MSC/VLR, which forwards the short message to the destination mobile station and receives back an SMS reception acknowledgment.

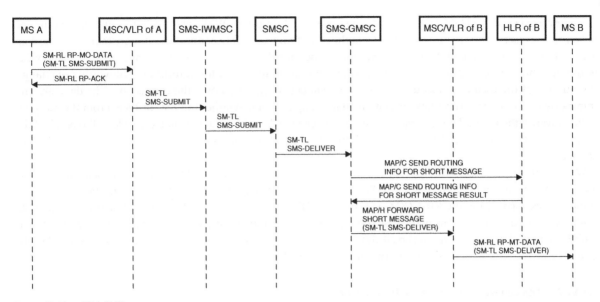

Figure 5.18 GSM SMS.

For further details about the messages used in these procedures, please refer to *Online Appendix I.2.5* (Figure 5.18).

If the mobile station of the SMS recipient is not reachable, either the HLR or MSC returns a negative acknowledgment to the SMS-GMSC for its MAP request. This causes the SMS-GMSC to inform the SMSC about the MS status. In such a case, the SMSC stores the short message so that it can be sent to the MS once it becomes reachable. The fact that there is a short message waiting for the MS is also stored into the HLR.

Later on, the mobile station registers back to the GSM network by sending a location update message to the MSC. The MSC checks if the mobile station has changed its location area. If it has, the MSC sends a MAP/D UPDATE LOCATION to the HLR; otherwise, it sends a MAP/D NOTE MS PRESENT to the HLR. As the HLR now becomes aware that the mobile station is reachable again, the HLR sends a MAP/C ALERT SERVICE CENTER message to the SMS-GMSC, which informs the SMSC accordingly. The SMSC completes the delivery of short message to the mobile station as described above.

5.1.11 Data Connections

5.1.11.1 Circuit Switched Data

Support of circuit switched GSM data has become obsolete with the introduction of packet switched GPRS technology, which is described in Section 5.2.3. The description of GSM CSD methods within Section 5.1.11 has only historical value. If the reader is not interested in such aspects of GSM evolution, the whole section could be skipped.

The original GSM Phases 1 and 2 standards supported only CSD over a single GSM TCH channel. The options for data transmission rate were 300, 600, 1200, 2400, or 9600 bps with a dedicated half- or full-speed channel. These data rates were on par with the data rates of early analog modems used within PSTN. When faster modem types were introduced in the 1990s, GSM CSD data rates fell behind.

The GSM Phase 2+ standards defined a new approach to reach higher data transmission rates up to 57.6 kpbs. GSM HSCSD used up to four GSM full- speed channels for a single data connection. The 144CC encoding method produced 14.4 kbps bitrate per timeslot; hence, 57.6 kbps rate could be reached with four timeslots. HSCSD is specified in 3GPP TS 22.034 [57] and TS 23.034 [58].

As described earlier, a GSM TDMA frame consists of eight timeslots. While voice call uses one single timeslot only, HSCSD reserves two to four timeslots from the GSM frame for a data connection. If more than two timeslots are used, the radio modem of the mobile station must be capable of simultaneous transmission and reception. It is also possible to reserve different numbers of timeslots asymmetrically for uplink versus downlink or change the number of the used timeslots dynamically while data is transported over the connection. The data streams transported over different timeslots are combined to a single stream at either the BSC or IWF. From the merging point onwards, the whole data stream is transported over a single 64 kbps E0 timeslot available in PSTN. The BSC must be able to perform a cell handover simultaneously for all the timeslots belonging to a single HSCSD data connection.

When setting up a HSCSD connection, a mobile station describes the needed transmission capacity and acceptable channel coding methods in a RIL3-CC SETUP message sent to the MSC. The MSC confirms the transmission resources reserved within the RIL3-CC CALL PROCEEDING message sent back to the mobile station. The BSC reserves the needed timeslots from the base station and sends an RRC ASSIGNMENT COMMAND message to the mobile station. After taking the reserved timeslots into use, the mobile station sends an RRC ASSIGNMENT COMPLETE message to the BSC to confirm the setup of the HSCSD channel.

5.1.11.2 Data Connectivity to External Data Networks

GSM data services have been designed to provide data connections from GSM mobile stations to external data networks. User data transport between two GSM mobile stations was an exceptional case, which the GSM standards do not specifically consider.

The data service provided by GSM was evolved from ISDN, which was specified on parallel with GSM in the 1980s. The aim was to support compatibility between GSM and ISDN so that both of those types of terminals could be connected to the same remote endpoints. When GSM was specified, it was expected that ISDN networks would gradually replace the traditional circuit switched PSTN networks. However, that did not eventually happen due to the strong adoption of IP packet-based technologies from the 1990s onwards. The main differences between ISDN and GSM come from the properties of GSM air interface:

- The transmission speed of one full-speed voice channel in GSM is just one- fourth of the 64 kbps speed as used for the basic rate ISDN transport channel.
- The transmission latencies are longer in the GSM network compared to the ISDN network.

As GSM has primarily been specified for voice, data adapters must be used on both the ends of the GSM connection. Adapters convert data formats as used on the external data networks and local computers connected to GSM terminals to formats usable over the GSM channel. The two types of adapters specified for GSM can be implemented as separate devices or functions of the mobile station or another GSM network element:

- Terminal adaptation function (TAF): An adapter connected to or inside of the mobile station, used to convert the data stream from a computer or telefax into a format used over the GSM channel.
- IWF: An adapter used at the edge of the GSM network to convert the data from a GSM channel to a format of an external data network.

The GSM data channel transports both the user data and adapter control data between the TAF and IWF functions. From the perspective of a GSM data user, the TAF and IWF functions are like external data modems used over traditional PSTN PCM audio channels.

Original GSM specifications defined a number of adaptation function types to interconnect GSM data with different types of external networks commonly deployed in the 1980s:

1) Connecting a GSM data flow to an audio modem via traditional fixed PSTN telephone network. During the 1980s and 1990s, audio modem connections to modem pools of universities and enterprises were the most common networking use cases for users of fixed PSTN networks. GSM was also expected to provide

connectivity between the audio modems at customer premises and behind the PSTN network. The setup used two conversion points: TAF was used to adapt the signal from the user's computer to the GSM channel and IWF to adapt the signal from the GSM channel to a PSTN PCM audio link of 64 kbps. In the original GSM specification, it was defined that IWF had to provide modem functionality as specified by CCITT for 300–9600 bps transmission speeds. The GSM connection between TAF and IWF adapters had to be able to transport both the user data and the control information between the remote data modem and the local computer connected to TAF over a serial interface. GSM adapters took care of conversions between synchronous GSM connections and asynchronous modem connections in cases where the clock frequency used for modems did not match with the GSM clock frequency.

2) Transmission of a telefax over a GSM connection to reach a remote telefax equipment connected to the PSTN network. Analog modems were also used for delivering telefaxes transported over the PSTN. At the mobile station, the TAF adapter shall be connected to a separate fax adapter when using an external telefax equipment with an integrated analog modem. In such a setup, the fax adapter and TAF would convert the audio signal from telefax to digital GSM format. A more straightforward setup would be to use telefax equipment with an integrated GSM mobile station rather than an integrated audio modem. Telefax data is transported over GSM to the IWF, which converts the data with its audio modem to the form used by remote telefax equipment.

3) Connecting GSM digital data to digital ISDN network. As the GSM channel speed is smaller than that of ISDN channel, a transmission rate adapter (RA) as specified in ITU V.110 is used for such a connection.

4) Connecting GSM data to a circuit switched public data network (CSPDN). The data connection is set up from the GSM network to CSPDN either directly or via the ISDN network. The mobile station uses the X.21 protocol for connecting to the CSPDN, while the GSM network uses a standard X.30 mechanism for transmission rate adaptation.

5) Connecting GSM data to a public X.25 packet data network (PSPDN):
   - Analog modems were used to create connections to X.25 packet data networks over PSTN. In the packet data network, the connection was supported using either a packet assembler disassembler (PAD) device, which composes the packets with X.28 protocol, or a packet handler (PH), which only forwards packets following the X.32 format from the mobile station to the packet data network.
   - The connection to the packet data network was created over the ISDN network using the X.32 protocol toward a packet handler in the PDN.
   - The connection to the packet data network was created directly from the GSM network so that the user does not need to select a telephone number within PSTN or ISDN networks for PDN connection. The communication between the mobile station and a PDN is done using either X.28 or X.32 protocols.

5.1.11.3 Data Transport within the GSM Network

Two different modes have been specified for GSM CSD transport: transparent and non-transparent:

- Transparent (T) mode could be used for real-time data flows with reasonable tolerance for errors. Transparent mode does not use error correction or retransmissions, but it provides a fixed transmission rate and latency. Data from upper layer protocols is transported as such over transparent mode connection.
- Non-transparent (NT) mode uses RLP protocol, which supports error checking and retransmission of corrupted data. The retransmission mechanism causes some variable delay for the data stream. To maximize the capacity available, the non-transparent mode may even drop upper layer error checksums from the packets and use only its own error correction mechanisms. Non-transparent mode was supported only for the following two upper layer protocols understood by TAF and IWF:
  - Asynchronous character-oriented protocol, where there is start and end bits around each character
  - LAPB protocol used in the context of X.25 packet networking

The RLP protocol of non-transparent mode was specified in 3GPP TS 24.022 [59]. RLP protocol is a variant from the HDLC protocol. RLP segments the upper layer protocol frames to blocks of 200 or 536 bits. Checksum is calculated for each block, and it can be used either for controlling retransmissions or forward error correction. The size of an RLP frame is either 240 or 576 bits, and it contains user data, checksum, and frame sequence number. The RLP frame has no frame start or end markers as the length of RLP block matches with a GSM block used at the radio interface.

Both the transparent and non-transparent mode rely on a variant of V.110 [60] protocol originally specified for ISDN. V.110 had the following responsibilities:

1) Adapt any asynchronous character-oriented data stream to synchronous GSM connection by delaying the transmission or removing some extra end bits with RA0 function.
2) Transport the clock frequency of the original data stream over the synchronous GSM connection. While GSM clock frequency is different from the one of the original data stream, V.110 uses specific commands used to adjust the clock phase of the receiver.
3) Pass modem control signals over GSM connection from a computer to the IWF with RA1' function.
4) Perform any needed rate adaptation for data streams between 600 and 9600 bps.

5.1.12 Mobility Management

5.1.12.1 PLMN and Cell Selection

According to GSM standards, a GSM network is operated by one single GSM service provider, within the territory of one country. GSM networks are identified with their **public land mobile network (PLMN)** IDs. Every GSM subscriber has a home network, which is the network of the operator that provides the GSM subscription and a SIM card for the subscriber. GSM subscribers are allowed to roam in GSM networks of other GSM operators and use the services of such visited networks, according to the roaming contracts between the operators. Every GSM operator has roaming partners practically in every country where GSM is deployed. In this way, GSM subscribers can use their GSM mobile stations in all the countries with GSM network support, if the subscription type just allows international roaming.

To enable roaming, the service providers have to provide the following inter-operator technical and administrative services:

- Exchange of subscriber information between networks
- Exchange of mobile station location information between networks
- Exchange of charging information between networks
- Definition of the services provided in the visited network, including support and tariffs

Network roaming may cause additional charging for the subscribers, settled between the home and visited operators. In the early days of GSM, the roaming charges were significant, but within Europe the EU regulation later pushed roaming tariffs down.

At the switch-on, a GSM mobile station starts GSM network selection process to enter the idle mode. The MS measures radio signals on the GSM frequency bands:

- To find those GSM PLMN networks that provide coverage at the location of the mobile station and select one of them for camping.
- To select the best cell of the chosen GSM network, providing the strongest radio signal for the mobile station. After choosing the cell, the mobile station makes a location update and registers itself to the location area of the cell.

There are a few partly contradictory requirements for the **cell search** and PLMN selection procedure:

- The mobile station should find and select the PLMN network as quickly as possible.
- The network search should not consume too much battery.
- The mobile station shall select the home network of the subscriber whenever under its coverage.

GSM network and cell selection process has the following steps:

1) After the mobile station is turned on, it starts searching FCCH and SCH of GSM cells. The mobile station starts the search over those frequencies it has used in its most recent home network access. Those frequencies cover the ones used by the most recently camped cell and its neighbor cells. If the mobile station is not able to find the home network in these frequencies, it shall continue the home network search by scanning all GSM frequencies supported by the mobile station. If the mobile station finds a GSM synchronization channel on these frequencies, it shall synchronize itself to the cell, measure the strength of the received radio signal, and listen to cell selection parameters sent on the BCCH channel. For each cell found, the mobile station calculates a special value called C1 based on the received BCCH parameter values, strength of the cell's radio signal, and the maximum transmit power of the mobile station. The mobile station initially camps on a cell with the biggest C1 value.
2) If the mobile station did not find home network cell on any frequency band, it can consider camping to a GSM cell of any other network found during the scan. The mobile station may select one of these networks either automatically or ask the user to select the network. The used method is chosen according to the mobile station settings.
   - Mobile stations using automatic selection check the found networks against two different lists stored to the SIM card:
     – Networks which have blocked access attempts earlier.
     – Networks which the home network provider or subscriber would prefer using when home network is not available. Among the available networks, the one with the highest preference on this list is chosen, which has not blocked an earlier access attempt.
   - In the manual search, the mobile station shows its user a list of networks found, except the ones which have earlier blocked an access attempt. The user can then pick the preferred network from that list.
3) After the network has been selected, the mobile station accesses a cell of that network. The mobile station listens to its BCCH to learn the frequencies used by the neighboring cells of the network. The mobile station then performs measurements for these cells and calculates the C1 values for them. Mobile station camps to the cell with the biggest C1 value.

After selecting the cell, the mobile station continues listening to the BCCH channel and starts listening to paging messages on the PCH channel. In idle mode, the mobile station also goes on measuring the neighboring cell frequencies learned from the BCCH channel SYSTEM INFORMATION TYPE2 messages. The purpose of continuing the measurements is to find out if the C1 values of other cells would suggest cell reselection, for instance, when the station is moving. Each cell sends a cell-specific value for the CELL_SELECT_HYSTERESIS parameter on its BCCH. This value is subtracted from the C1 value calculated for the cell when the mobile station does not yet use that cell. This adjustment is not done for the C1 value of the currently used cell. The hysteresis mechanism is to get the mobile station to slightly prefer its current cell and avoid unnecessary cell reselections when being close to the edge of two cells.

If the mobile station moves completely out of the GSM network coverage area, the mobile station starts to search for any other GSM network, as it does after being switched on. If no other network is found, the mobile station will periodically repeat the search over all the supported frequency areas until the network is found or the mobile station is switched off.

5.1.12.2 Location Update
After selecting the cell, the mobile station must tell the network its identity and location. The network responds to the mobile station, whether it is allowed to use all the GSM services of the network or only the emergency

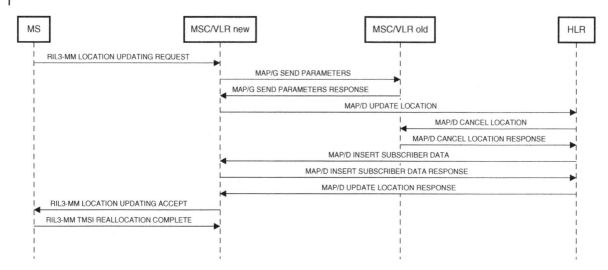

Figure 5.19 GSM location update.

call. In the latter case, the mobile station and its user must decide whether to stay in this network or camp to another network for full services. The location update procedure is performed as shown in Figure 5.19:

1) The mobile station requests a dedicated channel for the location update, as described in Section 5.1.7.1. After acquiring a dedicated standalone signaling channel, the mobile station sends a RIL3-MM LOCATION UPDATING REQUEST message. The message contains the TMSI code, which the mobile station earlier received from the MSC.

2) The MSC checks the old location area of the MS as encoded to the TMSI code. If the area belongs to another MSC, the new MSC/VLR sends a MAP/G SEND PARAMETERS message to the old MSC to retrieve the IMSI code of the user. The new MSC updates the location of the subscriber to the HLR of the subscriber's home network. If the MSC belongs to a visited network, the HLR checks if the subscriber has the right to use that network and returns the result to the new MSC. If the HLR accepts the new network, it stores the new location area and the new serving MSC/VLR to its subscriber location database and requests the old MSC/VLR to remove all data related to the subscriber from its VLR database. In the end, the HLR responds to the new MSC/VLR that the location update has been completed.

3) The new MSC/VLR may allocate a new TMSI code for the subscriber and send it to the mobile station within the location accept message. The new MSC/VLR stores into its database the state of the mobile station as reachable. The new MSC/VLR gets from the HLR the necessary subscriber security parameter values, such as encryption keys Kc, RAND random numbers, and the corresponding SRES authentication codes used to authenticate the user.

For further details about the messages used in these procedures and the case when the HLR rejects the new location, please refer to *Online Appendix I.2.6* (Figure 5.19).

If the mobile station has not changed its location area since it was turned off, the MSC/VLR already has the subscriber's authentication credentials when receiving the new location update message. If the location area has been changed, MSC/VLR gets the authentication credentials within a MAP/D INSERT SUBSCRIBER DATA message from the HLR. The MSC/VLR can thereafter authenticate the subscriber and start traffic encryption as described in Section 5.1.8.2. For further details about subscriber data stored in different network entities, please refer to 3GPP TS 23.008 [61].

The mobile station must regularly, at least once a day, send a location update message to the GSM network even if the location area has not been changed. The purpose of this is to ensure that both HLR and VLR databases of the network have the accurate location information of the mobile station regardless of any possible error situation. The mobile station in the idle mode may perform cell reselection as described in Section 5.1.12.1. Cell reselection triggers the mobile station to send a location update if the new cell belongs to another location area than that of the previously used cell.

When the mobile device is being switched off, the station sends a RIL3-MM IMSI DETACH message to the MSC to inform it about the upcoming switched off state. After receiving this message, the MSC stores to its VLR database the new state of the user as unreachable and informs the HLR about the same.

5.1.12.3   Handover in Dedicated Mode

In mobile networks, there are many conditions that cause a mobile station to lose its connection to a base station. If the station has an ongoing call, handover of the call between GSM base stations can be used to avoid call drop. The main reasons for handover decision are as follows:

1) When the mobile station moves farther away from the serving base station, its radio connection to the cell weakens. On the other hand, the movement may bring the mobile station closer to another base station and improves the quality of radio signal received from that cell. The call can be handed over between these cells when the radio connection to the new base station improves over the currently used connection. After the handover, the mobile station may use smaller transmission power to reduce the interference level of the network.

2) The used radio connection may experience very quick degradation if the radio signal is attenuated by an obstacle that suddenly appears between the MS and BTS. This may happen due to the mobile station itself moving behind an obstacle or other objects, like vehicles moving to positions where they block the signal. In such a case, an immediate handover would be needed to avoid the call being dropped.

3) When a cell experiences high load, there may be more mobile stations trying to camp on a cell and initiate calls than what the cell is able to support. In such a case, the GSM network may decide to hand over some of the ongoing calls to another nearby cell to balance the load between cells. The drawback of handovers done for load balancing is that the mobile stations are handed over to cells against which they do not have optimal radio connectivity. This causes the mobile stations to increase their transmission power, causing increase of interference level in the GSM network.

The GSM mobile station and base station measure frequently the quality of received signals to find out when handover would be either useful or necessary to maintain a call. Handover decision is made by the BSC or MSC based on the measurement data from the mobile station and base station. The measurement and handover decision process consists of the following steps:

1) The base station measures the error ratio and power loss of the dedicated channel.

2) The mobile station measures also the error ratio and power of the dedicated channel from its perspective. Additionally, during the idle unused timeslots, the mobile station measures power from all the BCH channels of any neighboring cells. To find out the frequencies of neighboring cells for measurement, the mobile station checks the SYSTEM INFORMATION TYPE2 messages sent over the BCCH of its own cell. The mobile station may also get more detailed information over its dedicated SACCH signaling channel about how the measurements should be performed.

3) The mobile station delivers its measurement data to the base station at most twice a second within the RRC MEASUREMENT REPORT messages sent over the active SACCH. In the RRC message, the mobile station tells the measured frequencies, measured signal quality values, and the BSIC codes detected on the SCH of the measured cells. This BSIC code identifies the measured cell to the network.

4) The base station forwards the measurement data to the BSC in a 48.058 MEASUREMENT REPORT message. The data is sent either as received from MS or pre-processed to minimize the traffic between the base station and BSC.

5) The BSC and MSC together know the load level of the cells under the BSC. When making handover decisions, the BSC takes into account all the above-mentioned data points as well as the maximum transmission power supported by each of the mobile stations.

Based on the measurement results, the BSC can send **power control** commands either to the base station or mobile station. Power control aims to ensure that neither of the stations would use any more transmission power than necessary, still keeping the quality of radio connections in acceptable levels. The BSC can send a 48.058 BS POWER CONTROL message to a base station for controlling its transmission power. The same message can also give the base station an order to adjust the power control parameter value sent to the mobile station over the SACCH.

When the quality of measured radio signals become significantly better for any other cell than the currently serving cell, the network may make a handover decision for the call. The main steps of handover process in GSM system are as follows:

1) The network (either a BSC or MSC) reserves dedicated channels for the mobile station into a new target cell.
2) The network creates new circuit switched connections from the MSC to the new base station, chosen as the target BTS of the handover.
3) The network instructs the mobile station to connect to the new cell and take the reserved dedicated channels into use, simultaneously releasing the channels used earlier with the old cell.
4) After the handover, the network releases the radio channels and corresponding circuit switched connections of the old cell, which is no longer used by the mobile station.

Depending on the relation of the old and new cells, there are three types of handovers in GSM as shown in Figure 5.20:

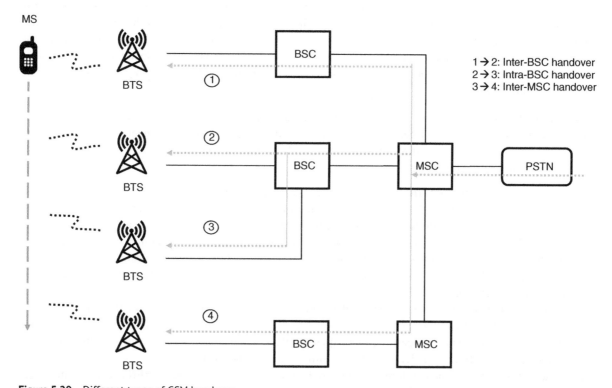

Figure 5.20 Different types of GSM handover.

- **Intra-BSC handover**, where both the old and the new cell are controlled by the same BSC.
- **Inter-BSC handover**, where old and new cells are controlled by two different BSCs connected to a single MSC. In this case, the currently serving BSC requests the MSC to perform handover to the new cell, as identified with its Cell ID and location area code LAC.
- **Inter-MSC handover**. If the handover is done between base stations under two different MSCs, the originally used MSC stays in control of the call until the call is released. That MSC is called **anchor MSC,** which maintains the call and collects the related charging data. The anchor MSC finds the other MSC based on the LAC code of the new cell, as given by the currently serving BSC.

In each of these cases, from the mobile station point of view the handover process looks the same. The mobile station receives a handover command from the network and moves to the new cell accordingly. The difference is in the steps needed to prepare for the handover in the network side. In the Intra-BSC case, the serving BSC can handle the handover autonomously with the base stations controlled by the BSC. In other cases, the involved BSC and MSC nodes must communicate with each other and allocate a new path for the voice signal between network elements before the handover command is sent to the mobile station. GSM handover procedures are described in 3GPP TS 23.009 [62].

The steps of GSM Inter-BSC **handover procedure** are as shown in Figure 5.21:

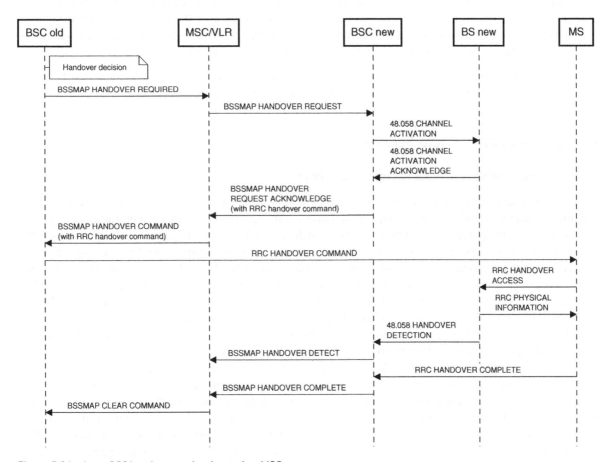

Figure 5.21 Inter-BSC handover under the anchor MSC.

1) The old BSC, which currently serves the mobile station, makes the handover decision. If the BSC itself does not control the target base station of the handover, the BSC sends a BSSMAP HANDOVER REQUIRED message to its currently serving MSC to inform it about the handover decision. The MSC sets up a new signaling connection control part (SCCP) connection to the new BSC, which is in control of the target cell. The MSC sends a BSSMAP HANDOVER REQUEST message to the new BSC.

2) The new BSC reserves dedicated radio channels and activates them at the target base station. The new BSC composes an RRC HANDOVER COMMAND for the mobile station and sends this RRC message to the MSC within its BSSMAP response message. The MSC establishes thereafter the needed circuit switched connections toward the new BSC. The MSC sends a handover command to the old BSC, which forwards the RRC message from the new BSC to the mobile station.

3) The mobile station learns the BSIC code and SCH channel frequency of the target cell from the RRC HANDOVER COMMAND. The message also describes the new dedicated channels, gives the MS an 8-bit handover reference number, and indicates whether synchronized or non-synchronized handover method shall be used. With the synchronized method, the message may also give a new timing advance value to be used for the target cell.

4) The mobile station connects to the target cell and starts right away receiving traffic over the new dedicated channels. The mobile station also sends RRC handover access bursts with the handover reference number over the dedicated channel to confirm the successful setup of the connection to the new cell.

5) After receiving access burst from the mobile station, the target base station sends a handover detection message to its BSC. The new BSC confirms the handover to the MSC so that the MSC can redirect the circuit switched connections from the old cell to the new one. Eventually, the mobile station declares the handover as complete so that the MSC can release any pending connections to the old cell.

For further details about the messages used in these procedures and the case when the old and new BSCs are connected to different MSC/VLRs, please refer to *Online Appendix I.2.7* (Figure 5.21).

5.2 General Packet Radio Service

5.2.1 Standardization of General Packet Radio Service

When the GSM network services were evolved further in the 1990s, the focus shifted from voice to data transport. The Internet became a major driver for data consumption, and engineers faced the need of transporting packet switched data efficiently over the GSM network. Packet switched data was earlier used in fixed networks for business purposes, but with the Internet it gradually became a mainstream consumer service for accessing email and browsing the Web. Fixed Internet access and GSM voice were forces behind the telecommunications boom of the 1990s, but the pressure increased for providing also mobile Internet access over proven GSM networks. Unfortunately, GSM mobile stations only supported CSD access, which is quite inefficient for transporting bursty packet data. Worse, GSM data service provided only symmetric and small data rates.

Even if the CSD connections enjoy benefits such as constant latencies, guaranteed bitrates, and no need in network level addressing in data packets, the rigid circuits waste capacity when used for Internet traffic. The typical packet data traffic pattern is highly variable. For instance, Web browsing mixes short periods of high-rate downlink data rate transport with long idle periods while the user is digesting the data. The traffic is also asymmetric as only small hypertext transfer protocol (HTTP) queries are sent an uplink while the big responses arrive by downlink. The GSM circuit switched connection wastes capacity for idle periods and slows down the high-rate data bursts, which exceed the constant bitrate of the connection. In theory, it would be possible to release the GSM data circuit for the idle periods, but this would cause extra delay and signaling load for releasing and reopening connections over the data consumption session.

To solve these problems, a new approach was used for GSM data transmission and its resource reservation. With the new GPRS service, the network dynamically reserves one or multiple GSM timeslots for the data channel as long as there is data to be transported and releases them when there is nothing to be sent [63]. The essential improvement compared to the circuit switched model is the dynamic, quick, and lightweight allocation of GSM timeslots for the radio channel to follow the pattern of user data packet flow. Compared to the long setup time of circuit switched end-to-end connection, activating an existing GPRS link can be done much quicker. While no data is transported, GPRS does not use network transmission capacity, which makes GPRS suitable for "always-on" and bursty IP connectivity. The GPRS system is also able to allocate different numbers of timeslots to uplink versus downlink when the use case requires asymmetric data rates. It is important to note that GPRS was not just a new way of using GSM air interface, but it introduced a completely new core network and protocol stack designs as well. In practice, both the network and mobile stations had to support separate CS voice and PS data systems, just sharing a common air interface L1 structure. Designing GPRS on top of GSM was a major operation for ETSI-driven standardization. Complete GPRS service description can be found from 3GPP TS 22.060 [64], which was published as part of GSM Phase 2+ release 98.

When the GPRS solution was being specified, there were a few competing packet switched protocols in common use within the fixed networks:

- Internet protocol IP
- X.25 packet data protocol for public packet switched networks
- Other vendor specific protocols

The GSM packet data transport solution was designed as generic to support all such protocols. Due to that, the solution was called **General Packet Radio Service (GPRS)**. When GPRS was eventually deployed after some years of specification, research, and development, the IP protocol had already become dominant. GPRS support for other protocols has not been widely used. In retrospect, it might be claimed that it would have been wiser to optimize GPRS for IP traffic rather than to create a protocol-independent general packet radio system.

The first GPRS networks entered commercial use around 2000, when the first GPRS mobile stations (or handsets) became available. The mobile stations were GPRS-enabled mobile phones, network cards that could be installed to computers, or personal digital assistant (PDA) devices supporting GPRS packet data connections. Typical GPRS use cases were multimedia messages (MMS), email, limited Web browsing capabilities, and using GPRS mobile station as a data modem for a portable computer.

GPRS is specified in 3GPP standardization series 40–55 and 21–35. These standards cover both the GSM voice system and the GPRS system as its extension. The most important GPRS specific 3GPP technical specifications are as follows:

- TS 22.060: GPRS service description – concepts and requirements
- TS 23.060. GPRS service description – functions and architecture
- TS 43.051: General description of GPRS protocols
- TS 43.064: General description of GPRS radio interface
- TS 44.060: Radio link control/Media Access Control (RLC/MAC) protocols
- TS 44.064: Logical link control (LLC) protocol
- TS 44.065: Subnetwork dependent convergence protocol (SNDCP) protocol
- TS 29.060: GPRS tunneling protocol (GTP) protocol
- TS 48.018: BSS GPRS protocol

GPRS mobility management (GMM) and session management (SM) procedures and messages are specified in 3GPP TS 24.008 [41].

At the time of this writing, GPRS has become a niche technology. Compared to newer cellular data systems, such as LTE, GPRS has many disadvantages: small bitrates, high latencies, and low spectral efficiency. GPRS and its EDGE enhancement are still supported in GSM networks for certain embedded devices that occasionally transfer only a small amount of non-time critical data.

5.2.2 Architecture and Services of GPRS System

GPRS network architecture and functionality is specified in 3GPP TS 23.060 [20].

5.2.2.1 GPRS System Architecture

The key constraint for GPRS design was reuse of existing GSM equipment and frequency bands. Consequently, GPRS uses GSM radio interface but manages the allocation of physical channels differently than the circuit switched GSM. An existing GSM network can be upgraded to support GPRS for all or a subset of its cells. Deployment of GPRS means the following activities for the network operator:

- Connecting new packet switched GPRS network elements to a GSM core network
- Software updates for base stations and BSCs to support GPRS
- Software updates for HLR, VLR, EIR, and AuC to support the additional parameters and information as used for GPRS

The GPRS network assigns a packet switched **tunnel** for each packet data flow of a GPRS mobile station, to carry the upper layer packet data protocol messages to the edge of GPRS network. The tunneling mechanism was designed to be able to transport different types of packet data protocols over the GPRS infrastructure. GPRS introduced a new mechanism, referred as **packet data protocol (PDP) context**, to manage the packet data protocol flows and related tunnels. The network sets up PDP contexts according to connectivity requests received from GPRS mobile stations. A mobile station sends a connectivity request toward a specific external **packet data network (PDN)** identified by its **access point name (APN)**. The request also defines the type of packet data protocol used. When receiving such a request the GPRS network creates a PDP context, assigns a packet data protocol address for the mobile station, and creates tunnels needed for routing and transporting the packets between the mobile station and the PDN. It must be noted that the tunnel routes the data packets only toward the correct destination PDN network, but to route packets to their ultimate destinations each packet must contain both source and destination addresses, such as IP addresses. These addresses are also used to map packets to the correct GPRS tunnels within the GPRS network.

The following types of network elements (see Figure 5.22) were added to GSM architecture to support GPRS (see 3GPP TS 03.02 [65] for further details):

- **Serving GPRS Support Node (SGSN)** with the following tasks:
  - Tracking the location of GPRS mobile stations and their GPRS service states within the GPRS tracking area managed by SGSN.
  - Querying the HLR database for GPRS subscriber authentication credentials and other subscription details to perform authentication and authorization when a mobile station attaches to the GPRS service.
  - Maintaining mappings of PDP contexts, GPRS tunnels, mobile stations and Gateway GPRS support nodes (GGSNs), which belong together.
  - Forwarding of packets between Packet Control Unit (PCU) and GGSN nodes via GTP tunnels.
  - Encryption of the packet data transported between the SGSN and mobile station.
- **Gateway GPRS Support Node (GGSN)** with the following tasks:
  - Management of PDP contexts and GTP tunnels to connect mobile stations to external packet data networks such as the Internet or an X.25 PSPDN.
  - Routing packets between external packet networks and SGSNs serving GPRS mobile stations. The GGSN is an edge router of the GPRS network. The GGSN routing mechanism is based on a database of PDP contexts, corresponding tunnels, and packet data protocol addresses.

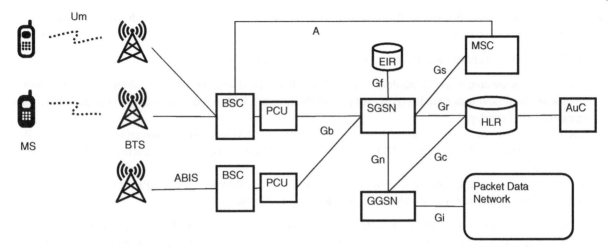

Figure 5.22 Architecture and interfaces of GPRS system.

- – Forwarding packet data between mobile stations and external packet networks within GTP tunnels between GGSN and SGSN nodes.
- – Assigning IP addresses for the PDP contexts and performing possible network address translation between the private IP address reserved for the mobile station and a public IP address owned by the GGSN, visible to the external networks.
- **Packet Control Unit (PCU)** is a function that forwards packet switched data from/to BSC to the SGSN of the GPRS network. The PCU segments long LLC frames into shorter RLC frames and takes care of both flow control and retransmissions. The PCU creates, supervises, and releases packet-switched calls. Toward the air interface, the PCU manages assignment of timeslots for both uplink and downlink GPRS data flows. A PCU can be implemented as part of a base station or BSC, or it may have its own dedicated PCU support node device connected to the BSC over the A_{gprs} interface.

GPRS core networks can be divided into the following parts:

- **The GPRS backbone** network connects the SGSN and GGSN devices within the operator network.
- **The Inter-PLMN GPRS backbone** or **IP roaming exchange (IPX)** network connects the SGSN of the visited network to the GGSN of the user's home network. This setup, known as home routing, is used to support users roaming abroad. Home routing allows the roaming user to exchange packet data with external packet data networks over the home GGSN, while being served by an SGSN of the visited network.

Like GSM specifications, GPRS specifications also define the interfaces as reference points between GPRS network elements. The GPRS network interfaces shown in Figure 5.22 are

- Gb: Interface between BSC/PCU and SGSN.
- Gs: Optional interface between SGSN and MSC.
- Gn: Interface between two GSN (either SGSN or GGSN) nodes, which belong to the same network
- Gp: Interface between SGSN and GGSN in different networks. This interface enables GPRS roaming where the SGSN is in a visited network and the GGSN in the home network.
- Gi: Interface between GGSN and external packet data network.
- Gf: Interface between SGSN and EIR.
- Gc: Interface between GGSN and HLR database. Gc interface may be omitted when the SGSN takes care of all necessary communication with the HLR.

- Gr: Interface between SGSN and HLR database, relying on the MAP protocol and an underlying SS7 stack.
- Gd: Interface between SGSN and MSC serving a SMSC.

GPRS mobile stations have been divided into three categories as follows:

- Type A stations are able to use GPRS data and GSM circuit switched voice connections simultaneously. This is enabled by the dual transfer mode (DTM) support in the network with which the circuit and packet switched timeslots do not overlap.
- Type B stations are not capable of simultaneous use of GPRS and GSM services. When starting a GSM voice call, the station automatically suspends its GPRS data connection. The data connection is reactivated after the voice call is closed.
- Type C stations let the user choose whether to use the device for GSM voice call or GPRS data. When used for GPRS data, the station is not even able to receive incoming GSM voice calls.

The service provided by the GPRS system does not depend only on the type of the mobile station but also the network operation mode. The networks supporting operation mode I are able to page the mobile stations for incoming mobile terminated calls while GPRS data is being transported. Mode II networks can page the mobile station for calls only when the GPRS functionality of the station is in the idle state.

3GPP standards specify 29 different **multislot classes** of GPRS mobile stations. The classes differ by the number of timeslots that can be used for a GPRS connection and whether simultaneous GPRS uplink and downlink data transfer is supported. The number of GPRS timeslots for a GPRS channel determine the maximum data rate of the GPRS device. Actual data rate depends also on the type of channel coding used on the radio interface. Maximum data rates of 70–150 kbps can be reached when using four to eight timeslots of a GSM TDMA frame for a single connection.

GPRS defines another packet temporary mobile subscriber identity (P-TMSI) identifier for the subscriber, in addition to the TMSI code as defined for GSM. P-TMSI is a code given by the SGSN to the subscriber located within the area controlled by the SGSN. P-TMSI is used in signaling messages sent as cleartext over radio interface between the mobile station and the network, like how the TMSI code is used for GSM.

5.2.2.2 GPRS Functions and Procedures

GPRS radio resource management takes care of allocation of GSM timeslots for GPRS data connections. GPRS extends GSM communication management with packet data **session management**. Session management is used to prepare the packet data connectivity by reserving a packet data protocol address for the mobile station and setting up GTP tunnels for transporting packet data flows through the GPRS network. The GGSN sets up its routing configuration so that any packets from the external packet data network toward the reserved address are routed to the tunnel toward the mobile station. These activities are performed at the **PDP context** activation. The concept of PDP context means pieces of configuration data shared between the GGSN, SGSN, and GPRS mobile station about the GPRS connection to the MS. The PDP context covers details of the packet data protocol and addresses used, the connected packet data network, the user data flows, and the GTP tunnels established to carry those within the GPRS network.

GMM is an extension of GSM mobility management MM. The main differences between MM and GMM are related to the mobility management states and accuracy of tracking the location of the mobile station:

- GPRS uses three mobility management states for the mobile station. When GPRS data transfer is going on, the mobile station is in ready state. In addition to idle and ready states, GPRS introduced the standby state. In the standby state, the mobile station is attached to GPRS service and has an active PDP context waiting for any new data to be transported. The network is able to initiate GPRS data transfer to a mobile station which is in the standby state by moving it to ready state. Also, the mobile station is able to move itself to ready state to transmit packet data. After the data has been transported or there is a long enough pause in the transport, the mobile station moves back to the standby state.

- For mobile station location tracking, the GPRS system uses a concept of **GPRS routing area (RA)**. Like the GSM location area, a GPRS routing area consists of a set of cells within which the mobile station may move without updating its location to the network. A routing area is always a subset of a GSM location area. The GPRS network tracks the mobile stations in ready state in the accuracy of a single cell and stations in standby mode within a routing area. Each routing area has a unique ID called routing area identifier (RAI).

Routing areas were introduced to balance the signaling traffic between the GPRS location messages and paging messages sent within a routing area. GSM and GPRS systems need areas of different sizes mainly because the GPRS system shall be capable of quicker and more frequent activation of the packet data connections compared to the GSM system with infrequent but potentially long circuit switched calls.

GMM consists of the following functions:

- GPRS attach and GPRS detach with which the mobile stations enter and leave the GPRS service. When attaching to GPRS, the mobile station informs the network of its location and capabilities so that the network can contact the mobile station when needed.
- Mobile station location updates in terms of a GPRS routing area, a GSM location area, and a cell. The accuracy of the location information stored into the network depends on the state of the GPRS service for the tracked mobile station. The location is tracked in the accuracy of a single cell while GPRS data is being transferred. Otherwise, when the mobile station is attached to GPRS service, its location is tracked in the accuracy of a routing area. Without GPRS attach, the network tracks the GSM location area of the mobile station.

5.2.2.3 GPRS Protocol Stack Architecture

GPRS protocol stack for Um and Gb interfaces is described in 3GPP TS 43.051 [26], and TS 23.060 [20] defines the complete GPRS protocol stack architecture. GPRS user plane protocol stacks shown in Figure 5.23 carry user data flows between the MS and GGSN.

The link layer of GPRS radio connection is divided into two protocols, MAC and RLC. These protocols, run between the base station and mobile station, multiplex the data flows and perform error correction over the radio interface. The base station subsystem GPRS protocol (BSSGP) takes care of similar link layer tasks between the base station and SGSN. The topmost LLC protocol of the link layer is used between the mobile station and SGSN for features such as encryption, flow control, and error correction. The SNDPC protocol is run between mobile station and SGSN on top of LLC. The main task of SNDPC is compression of the data to minimize the needed radio interface transport resources.

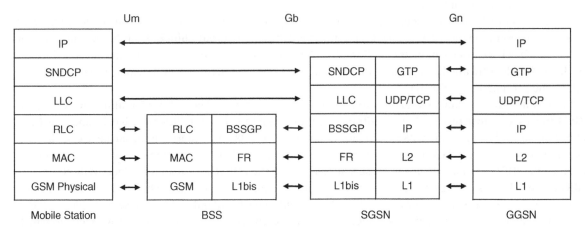

Figure 5.23 GPRS packet system user plane protocols.

The TCP/IP protocol stack is used to transport data between the SGSN and GGSN nodes. GTP tunneling protocol is the topmost protocol of the TCP/IP stack. The tunneling mechanism enables transporting of different packet data protocols (such as IP or X.25) over GPRS to/from the access points provided by GGSN toward external packet data networks.

Signaling between the SGSN and MSC is done with the SS7 protocols such as MTP, SSCP, TCAP, and MAP. Those protocols are used for GPRS in the same way as for GSM.

5.2.3 GPRS Radio Interface

Overview of the GPRS radio interface can be found from 3GPP TS 43.064 [66].

5.2.3.1 GPRS Radio Resource Allocation

The basic characteristics of GPRS radio interface come from the GSM air interface, which GPRS partly reuses but with certain extensions. Both the systems use shared TDM/FDM multiplexing and the common TDMA frame structure to support both GSM and GPRS services. The GPRS service uses the timeslots of the underlying GSM network. Individual bursts within the timeslots are used for both GSM and GPRS traffic. The base station can divide its timeslots between GSM and GPRS services either in a fixed way or dynamically based on the network load. In the latter case specific timeslots may be used for either of these services, depending on the instantaneous capacity demand per service.

The network grants GPRS transmission resources for a GPRS mobile station dynamically as **temporary block flows (TBF)**, based on the transmission needs that the MS has indicated to the network after a GPRS random access procedure. A mobile station may have one or multiple TBFs allocated. The uplink TBFs are separate from downlink TBFs, since the resource allocation is often asymmetric. Typically, more resources are granted to the downlink TBF than to the uplink TBF. A TBF has two dimensions for the dynamic resource allocation:

- GPRS packet data channel (PDCH)
- GPRS radio block

As we know, the GSM TDMA frame has eight timeslots. A PDCH is one of those timeslots allocated for GPRS traffic. Thus, the cyclic GSM frame structure may carry a maximum of eight PDCHs when none of the timeslots are used for voice. In the frequency space, PDCH uses the same frequency hopping scheme as a GSM TCH. The TBF allocated to the mobile station is linked to 1–5 PDCHs of a GSM frame. The number of PCDH resources that the mobile station can use simultaneously depends on the multislot class of the station and varies between 4 and 5 for downlink and 1–4 for uplink.

The basic unit of GPRS radio resource allocation is a group of four bursts used to carry one GPRS RLC/MAC protocol frame over a single PDCH channel. This group is called an **RLC data block**. Each of the four bursts of the RLC data block is carried within its own GSM TDMA frame, using the same timeslot number allocated for the PDCH. The **GPRS radio block** is a group of four consecutive GSM frames, capable of transporting eight RLC data blocks – one per timeslot (or PDCH).

GPRS has its own multiframe structure over the GSM TDMA frames. The GPRS 52-multiframe consists of 52 GSM frames, or rather two GSM 26-multiframes. GPRS multiframe has 12 GPRS radio blocks and four GSM frames that do not belong to any GPRS radio block. Frames 12 and 38 are used for timing advance calculations, while frames 24 and 51 can be used for measuring neighboring cells. Up to eight mobile stations may be multiplexed to one PDCH so that only one of those stations can use the PDCH within a radio block. Within the next GPRS radio block, the same PDCH may be assigned to another mobile station. Thus, when a mobile station gets a PDCH assigned to it, the station may not use that PDCH continuously but only within those radio blocks where the PDCH is allocated to that station (see Figure 5.24). For each radio block, the network indicates to which mobile station (and its TBF) the PDCH belongs.

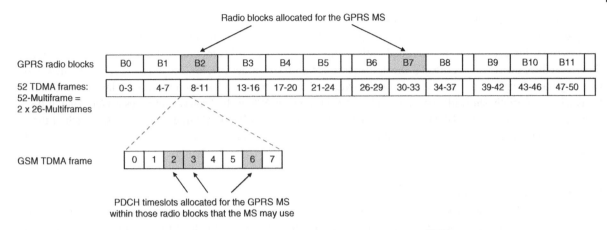

Figure 5.24 PDCHs and radio blocks allocated to the GPRS mobile station within the GPRS multiframe.

A temporary block flow (TBF) is the temporary allocation of PDCH timeslots within a number of non-consecutive GPRS radio blocks for a mobile station. As its name indicates, the allocation of TBF is not permanent, and thus it is bound in time for a number of radio blocks. When the mobile station has to send or receive user data over GPRS, the network grants the station with one or multiple PDCH timeslots for as many radio blocks needed until the data transport has concluded. Thereafter, the TBF allocation is released. For bidirectional traffic, a TBF is allocated for either of the directions. The TBF is identified by a **temporary flow identifier (TFI)**. At the allocation of uplink TBF, the network gives the mobile station an **uplink state flag (USF)** value separately for each allocated PDCH. For that PDCH, the USF value uniquely identifies one mobile station among those eight stations or less multiplexed to the same PDCH.

While allocation of PDCHs for a TBF is fixed when the TBF is created, the same does not apply to the allocation of radio blocks. GPRS uses a few different mechanisms to indicate the mobile station when the assigned PDCH(s) of a radio block belongs to the MS. One of those mechanisms is used for downlink and others for uplink. The four timeslots of one PDCH within a radio block carry a single RLC protocol frame. The downlink RLC frame header contains the TFI of the TBF to which the frame belongs to. The mobile station reads all the RLC headers of downlink PDCHs of its downlink TBF. When the header contains the TFI of its own TBF, the mobile station reads the complete RLC protocol frame that was sent on that PDCH over the radio block. The uplink radio block allocation of PDCH can be indicated to the mobile station in the following ways:

1) Fixed allocation. When the TBF is created, the network sends the mobile station a bit map per PDCH indicating those radio blocks in which the PDCH belongs to the mobile station. A single bit of the map tells if the PDCH can or cannot be used in the radio block corresponding to the bit. The sequence of bits of the map means the consecutive radio blocks after the TBF start time, one bit per radio block. If the mobile station still has more data to send when the radio blocks described within the bit map have been transmitted, the mobile station may request the network to send a new bitmap for additional radio blocks.

2) Dynamic allocation. The network manages the radio block allocation between mobile stations dynamically, one radio block at a time. The MAC header sent with the RLC block contains a USF, which identifies the mobile station that may use the PDCH in the next uplink radio block. The USF has eight possible values, each indicating a specific mobile station multiplexed to that PDCH. With extended dynamic allocation, the network may use a single USF to assign several timeslots for mobile station with a high multislot class.

In the network side, the allocation is managed by the PCU. For downlink, the PCU allocates as many blocks as needed for the downstream traffic. For uplink, the PCU allocates blocks based on the requests received from mobile stations. The uplink MAC header has a countdown field indicating the number of radio blocks needed to complete the data transfer. The original design of GPRS was to release the uplink TBF when the countdown reaches the value zero. With the extended uplink TBF method, the TBF is maintained until the expiry of an idle timer to reduce delay and overhead if the uplink transmission continues in a few seconds. Creation of a new TBF typically takes longer than half a second, so the extended uplink tries to avoid such kinds of delays for traffic patterns with potentially short idle periods, such as Web browsing. When the browser retrieves a Web page, it typically needs to send multiple HTTP requests after each other, as described in Chapter 3, Section 3.6. Each of those requests is sent separately but very soon after each other, so the extended uplink TBF method may keep the TBF alive long enough to retrieve a complete page.

5.2.3.2 GPRS Logical Channels

The GPRS mobile station uses the FCCH and SCH of the GSM frame structure to synchronize itself to the GSM frame cycle. Additionally, GPRS mobile station listens to SYSTEM INFORMATION TYPE13 messages sent on the BCCH. These messages tell the mobile station how it can find the packet broadcast control channel (PBCCH) channel providing more information about the GPRS service and its shared channels in the cell. If the cell does not support the PBCCH channel, the additional information is given directly in the SYSTEM INFORMATION TYPE13 message.

The following logical channels were originally defined for GPRS:

1) Shared signaling channels used by all GPRS mobile stations camping in a cell:
    - Packet broadcast control channel (PBCCH): The channel on which six different types of PACKET SYSTEM INFORMATION messages are broadcast within the cell. These messages provide mobile stations with information about the GPRS network and various parameter values used to control mobile station functionality, such as
        - How PCCCH logical channels are mapped to PDCH channels of the cell
        - The number of information bits the mobile stations shall add to their PACKET CHANNEL REQUEST messages
    - Packet common control channels (PCCCH) are in shared use of all mobile stations within the cell, but without broadcast mechanism. Only one mobile station may use an uplink channel at a time, and the messages on the downlink channel have the address of the destination mobile station.
        - The packet paging channel (PPCH) on which the network sends a paging message to a mobile station for either incoming packet data or a circuit switched voice call. Paging messages are sent in all the cells that belong to the routing area (RA) where the mobile station has most recently sent a routing area update message. Like the GSM PCH channel, the GPRS PPCH channel can be divided into multiple subchannels for DRX. When using DRX, the mobile station has to listen to only one of the PPCH subchannels and save its battery the rest of the time.
        - The packet random access channel (PRACH) on which the mobile station can request itself a dedicated channel for packet switched data transport.
        - The packet access grant channel (PAGCH) is used by the network to inform the GPRS mobile station about the dedicated GPRS TBF allocations for packet data traffic channel (PDTCH) data and packet associated control channel (PACCH) signaling channels.
        - The packet notification channel (PNCH) used by the network to send PTM-M message to a group of GPRS mobile stations. This message is used to inform the mobile stations about point-to-multipoint packet data arriving to the members of the group.
2) Packet data channels dedicated to a single mobile station
    - Packet data traffic channel (PDTCH) is used to transport packet switched user data. The PDTCH channel consists of the PDCH timeslots within radio blocks allocated to the mobile station.

3) Signaling channels dedicated to a single mobile station
   - Packet associated control channel (PACCH) is used to transport signaling messages between the network and a single mobile station. The PACCH channel is used for packet resource requests and assignment messages, packet data acknowledgments, and power control commands. The PACCH may be used to notify mobile stations about incoming circuits switched calls, when the network supports such an operation mode. The TBF resource allocation of the mobile station is divided between PDTCH and PACCH. The payload type field of the MAC headers sent with the RLC block tells if the block carries PDTCH user data or PACCH signaling.
   - Packet timing control channel (PTCCH) to which GPRS mobile stations send uplink bursts to allow the base station to determine the timing advance needed by the mobile station. The network sends the timing advance commands to the mobile stations on the downlink of the PTCCH channel.

According to the original specifications, the GPRS network was able to multiplex the following combinations of logical channels to a single PDCH over the GPRS multiframe:

1) PBCCH + PCCCH + PDTCH + PACCH + PTCCH
2) PBCCH + PCCCH
3) PCCCH + PDTCH + PACCH + PTCCH
4) PDTCH + PACCH + PTCCH

In other words, one PDCH timeslot (in 52 consecutive GSM frames) could carry many types of signaling channels together with a dedicated packet data channel. Multiplexing of the logical channels to the PDCH is done in the following way: Among the available combinations of logical channels to be multiplexed to the PDCH, one is chosen. Each of the channels in the combination is granted with a certain fraction of the 52 PDCH bursts of the multiframe. The channels are mapped to the sequence of 52 bursts in the order as stated above for each of the four multiplexing options. To limit impact of short disturbances on the radio channel, the sequence of 52 PDCH bursts for logical channel multiplexing is defined in a very specific way, instead of just concatenating the bursts in the order of TDMA frames transmitted. As described earlier, the GPRS multiframe consists of 52 GSM TDMA frames, grouped to 12 GPRS radio blocks B0–B11 of four frames. The four GSM frames that do not belong to any GPRS radio block are used for the PTCCH. PDCH bursts of the radio blocks are used for the other channels. The bursts of different radio blocks are concatenated to one contiguous stream of bursts with the following order of radio blocks: B0, B6, B3, B9, B1, B7, B4, B10, B2, B8, B5, and B11.

As an example, we use the PDCH to which the following logical channels have been multiplexed according to option 4: PDTCH (ten radio blocks B0, B6, B3, B9, B1, B7, B4, B10, B2, and B8), PACCH (two blocks B5 and B11), and PTCCH (four individual GSM frames, one after each group of three radio blocks). In this case, the PDCH timeslots of the multiframe are divided between the logical channels as follows:

- PDTCH: frames 0–3, 26–29, 13–16, 39–42, 4–7, 30–33, 17–20, 43–46, 8–11, 34–37
- PACCH: frames 21–24, 47–50
- PTCCH: frames 12, 25, 38, 51

While 3GPP has evolved GSM and GPRS specifications, the packet broadcast and common control channels PBCCH and PCCCH have been deprecated. Their services have been merged with the GSM broadcast and common control channels BCCH and CCCH. The latest versions of TS 44.018 [42] say: "Independently of what is stated elsewhere in this and other 3GPP specifications . . . the network shall never enable PBCCH and PCCCH." This is due to optimizing the service. These GPRS channels were eventually deemed redundant to GSM broadcast and common channels, so it was decided to merge those to simplify the system. Consequently, from the above-mentioned four specified logical channel combinations for PDCH, only options 3 and 4 are still deployed.

Table 5.2 GPRS coding schemes.

	PDCH data rate (kbps)	Coding ratio (%)	Coding method
CS-1	8	50	A checksum of 40 bits is added to the 181 bits of RLC/MAC frame and the three USF bits. The result is encoded with 1/2 convolutional code.
CS-2	12	66	A checksum of 16 bits is added to the 268 bits of RLC/MAC frame and the 6 precoded USF bits. The result is encoded with 1/2 convolutional code and thereafter predefined 132 bits are removed (puncturing).
CS-3	14.4	75	A checksum of 16 bits is added to the 312 bits of RLC/MAC frame and the 6 precoded USF bits. The result is encoded with 1/2 convolutional code and thereafter predefined 220 bits are removed (puncturing).
CS-4	20	100	A checksum of 16 bits is added to the 428 bits of RLC/MAC frame and the 3 USF bits. Thereafter, precoding is applied to USF bits to generate a sequence of 12 bits.

5.2.3.3 GPRS Channel Coding and Transmitter Design

GPRS uses the same methods for dividing the user bit data stream to bursts as GSM. However, GPRS supports the four **coding schemes** as in Table 5.2 to generate a sequence of 456 bits from RLC/MAC frame, to be carried as an RLC data block within the four PDCH bursts of a GPRS radio block:

Note that the PDCH data rate means the data rate for using one timeslot per GPRS frame. The full data rate of GPRS connection is the PDCH data rate multiplied with the number of PDCHs (timeslots) used per GSM TDMA frame. As can be seen in the table, the size of the RLC/MAC frame is different in these encoding methods. The CS-4 coding method provides the highest bit rate for traffic tolerant to errors or delay due to retransmissions. CS-4 is typically used over channels with good radio signal quality. CS-1 minimizes the latency since the strong forward error correction reduces the need for any retransmissions.

The **coding ratio** is number of original data bits compared to the user data bits used in the transmission. Coding schemes with smaller coding ratio are more robust against transmission errors as 1 user data bit is encoded with multiple redundant bits. Coding ratio can be adjusted with the puncturing technique, where some of the data bits in fixed bit positions are not sent at all. The receiver adds 0 bits to these positions and then applies convolutional decoding. If the punctured bit was 1 rather than 0, then these bits would appear to the decoder as bit errors. Puncturing increases effective data rates but decreases efficiency of the forward error correction. Puncturing ratios are chosen to find a reasonable balance between those.

The sequence of 456 bits generated in the channel coding process is interleaved to four bursts of the radio block with GSM interleaving methods. The structure of the burst sent on the PDTCH channel is the same as used on GSM TCH channel. The stealing bits of the burst are used in GPRS to indicate the coding scheme used for the burst. The burst sent on the PRACH channel had the same structure as a burst on the GSM RACH channel. Figure 5.25 shows the block diagram of GPRS transmitter.

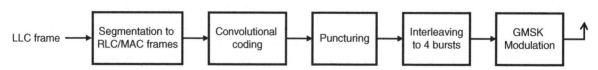

Figure 5.25 GPRS transmitter design.

5.2.4 Protocols between MS and GPRS Network

5.2.4.1 MAC Protocol

GPRS medium access control (MAC) protocol takes care of multiplexing PDTCH data channel and PACCH signaling channel to the PDCH channel allocated to the mobile station. The GPRS MAC protocol is specified in 3GPP TS 44.060 [67] for mobile stations that use traditional GSM A/Gb interfaces and in 3GPP TS 44.160 [68] for those that rely on the Iu interface, which was introduced originally for UMTS but reused later also for GPRS.

Uplink and downlink GPRS MAC frames use different structures and header fields as shown in Figure 5.26. The uplink MAC frame sent by the mobile station has the following fields:

- Payload type tells if the frame contains the user data (PDTCH) or signaling (PACCH).
- Countdown value (CV) is the number of radio blocks still needed for the TBF. This value is decreased after each radio block transmitted. The network uses this value to check the number of radio blocks to be allocated before releasing the TBF.
- Stall indicator tells if the RLC transmit window of the mobile station is able to move or not.
- Retry bit tells if the mobile station has sent a CHANNEL REQUEST message only once or multiple times during its most recent random access attempt.
- Payload data is inside of the uplink MAC frame, which is an RLC frame.

The downlink MAC frame sent by the base station has the following fields:

- Payload type tells if the frame contains the user data (PDTCH) or signaling (PACCH).
- Relative reserved block period (RRBP) indicates the mobile station the radio block uses for RLC/MAC PACKET CONTROL ACKNOWLEDGE messages or other PACCH channel signaling.
- Supplementary/polling (S/P) bit tells if the mobile station shall send the acknowledgment on the radio block identified by RRBP or not.
- USF tells which of the mobile stations listening to this timeslot may use the next uplink radio block for its transmission.
- Payload data is inside of the downlink MAC frame, which is an RLC frame.

5.2.4.2 RLC Protocol

GPRS Radio Link Control (RLC) protocol transports LLC protocol frames between the base station and mobile station. It additionally has the task to segment the LLC frames to blocks that fit into a single radio block and reassembly of those frames on the receiving end of the link. As a result of the segmentation process, a single RLC frame may contain only a fraction of a single LLC frame, or it may have the end of one LLC frame and the start of the next one. After channel coding, the length of one RLC/MAC frame equals the length of four GPRS bursts. Thus, one GPRS PDCH can carry the frame within a single radio block. GPRS RLC protocol is specified in 3GPP TS 44.060 [67].

Uplink	Payload type	Countdown value	Stall indicator	Retry bit	Payload

Downlink	Payload type	RRBP	S/P bit	USF	Payload

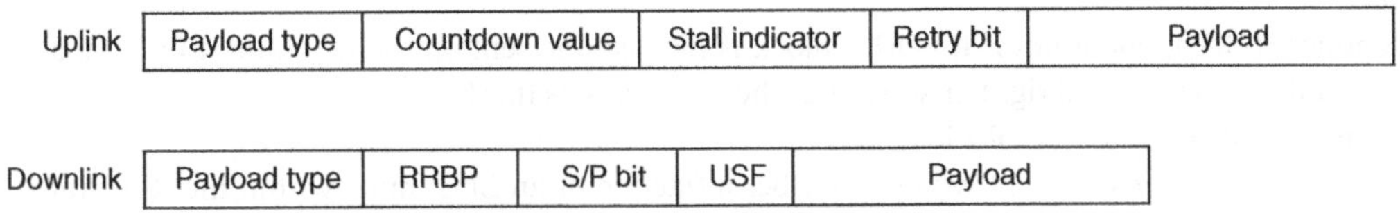

Figure 5.26 Structures of GPRS MAC uplink and downlink frames.

Figure 5.27 Structures of GPRS RLC uplink and downlink data frames.

RLC protocol uses one of the following modes for transporting the frames:

- Acknowledged mode is where the received frames are acknowledged and retransmitted if the frame was corrupted on the radio path. A retransmission window of 64 frames is used so that transmission of new frames stops only when no acknowledgments are received for 64 consecutive frames. The acknowledge message tells the sequence number of the frame up to which all frames have been successfully received. Retransmission of a frame can be requested with a negative acknowledgment. While RLC data frames are transmitted over the PDTCH channel, RLC acknowledgments are sent on the PACCH.
- Non-acknowledged mode is without any RLC level retransmissions. The transmission delay of RLC frames over radio interface is fixed, but the transmission is less reliable than in the acknowledged mode.

The RLC mode used is chosen when the mobile station requests the network to allocate a new packet data protocol context (PDPC) for it.

Like MAC frames, the uplink and downlink RLC have different structures, as shown in Figure 5.27. The uplink RLC frame has the following fields:

- Packet flow identifier (PFI) indicator (PI) bit tells if the frame has PFI fields.
- Temporary flow identifier (TFI) is the identifier of the TBF to which the RLC block belongs. The TBF identifies the mobile station that has sent the frame.
- Temporary logical link identifier (TLLI) indicator (TI) bit tells if the frame has a TLLI field.
- Block sequence number (BSN) is the sequence number of the RLC block.
- There are one or multiple records, each of which describes a fraction of an LLC frame within the RLC frame payload:
  - Length indicator (LI) is the number of octets of the LLC frame.
  - More (M) bit tells if the payload has a fraction of another LLC frame.
- Temporary logical link identifier is the identifier of LLC protocol link between the mobile station and SGSN.
- Packet flow identifier.
- Payload of the uplink RLC frame carrying the LLC fragments.

The downlink RLC data frame has the following fields:

- Power reduction (PR) tells if the mobile station shall adjust its transmission power for upcoming radio blocks.
- TFI is the identifier of the TBF to which the RLC block belongs. The TBF identifies the mobile station that will receive the frame.
- Final block identifier (FBI) bit identifies if this RLC block is the last one sent on the downlink to the mobile station. The downlink TBF is released right or soon after the network sets this bit.
- BSN is the sequence number of the RLC block.
- There are one or multiple records, each of which describes a fraction of an LLC frame within the RLC frame payload:
  - LI (length indicator) is the number of octets of the LLC frame.
  - M (more) bit tells if the payload has a fraction of another LLC frame.
- The payload of the downlink RLC frame carries the LLC fragments.

The downlink RLC signaling frame has the following fields:

- Reduced block sequence number (RBSN): the sequence number of a signaling block.
- Radio transaction identifier (RTI) has identical value in those radio blocks that have a segment of one single upper layer signaling message.
- Final segment (FS) bit identifies the block which contains the last segment of a single upper layer signaling message.
- Power reduction (PR) tells if the mobile station shall adjust its transmission power for upcoming radio blocks.
- TFI is the identifier of the TBF to which the RLC block belongs. The TBF identifies the mobile station that will receive the frame.
- Reduced block sequence number extension (RBSNe) is the sequence number when extended RLC/MAC control message segmentation is used.
- Final segment extension (FSe) bit identifies the block that contains the last segment of one single upper layer signaling message when extended message segmentation is used.
- The payload of the RLC frame.

5.2.4.3 LLC Protocol

Logical Link Control (LLC) protocol transports upper layer protocol messages over the logical link between the mobile station and SGSN. GPRS LLC protocol is specified in 3GPP TS 44.064 [69]. The LLC protocol provides the following services:

- Acknowledged and non-acknowledged modes like the RLC protocol. Acknowledged mode uses frame checksums and retransmissions for error control. In the non-acknowledged mode, LLC can be configured to either drop corrupted frames or pass them to the upper layer protocol, which would take care of correcting the errors.
- Flow control and message sequence control over the logical link.
- Encryption and integrity protection of transmitted data using the keys created during the authentication process.
- Point-to-multipoint messages from an SGSN to multiple mobile stations.

Temporary logical link identifier (TLLI) identifies a logical link between the GPRS mobile station and the SGSN. More precisely, the SGSN assigns a TLLI to the mobile station to identify it. As long as the mobile station stays in the cells within the routing area served by the SGSN, the link and its TLLI identifier remain the same. But when the mobile station moves to a cell under another SGSN, a new logical link is opened and the old link is closed. The MS gets a new TLLI from the new SGSN. The link is kept alive over any idle periods when no user data is transferred.

Multiple parallel data flows may be multiplexed to the logical link. Those data flows are identified with SAPI code of the LLC protocol. The following SAPI values are defined in GPRS:

- SAPI = 1: the flow is used to transport GMM messages
- SAPI = 2: tunneling of messages
- SAPI = 3: messages using QoS level 1
- SAPI = 5: messages using QoS level 2
- SAPI = 7: short messages over GPRS
- SAPI = 8: tunneling of messages
- SAPI = 9: messages using QoS level 3
- SAPI = 11: messages using QoS level 4

The QoS levels are used to assign priorities between different types of messages when the network is under high load. There are no fixed QoS parameter values or for any SAPI, but those are negotiated when taking SAPI into use. Data link connection identifier (DLCI) is the combination of TLLI and SAPI to identify a data flow over the logical link

Address	Control	Information	FCS

Figure 5.28 Structure of the GPRS LLC frame.

The LLC frame structure shown in Figure 5.28 resembles HDLC protocol frames. The LLC frame has the following fields:

- Address field, consisting of the following subfields:
  - The protocol discriminator (PD) has value 0 for GPRS LLC.
  - The command/response (C/R) bit that tells if the LLC frame contains a command or response.
  - SAPI identifies the flow and defines the QoS level of the message. The specific upper layer protocol GMM, SMS, or SNDCP transported by the LLC frame can also be deduced from the SAPI.
- Control field, which describes the type of the frame:
  - Acknowledged or non-acknowledged mode
  - Supervisory function or control function
- Information field, which contains the upper layer protocol data payload.
- The frame check sequence (FCS), a checksum of 24 bits calculated over the frame.

The maximum size of an LLC frame in GPRS is 1600 octets.

5.2.4.4 SNDCP Protocol

GPRS subnetwork dependent convergence protocol (SNDCP) is specified in 3GPP TS 44.065 [70]. The SNDCP protocol has the following tasks:

- Compress the headers and data of the transported packet data protocol before segmentation. Usage of compression is negotiated when opening the PDP context for the mobile station.
- Splitting the transported data to segments that can be carried as payload of LLC frames.
- Adapt different packet data protocols to the LLC protocol used over the link. Multiplex the packet data protocol messages with the same QoS level to one single SAPI flow of the LLC layer.
- Use either acknowledged or non-acknowledged LLC mode on the logical link for data transfer. Establish and release LLC connections for acknowledged mode.
- Reassemble and decompress data received from the LLC and forward it to the correct packet data protocol entity.

The SNDCP frame has the following fields shown in Figure 5.29:

- X-bit, always 0
- F-bit (first segment indicator) tells if the SNDPC frame contains the first segment of an upper layer protocol message
- T-bit protocol data unit ([PDU] type) tells if the frame uses acknowledged or non-acknowledged mode.
- M-bit (more) tells if the SNDPC frame contains the last segment of an upper layer protocol message.
- NSAPI identifier tells the type of the packet data protocol carried as the payload of SNDCP.
- DCOMP tells the type of compression applied to the upper layer data.
- PCOM tells the type of compression applied to the addresses of the carried packet data protocol.
- Segment number used for non-acknowledged SNDCP mode.

X	F	T	M	NSAPI	DCOM	PCOM	Segment number	N-PDU number	Payload

Figure 5.29 Structure of GPRS SNDCP frame.

- N-PDU number is the sequence number of the upper layer protocol frame.
- Payload of the SNDCP frame for carrying upper layer data segments.

5.2.5 Protocols of GPRS Network

5.2.5.1 NS Layer Protocols

Network Service (NS) layer provides a link layer connection between the PCU and SGSN. For the link implementation, 3GPP has defined two options:

- The original GPRS design relied on E1 links on top of which virtual connections were created with the frame relay protocol, described in Chapter 3, Section 3.2.1.
- The newer design was to use the internet protocol over Ethernet fiber links. This option has gradually replaced the frame relay in GPRS network deployments. Similar links are used also between the GGSN and SGSN.

With frame relay, a link of the NS layer is a virtual connection created over all the physical cable links and other pieces of network equipment between the PCU and SGSN. Every single piece of equipment along the virtual connection path shall support the FR protocol and have a static routing configured to forward FR frames between correct physical links. This kind of connection is called **permanent virtual circuit (PVC)**. The frame relay provides the following features:

- Permanent end-to-end virtual circuits that can be set up by the network management system
- Statistical multiplexing of data packets from different virtual circuits to physical links

The PCU has one PVC circuit to the SGSN while the SGSN has one PVC to every PCU connected to it. One virtual connection to a PCU carries all the GPRS data flows for mobile stations served by the PCU.

With IP over Ethernet, the frame relay switches have been replaced by IP routers. No virtual connections are used, but the Ethernet frames are terminated at the end of each link. The IP packets are then routed between the PCU and SGSN.

5.2.5.2 BSSGP Protocol

Base station subsystem GPRS protocol (BSSGP) is used on top of the NS layer between the SGSN and PCU. The GPRS BSSGP protocol is specified in 3GPP TS 48.018 [71]. When using frame relay on the NS layer, BSSGP maps traffic from BSSGP virtual circuits (BVC) to the PVCs of the NS layer. The BSSGP protocol takes care of BVC packet flow control and buffering of data packets received from different sources before forwarding them to the links. BSSGP has three layers of data buffers on top of each other, as shown in Figure 5.30. The buffers are called as buckets, as they behave like ATM buckets described in *Online Appendix H.5.4*.

- The bucket for an individual packet flow context (PFC), which carries an individual data flow for mobile station. Usage of the PFC flow control and buckets is optional.
- The bucket for the complete aggregate packet data flow toward a mobile station.
- The bucket for the packet data flow passing through the BVC to the NS layer.

Buffering is used for statistical multiplexing, since the amount of data sent or received by mobile stations may temporarily exceed the capacity reserved for the BVC. The excess packets are buffered to queue for transmission. If a buffer would overflow, some of the packets are dropped and an LLC-DISCARDED PDU message is sent to the source of the dropped packets to inform it about the buffer overflow. The source could then reduce its data rate.

The BSSGP protocol uses FLOW-CONTROL PDU messages specific to each bucket to adjust the sizes of buffers in the end of the connection and the output data rate from the buffer. The user data packets (LLC PDUs) are carried as the payload of BSSGP UNITDATA PDU messages.

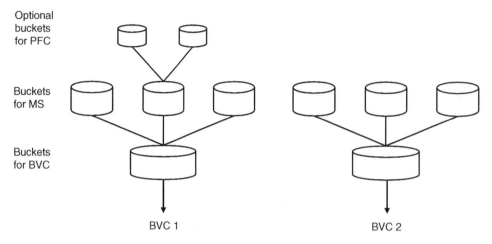

Figure 5.30 Buffers of BSSGP protocol.

In addition to user data transport, the BSSGP protocol supports also various GPRS mobility management and PDP context management signaling procedures. An SGSN uses the BSSGP protocol to instruct the BSS subsystem to page a mobile station to activate the data connection when a data packet arrives to the mobile station. Also, the PS handover control messages are carried over BSSGP.

BSSGP protocol messages start with the PDU type field, the value of which determines the composition of other fields within the message.

5.2.5.3 GTP Protocol

GPRS Tunneling Protocol (GTP) is used between the GGSN and SGSN to transport both user data and signaling messages. Basically, it would be possible to transport both types of data directly on top of the IP network used between the GGSN and the base station. The drawback is that any move of a mobile station between cells would cause the routing tables of the IP routers to reflect the new location of the IP address of the mobile station. Such a network-wide routing table update is both a slow and heavy operation. The GTP tunneling approach solves this problem by hiding the movements of the mobile station from the underlying IP network. Instead, only SGSNs track the locations of mobile stations. Other routers within the GPRS network see only IP addresses of tunnel endpoints, rather than IP addresses assigned to the mobile stations. The SGSNs inform GGSNs via which SGSN and related tunnel the mobile stations can be reached. The IP routing tables are then modified only at GGSN per MS but not in the rest of the IP network infrastructure between the GGSN and mobile station. At the GGSN, the endpoint of routing is the SGSN that servers the mobile station, rather than the mobile station itself.

GTP is divided to the GTP-U and GTP-C subprotocols, where U stands for user and C for control. User data flows between the GPRS mobile station and an external packet data network are carried within GTP-U tunnels created between the GGSN and SGSN. The GPRS signaling messages between the GGSN and SGSN and are sent with GTP-C protocol. PDP context management messages, tracking area update messages or GTP tunnel management messages are examples of GPRS signaling messages. The GPRS GTP protocol is specified in 3GPP TS 29.060 [72].

The Gn interface between GSN nodes uses the packet switched IP protocol stack. GTP protocol enables the GPRS network to transport multiple different user packet data protocols (such as X.25 and IP) so that the GGSN and SGSN nodes do not need to understand and interpret those protocols. The user data packets are encapsulated into GTP-U packets, which form tunnels maintained with the GTP-C protocol. The GTP-U packets are carried over the network with underlying TCP/UDP/IP protocols.

User data packets from external networks are routed via GGSNs to GPRS mobile stations. In those packets, the destination is identified with the address that the GGSN has given for the mobile station when the PDP context was created. The GGSN routes those packets toward the mobile station and its SGSN over the GTP-U tunnel of the PDP context. The same tunnel is also used for uplink data packets from the mobile station to the external packet data network. The tunneling mechanism encapsulates the inner user data packet into an outer GTP-U packet of the tunnel. The source and destination addresses of the outer GTP-U packet are IP addresses of the GGSN and SGSN. In this way, GTP relies on IP routing mechanisms. The IP address that the mobile station has for its PDP context is used in the downlink packet routing process as follows:

- When a user data IP packet arrives from an external PDN network to the GGSN, its destination IP address is used to select the PDP context and GTP tunnel to transport the user data packet to the correct SGSN node.
- When the SGSN receives the user data packet, it checks the destination IP address of the packet to route it to the destination mobile station. The packet arrives to the mobile station via the base station under which the mobile station currently camps.

If GGSN receives packets from an external data network with destination addresses not mapped to any of its PDP contexts, those packets shall be processed in either of the following two ways:

- If the GGSN finds from its database that the address is fixedly reserved for a mobile station, the GGSN tries to find out which SGSN is currently serving the mobile station. If the mobile station has attached to GPRS service, the GGSN opens a PDP context to the mobile station and forwards the packets to it.
- If the GGSN does not find the destination address from its database, the packet is dropped.

The structure of GTP frame depends on the purpose to which the frame is used. There are always the following fields in the frame:

- Version of the GTP protocol
- Protocol type: GTP or GTP'. GTP' is a charging protocol based on GTP.
- E, S, and PN-bits, which tell if the frame contains extension headers, sequence number and/or N-PDU number of the upper layer protocol frame.
- Message type
- Payload length
- Tunnel endpoint identifier (TEID), which identifies the endpoint of the GTP tunnel. TEID is generated from the user's MCC, MNC, and MSIN numbers at the PDP context activation.

The frame may additionally have also the following fields, as shown in Figure 5.31:

- Sequence number of the frame
- N-PDU number of the upper layer protocol, such as LLC
- Other additional fields
- The payload of the GTP frame

GTP version	Protocol type	E	S	PN	Message type	Payload length	TEID	Sequence number	N-PDU number	Payload

Figure 5.31 Structure of the GPRS GTP data frame.

5.2.6 Radio Resource Management

5.2.6.1 Opening and Releasing of Dedicated GPRS Radio Channels

The GPRS mobile station is in either of the two following modes, depending on whether it has a dedicated packet data channel or not:

1) **Packet idle mode:** No dedicated packet data channel has been granted for the mobile station in packet idle mode. The mobile station listens to the messages sent to it over the PCH and PPCH channels as well as the SYSTEM INFORMATION messages broadcasted on BCCH (and earlier PBCCH) channels. The mobile station may also send a random access request over the RACH channel. The network responds to such access requests on the AGCH channel, after which the mobile station moves to the dedicated mode. The mobility management state of the MS is either **idle** or **standby**.

2) **Dedicated mode:** The mobile station has an active packet data connection with the network in the dedicated mode. The network has granted the station with a PDTCH user data channel and a PACCH signaling channel. Additionally, the network has allocated the mobile station a PTCCH channel, which is used to adjust the timing advance of the mobile station. The mobility management state of the MS is **ready**.

When the mobile station moves from idle to dedicated mode, a TBF is created. The TBF is released when moving back to idle. In the dedicated mode, the mobile station has a packet data connection open at least up to the SGSN at GPRS attach, but up to the GGSN when the PDP context has been activated. The packet data connection is opened separately from the mobile station toward the network (uplink) and from the network to mobile station (downlink). The request to activate the connection may be initiated either by the mobile station or network.

The mobile station moves from the packet idle mode to the dedicated mode by performing an access procedure, because of one of the following reasons:

1) The mobile station has to send a signaling message or user data.
2) The mobile station should receive an incoming data packet. In this case, the mobile station requests access after the network has paged the station.
3) The mobile station sends an RLC/MAC control message.

There are three ways to perform the GPRS access procedure:

- One-phase access: After receiving the random access request, the network immediately grants the mobile station with a single PDCH. One-phase access can only be used if the mobile station responds to paging or wants to use the RLC acknowledged mode over one single PDCH.
- Two-phase access: After receiving the random access request, the network grants the mobile station with PDCH access to one or two radio blocks only. The mobile station uses the grant to send a packet resource request, based on which the network grants the mobile station with a packet data channel. Two-phase access must be used if the mobile station either requests multiple PDCHs for higher data rates or wants to use the non-acknowledged RLC mode.
- Single block access: After receiving the random access request, the network grants the mobile station with PDCH access to one radio block. The mobile station uses the grant to send an RLC/MAC control message or to initiate the two- phase access.

The complete GPRS access procedure is done as shown in Figure 5.32:

1) The mobile station requests a dedicated packet data channel by sending either an RRC CHANNEL REQUEST or RLC/MAC EGPRS PACKET CHANNEL REQUEST random access message to the BSC over the GSM RACH channel. With a CHANNEL REQUEST message, the mobile station may either request one-phase access or single block access. With an EGPRS PACKET CHANNEL REQUEST message, the mobile station

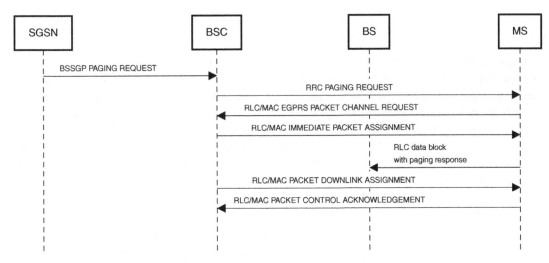

Figure 5.32 GPRS downlink channel activation for MT data.

may request either a one- or two-phase access procedure and provide further information about the request type specific options.

2) The BSC responds to the mobile station over the GSM AGCH channel with either an RRC IMMEDIATE ASSIGNMENT or a MAC/RLC IMMEDIATE PACKET ASSIGNMENT message. If one-phase access is used, the message describes the PDCH channel, which is continuously allocated to the mobile station and provides the TFI and USF identifiers of the TBF. The message may also have an optional bit vector indicating the radio blocks allocated to the mobile station. The message completes the one-phase or single block access procedure. Instead of the one-phase PDCH grant, in its response the network may alternatively provide only single or multiple PDCH radio blocks for the following cases:

   - The mobile station requested single block access.
   - The mobile station requested two-phase access.
   - The mobile station requested one-phase access but the network decided to use the two-phase access.

3) In the two-phase access, the mobile station uses the granted radio block to send an RLC/MAC PACKET RESOURCE REQUEST message over the PACCH to the BSC. This message describes the details of the channel request, the radio access capabilities of the mobile station, the temporary logical link identity (TLLI) identifier of the logical link, and the TFI, if the request is related to an existing TBF. The packet channel can be reserved either for transporting a defined amount of data or until further notice.

4) The BSC responds by sending either an RLC/MAC PACKET UPLINK/DOWNLINK ASSIGNMENT or a MULTIPLE TBF UPLINK/DOWNLINK ASSIGNMENT message to describe the PDCH allocation. The former type of message is used for a single TBF and the latter for multiple TBFs and data flows. The message contains the following pieces of information related to the dedicated packet data channel:

   - The TFI identifier of the TBF, used to identify the downlink GPRS radio blocks allocated for the TBF and the TBF itself for any further requests
   - PDCH timeslots allocated for the TBF and the parameters to define the frequency hopping schemes on them
   - Array of USF identifiers, one per PDCH timeslot, used for identifying uplink PDCH radio blocks allocated dynamically for the mobile station
   - The coding scheme, MAC, and RLC modes to be used on the TBF
   - Timing advance and power control parameters

When the GPRS mobile station has a PDP context in the standby state, the SGSN activates the downlink data channel for a mobile terminated data packet as follows:

1) The SGSN sends a BSSGP PAGING REQUEST to the BSS, causing the BSC to send an RRC PAGING REQUEST to the mobile station. The MS is identified by the P-TMSI identifier of the subscriber. The MS responds with a random access message, which tells the reason of the request as one-phase access for a paging response. The BSC then sends an immediate assignment message to allocate a single signaling PDCH for the mobile station. The mobile station then sends its response to the paging request over the newly allocated signaling channel.
2) After completing paging, the BSC sends a packet downlink assignment message to the mobile station. This message describes the PDCH timeslot, power control, and timing parameters to be used for the packet data channel over which the MT data packet will arrive. After taking the downlink channel into use, the mobile station may at any moment reserve also an uplink data channel by sending a channel request within the packet data acknowledgment messages.

For further details about the messages used in this procedure, please refer to *Online Appendix I.3.1* (Figure 5.32). The network may release the uplink connection when the mobile station decrements the countdown value counter of the MAC frame to zero to tell that all user data has been sent. The mobile station may start using the countdown mechanism either spontaneously or after the network has requested the connection release by sending an RLC/MAC PACKET PDCH RELESE message to the mobile station. To finally release the uplink connection, the BSC sends an RLC/MAC PACKET UPLINK ACK message with Final_Ack value 1.

The downlink connection is released so that the network sends an RLC frame with value 1 in its FBI field. The mobile station acknowledges the release by setting the Final_Ack value as 1 to the RLC/MAC PACKET DOWNLINK ACK response. If the connection uses unacknowledged RLC mode, the mobile station may send a separate RLC/MAC PACKET CONTROL ACK message. The mobile station may also request the network to release the connection by sending a message with its TBF_RELEASE bit set.

5.2.7 Mobility Management

5.2.7.1 GPRS Attach

In the GPRS attach, the mobile station announces itself to an SGSN. The purpose of GPRS attach is to get a SGSN to start tracking the location of the mobile station so that the network could route mobile terminated packet data to the station. The mobile station attaches to one single SGSN node at any time. This SGSN serves the routing area within which the mobile station is located. At the GPRS attach, the SGSN authenticates the subscriber and checks which services the subscriber is entitled to use.

The GPRS mobile station may be in one of the three GPRS mobility management states:

- **Idle:** The mobile station is not attached to a GPRS network.
- **Ready:** The mobile station is attached to a GPRS network and has a PDP context. An MS enters ready state whenever signaling or user data is sent to/from it over GPRS. After the data transmission stops, the mobile station stays in ready state until expiration of timer T3314. In the ready state, the mobile station must report any cell reselection performed.
- **Standby:** The mobile station enters the standby state when it exits the ready state by the timer expiration. In the standby state, the mobile station reports cell reselections only when the routing area changes. The network can move the mobile station from the standby to the ready state by paging MS for incoming data.

The GPRS attach procedure is performed as shown in Figure 5.33:

1) The mobile station searches the GSM network and synchronizes itself to the FCCH and SCH channels of the cell. The mobile station starts reading system information messages. From the SYSTEM INFORMATION TYPE13 message, the mobile station acquires parameters needed for GPRS attach, such as routing area code, timer values, and supported optional GPRS features.

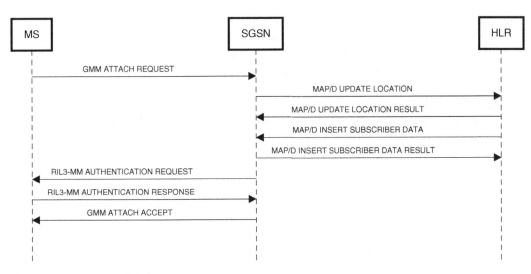

Figure 5.33 GPRS attach procedure.

2) The mobile station opens a GPRS dedicated packet signaling channel as described in Section 5.2.6.1. The mobile station uses the channel to send a GMM ATTACH REQUEST to the SGSN, in order to register itself to the GPRS service. Additionally, the mobile station may request registration to the GSM service for SMS and voice calls. The mobile station identifies itself with its P-TMSI identifier allocated to the station and gives the RAI identifier of the routing area where that P-TMSI was most recently used.

3) The SGSN updates the new location of the subscriber to the HLR of the subscriber's home network. The HLR confirms if the subscriber is entitled to use the GPRS service of the attached network and sends the GPRS subscriber information back to the SGSN. Thereafter, the SGSN authenticates the subscriber and allocates a new P-TMSI identifier for the mobile station. The SGSN completes the sequence with an attach accept response to the mobile station, referring to the new P-TMSI.

For further details about the messages used in these procedures and protocols below GMM, please refer to *Online Appendix I.3.2.*

The mobile station may also perform a combined GSM/GPRS attach procedure in which the SGSN informs an MSC over the Gs interface about the location of the mobile station. Both the SGSN and MSC inform the HLR about the location of the MS within the packet switched and circuit switched domains, respectively. If the HLR knows any old SGSN and MSC, which have previously served the mobile station, the HLR sends cancel location messages to them (Figure 5.33).

The mobile station may disconnect from the GPRS service by sending the SGSN a GMM DETACH REQUEST. The SGSN acknowledges this message with a GMM DETACH ACCEPT response. Thereafter the SGSN will tear down any related active PDP contexts by sending a DELETE PDP CONTEXT REQUEST messages to all the GGSN nodes that had active PDP contexts for the disconnected mobile station.

5.2.7.2 Cell Reselection

Like a GSM station, the GPRS mobile station selects the serving cell based on the signal quality measurements. The GPRS mobile station may perform cell reselection either autonomously or with the GPRS specific network-assisted cell change (NACC) procedure. These procedures work as follows:

- Autonomous cell selection works as in GSM. The mobile station learns frequencies used by the neighboring cells from the SYSTEM INFORMATION 2 message. After doing measurements, the MS may decide to change the serving cell. The MS synchronizes to the new cell and reads its system information messages.

- With NACC, the difference is that after making measurements and deciding to change the cell, the MS may send an RLC/MAC PACKET CELL CHANGE NOTIFICATION message to the serving base station. The base station responds with an RLC/MAC PACKET NEIGHBOR CELL DATA message with relevant parts of the system information of the new cell to speed up the MS to connect to the new cell.

5.2.7.3 Routing Area Update

When moving from one cell to another in the standby state, the GPRS mobile station checks if the new cell belongs to a different routing area than the old cell. If that is the case, the mobile station shall acquire a dedicated channel and send a routing area update message to announce its new location to the network. Since the dedicated channel is used to send a single signaling message, the mobile station uses the one-phase access method to get a PACCH allocation for those radio blocks needed to send the update.

Routing area update messages are not used when the mobile station is using TBF. In that case, the network tracks the location of the mobile station in the accuracy of a cell as follows. The mobile station in the dedicated mode moves between two cells within a single routing area. The mobile station releases its TBF connection toward the old cell and opens a new TBF for the new cell. When any packets sent by the mobile station reach the SGSN over the new TBF, SGSN detects the location of the mobile station from the global cell identifier added by the PCU to the packets.

These are the following major scenarios for the GPRS routing area update procedure:

- The mobile station moves from a routing area to another. In this case, the mobile station sends a routing area update message.
- The mobile station moves simultaneously between two routing areas and GSM location areas. In this case, the mobile station updates both its routing area and location area with a combined update procedure.
- The mobile station sends a periodic routing area update even if it has not changed the cell.
- The mobile station sends a routing area update right after attaching to GPRS.

The GPRS routing area update procedure is performed as shown in Figure 5.34:

1) The mobile station sends a GMM ROUTING AREA UPDATE REQUEST message to the SGSN that serves new cell. The MS identifies itself with its P-TMSI identifier and the identifier of its old routing area in case the SGSN was changed. The new SGSN retrieves the mobility management and PDP contexts of the mobile station from the old SGSN. After providing the context data, the old SGSN establishes a GTP tunnel to the new SGSN and sends any data packets buffered for the mobile station via the tunnel. The new SGSN forwards those packets to the mobile station.
2) The new SGSN contacts all the GGSN nodes that have active PDP contexts for the mobile station. Those GGSN nodes establish new GTP tunnels to the new SGSN and tear down the old ones toward the old SGSN.
3) The new SGSN updates the location of the mobile station to the HLR, which now tells the old SGSN to purge any expired location and subscriber data records for the subscriber. The HLR thereafter sends the subscriber data records to the new SGSN, so that the new SGSN can authenticate the subscriber. Finally, the new SGSN acknowledges the routing area update procedure.

For further details about the messages used in these procedures, please refer to *Online Appendix I.3.3*.

5.2.8 Packet Data Connections

5.2.8.1 PDP Context Management

To transport packet data over GPRS, the mobile station must open **PDPC** after attaching to the GPRS service. GPRS attach enabled the network to track the routing area of the mobile station, but a PDP context is needed for GPRS data transfer. The mobile station gets a packet data protocol address at the PDP context activation.

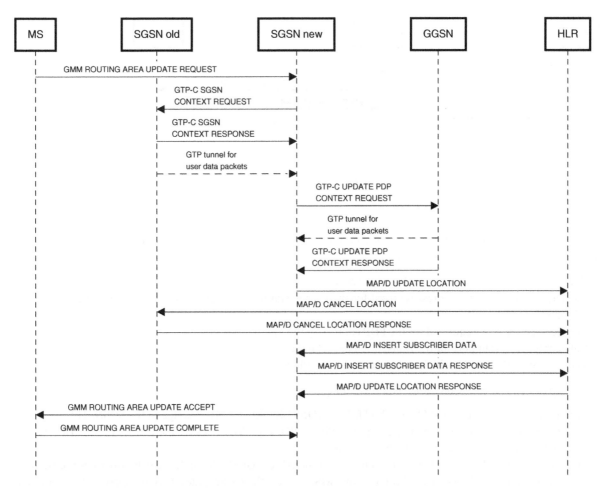

Figure 5.34 Routing area update for GPRS.

This address is used as a source or destination address within the transported user data packets. The PDP context activation involves both the GGSN and SGSN for setting up GTP tunnels for the mobile station and updating their internal routing tables for the new packet data protocol address granted to the mobile station. The packet data protocol used by the mobile station is in practice IP protocol while GPRS also supports X.25 protocol. The X.25 protocol was still in use when the GPRS protocol was designed but has since become obsolete. If the mobile station supports simultaneous GPRS connectivity toward different external data networks, the mobile station may open multiple PDP contexts toward different GGSN nodes. In this case, the mobile station gets a new PDP address for each of the contexts opened.

An active PDP context provides the mobile station with a packet data connection toward a single **access point (AP)** of an external data network and its packet data service. Every packet data service, such as MMS multimedia messaging or generic Internet access, has its own access point. The access point is identified with its unique APN. When using the IP protocol, the APN is a domain name, which globally identifies both the GGSN node, the specific packet data service, and its provider. The format of the APN follows the convention of <service name>. mnc<MNC>.mcc<MCC>.gprs where any part limited with <> would be replaced with the corresponding name or numeric code of the entity. MNC and MCC are the mobile network and country codes of the operator which owns the GGSN. The mobile station uses the APN when requesting activation of a PDP context for the packet data service.

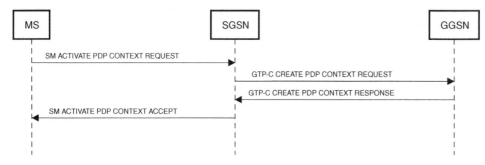

Figure 5.35 GPRS PDP context activation by mobile station.

The PDP context is represented as a data structure stored to the mobile station, SGSN and GGSN nodes. This structure has the following key elements:

- The type of the packet data protocol used: IP or X.25
- The packet data protocol address reserved for the mobile station either permanently or for the lifetime of the PDP context
- The QoS service level requested for the PDP context by the mobile station
- Data compression support for the transported packet data
- The address of the GGSN, which is the anchor for the PDP context

Activation of a PDP context is done as shown in Figure 5.35:

1) The mobile station sends an SM ACTIVATE PDP CONTEXT REQUEST message to the SGSN. This message identifies the PDP type, APN name, QoS parameters, SNDCP protocol, NSAPI identifier, and the packet data protocol address requested by the mobile station.
2) The SGSN checks the subscriber data received from the HLR at GPRS attach. If the subscriber has the right to activate the requested type of PDP context, the SGSN selects the GGSN gateway that supports the requested PDP type and APN. For instance, when the mobile station wants to access the Internet, the SGSN may use a local DNS server to return the IP address of the GGSN whose domain name matches with the APN received from the mobile station.
3) The SGSN sends a GTP-C CREATE PDP CONTEXT REQUEST message to the selected GGSN, which allocates a PDP address for the mobile station. The GGSN may use an address permanently assigned for the station or use a dynamically allocated address retrieved from a local DHCP server.
4) The GGSN creates a GTP protocol tunnel toward the SGSN and gives a TEID for the tunnel. The GGSN stores the mapping between the TEID and the packet data address allocated to the MS, to be used for routing user data packets over the correct tunnel. The GGSN sends a GTP-C CREATE PDP CONTEXT RESPONSE message to inform the SGSN about the TEID of the GTP tunnel and the PDP address allocated for the mobile station.
5) SGSN finally sends an SM ACTIVATE PDP CONTEXT ACCEPT message to the mobile station. This message provides the mobile station with the allocated PDP address and other parameters of the PDP context (Figure 5.35).

If the mobile station has a permanently allocated PDP address, other packet data protocol endpoints may send data to it even when the station does not have an active PDP context. In such a case, the GGSN has to activate the PDP context. After receiving a data packet, the GGSN shall find out the routing area where the mobile station is located and thereafter create GTP tunnels to the SGSN serving the station. The PDP context is activated by the GGSN as follows:

1) The GGSN sends a MAP/D SEND ROUTING INFO FOR GPRS to the HLR, identifying the subscriber with the IMSI code mapped to the permanent destination PDP address. In its response, the HLR provides the address of the SGSN currently serving the mobile station.

2) The GGSN sends a GTP-C CREATE PDP CONTEXT REQUEST message to the SGSN, which sends a SM REQUEST PDP CONTEXT ACTIVATION message to the mobile station. This message triggers the mobile station to initiate the PDP context activation procedure for the given packet data protocol address and APN. The activation procedure is like that described earlier.

The mobile station may request deactivation of a PDP context by sending an SM DEACTIVATE PDP CONTEXT REQUEST message to the SGSN. Consequently, the SGSN sends a GTP-C DELETE PDP CONTEXT REQUEST message to the GGSN. The GTP tunnel is torn down, and after getting acknowledgment from the GGSN, the SGSN returns an SM DEACTIVATE PDP CONTEXT ACCEPT response to the mobile station.

5.2.8.2 Transfer of Packet Data in GPRS System

The path of packet data in GPRS network can be divided into three segments, each relying on a specific protocol stack for relaying user data packets. These stacks were described in Section 5.2.2.3.

- The segment between the GPRS mobile station and the BSS subsystem
- The segment between the BSS and SGSN
- The segment between the SGSN and GGSN

When forwarding user data packets from the mobile station over these segments, the PCU, SGSN, and GGSN nodes behave as follows:

1) The mobile station sends user data within RLC/MAC frames over the allocated PDTCH channel. The TLLI identifier within the RLC frame identifies the LLC link of the PDP context. The RLC/MAC frames contain segments of LLC frames, which the PCU forwards over the trunk connection to the SGSN.

2) The SGSN checks the SAPI identifier of the received LLC frames to find out the upper layer protocol to which the payload of the LLC frame belongs. The LLC payload contains user data packets encapsulated into SNDCP packets. The NSAPI field of the SNDCP header tells the type of packet data protocol used. The SGSN terminates LLC and SNDCP protocols and unwraps the user data packet from the SNDCP packet. The SGSN then accesses the header of the user data packet to check the PDP destination address. The SGSN uses this address to route the user data packets over the correct GTP tunnel, which has been set up for the PDP context. The SGSN encapsulates the user data packets to GTP-U packets sent to the GGSN node in the end of the tunnel.

3) The GGSN unwarps the received GTP-U packets and forwards the user data packets to the external packet data network identified within the PDP context.

User data packets arriving from the external data network toward the mobile station are processed in the reverse manner.

5.3 EDGE

While GPRS provided a solution for delivering variable-rate packet switched data flows over the GSM air interface, the provided data rates were modest in comparison to fixed Internet access technologies, such as ADSL. To increase mobile data rates, 3GPP enhanced the GPRS solution with a few techniques, such as advanced modulation and coding scheme (MCS)s.

Enhanced data rates for GSM evolution (EDGE) means a set of techniques used to increase bitrates of circuit switched GSM and packet switched GPRS data connections. EDGE covers two separate solutions: Enhanced

Circuits Switched Data (ECSD) and Enhanced General Packet Radio Service (EGPRS). EGPRS can be considered as an essentially new air interface toward the GPRS system. Both ECSD and EGPRS rely on modulation methods specified in 3GPP TS 45.004 [31].

EDGE was introduced into 3GPP GSM and GPRS standards in GSM Phase 2+ release 98. After EDGE was launched, 3GPP introduced a new acronym, GERAN, to cover radio access networks of the GSM family: GERAN stands for GSM-EDGE Radio Access Network. 3GPP has continued to evolve GERAN specifications and introduce various enhancements for GERAN over successive 3GPP specification releases.

5.3.1 ECSD

Enhanced circuit switched data (ECSD) brings new modulation and line coding methods for GSM CSD bursts. The new methods support improved data rates up to 43.2 kbps per timeslot. The maximum transfer rate of an ECSD connection is 64 kbps with two GSM timeslots, each providing 32 kbps data rate. Compared to HSCSD, the maximum data rate of ECSD is the same, but it is achieved with a smaller number of timeslots. That saves GSM transmission resources for other connections.

5.3.2 EGPRS

Enhanced general packet radio service (EGPRS) is an improved version of GPRS to provide higher data transfer rates, theoretically up to 400 kbps. In practical network conditions, the maximum achievable bitrate is limited approximately to 100 kbps. The most important new EGPRS methods are as follows:

- New modulation and channel coding methods for the PDCH channel to increase the transmission rates up to 60 kbps per GSM timeslot.
- Merging of RLC and MAC protocols so that each radio block has one single RLC/MAC header area instead of separate RLC and MAC headers. This reduces the amount of RLC/MAC overhead. Note that the RLC/MAC messages referred to in Section 5.2.7 already followed this EDGE convention.
- Usage of two new complementary techniques to improve the retransmission process:
  - **Incremental redundancy (IR):** When an RLC/MAC frame is retransmitted, the punctured bits removed after the convolutional coding are selected differently for the retransmitted frame compared to the earlier transmissions of the frame. This allows the receiver to combine information from different radio blocks carrying copies of the frame and to reconstruct the original frame even if every copy would have its own bit errors.
  - **Link adaptation (LA):** When the signal quality of the channel is bad enough to cause many retransmissions, a more robust but slower coding scheme can be used to improve the probability of forward error correction to catch and fix the bit errors.

EGPRS supports nine different **MCS** options to be used for link adaptation. All of these methods share the following features:

- The RLC/MAC header is encoded separately from the payload data. Both the header and payload have their own checksums calculated before the convolutional coding. For headers, stronger protection is applied than for the payload by allocating a higher number of bits to the checksum per the number of header bits being protected.
- **Convolutional coding** with 1/3 rate (three output bits per one input bit) is used by all of the MCS schemes.
- **Puncturing** is used after convolutional coding to adjust the bitrate. Some of the bits in predefined positions are dropped before transmission to increase bitrate with the cost of making the forward error correction weaker. Each MCS scheme has a few optional puncturing schemes that can be used to achieve specific data rate.
- A new puncturing scheme (CPS) indicator field of the RLC/MAC header is used to communicate the chosen combination of MCS and puncturing to the remote end.

Table 5.3 EGPRS modulation and coding schemes.

Coding scheme	Modulation method	PDCH data rate (kbps)	Coding ratio (%)	User data bits per radio block	Encoded header bits	Encoded data bits
MCS-1	GMSK	8.8	53	176	68 DL / 80 UL	372
MCS-2	GMSK	11.2	66	224	68 DL / 80 UL	372
MCS-3	GMSK	14.8	85	296	68 DL / 80 UL	372
MCS-4	GMSK	17.6	100	352	68 DL / 80 UL	372
MCS-5	8PSK	22.4	37	448	100 DL / 136 UL	1248
MCS-6	8PSK	29.6	49	592	100 DL / 136 UL	1248
MCS-7	8PSK	44.8	76	896	124 DL / 160 UL	1224
MCS-8	8PSK	54.4	92	1088	124 DL / 160 UL	1224
MCS-9	8PSK	59.2	100	1184	124 DL / 160 UL	1224

- USF code is separately precoded to make sure that the uplink blocks are received by the correct mobile station. MCS schemes 1–4 encode USF with 12 bits while schemes 5–9 are with as many as 36 bits.

Properties of the EGPRS coding schemes [28] are summarized in Table 5.3.

MCS 1–4 use GMSK modulation, with which a burst carries 114 information bits and a radio block of four bursts totals 456 bits. The 8PSK modulation used in schemes 5–9 is able to represent 3 bits in one coding symbol; thus, the number of information bits per radio block is tripled to 1348 downlink and 1384 uplink. In the preceding table, the max data rates are given per timeslot or PDCH. Higher bitrates can be reached by granting the mobile station with multiple slots or PDCHs to be used in parallel. For MCS coding schemes 7–9, the RLC block size has been reduced to two bursts instead of four. This reduces the amount of data to be retransmitted if an RLC block is corrupted by sending one of its bursts over a badly disturbed RF channel.

A base station is able to serve multiple mobile stations with different GPRS and EGPRS coding schemes in different PDCH timeslots. The coding scheme applied to a PDCH can be selected individually for every radio block, based on the bit error rates (BER) measured for the earlier blocks. The base station itself measures BER of uplink traffic and relies on the reports received from mobile station about the downlink traffic. The new EGPRS-specific MCS schemes are used only on dedicated packet channels. The common and shared channels that are used for both GSM and GPRS still rely on the GSM coding schemes. Otherwise, mobile stations without EGPRS support could not use the cell.

5.3.3 EGPRS2

The final evolution step of EGPRS is EGPRS2, which brought the following enhancements to the spec:

- New types of modulation: Quadrature phase shift keying (QPSK), 16-QAM, and 32-QAM, in addition to 8PSK.
- New coding schemes, which are different for uplink and downlink.
- **Turbo coding** instead of the traditional GPRS convolutional coding. An overview to turbo coding is given in *Online Appendix A.6.4*.
- A new reduced transmission time interval (RTTI) in addition to the basic transmission timer interval (BTTI). BTTI means the traditional GPRS case where a radio block consists of a single PDCH timeslot of four consecutive TDMA frames. With RTTI, a radio block is transmitted with a pair of PDCH timeslots within two consecutive TDMA frames. The RTTI decreases the GPRS transmission latency from 20 to 10 ms.

Table 5.4 EGPRS2 coding schemes.

Coding schemes	Direction	Modulation method	PDCH data rate	Coding ratios (%)
DAS-5 . . . DAS-7	Downlink	8-PSK	22.4–32.8	37–54
DAS-8 . . . DAS-9	Downlink	16-QAM	44.8–54.4	56–68
DAS-10 . . . DAS-12	Downlink	32-QAM	65.5–98.4	64–96
DBS-5 . . . DBS-6	Downlink	8-PSK	22.4–29.6	49–63
DBS-7 . . . DBS-9	Downlink	16-QAM	44.8–67.2	47–71
DBS-10 . . . DBS-12	Downlink	32-QAM	88.8–118.4	72–98
UAS-7 . . . UAS-11	Uplink	16-QAM	44.8–76.8	55–95
UBS-5 . . . UBS-6	Uplink	8-PSK	22.4–29.6	47–62
UBS-7 . . . UBS-9	Uplink	16-QAM	44.8–67.2	46–70
UBS-10 . . . UBS-12	Uplink	32-QAM	88.8–118.4	71–96

Table 5.4 gives an overall summary of EGPRS2 coding schemes.

EGPRS2 mobile stations are divided into two categories: EGPRS2-A and EGPRS2-B. Category A devices support only DAS and UAS schemes, while category B devices support all the EGPRS2 coding schemes.

5.4 Questions

1 Please list the services provided by GSM for its users.

2 Which are the types of elements and their roles in the GSM BSS subsystem?

3 What is IMSI?

4 What are the main advantages of GSM frequency hopping?

5 What kind of goals does a good voice codec design meet?

6 How does a GSM phone select the used GSM network?

7 In which ways is GPRS better for packet switched data compared to GSM?

8 What are the three types of GPRS network elements added on top of the GSM architecture?

9 What is a temporary block flow and how is it created?

10 What does "modulation and coding scheme" mean?

11 Why does GPRS use tunneling?

12 What does GERAN mean?

References

1 Mouly, M. and Pautet, M.-B. (1992). *The GSM System for Mobile Communications*. Palaiseau: Cell & Sys.

2 Sauter, M. (2021). *From GSM to LTE-Advanced Pro and 5G : an introduction to mobile networks and mobile broadband*. West Sussex: Wiley.

3 ITU-T Recommendation T.30 Procedures for document facsimile transmission in the general switched telephone network.

4 3GPP TS 23.081 Line Identification Supplementary Services; Stage 2.

5 3GPP TS 23.091 Explicit Call Transfer (ECT) Supplementary Service; Stage 2.

6 3GPP TS 23.082 Call Forwarding (CF) Supplementary Services; Stage 2.

7 3GPP TS 23.083 Call Waiting (CW) and Call Hold (HOLD) Supplementary Services; Stage 2.

8 3GPP TS 23.088 Call Barring (CB) Supplementary Services; Stage 2.

9 3GPP TS 23.086 Advice of Charge (AoC) Supplementary Services; Stage 2.

10 3GPP TS 23.084 Multi Party (MPTY) Supplementary Services; Stage 2.

11 3GPP TS 23.085 Closed User Group (CUG) Supplementary Services; Stage 2.

12 3GPP TS 23.089 Unstructured Supplementary Service Data (USSD).

13 3GPP TS 02.03 Teleservices Supported by a GSM Public Land Mobile Network (PLMN).

14 3GPP TS 03.20 Security-Related Network Functions.

15 3GPP TS 23.002 Network Architecture.

16 3GPP TS 42.017 Subscriber Identity Module (SIM); Functional Characteristics.

17 3GPP TS 48.052 Base Station Controller - Base Transceiver Station (BSC - BTS) Interface; Interface Principles.

18 3GPP TS 48.002 Base Station System - Mobile-services Switching Centre (BSS - MSC) interface; Interface principles.

19 Anttalainen, T. and Jääskeläinen, V. (2015). *Introduction to Communications Networks*. Norwood: Artech House.

20 3GPP TS 23.060 General Packet Radio Service (GPRS); Service description; Stage 2.

21 3GPP TS 23.003 Numbering, Addressing And Identification.

22 3GPP TS 29.010 Information Element Mapping Between Mobile Station - Base Station System (MS - BSS) and Base Station System - Mobile-Services Switching Centre (BSS - MSC); Signalling Procedures and the Mobile Application Part (MAP).

23 3GPP TS 44.001 Mobile Station - Base Station System (MS - BSS) interface; General Aspects And Principles.

24 3GPP TS 48.006 Signalling Transport Mechanism Specification for the Base Station System - Mobile Services Switching Centre (BSS - MSC) interface.

25 3GPP TS 48.058 Base Station Controller - Base Transceiver Station (BSC - BTS) Interface; Layer 3 specification.

26 3GPP TS 43.051 GSM/EDGE Overall description; Stage 2.

27 3GPP TS 48.008 Mobile Switching Centre - Base Station System (MSC-BSS) Interface; Layer 3 specification.

28 3GPP TS 45.001 GSM/EDGE Physical Layer on the Radio Path; General Description.

29 3GPP TS 45.002 GSM/EDGE Multiplexing and Multiple Access on the Radio Path.

30 3GPP TS 45.003 GSM/EDGE Channel Coding.

31 3GPP TS 45.004 GSM/EDGE Modulation.

32 3GPP TS 45.005 GSM/EDGE Radio Transmission and Reception.

33 3GPP TS 45.008 GSM/EDGE Radio Subsystem Link Control.

34 3GPP TS 45.009 GSM/EDGE Link Adaptation.

35 3GPP TS 05.02 Multiplexing and Multiple Access on the Radio Path.

36 3GPP TS 44.003 Mobile Station - Base Station System (MS - BSS) Interface Channel Structures and Access Capabilities.

37 3GPP TS 44.004 GSM/EDGE Layer 1; General Requirements.

38 3GPP TS 43.013 Discontinuous Reception (DRX) in the GSM system.

39 3GPP TS 44.005 GSM/EDGE Data Link (DL) Layer; General aspects.

40 3GPP TS 44.006 Mobile Station - Base Station System (MS - BSS) interface; Data Link (DL) layer specification.

41 3GPP TS 24.008 Mobile Radio Interface Layer 3 Specification; Core Network Protocols; Stage 3.

42 3GPP TS 44.018 Mobile Radio Interface Layer 3 Specification; GSM/EDGE Radio Resource Control (RRC) protocol.

43 3GPP TS 48.004 Base Station System - Mobile-services Switching Centre (BSS - MSC) interface; Layer 1 specification.

44 3GPP TS 48.054 Base Station Controller - Base Transceiver Station (BSC - BTS) interface; Layer 1 Structure of Physical Circuits.

45 3GPP TS 48.056 Base Station Controller - Base Transceiver Station (BSC - BTS) interface; Layer 2 Specification.

46 3GPP TS 29.002 Mobile Application Part (MAP) Specification.

47 3GPP TS 23.096 Name Identification Supplementary Services; Stage 2.

48 3GPP TS 24.010 Mobile Radio Interface Layer 3; Supplementary Services Specification; General Aspects.

49 3GPP TS 06.10 Full Rate Speech Transcoding.

50 3GPP TS 46.010 Full Rate speech; Transcoding.

51 3GPP TS 06.20 Half Rate Speech Transcoding.

52 3GPP TS 46.020 Half Rate Speech; Half Rate Speech Transcoding.

53 3GPP TS 46.001 Full Rate Speech; Processing Functions.

54 3GPP TS 46.002 Half Rate Speech; Half Rate Speech Processing Functions.

55 3GPP TS 23.040 Technical Realization of the Short Message Service (SMS).

56 3GPP TS 24.011 Point-to-Point (PP) Short Message Service (SMS) Support on Mobile Radio Interface.

57 3GPP TS 22.034 High Speed Circuit Switched Data (HSCSD); Stage 1.

58 3GPP TS 23.034 High Speed Circuit Switched Data (HSCSD); Stage 2.

59 3GPP TS 24.022 Radio Link Protocol (RLP) for Circuit Switched Bearer and Teleservices.

60 ITU-T Recommendation V.110 Support by an ISDN of data terminal equipments with V-series type interfaces.

61 3GPP TS 23.008 Organization of Subscriber Data.

62 3GPP TS 23.009 Handover Procedures.

63 Bates, R. (2002). *GPRS General Packet Radio Service*. New York: McGraw-Hill.

64 3GPP TS 22.060 General Packet Radio Service (GPRS); Service Description; Stage 1.

65 3GPP TS 03.02 Network Architecture.

66 3GPP TS 43.064 General Packet Radio Service (GPRS); Overall Description of the GPRS Radio Interface; Stage 2.

67 3GPP TS 44.060 General Packet Radio Service (GPRS); Mobile Station (MS) - Base Station System (BSS) Interface; Radio Link Control / Medium Access Control (RLC/MAC) protocol.

68 3GPP TS 44.160 General Packet Radio Service (GPRS); Mobile Station (MS) - Base Station System (BSS) Interface; Radio Link Control / Medium Access Control (RLC/MAC) Protocol Iu Mode.

69 3GPP TS 44.064 Mobile Station - Serving GPRS Support Node (MS-SGSN); Logical Link Control (LLC) Layer Specification.

70 3GPP TS 44.065 Mobile Station (MS) - Serving GPRS Support Node (SGSN); Subnetwork Dependent Convergence Protocol (SNDCP).

71 3GPP TS 48.018 General Packet Radio Service (GPRS); Base Station System (BSS) - Serving GPRS Support Node (SGSN); BSS GPRS Protocol (BSSGP).

72 3GPP TS 29.060 General Packet Radio Service (GPRS); GPRS Tunnelling Protocol (GTP) across the Gn and Gp Interface.

6

Third Generation

The third-generation networks were designed to support both voice centric and data centric use cases. Voice services were provided over the traditional circuit switched technology, and support for packet switched data was also built in to the systems. Third-generation networks used code division multiplexing in their air interfaces. Code division provided the capabilities needed to increase bitrates and adjust them for data connections that required smaller, larger, or even variable bitrates.

6.1 Universal Mobile Telecommunications System (UMTS)

6.1.1 Standardization of Third-Generation Cellular Systems

When the third-generation (3G) mobile phone system standardization was carried out, it had become evident that mobile telephony and messaging was not enough. The earlier global system for mobile (GSM) system had to be complemented with the packet switched general packet radio service (GPRS) extension. Use cases such as mobile email, network-connected calendaring, and Web browsing required mobile phones to support seamless data connections. The 3G systems should support telephony and data use cases in an equal way. Consequently, the radio interface design was no longer optimized for voice, but it had to be generic enough to support both voice and data, with variable bitrates higher than those provided by earlier GSM EDGE radio access network (GERAN) networks.

ITU-R defined the following goals for IMT-2000 cellular systems in recommendation M.687 [1]:

- Services with their Quality of Service (QoS) comparable to that of fixed networks shall be supported for mobile users, irrespective to their location.
- The standard was to be globally deployable and support international roaming.
- The system had to enable seamless integrated usage of telephony and data services.
- Services provided by the system should be compatible with fixed networks, such as ISDN or PSTN.
- The system had to support data rates up to 2 Mbps at phase 1 and higher rates later on.
- The services provided by the system had to be decoupled from the radio technology used. The choice of the radio technology should not constrain the types of new services and applications, which could be created on top of the 3G radio.
- The system was expected to provide high-spectral efficiency to maximize the system capacity provided with the available frequency band.
- The system should support mobile terminals of various sizes such as handheld or vehicle mountable ones.
- The system should support an improved level of security in comparison to GSM.

Converged Communications: Evolution from Telephony to 5G Mobile Internet, First Edition. Erkki Koivusalo.
© 2023 The Institute of Electrical and Electronics Engineers, Inc. Published 2023 by John Wiley & Sons, Inc.
Companion website: www.wiley.com/go/koivusalo/convergedcommunications

Additionally, ITU-R recommendations M.816 [2], M.1034 [3], and M.1079 [4] stated the following goals for IMT-2000:

- The new system was expected to support simultaneous transmission of multiple different types of media to support multimedia sessions consisting of sound, pictures, video, and other data.
- The system had to support asymmetric and variable data rates.
- The system had to support smaller latencies and error ratios compared to the GPRS system. In GPRS, the round-trip time could be as long as half a second, which is problematic for TCP connections using a slow start. As described in Chapter 3, Section 3.3.5, a TCP connection starts with low bitrate and increases it against the acknowledgments received. Longer latencies mean a longer time for receiving the acknowledgments, which causes the data rates of new TCP connections to increase slowly. When Web browsers open multiple TCP connections for fetching small amounts of information per connection, loading Web pages over GPRS is quite slow [5].

Considering the practical deployments, 3G standardization had the following goals:

- To be commercially viable, 3G systems had to provide clear value added and higher data rates than what GSM and GPRS data services did.
- The 3G system had to support handovers to/from the 2G systems.
- The system architecture and its air interface had to be entirely well defined. The most important system interfaces had to be open to support interoperability between different vendors.

In the global scale, the following 3G cellular systems were specified:

- WCDMA UMTS relying on the wideband CDMA technology specified in 3GPP for global markets.
- CDMA2000 and EV-DO relying on another CDMA technology specified in 3GPP2 for primarily American markets. These systems were also deployed on other continents, but not as widely as WCDMA UMTS systems.
- Mobile WiMAX relying on OFDMA multiplexing and common IEEE 802 architecture originally defined for local area networks. WiMAX was specified in IEEE originally as a fixed wireless access system. In the specification version 802.16e [6], WiMAX took OFDMA into use and become a mobile cellular data system with handover support between WiMAX cells. WiMAX was a closed system as no interoperability with other cellular networks was provided. WiMAX deployments were established in many countries around of the world, but most of the WiMAX networks were rather small. Eventually, WiMAX system design was aligned with the 4G LTE design. For more information about WiMAX, please refer to *Online Appendix G.2*.

Third-Generation Partnership Project (3GPP) was a new global standardization forum established to specify the next generation cellular system as an evolution from the second-generation GSM and GPRS systems. In addition to its core task, 3GPP took over the maintenance of the GSM and GPRS standards. Later on, 3GPP continued its work with 4G and 5G cellular system standardization.

3GPP2 was a North American standardization forum working on parallel to 3GPP to specify 3G cellular system based on the CDMA technology. CDMA (IS-95) was the mainstream 2G cellular technology deployed in North America while **Evolution-Data Optimized (EV-DO)** was the 3G technology evolved from it. EV-DO revisions 0, A, and B were the evolutionary steps of the standard. Revision 0 supported downlink bitrates up to 2.4 Mbps but uplink only 153 kbps. Revision A upgraded the downlink up to 3.1 Mbps and uplink to 1.8 Mbps. Revision B was a multicarrier solution capable of achieving 14.7 Mbps downlink and 4.9 Mbps uplink bitrates [7].

Universal Mobile Telecommunications System (UMTS) was the name chosen by 3GPP for the new 3G system to replace the legacy GSM. As its name indicates, the target was to create a global standard, instead of multiple regional incompatible 3G cellular system standards. UMTS was a step in this direction, but not the final breakthrough as American standardization forums managed to create their own 3G system. While both global

UMTS and American EV-DO standards use CDMA technology and 3GPP2 decided to reuse relevant 3GPP standards, the systems stayed technically incompatible with each other as the air interface specifications were fundamentally different.

The specification work for 3G mobile networks was officially launched already in 1992 at the WARC conference held by the International Telecommunications Union (ITU). In WARC radio bands, around 2 GHz frequencies were defined to be used for 3G networks [5]. The original aim of ITU was to create a single cellular system standard for global use and avoid regional fragmentation of technologies. The European and Asian standardization organizations were able to agree on a shared 3G technology to be developed, but the North American organizations took their own way toward 3G as an evolution of their legacy 2G CDMA systems.

The standardization work for UMTS was prepared in the RACE and ACTS/FRAMES research projects financed by EU during the 1990s. These projects studied various types of cellular technology radio interface candidates against the goals set for 3G cellular systems. The evaluated technologies were more advanced compared to the relatively simple GSM radio interface. They had become feasible by the recent increase of processing power in mobile computing hardware platforms. Eventually, in 1998 the European Telecommunications Standards Institute (ETSI) selected the **wideband code division multiplexing (WCDMA)** technology for the new 3G system radio interface. This choice was also supported by the biggest Japanese mobile operator NTT DoCoMo. In the same year, ETSI, Japanese standardization forums TTC and ARIB, Korean TTA, and T1P1 from the United States agreed to officially start 3G UMTS standardization in the newly established 3GPP standardization forum. Later on, other Asian and American standardization forums, major system vendors, and cellular operators joined 3GPP. The enterprises were allowed to join 3GPP under an authorization given by any of the regional standardization organizations.

While the WCDMA radio interface of the emerging UMTS standard was completely new, the same cannot be said about UMTS core network architecture. 3GPP made the decision to reuse as much as possible from the legacy GSM and GPRS core networks for UMTS. The approach was to redefine a new **UMTS terrestrial radio access network (UTRAN)** to replace the GERAN BSS and support WCDMA access for both circuit and packet switched services.

The first UMTS networks, compliant to 3GPP standards, were launched in Japan in 2001 and in Europe in the following years, 2002–2003. In those years, 3G networks were built only for the densest cities. Operators did not proceed with building country-wide WCDMA coverage too quickly, partly due to the high price already paid for the 3G licenses in Europe. Coverage areas of WCDMA networks were expanded gradually, so that some sparsely populated rural areas were still covered only by 2G networks. Building coverage for 3G networks was more difficult than for 2G networks, since the 3G higher frequency radio signals could not reach as long distances as the lower frequency GSM bands. Because of the shorter signal reach, operators were not able to build country-wide 3G coverage simply by adding 3G support to their existing 2G base station sites. Over time,2004–2011, WCDMA became the mainstream cellular technology. The WCDMA data service was enhanced with many steps over those years. At the time of this writing in 2022, many WCDMA networks have been or are being decommissioned since newer data-centric 4G and 5G cellular technologies provide better service and spectral efficiency. It is advantageous for cellular operators to refarm frequencies earlier used for WCDMA networks to those more modern systems, as described in Chapter 7, Section 7.1.4.1. This book has an extensive description of the 3G WCDMA system, since many of its architectural, protocol design, and security solutions are either inherited from or at least had a major impact on 4G LTE/SAE system design, even if the air interface of the 3G WCDMA system is totally different than that of the 4G LTE.

In theory, the original 3GPP Rel-4 compliant WCDMA system supports 2 Mbps data rates for a single data connection. In practice, such maximal rates were not initially achieved anywhere else than in optimal laboratory conditions. In practice, the first 3G networks were limited to 384 kbps maximum and 64–128 kbps sustainable downlink bitrates, which was only comparable to or slightly better than what was available over GPRS at that time but far from astonishing compared to 1 Mbps ADSL. Later on, UMTS data rates were increased by enhancing the 3G WCDMA networks with new modulation and channel coding methods. Latencies were decreased by giving the 3G base stations a wider area of responsibilities compared to the original UMTS design. These enhancements, referred as the **high-speed packet access (HSPA)** technology, were deployed to existing 3G networks a few years after the initial launches of 3G services. After several iterations, HSPA downstream data rates were eventually increased up to 14 Mbps, well above the original 3G target.

3GPP WCDMA UMTS standards have been developed and published as subsequent standards releases. Its major features covered in Chapter 6, Section 6.1, were as follows:

- 3GPP R99 was completed during 1999 for its major parts. Its focus areas were the WCDMA radio interface and UTRAN radio network. R99 introduced also various UICC and USIM features.
- 3GPP Rel-4 was completed in 2001. Its main focus area was redefinition of the circuit switched core network.
- 3GPP Rel-5 introduced various small enhancements for WCDMA radio interface and radio resource management, such as the site selection diversity described in Chapter 6, Section 6.1.4.3.

A 3GPP specification release is bound by its technical scope and contents rather than its publication time. At first, the 3GPP had a goal to provide one standards release per year, and the release naming was based on the target year of publishing the release. This approach turned out to be too difficult. 3GPP did not manage to force the working groups completing specifications with agreed scope in a fixed schedule. From Rel-4 onwards, the release names simply had a running number. While a specification release is declared complete at a certain point and its scope is frozen, the specifications stay under constant maintenance and new versions are released on quarterly basis to fix any errors and mistakes found from the earlier versions. Chapter 6, Section 6.1, describes WCDMA UMTS design as per 3GPP Rel-4/5 specifications and the later HSPA enhancements of Rel-5 to Rel-7 are described in Chapter 6, Section 6.2.

As the earlier GSM standards were divided to different standards series, so to were the 3G UMTS standards. Each of the series has a specific scope. The 3GPP UMTS standardization series are based on and extended from those that ETSI had defined for GSM and GPRS standards. When the 3G standards were initially published, the 3GPP UMTS standardization series were as follows:

- TS 21: General, terms and acronyms
- TS 22: Services and requirements of the UMTS system
- TS 23: Architecture, functions and general description of UMTS network
- TS 24: UMTS radio interface and its protocols
- TS 25: Physical layer of UMTS radio interface
- TS 26: Voice encoding in UMTS networks
- TS 27: Terminal adaptors for data terminals
- TS 28: Interfaces and signaling protocols between the base station subsystem and the mobile switching center
- TS 29: Interoperation of UMTS and fixed networks
- TS 31: UMTS SIM card (USIM)
- TS 32: Operation and maintenance (OAM) as well as charging in the UMTS network
- TS 33: UMTS information security
- TS 34: UMTS equipment and type approvals
- TS 35: Algorithms related to UMTS information security

6.1.2 Frequency Bands Used for WCDMA UMTS

The following frequency areas were reserved globally for 3G systems:

- WCDMA frequency division duplex (FDD):
  - FDD means the way of providing separate dedicated radio bands for downlink and uplink traffic; see *Online Appendix A.9.1* for further details.

- Two 60 MHz bands were allocated for WCDMA FDD:
  - ○ Frequency band 1920–1980 MHz uplink channels from the mobile station to the base station.
  - ○ Frequency band 2110–2170 MHz downlink channels from the base station to the mobile station.
- WCDMA time division duplex (TDD)
  - TDD means the way to use one single radio band for both downlink and uplink traffic and separate the directions with time division multiplexing.
  - Two bands were allocated for WCDMA TDD:
    - ○ Frequency band 1900–1920 MHz
    - ○ Frequency band 2010–2025 MHz
- In 2000, ITU reserved also the following bands for IMT-2000 standard:
  - Frequency band 1710–1885 MHz
  - Frequency band 2500–2690 MHz
  - Frequency band 806–960 MHz

In one region, a frequency band can be used by one network only. To support multiple operators per country, the countries have split these bands to subbands and given licenses for these subbands to different network operators. Many European countries arranged auctions for granting the licenses while others granted the licenses without extra cost for those operators who were able to fulfill predefined acceptance criteria. The high prices paid by operators in some of these auctions eventually slowed down building 3G networks, as the initial investment for acquiring the frequency bands was already very high.

6.1.3 Architecture and Services of UMTS Systems

6.1.3.1 UMTS Services
An UMTS network can be divided into two service domains:

- **Circuit switched domain**, which has been developed further from the GSM network architecture
- **Packet switched domain**, which has been developed further from the GPRS network architecture

UMTS services are essentially a combination of GSM circuit switched and GPRS packet switched services.

6.1.3.2 UMTS System Architecture
The UMTS cellular system on its highest level can be divided into the following subsystems:

- **User equipment (UE)** is the mobile station and a UMTS SIM card within it.
- **UMTS terrestrial radio access network (UTRAN)** is the radio network consisting of WCDMA base stations and radio network controller (RNC) nodes to manage the base stations.
- **Core network (CN)** is the network core consisting of mobile switching centers, packet data routers, and different kinds of databases to store information about the subscribers and their mobile stations. Additionally, the 3G IP Multimedia Subsystem (IMS) core is part of the network core. Very few IMS services were yet taken into use within 3G networks, but with the next generation, 4G technologies were proven a solid platform for IMS. The structure and functions of the IMS core are described in Chapter 9, Section 9.3.

One leading principle of UMTS network design was to make the radio network and core network only loosely coupled and as independent from each other as possible. The aim was to minimize the need to modify core network while improving the radio network design. Loose coupling was meant to support introduction of new radio technologies and getting them deployed on parallel to the earlier ones. All the functions of WCDMA radio resource management are centralized to the UMTS radio access network (UTRAN). One example of this is the cooperation between peer radio network controllers to support cell reselections and handovers without involving mobile switching centers.

The UMTS cellular system consisted of the following physical elements, shown in Figures 6.1 and 6.2:

- **Mobile equipment (ME)** was the UMTS mobile station, which was either a mobile phone or UMTS capable accessory for a computer. While GSM mobile phones had rather limited text-based user interfaces, UMTS smartphones already had larger screens with graphical user interfaces for advanced functions. Two different microprocessors could be used to support the processing needs of such smartphones. A baseband processor was used to support the radio modem and an application processor to run the user interface software and an operating system to support various end-user applications.

- **Universal integrated circuit card (UICC)** was a smart card to store UMTS subscriber information. The UICC card was equipped to the slot reserved for the card within the UMTS mobile equipment. The UICC card was an improved version of the SIM card specified for GSM. The UICC card contains a set of separate applications, the most important of which is the **universal subscriber identity module (USIM)** application. The USIM provides the secret keys and algorithms used for authentication. It also has subscriber identity codes, such as IMSI, and identifiers of the home network. The combination of the ME and its UICC is often referred to as the **user equipment (UE).**

- **Base station (BS)** was also known as **NodeB**. UTRAN base stations consisted of radio transmitters, receivers, and the antennas serving UTRAN cells. Typically, one NodeB was able to control two to three cells as sectors created with directional antennas. Additionally, the NodeB had a computer system, known as the baseband unit, for controlling its functions and a data link to the radio network controller (RNC) node. Most typically, the baseband unit was located at the equipment room at ground level while the radio units were located high on the mast close to antennas. These units were connected with coaxial cables or optical fibers. The base station had the following tasks:
  - Modulation and multiplexing of WCDMA radio signals
  - Management of physical channels and frame structures used on the radio path
  - Synchronization, channel coding, interleaving, and rate adaptation of transmitted data
  - Management of the transmission power of the UE based on the SIR (signal to interference ratio) target value received from the RNC

- **RNC** was the UTRAN radio network controller, which managed a number of UTRAN base stations and connected them toward the mobile switching center or SGSN node of the packet switched core network. While a GSM BSC typically controls tens of base stations, one RNC was used to control even hundreds of NodeBs. The RNC had the following tasks:
  - Manage connections between the core network, base stations, and user equipment behind the radio interface
  - Manage WCDMA radio resources to maximize capacity, stability, and quality of the radio connections for all user equipment being served

  The RNC may have one of the following roles toward a specific user equipment:
  - Serving RNC (SRNC) was responsible of the connection between the radio network and core network to communicate with a single user equipment (UE). SRNC decided the set of cells which worked together to communicate with the UE. SRNC was the endpoint of various WCDMA protocols such as RRC used for UE radio resource management and RANAP used between the RNC and core network.
  - Drifting RNC (DRNC) managed connections toward the UE via those base stations which were not directly connected to the SRNC. The SRNC may ask a DRNC to open a parallel radio connection toward the UE as part of a WCDMA soft handover (see Chapter 6, Section 6.1.4.3).
  - Controlling RNC (CNRC) was responsible for managing the set of radio resources used by a single base station. The CNRC contributed to the radio resource management from the perspective of a cell rather than of a UE. The CRNC took care of the cell load and congestion control. With **admission control** algorithms, the CRNC decided if a new radio connection could be opened for a UE. The CRNC allocated WCDMA spreading codes for the radio connections.

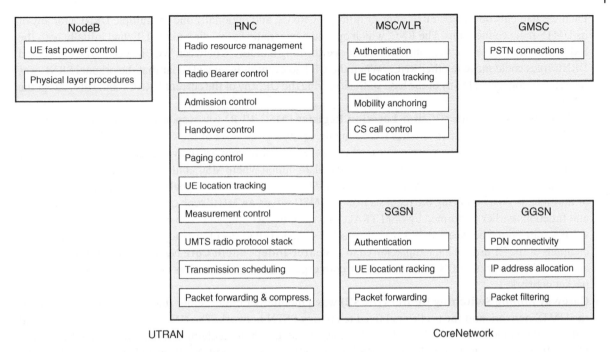

Figure 6.1 Functional split between UTRAN and core network elements.

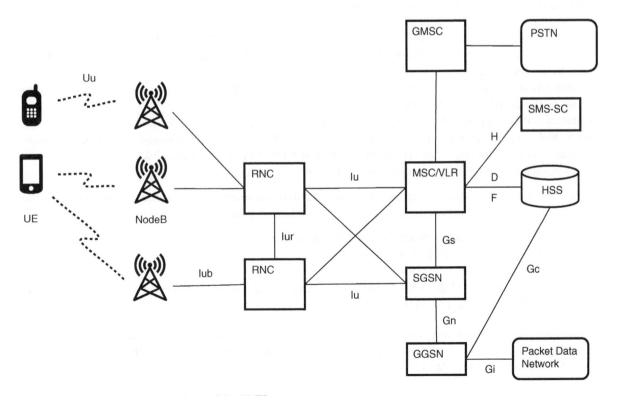

Figure 6.2 Architecture and interfaces of the UMTS system.

Each UE had only one single SRNC, one or multiple CRNCs, and optionally one or multiple DRNCs for any of its UMTS radio connections. The RNC, which opened a new radio connection to the UE, had initially both the SRNC and CRNC roles for the connection. After performing handovers between WCDMA cells, the CRNC and DRNC roles could move to other RNCs. When the UE had eventually moved so far that the original SRNC was not directly connected to any of the base stations serving the UE, any of the current DRNCs could also adopt the SRNC role.

- **Mobile Switching Center/Visitor Location Register (MSC/VLR)** was responsible for connecting the circuit switched calls. Its responsibilities were roughly equivalent to those of the GSM MSC and VLR elements. The UMTS MSC had all the voice codecs, which were needed to connect the voice call from UTRAN to the traditional fixed PSTN network. 3GPP Rel-4 specified an option where MSC could be divided into two entities: an MSC server and a media gateway (MGW). The MSC server took care of signaling, and it would control one or multiple MGWs. Following the requests from the MSC server, an MGW took care of CS bearer management and functions earlier performed by GSM TRAU, like voice transcoding and echo cancellation [8]. The encoded voice is transported between MGWs over the packet switched IP protocol instead of the traditional 64 kbps E0 channels of the E1 links. This setup, referred to as **Bearer-Independent Core Network (BICN),** was applied also to GSM networks. BICN is an example of **convergence,** where all services are delivered over a single IP packet network.
- **Gateway Mobile Switching Center (GMSC)** was a mobile switching center that connected calls between the UMTS network and the traditional fixed telephony PSTN network. The GMSC had a special task to find out the MSC/VLR under which the UMTS UE was located, when receiving a mobile terminated call from PSTN. It was possible to divide the GMSC into a GMSC server and MGW switches controlled by the GMSC server.
- **Serving GPRS Support Node (SGSN)** had the following tasks in UMTS:
  - Track the location and mode of the UMTS UE for the packet switched service.
  - Forward packets between the UEs in the routing areas served by the SGSN and the UMTS packet core network. Unlike 2G SGSN, the UMTS SGSN did not terminate the GTP tunnel from GGSN but passed it up to the SRNC serving the UE.
  - Authenticate the subscriber for packet switched services and check the services the subscriber is entitled to use during the UMTS GPRS attach.
- **Gateway GPRS Support Node (GGSN)** was as described in Chapter 5, Section 5.2.2.1.
- **HLR, AuC,** and **EIR** databases had the same tasks as in a GSM system. In UMTS system, these databases were typically combined into one single **Home Subscriber Server (HSS)** entity, responsible for subscriber authentication, subscription data management, and UE location management.

Like GSM and GPRS, the UMTS specifications defined standard interfaces, or reference points, between the network entities. The most important new UMTS interfaces, also shown in Figure 6.2, were:

- Uu: WCDMA radio interface between UE and NodeB base station
- Iub: The interface between UTRAN NodeB and RNC
- Iur: The interface between two UTRAN RNCs
- Iu: The interface between UTRAN RNC and the core network. The Iu interface has two different variants: Iu(cs) toward the MSC and Iu(ps) toward the SGSN.

Within the UMTS network core, the names of reference points were mostly the same as in GSM and GPRS. In some of those interfaces, the UMTS core network had its own protocols and for the rest of the interfaces the GSM protocols were enhanced for UMTS.

In the UMTS system, the subscriber and user equipment identifiers were the ones already defined for GSM and GPRS systems: IMSI, TMSI, P-TMSI, MSISDN, MSRN, and IMEI. Additionally, there was one new major identifier defined for the UMTS radio network control purposes:

- **Radio Network Temporary Identity (RNTI)** was an identifier with which RNC identified a single UE that owned a certain UMTS RRC connection. The RNC reserved one of the following identifier types for the UE, depending on the role of the RNC for the RRC connection management:
  - UTRAN RNTI (U-RNTI) used to identify the UE for handovers or paging. U-RNTI could also be used in downlink MAC frames sent over common or shared channels to identify the destination UE. The U-RNTI is a combination of SNRC identifier and S-RNTI given for the UE.
  - Serving RNC RNTI (S-RNTI) was allocated to the UE by the SRNC when opening an RRC connection. The UE used S-RNTI for messages it sent to the SRNC.
  - Drift RNC RNTI (D-RNTI) was allocated to the UE by the DRNC when the DNRC engaged itself to an existing RRC connection. The UE used D-RNTI for messages it sent to the DRNC.
  - Cell RNC RNTI (C-RNTI) was allocated to the UE by the CRNC that controlled the single NodeB, which carried RRC connection to the UE. The UE used C-RNTI for messages it sent to the CRNC.

For location management, UMTS system used the same identifiers as defined for GSM and GPRS systems: CGI, LAI, and RAI. In UTRAN the routing area was split to UTRAN registration areas (URA) so that the registration area was under control of a single RNC.

A detailed description of UMTS identifiers and network architecture can be found from 3GPP TS 25.401 [9].

6.1.3.3 UMTS Bearer Model

UMTS system specifications introduced the concept of **bearer** for UMTS connectivity and QoS model. Bearers are used to transport either signaling or user data, whether circuit or packet switched. Every bearer comes with QoS requirements, specific to the bearer type. UMTS bearers of different types compose a hierarchical structure shown in Figure 6.3. The link level bearers on the bottom of the diagram are used over a single link of UMTS Iub and Iu reference interfaces. Bearers on the top span over multiple interfaces of the UMTS reference model. Bearers on the higher levels of the hierarchy use the underlying bearers to transport data across the network interfaces. In the diagram, the interfaces are shown as gaps between the gray boxes, which depict different parts of the UMTS network. On each

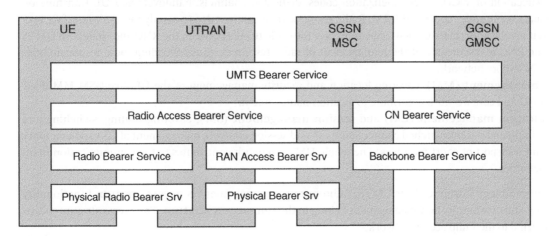

Figure 6.3 Bearer architecture of UMTS system. *Source:* Adapted from 3GPP TS 23.107 [10].

UMTS system interface the bearers are implemented with a protocol stack specific to the interface. The purpose of the abstract and hierarchical bearer model is to provide well-defined terminology for referring to the connections without being forced to instead refer to implementation-specific protocol details. A well-structured model also supports protocol design with which it is possible to do local changes to the implementation of a reference interface without impacting the others. The UMTS QoS concept and bearer architecture are described in 3GPP TS 23.107 [10].

The basement of the bearer hierarchy is built with the following link level bearers used between different UMTS network subsystems:

- The **radio bearer service** supports connections between the UE and RNC. Radio bearers consist of WCDMA air interface radio channels between the UE and NodeB and protocol connections between the NodeB and RNC. A radio bearer may be used either as a signaling bearer or a user data bearer. The implementation consists of protocol stacks underlying to the RLC protocol (see Section 6.1.3.5).
- The **RAN access bearer service,** also known as Iu bearer service, supports connections between the RNC and UMTS core network inner edge. The core network edge elements are MSC/VLR for the circuit switched domain and SGSN for the packet switched domain. The implementation consists of protocol stacks specified for the Iu interface, which are also described in Section 6.1.3.5.
- The **backbone bearer service** is implemented with the trunk network connections within the UMTS network core. In the circuit switched domain, the backbone bearers connect MSC/VLR and GMSC, and in packet switched domain, they connect the SGSN and GGSN.

From the UE point of view, the three different layers of bearers are as follows:

- **Radio bearers** connect the UE to the RNC.
- **Radio access bearers (RAB)** connect the UE to the UMTS core network inner edge.
- **UMTS bearers** connect the UE to the GGSN and GMSC gateways at the UMTS core network outer edge.

6.1.3.4   UMTS Functions and Procedures

The UMTS system specifications define the following areas of interaction or functions between the UE and UMTS network:

- **Radio resources management (RRM)**, covering management of radio bearers between the RNC, NodeB, and UE. Allocation of WCDMA channelization codes, dedicated channels, handovers and UE transmission power management under control of RRM. Radio resource management does not only mean air interface procedures as it involves also management of bearers over the cable trunks between the RNC and NodeB. In UMTS network the RNC is responsible of the management of radio resources, thus offloading these responsibilities from the UMTS core network.
- **Mobility management (MM)**, covering location and security management of the UE, like GSM MM does. Both the RNC and core network contribute to location management.
- **Communication management (CM)** and **session management (SM)**, covering routing, switching and release of circuit and packet switched connections (call and session control), management of CS supplementary services, and transport of short messages. The whole UMTS network contributes to the communication management activities.

With the terminology introduced for UMTS, radio resource management is **access stratum** functionality while mobility and communication management belong to the **non-access stratum**. The access stratum means functions internal to the radio access network.

The life cycle of a UMTS radio connection can be described as follows:

1) If the creation of the connection is initiated from the network side, the network at first pages the UE. After receiving the paging message, the UE starts random access and proceeds as when the connection was initiated by the UE.

2) The UE and radio access network proceed to set up a radio bearer via which further signaling messages can be exchanged. This phase is called the RRC connection setup. The established radio bearer (RRC connection) can be used for managing CS or PS radio access bearers.

3) After the RRC connection has been opened, the UE sends its first message over it to the network core. In this message, the UE tells the purpose for which radio access bearers would be needed. The network consequently allocates sufficient radio resources for the UE, over either a shared or a dedicated channel.

4) Before the UE is allowed to continue, network authenticates the subscriber and starts encryption of the traffic.

5) After completing the authentication, the network opens further radio access bearers for the UE, based on the request from the UE.

6) The UE can use the radio access bearers for one of the following MM, CM, and SM procedures:
   - Sending location update message to the network
   - Sending measurement data about the radio connection or perform a handover
   - Opening a circuit switched connection for CS call establishment
   - Activating a PDP context to be used for packet switched data transport

7) After the necessary MM or CM interactions and the user data transfer have been completed, the radio access bearers are released. The UE may also deactivate the PDP context, if it aims to stop communicating with a packet switched network.

8) Eventually, the RRC connection is released as it is not needed for either signaling or user data. The radio resources now become available for other connections.

Examples of UMTS signaling procedures can be found from TS 25.931 [11].

6.1.3.5 UMTS Protocol Stack Architecture

UMTS network protocols can be categorized by the following dimensions:

- The type of information carried by the protocol: user data or signaling
- The type of the related connection: circuit or packet switched.
- Role of the protocols in the protocol stack layer structure: generic data transfer protocols, radio network control protocols, and system level protocols
- Network subsystem for the protocol: radio interface, radio network or core network
- The specific network reference point over which the protocol is run

The first of these dimensions is the most fundamental one: Control plane protocols are used for signaling, which means messages used to control and manage UMTS functions. User plane protocols are used to transport user data or voice. Two different protocol architectures have been defined for UMTS control and user planes. The control plane protocol architecture is depicted in Figure 6.4 and the user plane architecture in Figure 6.5.

The 3GPP specifications divide the protocols to subsets based on the network subsystems within which the protocols are used:

- Specification series TS 24 covers detailed specifications of the UMTS radio interface protocols. The UMTS radio interface protocol architecture is described in 3GPP TS 25.301 [12], which belongs to another specification series.
- Specification series TS 25 and 29 cover detailed technical specifications of protocols between the radio network and core network.

These protocols closely related to the UMTS radio service have been defined specifically for UMTS purposes, while protocols used within the core network have evolved from GSM and GPRS core network protocols. The link and network layer protocols used within radio and core networks – such as SDH, ATM, and IP – have been chosen rather than developed for the UMTS protocol architecture.

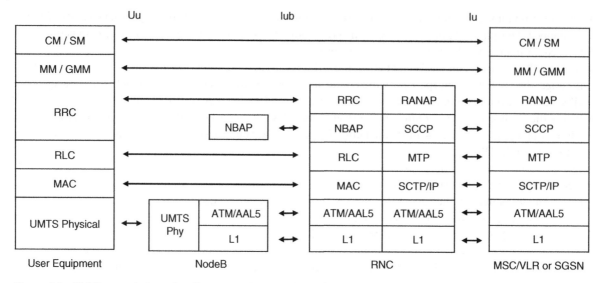

Figure 6.4 UMTS control plane signaling protocols with ATM option.

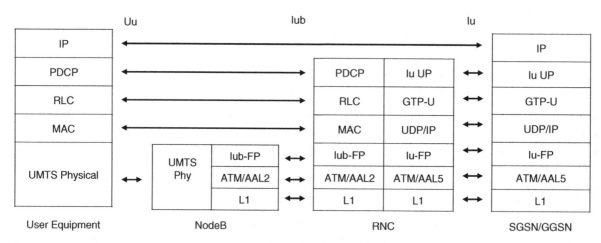

Figure 6.5 UMTS user plane protocols in packet switched domain with ATM option.

Protocols in the different layers of the protocol stacks have the following roles:

- Transport protocols used to transfer both user data and signaling messages over each of the UMTS network interfaces between the endpoints of the physical or virtual link.
- Radio network control protocols, the endpoints of which are the RNC and either the UE or NodeB. These signaling protocols are used for radio access bearer management.
- System level signaling protocols with which the UMTS network core communicates with either the UE or RNC.

6.1.3.6   UMTS Radio Channel Architecture

GSM logical channels were introduced in Chapter 5, Section 5.1.4.2. UMTS WCDMA system specifications reuse the concept of **logical channel** to describe various types of information flows between the UE and network. The structure and naming of UMTS logical channels follow closely to those of GSM. But the way of mapping the logical channels to the **physical channels** of the radio interface is very different in these two systems. A physical channel of GSM consists of a sequence of bursts, each of which has well-defined coordinates in the two-dimensional space of GSM carrier frequencies and GSM TDMA frame timeslots. The GSM system has a single physical channel reserved for each separate logical flow. The physical channels of the WCDMA system are defined with **channelization (spreading) codes** used in the code division multiplexing space. Allocation of the WCDMA spreading codes can be done in various ways. Due to the flexibility of code division multiplexing, WCDMA supports many different types of physical channels that can be used in diverse ways as fit to the context. A single UMTS logical channel can be transported to the UE over one or another type of WCDMA physical channel, depending on the combination of logical flows that the UE needs.

Figure 6.6 is a representation of the WCDMA radio protocol stack. It shows how the logical and physical channels are related to the layers of the stack. Within the RAN side, the MAC protocol of the RNC maps different RLC protocol flows to the UMTS logical channels and adapts those to **transport channels** toward the NodeB. At NodeB, the data is mapped from the transport channels to the WCDMA physical channels. The RNC is not able to access WCDMA physical channels directly, since those are terminated at the NodeB base station. That is why the data is forwarded between RNC and NodeB over the transport channels of Iub interface. Transport channels are implemented with the FP protocol described in Section 6.1.7.3. For the protocol design point of view, a transport channel is an abstraction layer that hides any possible changes to the WCDMA physical channel structure from the RNC. In case WCDMA physical channels would be restructured, the changes would impact only NodeBs but not RNCs as long as the transport channel design would stay as is.

The specific types of WCDMA logical, transport and physical channels and mappings between them are introduced in Sections 6.1.4.6 and 6.1.4.5. Detailed specification of WCDMA logical and transport channel design can be found from 3GPP TS 25.301 [12].

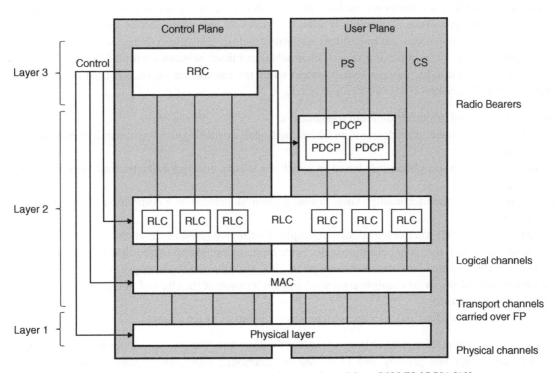

Figure 6.6 WCDMA radio protocol stack and channels. *Source:* Adapted from 3GPP TS 25.301 [12].

6.1.4 WCDMA Radio Interface

This chapter provides an overview of the WCDMA physical layer. The 3GPP provides a more detailed summary of WCDMA physical layer functions and other related 3GPP specifications in 3GPP TS 25.201 [13].

6.1.4.1 Modulation and Multiplexing

UMTS WCDMA uses two different modulation methods, one for each direction. QPSK is used for downlink and OQPSK for uplink. As described in *Online Appendix A.4.5*, QPSK uses four values of phase difference to encode any combination of two data bits. When both the bits are different than the previous two bits, the transmitted signal has a phase change of 180°. This causes a very steep change in the amplitude of the signal, which is challenging for the amplifier of the radio transmitter. Such a change is yet feasible for the transmitters of the NodeB, but it was not considered feasible for the user equipment. Because of this, another OQPSK modulation method was selected for WCDMA uplink signals. Like QPSK, OQPSK uses four values of the phase difference, but any change of the phase is limited to 90° only. With QPSK there is just one phase change per the symbol of two bits, but with OQPSK the phase is changed twice per symbol. Against the 180° phase change of QPSK, the OQPSK method produces two changes of 90° to the transmitted signal. As a consequence, OQPSK results are twice as frequent, but there are smaller changes to signal phase than QPSK.

WCDMA uses DSSS spread-spectrum technique to multiplex the UEs to a cell. The DSSS method is described in *Online Appendix A.9.1*. All the WCDMA cells of a network may use the same 5 MHz wide system bandwidth for transmission; thus, WCDMA uses a single-frequency network design. DSSS spreading codes are used to distinguish different transmission streams from each other and what appears to be the spectrum of white noise. With DSSS every information bit to be transmitted is encoded with a code word of N chips. WCDMA transmitter sends 3.84 million chips per second; thus, the **chip rate** of Uu radio interface is 3.84 Mcps.

The WCDMA system uses two different types of multiplexing codes: a scrambling code and a channelization code. Technically, **channelization codes** are used as DSSS spreading codes while **scrambling codes** are used to scramble the signals after spreading. Scrambling is often used to avoid long sequences of single bit values, as created by some channelization codes that have rather uniform structure. UMTS uses these codes for two logical purposes: as identifiers and for multiplexing various channels transmitted between multiple UEs and NodeBs. The scrambling code is used to identify the source system while the channelization code is used to identify a specific channel or the destination UE for a downlink signal.

- Scrambling codes are used to identify different transmitters:
  - Each UE has its own scrambling code based on which the NodeB can distinguish the transmissions from different user equipment.
  - Each NodeB has its own scrambling code based on which the UE can distinguish the transmissions of different WCDMA cells.
- Channelization codes are used for multiplexing, so that a physical channel or a destination UE has its own channelization code:
  - For uplink, the UE uses channelization codes to multiplex its different physical channels.
  - For downlink, the NodeB uses channelization codes to multiplex common and shared channels as well as the dedicated channels allocated for individual destination UEs. Multiplexing of the physical channels is done with a combination of the channelization code used and the location of the channel in the WCDMA downlink TDMA frame structure.
  - In both directions, channelization codes are used to adjust the bitrate of the channel.

WCDMA system has 8191 different scrambling code values for downlink. These codes are divided into 512 code groups, each of which has one primary and 15 secondary scrambling codes. Every NodeB has one of the code groups. The 512 **primary scrambling codes (PSC)** are further divided into 64 PSC groups of eight

PSC codes. The NodeB must always use the primary scrambling code but it may also use a set of secondary scrambling codes from the same group, as allocated to the NodeB based on the total capacity need of the cell. The primary scrambling code is also used as non-unique local cell identifier used by the UEs for WCDMA cell search. In this way, any UE that performs a blind cell search has a limited search space over the 512 globally used primary scrambling codes. For uplink, the UE uses either a short scrambling code or 256 bits or a longer gold code of 38 400 bits. Because both types of uplink scrambling codes have millions of values, there is no need for any sort of code planning, and the RNC can just allocate any unused scrambling code for a UE.

The channelization code word length of WCDMA ranges between 4 and 256 chips, derived as 2^k, where $k = 2..8$. The length of the channelization code is also called its **spreading factor (SF)**, meaning the number of chips produced by encoding one data bit with the code. User data rate of WCDMA channel can be calculated as follows [7]:

Chip rate * bits per QPSK symbol * parallel codes * coding rate / spreading factor

As the spreading factor appears as a divisor in this equation, a large spreading factor of a long channelization code decreases the data rate but makes the transmission more robust against errors. High spreading factor values, such as 128 or 256 chips per code word, are used for low bitrate voice connections and critical signaling messages. The additional robustness gained by high SF is called **processing gain**.

The available channelization codes are organized as **code trees**, like the one shown in Figure 6.7. In this tree, the shortest codes are close to its root while the longest codes of 256 bits are its leaves. As the transmitted code words are produced by applying both channelization and scrambling codes to the data, there is a relation between those. For each scrambling code (one per UE and NodeB) there is a channelization code tree from which channelization codes can be picked. This means every transmitter is able to allocate its channelization codes independently of the other transmitters. If two devices happen to allocate the same channelization codes, their transmissions are eventually distinguished from each other by the scrambling process.

When a channelization code of the tree is taken into use, it blocks using those codes of the same tree, which the receiver could mix with the chosen one. When a code (C) is taken into use from a tree, the following rules must be obeyed from taking any other code into use from the same tree:

- No channelization code can be used from the path of the tree between the code C taken into use and the root of the tree. If one of those codes would be used, its bit sequence would match the first bits of code C, making it impossible for the receiver to distinguish between those.

Figure 6.7 WCDMA channelization code tree structure. *Source:* Adapted from 3GPP TS 25.213 [14].

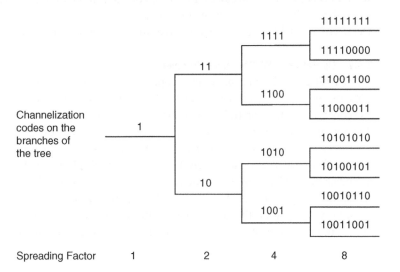

- No channelization code can be used from the branches of the tree below the selected code C. If one of those codes would be used, the bit sequence of code C would match the first bits of the other longer code, making it impossible for the receiver to distinguish between those.

This introduces a tradeoff for channelization code selection. The base station can either support a few connections with high data rates or many connections with low data rates. Taking into use a short channelization code with low spreading factor for a high-speed connection blocks the possibility of picking multiple longer codes from the same branch of the tree, which otherwise could support many slower connections. On the other hand, if the RNC allocates NodeB with a number of long channelization codes with high spreading factors from the parallel branches of the tree for many connections, those will block usage of any shorter code from the same branch for a high-speed connection.

This tradeoff is alleviated with a dynamic way of using the channelization codes. Short channelization codes are allocated for the UEs as long as needed for active high-bitrate data channels. Whenever there is a gap or idle period in the user data transmission, the RNC reallocates a longer code with bigger spreading factor for the channel. With this way, the UE may still use the channel quickly when new data is to be transmitted and the bitrate of the channel can be increased just by swapping its spreading code to a shorter one. On uplink, the channelization codes and spreading factors can be dynamically changed for each WCDMA frame sent by the UE. On downlink, to conserve channelization codes it is possible to send data to the UE over a shared channel with a single channelization code instead of setting up a new dedicated channel toward every UE. Spreading factors cannot be changed for every WCDMA frame of dedicated downlink channels, but they can be dynamically adjusted frame-by-frame for the shared downlink channels.

Channelization code management is not the only factor to limit capacity of WCDMA cell and data rates available for UEs. The other aspect is related to the fact that all the UEs within the cell use the same frequency band. When only few parallel transmissions exist, a CDMA system can easily recognize each separate CDMA encoded signal. But when the number of simultaneous users or data rates per users grow, then the noise power created by the other interfering signals increases compared to the power of the specific CDMA signal tried to be recognized. A single UE may try to overcome this by increasing the transmit power to improve SNR. That, however, increases the noise toward the other signals. DSSS signals are more robust against the noise when using high spreading factors and lower data rates. In the end, maximum capacity of a cell is reached when many of the UEs within the cell are using high transmit power and the resulting noise starts to deteriorate data flows of the UEs located at the cell edge even when using maximum power. This phenomenon where the UEs at the cell edge lose their connections is called **cell breathing** [15]. To control cell breathing, UMTS uses admission control algorithms to prevent the cell load from increasing too high.

Details of WCDMA modulation and spreading methods can be found from 3GPP TS 25.213 [14].

6.1.4.2 Operation of WCDMA Rake Receiver

WCDMA **rake receiver** is able to pick copies of the same signal received at slightly different moments. Those copies are reflections or refractions of the transmitted signal. Since the lengths of signal propagation paths of those copies differ, their arrival times at the receiver differ slightly as well. While copies of the signal would cause fading for an ordinary receiver, a rake receiver is able to take advantage of those. A rake receiver is able to combine the received copies into one single signal image, which reflects the transmitted signal better than any of the individual copies received. A rake receiver consists of multiple parallel fingers, each processing one copy of the received signal, and a combiner, which creates the received signal as a combination of those processed by rake fingers. Rake fingers are able to pick from the apparently random signal the known code words and any of their copies, provided that all the copies arrive within a relatively short period. WCDMA utilizes rake technique to combine multiple received instances of the same signal created in the following ways

- Copies of signal sent from a single antenna but propagated either directly, reflected or refracted over different paths as described in *Online Appendix A.3.3.*
- Copies of signal sent simultaneously by two to three different base stations within a soft handover, as explained in Section 6.1.4.3. In this case, the signal copies use different scrambling codes, so each rake finger is configured to use the scrambling code of one of those base stations within the active set.
- Copies of the signal sent simultaneously by two antennas of a single base station. On a dedicated channel, the base station is able to adjust the phase and amplitude difference between the antennas to maximize the strength of the signal received. That technique is called **closed loop transmit diversity**.

In comparison to GSM, the WCDMA is more efficient radio technology. Since the WCDMA is able to multiplex many connections to the same frequency band, it is possible to make the WCDMA base station much smaller than GSM base station. The GSM base station needs to have multiple radio frequency units and a WCDMA base station needs only a single unit. To separate different channels from each other, the WCDMA base station uses digital signal processing circuits, which take less space than the analog radio frequency units. Further, the WCDMA has higher spectral efficiency compared to GSM so that with given radio bandwidth a WCDMA cell is able to support twice as many calls as a GSM cell.

6.1.4.3   UMTS Handover Types

A WCDMA connectivity model is fundamentally different from the GSM model. A GSM mobile station is always connected to one single cell at the time. Handovers between GSM cells are **hard handovers,** where any connections to the previously used cell are cut while moving the radio connections to the new cell. WCDMA, on the other hand, introduced a new **soft handover** concept where the UE is connected to multiple base stations at the same time. As all the UEs and cells of a WCDMA network always use the same wide band frequency area, such multiconnectivity is possible for WCDMA. Even if soft handover is an important WCDMA mechanism, it is not the only way for the UE to perform handovers. WCDMA supports the following three handover approaches:

- **Hard handover:** The UE must break its connection to the old base station when switching over to the new one if the two RNCs controlling the old and new base station do not have a direct connection between each other over the Iur interface. Any inter-RAT handover between WCDMA and another radio technology is also a hard one. Hard handover of a circuit switched call is always done under control of the anchor MSC, which stays in the call path despite the handover. Connections over common and shared channels always use hard handover as the UE uses those always toward one single cell only. In the hard handover, the contributing RNCs and the core network prepare connections to the new cell prior to the handover execution at the UE. Eventually, the SNRC sends a handover command to the UE, instructing it about parameters such as channelization and scrambling codes to be used in the new cell. The connections experience a short interrupt of approximately 100 milliseconds during the handover. Hard handover always involves the SNRS relocation process.
- **Soft handover:** If the handover is performed between two cells controlled by one single RNC or by two RNCs interconnected over Iur interface, it is possible just to update the **active set** of cells with which the UE is engaged on communication over dedicated channels. In this case, the UE starts communicating with the new cell while it still maintains its connection to the old cell(s). No disruption is caused for existing connections by soft handover. UE may have up to six cells in its active set, but typically no more than two or three cells are used. It is possible to remove any cell from the active set when the connection quality toward the cell has degraded under a predefined threshold.
- **Softer handover:** When forming the active set with cells being sectors of one NodeB, the copies of signals can be combined already at NodeB rather than RNC. At softer handover, only one of the sectors transmits to downlink while all the sectors contributing to softer handover listen to the UE for uplink.

The operation of the soft handover is as follows.
Downlink:

- WCDMA rake receiver of the UE is able to combine signals received from multiple base stations, even if those signals use different scrambling codes and their arrival timing is slightly different. The SRNC, which is in control of the soft handover, picks base stations to transmit signals for a dedicated channel toward the UE. The set of cells contributing to soft handover is known as the active set for the UE. To minimize the arrival time differences between the signals from different NodeBs of the active set, the RNC adjusts the timing of the transmissions of the NodeB. The RNC tells the UE the scrambling codes of NodeBs in the active set, so that the UE can configure the fingers of its rake receiver with those scrambling codes. Thereafter, each finger is focused to receive transmission from a different cell in the active set. The downlink soft handover method has the following benefits:
  - The total signal power received by the UE located near to the edge of the cells is higher than in a case where only a single NodeB would transmit the signal. Soft handover increases the network coverage area per base station and allows a geographical area to be covered by a smaller number of base stations.
  - The overall interference within the WCDMA network can be decreased, when the UE and NodeBs in the active set can use smaller transmission power than when relying only on a single cell.

Uplink:

- WCDMA rake receiver of NodeB is able to combine signal copies received from a single UE via different propagation paths and arrival timing. This WCDMA property is called **micro diversity**, which a single NodeB may take advantage of even without a soft handover. Due to the micro diversity, the NodeB can provide the RNC with a higher-quality version of received signal compared to any single copies of it available to the NodeB. An additional benefit is that there is no need for UE timing advance in UMTS. The NodeB simply does not need to align the signals received from multiple UEs to a rigid TDMA timeslot structure. WCDMA uses TDMA frame structure within the uplink transmission from a single UE but not a common frame for multiple UEs.
- At soft handover, the RNC is able to combine information received from multiple base stations in the active set. Combining the signals at RNC is called **macro diversity**. To use macro diversity, the RNC configures NodeBs in the active set with the channelization and scrambling codes of the dedicated channel given to the UE. Thereafter, each NodeB in the active is able to receive copies of the radio block transmitted by the UE and provide a version of it to RNC. At the RNC, macro diversity is implemented so that the RNC gets the received radio blocks (interleaved over 1–8 frames) and related **signal-to-interference ratios (SIR)** from all the NodeBs within the active set. From those multiple copies, the RNC simply uses the one with the best SIR.

The target of the soft handover mechanism is to minimize total transmission power and the overall interference between all the radio connections within the network. This is achieved by UEs at the edge of multiple cells relying on multiple low-power signals from its active set rather than on a single high-power signal from one NodeB. There is, however, a tradeoff as decrease of the interference level depends on how many UEs apply the soft handover simultaneously. When adding more cells to active sets, every new connection uses a fraction of available capacity and makes a small addition to the interference level. The gain can be adjusted by tuning the threshold values of signal strength differences for updating the active set.

The new cells and their NodeBs added to the active set may be controlled by one or multiple RNCs. In every case, the connection to the UE is controlled by the serving SNRC and possibly supported by one or multiple DNRCs as shown in Figure 6.8. When the UE moves across the network, it may perform a number of successive soft handover active set updates. Eventually, the connection may rely on cells, none of which is under control of the original SNRC. In this case, a **serving radio network subsystem (SNRS) relocation procedure** is needed to upgrade one of the current DRNCs to a new SNRC.

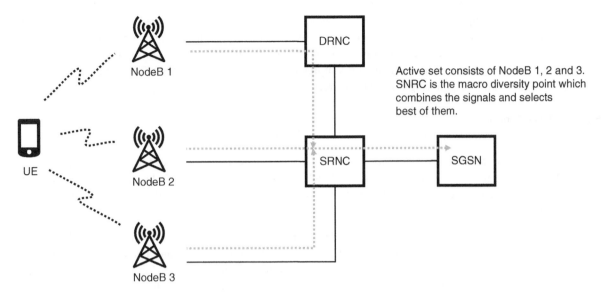

Figure 6.8 WCDMA soft handover.

In a basic soft handover scenario, all the NodeBs in the active set transmit the whole dedicated channel to the UE. The dedicated channel consists of two physical channels, DPDCH data channel and DPCCH control channel, as described in Section 6.1.4.6. 3GPP Rel-5 specifications introduced another approach called **site selection diversity (SSDT)**. In SSDT, only the base station that has the best connection to the UE would transmit the complete dedicated channel. Other NodeBs in the SSDT active set transmit only DPCCH control data and keep DPDCH in a permanent DTX state. The RNC selects the NodeB for active DPDCH user data channel based on the UE measurements of active set pilot signals. The SSDT may lead to a smaller overall interference compared to the basic soft handover when the UE does not move too fast. For UEs with high mobility, SSDT may cause too frequent handovers for the DPDCH traffic.

The WCDMA UE and base station measure the quality of the connection continuously to support the RNC handover decision making. The handover may be initiated by either the UE or network, but in any case, the final handover decision is made by the RNC. After making a handover decision, SRNC initiates the radio resource reservation process for the NodeBs in the active set. The UE measures the quality of the current connection, but also the strength of the signals received from other nearby cells that use either WCDMA or another cellular radio technology. The latter is used to perform **intra-system handovers** in cases when UTRAN deems the change of the radio technology as the best way to continue supporting the UE. As the WCDMA radio interface relies on continuous transmission and the UE has only one receiver, special arrangements are needed to allocate some time for the UE to measure other frequencies. This is done with the WCDMA **compressed or slotted mode**. For compressed mode, the NodeB uses DTX to provide the UE with idle periods which the UE can use for measurements, as defined in 3GPP TS 25.215 [16]. The applied DTX reduces the transmission rate of the channel. If there is a need to avoid bitrate fluctuation, it is possible to use reduced spreading factor for some frames of the dedicated channel to compensate the capacity lost for the measurement slots. Another option is to modify the puncturing of the channel coding to reduce the needed physical bitrate. In both the cases, the reliability of the signal is decreased and the UE may need to increase its transmission power to avoid transmission errors.

The usage of compressed mode either increases the transmit power of the UE or the error ratio of the channel, especially near the cell edge, if the UE operates already with high transmission power. When the transmission power is increased, the total cell interference is also increased and the cell total capacity decreased. Because of these side effects the compressed mode should be used only when justified by the cell load or when measurement slots are needed for measuring other cells when approaching handover thresholds of current signal quality. The UE switches to the compressed mode as instructed by the RNC to start preparations for intra- or inter-RAT handover to another frequency band. When starting the compressed mode, the RNC tells the UE which frequencies and radio technologies of the neighboring cells to measure.

6.1.4.4 Power Control

In WCDMA networks where all UEs use the same frequency band, power control is an essential feature:

- If a UE uses too low transmission power, the NodeB is not able to correctly receive the radio signal sent by the UE.
- If a UE uses too high transmission power, it will block transmissions of other UEs located farther away from NodeB. This is known as **near-far effect**, which decreases the number of UEs that the cell would otherwise be able to serve.

A WCDMA **fast power control** mechanism seeks the balance between these two unwanted cases. The NodeB sends transmission power commands to UEs significantly more frequently (1500 times per second) compared to the GSM base station (1 to 2 times per second). Such a continuous UE power control aims at compensating even minimal changes of the radio channel quality caused by small changes of the UE position. The frequency sharing between WCDMA UEs may cause signal fading even with small changes of UE location when opposite phases of two signals cancel each other out. The fast power control of WCDMA is able to compensate such fading when the UE moves at most with walking speed. With higher speeds, when a UE is used in a vehicle, even the fast power control is not able to cope with the quick changes of the radio connection.

Before opening a radio connection to the base station, the UE calculates the initial transmission power by measuring the strength of CPICH pilot signal received from the NodeB. The UE uses pilot signal to estimate its distance from the NodeB and the initial transmission power to be used. The UE limits its transmission power to the maximum allowed power in the cell, which the NodeB broadcasts in system information SIB3 and SIB4 blocks. Because the amount of attenuation experienced by uplink and downlink signal is different, such an **open loop power control** method gives only a rough estimate of transmission power level needed to reach the base station.

After the UE has an RRC connection to the NodeB and uses a dedicated channel, it gets a **transmit power control (TPC)** command from the NodeB once for each WCDMA timeslot. The NodeB measures the signal-to-noise and block error ratios of the transmission from the UE and tells the UE to either increase or decrease its transmission power in steps of 1 to 2 decibels. With this kind of **closed loop power control** with feedback from NodeB the network is able to control the UE power quite precisely. To maximize the likelihood of the UE to receive power control commands correctly, the NodeB can send them with higher power than it uses to transmit user data on the dedicated channel.

When applying the soft handover for the dedicated channel, UE gets power control commands from all NodeBs in the active set. In that case, the UE decreases its transmission power if any NodeB in the active set tells the UE to decrease the power. The UE increases its power only when all the NodeBs in the active set tell the UE to increase its power. When using SSDT, only the single base station with the active DPDCH sends transmit power control commands, and other base stations send just the rest of DPCCH signaling to the UE.

As described earlier, the WDCMA network uses **macro diversity** where the RNC picks the best copies of radio blocks delivered by the NodeBs in the active set. As mentioned earlier, those radio blocks are given with measured SIR values. Because of that, the RNC has the best knowledge about acceptable signal-to-noise ratio of the UE considering the whole active set, to keep the block error ratio in control. The RNC provides feedback for NodeBs in the active set for their power control decisions. With such **outer loop power control,** the NodeBs in the active set get target signal-to-noise ratios for the UE and adjust their power control commands accordingly.

The UE also sends power control commands to the NodeBs in the active set. The UE compares the signal-to-noise and block error ratio of transmissions from NodeBs in the active set. The UE selects the value of NodeB transmission power based on the block error ratio of combined signal. All the NodeBs in the active set adjust their transmission power of code words sent to the UE based on the UE power control commands. The power control can be used for both the dedicated channels as well as for the shared channels, but not for common channels. WCDMA power control is specified in 3GPP TS 25.214 [17].

6.1.4.5 Logical and Transport Channels

The WCDMA MAC protocol layer provides the following **logical channels** as carriers for different RLC flows:

- Broadcast control channel (BCCH) is a downlink channel, which broadcasts data for all the UEs camping in the cell. The SIB blocks within the SYSTEM INFORMATION messages sent over BCCH describe the network environment and give values for various UE control parameters such as network identity, location area, cell access restrictions, allowed range of UE transmission power within the cell, scrambling codes, and frequencies of neighboring cells to be used for measurements.
- Paging control channel (PCCH) is a downlink channel on which the NodeB sends paging messages to notify UEs about incoming calls or data packets.
- Common control channel (CCCH) is a bidirectional channel shared by all UEs camping on the cell. Any message sent by NodeB on CCCH has only one destination UE.
- Common traffic channel (CTCH) is a downlink channel used to carry data to all or a subset of the UEs camping on the cell.
- Dedicated control channel (DCCH) is a bidirectional signaling channel dedicated for a single UE. DCCH is used for transporting mobility, session, and communications management messages.
- Dedicated traffic channel (DTCH) is a bidirectional user data channel dedicated for a single UE. The DTCH may carry either a voice or data stream.

At the RNC, the WCDMA MAC protocol maps different RLC protocol flows to logical channels and adapts those to the **transport channels** toward the NodeB. The following transport channels have been defined for UMTS:

1) Downlink channels:
    - Broadcast channel (BCH) is used to broadcast system information messages describing the UMTS cell configuration to all UEs camping in the cell. The BCH carries BCCH logical data flows to the NodeB, which forwards the broadcast messages to the P-CCPCH physical channel.
    - Paging channel (PCH) is used for paging messages to the UEs. The PCH carries PCCH logical data flows to NodeB, which forwards the paging messages to the S-CCPCH physical channel.
    - Forward access channel (FACH) is used by the RNC to send RRC CONNECTION SETUP messages to the UEs. The FACH channel is used also for small amounts of data to a single or multiple UEs which camp in a cell without a dedicated channel. The FACH carries logical CCCH, DCCH, CTCH, and DTCH data flows (downlink control or user data messages) to the NodeB, which forwards the messages to the S-CCPCH physical channel.
    - Downlink shared channel (DSCH) is used to carry messages of dedicated DCCH control or DTCH user data flows to the UEs over the PDSCH physical channel.
2) Uplink channels:
    - Random access channel (RACH) is used to carry RRC CONNECTION REQUEST random access requests when the UE wants to initiate a CS call or send data or location update message. The NodeB forwards messages from the PRACH physical channel to the RACH transport channel.
    - Common packet channel (CPCH) is shared by multiple UEs, which can use the channel for sending data packets toward the network. The NodeB forwards any messages from the PCPCH physical channel to the CPCH transport channel.

3) Bidirectional channels:
  - Dedicated channel (DCH) is used to transfer large amounts of data and signaling during the active data transfer. A DCH is dedicated to a single UE to transport either DTCH user data and DCCH control data flows. In the radio interface, these data flows are split to two different physical channels: the user data flow is sent over a DPDCH channel while the control data flow is over a DPCCH channel.

The RNC dispatches the data flows from/to the UE over various types of transport channels. The transport channels are carried over the Iub interface between RNC and NodeB either over fiber optic transmission cables or with microwave radio links. Small capacity links could be built with E1 or PDH trunks while SDH was used for higher capacity links. In the original UMTS design, ATM protocol was used for dynamically sharing the link capacity between different connections. Later, when ATM was losing its popularity, another IP transport option was introduced for UMTS. In this option, UDP/IP protocols were used on top of the PPP/HDLC link layer (see IETF RFC 1661 [18] and 1662 [19]) to carry frame protocol messages. 3GPP TS 25.434 [20] describes these options in detail. Above the deployed generic transport layer based on ATM or UDP/IP, the frame protocol is used to carry the data specific to different transport channels. The frame protocol is introduced in Section 6.1.7.3 and specified in 3GPP TS 25.435 [21] and TS 25.427 [22].

Data flows that are carried over one single transport channel are dynamically mapped in the NodeB to different types of physical channels, as described in 3GPP TS 25.211 [23]. The types of used physical channels are chosen based on the total load of the cell and channelization codes already reserved for dedicated channels. With small amounts of data, shared channels may be used while dedicated channels are allocated for larger amounts of data. A dedicated physical channel granted to the UE may transfer data from one or multiple DCH transport channels. Multiple DCH channels are used when the UE has data flows with different QoS requirements.

GPRS radio blocks were the basic units radio resource allocation, and UMTS uses **transport blocks** for the same purpose. As the name suggests, the UMTS transport blocks are related to the transport channels. Each transport channel has a set of transport formats specific to the channel type. The **transport format** specifies the following properties of the transport block:

- Transport block size.
- Transport block set size, which is the number of transport blocks that can be sent to the UE within one transmission time interval (TTI).
- **Transmission time interval (TTI),** which is the length of the window used for sending the transport block set and changing the physical channel mapping used for the UE. The MAC protocol provides the physical layer with a new transport block set once per TTI. A transport block set is scheduled to a single physical channel. Transport blocks from one transport channel can be switched from one type of physical channel to another as complete transport block sets at TTI boundaries. In UMTS, the TTI length is either 10, 20, 40, or 80 milliseconds, capable of delivering one to eight WCDMA frames.
- The type of the error correction scheme and length of the CRC checksum used on the transport channel.
- **Rate matching** parameters used to control the bitrate of the channel.

A physical channel is able to multiplex transport blocks from many transport channels. When each of those transport block sets may use its own transport format, the physical channel carries a combination of those transport formats. When opening a dedicated channel for a UE, the RNC tells the acceptable transport format combinations that can be used for its uplink. For this purpose, the RNC uses the **transport format combination identifier (TFCI)** parameter sent to the UE over the DPCCH physical channel.

6.1.4.6 Frame Structure and Physical Channels

The WCDMA radio interface uses the TDMA frame structure for synchronization and to multiplex different physical channels to one spread spectrum channel. The most important purpose of the downlink frame structure is to enable the UE to find the starting positions of the code words for each physical channel. The base station sends a fixed code word on its synchronization channel. After the UE has detected the synchronization code word, it can derive the frame timing and synchronize itself to the frame cycle of the cell as used for all physical channels. Unlike GSM, no multiframe structure has been defined for WCDMA frames.

The WCDMA TDMA frame of 10 milliseconds is divided into 15 timeslots, each of 2/3 milliseconds. In each timeslot, a total of 2560 chips are transmitted so that the whole frame has 38400 chips, which is the length of uplink gold scrambling code. As mentioned earlier, the spreading factor of the WCDMA uplink data channel can be changed for each frame to adjust data rate every 10 milliseconds. For downlink, the spreading factor is fixed and the data rate is adjusted either with discontinuous transmission or by modifying the channel coding for the amount of redundancy used for forward error correction.

WCDMA has three categories of physical channels on its radio interface:

- **Common downlink channels** with which the UE is able to detect the WDCMA cell and its physical channels to be able to register to the WCDMA service.
- **Shared bidirectional channels** which carry small amounts user data and signaling to/from multiple UEs.
- **Dedicated channels** allocated for individual UEs for carrying large amounts of user data and signaling.

The WCDMA radio interface has a very large set of physical channels:

1) Downlink common and shared channels:
   - Primary synchronization channel (P-SCH) is a synchronization channel with which the UE is able to find the cell and detect its timeslot boundaries. P-SCH code words are sent within the first 256 chips of every WCDMA timeslot. The WCDMA system uses the same P-SCH code word – WCDMA primary **synchronization code** – in every cell so that the UEs could easily detect slot timing of the cell.
   - Secondary synchronization channel (S-SCH) is the secondary synchronization channel from which the UE can derive the WCDMA frame timing and scrambling code group of the cell. S-SCH code words are sent on parallel to the P-SCH code words within the first 256 chips of every WCDMA timeslot. While P-SCH uses the same code word for every timeslot, S-SCH code words are different for every timeslot of a frame. The S-SCH channel of a cell uses a fixed sequence of 15 code words so that this sequence repeats in every frame. There are 16 values available for S-SCH code words, but by arranging those values in different ways as a sequence, 64 different sequences exist. When the UE detects the sequence used for the frame, it finds out which of the 15 timeslots is the first one within the frame and which of the 64 **primary scrambling code (PSC)** groups is used by the cell.
   - Common pilot channel (CPICH) is used by the UEs to detect the primary scrambling code of the cell, from the code group identified with the S-SCH channel. CPICH code words of 2560 chips are sent within every timeslot of the frame. CPICH code words of a cell have one fixed value as they created by encoding a predefined 20-bit sequence with a known CPICH channelization code and the primary scrambling code of the cell. As the CPICH bit sequence and channelization codes are known to the UE in advance, the specific primary synchronization code is the only item to be detected. The UE, which already has the slot and frame synchronization with the cell and knows the PSC group used, shall just try correlating the received CPICH code words with the eight PSC options within the PSC group. By doing measurements of the signal power on the CPICH channel, the UE can figure out which cells are the best to communicate with and calculate its first estimate of the transmission power used. By adjusting the power of the CPICH channel of different cells, the RNC can control how the total load of the UMTS network is distributed between different cells of the routing area.
   - Primary common control physical channel (P-CCPCH) carries BCCH information broadcast to all the user equipment within the cell. The primary scrambling code is used on the P-CCPCH channel. The transmission rate of the channel is a modest 30 kbits in order to minimize the fixed transmission power needed for it. P-CCPCH is time-division multiplexed with the synchronization channel; thus, it is sent on the first 2304 chips of every timeslot of the frame, which are not used for synchronization channels.

- Secondary common control physical channel (S-CCPCH) carries signaling, such as paging and connection setup messages, toward those UEs that do not have a dedicated channel. The cells may have one or many S-CCPCH channels, each of them having a fixed-length channelization code. A spreading factor of each S-CCPCH channel depends on the bitrate allocated for the channel.
- Physical downlink shared channel (PDSCH) carries data packets to multiple UEs sharing this channel. With small amounts of data, the average transmission rate needed by the UE may stay low even if the peak rate would be high. In such a case, it is better to send the data over a shared channel rather than reserve a dedicated channel for the UE. The dedicated channel needs its own channelization code from the code tree of the NodeB, and as described earlier any code reserved from the tree blocks other codes to be used. Instead of allocating a high bitrate dedicated data channel, the RNC could give the UE a slow data rate signaling channel for uplink transmission power commands and to tell the UE how it can identify its own packets from the shared PDSCH channel. The channelization code and spreading factor of the PDSCH channel may be changed for every frame to adjust the data rate of the channel.
- Acquisition indication channel (AICH) is used by the NodeB to reply to random access requests by repeating the preamble sent by the UE over the PRACH channel.
- Paging indication channel (PICH) carries a set of PI (paging indicator) bits to announce the UEs about a paging message being delivered over the S-CCPCH channel. Every PI bit identifies a subset of UEs using the cell. When detecting such a bit set for its own UE group, the UE shall start listening to the S-CCPCH channel for paging messages. To save its power, the UE in idle state does not listen to the PICH channel continuously but only periodically.
- CPCH status indication channel (CSICH) carries information about the CPCH and PCPCH channel configurations in the cell.
- Collision detection indicator channel (CD-ICH), channel assignment indicator channel (CA-ICH), and access preamble acquisition channel (AP-AICH) are used for the random access, contention resolution, and access control procedures of the shared PCPCH uplink channel. **Contention resolution** means a process used to make sure that only one UE uses the radio resources granted, if multiple UEs tried out random access at the same time. The WCDMA random access and contention procedures work as follows: The UE may try to access the PCPCH channel by sending a random access preamble on it. The network will respond to the random access request on the AP-AICH channel. After that, the UE continues by sending a collision detection/channel access message over the PCPCH, to which the network responds on a CA-ICH channel. If there is no collision and access is granted, the UE can start sending its data over a PCPCH channel.

2) Uplink shared channels:
- Physical random access channel (PRACH) is used by the UE to request itself a dedicated channel when initiating a circuit switched call, user data transmission, or a location update message.
- Physical common packet channel (PCPCH) is shared by multiple UEs for sending data packets toward the network, one UE at a time. This channel supports fast power control and a collision detection mechanism for packets from different UEs.

3) Bidirectional dedicated channels:
- Dedicated physical data channel (DPDCH) is the dedicated channel for one UE to transmit user data (DTCH) and access stratum messages for radio resource management and non-access stratum messages for mobility, communications, and session management. A single UE may have one or multiple parallel DPDCH channels allocated to itself.
- Dedicated physical control channel (DPCCH) is the signaling channel dedicated for one UE to transmit control data (DCCH). Only one DPCCH is allocated per UE. This channel may carry the following types of control data:
  - Pilot bits, which the receiver uses to estimate the signal-to-noise ratio of the channel, which is used for determining the transmission power control commands

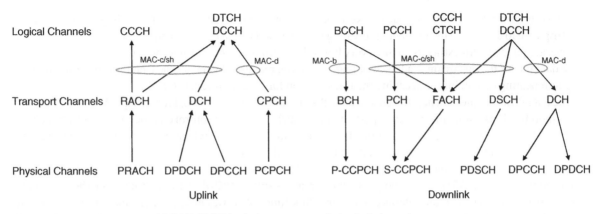

Figure 6.9 Mapping between WCDMA UMTS logical, transport, and physical channels.

- Transmission power control (TPC) commands to control the UE transmission power on the DPDCH channel
- Transport format combination indicator (TFCI) to describe the combination of data flows mapped to the dedicated DPDCH channel. This mechanism was described in Section 6.1.4.5.
- Feedback information (FBI) bits used to control the phase and frequency difference of multiple antennas used in parallel

A dedicated channel uses one channelization code allocated for the UE. The data and control channels are multiplexed on the downlink using time division and on the uplink using the I/Q modulation of the radio interface. In the latter, each modulated uplink symbol has information from both the channels. The benefit of the I/Q modulation is that there are no long breaks in the transmission even when there are pauses in the user data transmission. This will prevent any disturbances for nearby loudspeakers on the acoustic frequencies when the DPCCH channel is used to send TPC commands in 1.5 kHz frequency. Compared to the shared PCPCH channel, higher uplink data rates can be used on the dedicated DPDCH channel, which has more precise control of the UE transmission power. On the DPDCH, the network can adjust the UE transmission power carefully to balance the transmission quality and interference caused, taking into account both the location and mobility state of the UE.

The NodeB connects the transport channels of the MAC layer (from RNC) to the physical channels of WCDMA radio interface. Figure 6.9 shows a subset of all the mapping options that WCDMA has between and among its logical, transport, and physical channels. Further details about mapping between physical and transport channels can be found from 3GPP TS 25.211 [23].

Note that Figure 6.9 refers to MAC layer entities MAC-b, MAC-D, and MAC-c/sh, which are introduced in Section 6.1.5.1.

6.1.4.7 WCDMA Transmitter Design

The WCDMA transmitter at the NodeB takes the transport blocks as its input and generates the signals for the antennas as the output. Figure 6.10 shows the NodeB transmitter design as a block diagram. The complete operation of the transmitter is as follows:

1) A CRC checksum is calculated and attached to every transport block received from the transport channel.
2) Transport blocks are either concatenated or segmented to code blocks, depending on the sizes of transport blocks as in comparison with the code block size.

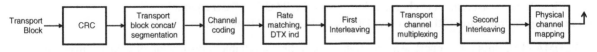

Figure 6.10 WCDMA downlink transmitter design.

3) For forward error correction, the code blocks are encoded with either Viterbi or Turbo coding, using a specific **coding rate** to determine the number of resulting bits against the data bits supplied for encoding. The encoded blocks are padded with extra bits so that every block delivered to interleaver has the same size.

4) The data rate of the bit stream is adapted or matched to the rate used by the physical channel by either removing or repeating certain bits or disabling the transmission for certain timeslots (DTX).

5) If the data flow tolerates latency, the encoded blocks are interleaved over two, four, or eight WCDMA frames.

6) Transport block sets from different transport channels (such as data and control channels) are multiplexed into a single data stream. If the UE has multiple dedicated channels (each of which has its own spreading code) to achieve high data rates, the bit stream is divided evenly to those channels.

7) Finally, a second interleaving round is done for data within a frame, separately for each physical channel multiplexed to the frame. After interleaving, the data stream is split to WCDMA frame timeslots to be scrambled and transmitted over the radio interface. One downlink timeslot carries both DPDCH and DPCCH channels multiplexed with TDM.

The UE processes its transmissions in nearly the same way. The main differences are that the UE performs the first interleaving before the transport rate matching and that the UE does not use discontinuous transmission (DTX). WCDMA multiplexing and channel coding is specified in 3GPP TS 25.212 [24].

6.1.5 Protocols between UE and UMTS Radio Network

The UMTS radio interface protocol architecture is described in 3GPP TS 25.301 [12]. This chapter provides brief overviews of protocols within the UMTS radio protocol stack.

6.1.5.1 MAC Protocol

The main task of **WCDMA medium access control (MAC)** protocol is to map the logical channels to WCDMA transport channels as described in Section 6.1.4.5. The WCDMA MAC protocol is specified in 3GPP TS 25.321 [25]. WCDMA Rel-5 specifications defined the following protocol entities of MAC:

- MAC-b takes care of mapping the common BCCH logical channel to the BCH transport channel.
- MAC-c/sh manages traffic of the common and shared channels of the cell. For shared channels, the MAC-sh may build transport block sets from individual transport blocks to different UEs. To identify the destination UEs, MAC adds either a U-RNTI or C-RNTI identifier to the MAC frame of the transport block.
- MAC-d manages traffic of dedicated channels. MAC provides the DCH with a transport block set once per TTI. All the blocks within the set belong to one single UE and a single RLC radio bearer of the DCH.

In addition to the above-mentioned tasks, the WCDMA MAC protocol has the following responsibilities:

- Transport channel selection. MAC decides if the data from a logical channel is delivered over a common, shared, or dedicated channel. The decision is done based on the properties of the data, needed bitrates, and the set of channels currently allocated for the UE.
- Multiplexing different logical data flows into a single transport channel, which may carry both signaling and multiple user data flows.
- Selecting a transport format for the transport blocks. MAC does its choice based on the data rate of the logical channel and the chosen transport channel.
- Scheduling the transmissions over common and shared channels based on the priorities of different connections between the network and UE.
- Traffic volume measurement to choose delivery over the RACH or DCH channel.
- Encryption of transparent RLC data over DCH with the KASUMI algorithm.

WCDMA MAC does not support segmentation, reassembly, or reliable transmission of data as those functions belong to the RLC protocol.

TCFT	UE identification type	UE RNTI	C/T	User data payload

Figure 6.11 Structure of the WCDMA MAC frame.

The WCDMA MAC frame has the following fields, as shown in Figure 6.11:

- Target channel type field (TCFT) is the identifier of the logical channel to which the frame belongs.
- UE identification type bits describe the type of the RNTI identifier within the UE identifier field.
- UE identifier is either the C-RNTI or U-RNTI identifier of the UE.
- C/T field tells the instance of the logical channel if the transport channel is mapped to multiple logical channels of the same type.
- User data field for the upper layer data frame.

6.1.5.2 RLC Protocol

WCDMA radio link control (RLC) protocol provides a link level connection (radio bearer or signaling radio bearer) between the UE and RNC. The WCDMA RLC protocol is specified in 3GPP TS 25.322 [26].

The WCDMA RLC protocol provides three different service types:

- Acknowledged mode (AM), where encrypted data frames are acknowledged and retransmitted, when necessary. The acknowledged RLC mode supports reliable transmission of data packets, where the correct contents and order of the received data packets are ensured. The acknowledged mode is used for transporting packet switched IP protocol user data.
- Unacknowledged mode (UM) does not support acknowledgments or retransmissions but supports encryption. Error checking is left for protocols above RLC.
- Transparent mode (TM) does not use separate RLC frame or encryption. Those tasks are left for the underlying MAC protocol. The RLC just passes the data from upper layers to MAC as such. Transparent mode is used for transporting circuit switched voice and BCCH broadcast messages.

The WCDMA RLC protocol performs the following tasks:

- Segmentation and reassembly of the upper layer protocol frames so that the segments would fit to the transport channel blocks.
- Composing the user data field of the RLC frame from one or multiple segments and padding the field with extra bits when needed.
- Acknowledgments, retransmissions, and flow control of RLC frames in AM mode using automatic repeat request (ARQ) mechanism; see *Online Appendix A.6.3*. RLC retransmissions are triggered by notifications from the WCDMA physical layer about corrupted data blocks that could not be recovered by forward error correction. Acknowledged mode uses flexible acknowledgment window size between 1 and 2^{12}. Large windows are used in bad network conditions to avoid interruption of sending new frames while waiting for acknowledgments for the earlier frames. This was a major improvement compared to the fixed size retransmission window used in GPRS [15].
- Encryption of user data in AM and UM modes with the KASUMI algorithm.

RLC/MAC protocol of GPRS is used for packet switched data. The WCDMA RLC is used for both packet switched data or circuit switched voice. WCDMA RLC frames are different for signaling and user data. The structure of a user data frame depends on the RLC mode. The first field of the frame tells if the frame is a signaling or data frame. The second field of a signaling frame describes the type and structure of the signaling message. Depending on the RLC mode, the user data frame has the following header fields:

- Sequence number (SN) of the RLC frame (UM and AM).
- Poll bit (P) requesting STATUS information from the other end (AM).

- Length of the RLC payload (UM and AM).
- RLC payload (TM, UM, and AM) for the user data.
- Possible padding bits (UM and AM) or the RLC STATUS message in the end of the frame (AM).

6.1.5.3 Packet Data Convergence Protocol

WCDMA packet data convergence protocol (PDCP) is used to transport packet switched data between the SRNC and UE. PDCP packets are transported over the RLC protocol. WCDMA PDCP protocol is specified in 3GPP TS 25.323 [27].

The most important task of the PDCP protocol is header compression of TCP/IP user data packets. Compression is used to minimize the number of overhead bits sent over the radio interface. The task is done with two protocols defined by IETF: IP header compression (RFC 2507 [28]) and robust header compression (RoHC, RFC 3095 [29]). The basic idea of header compression is not to transmit those parts of IP headers that repeat identically in a stream of packets. Those parts are sent only once and the omitted headers are restored accordingly in the other end. RoHC profiles and compact RoHC signaling are used to control this process.

In SRNC relocation, PDCP ensures that no user data packets are lost or duplicated due to the switchover and connection rerouting processes.

6.1.5.4 Radio Resource Control Protocol

WCDMA radio resource control (RRC) protocol is used for radio resource management. The RRC has a state model to describe the status of network connectivity and usage of radio resources at the UE. In the **RRC state model** the UE can be in one of the five states shown in Figure 6.12:

- **IDLE:** The UE is attached to the network and listens to cell broadcast messages but does not have an RRC connection with the network. The UE may have a PDP context but it is currently not in active use. The network has released the radio connections related to the PDP context. The network has not given any RNTI identifier for the UE. The UE does not perform any location updates and does not try listening to paging messages. To move out from the RRC idle state, the UE must gain an RRC connection to perform authentication, start encryption, and send a location update.
- **Cell DCH:** The UE is connected with the RNC over a dedicated channel and has RNTI identifiers. The UE has been given with spreading codes for the dedicated channel and a scrambling code to be used for the uplink. UE measures the reception quality of the radio link versus the signals received from neighboring cells, according to the instructions received from the RNC. The UE can change the serving cell only with a handover under control of the RNC.
- **Cell FACH:** The UE is connected with the RNC over common and shared channels. In the FACH state, small amounts of data or signaling can be exchanged over the low bitrate downlink FACH or uplink RACH transport channels. The UE is able to identify a MAC frame sent to it with help of the U-RNTI or C-RNTI identifier within the MAC header. In the Cell FACH state, the UE may perform cell reselections autonomously.
- **Cell PCH:** The UE is not connected to any RNC but listens regularly to the paging channel. In a Cell PCH state the UE informs the RNC about all its cell changes, so the network knows in which cell the UE camps and can

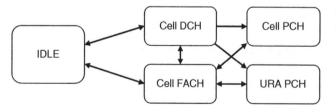

Figure 6.12 WCDMA RRC state model and possible state transitions.

page the UE in that cell. To receive incoming data, the RNC must page the UE and move it to the Cell FACH state for the data transfer.

- **URA PCH:** The UE is not connected with any RNC, but listens regularly to the paging channel. In the URA PCH state, the UE informs the RNC about changes of its UTRAN registration area, so the network can page the UE in that URA. Otherwise, the state is similar to Cell PCH state.

The WCDMA RRC protocol module controls the functions of the underlying RLC and MAC protocols. The WCDMA RRC protocol is specified in 3GPP TS 25.331 [30].

The RRC protocol messages are transported as payload of the RLC protocol. The RRC protocol selects the RLC mode used. The chosen mode depends on the purpose of the RRC message, as some messages need minimal latencies while maximal reliability might be the preference for others.

The RRC protocol has the following tasks:

- Broadcasting SYSTEM INFORMATION messages of the BCCH logical flow to all UEs camping on the cell
- Paging a UE for one of the following reasons:
  - Incoming call
  - Mobile terminated data message
  - Notification about changes in the SYSTEM INFORMATION messages
- Cell reselection when the UE does not have an active connection with the network
- Establishment of radio bearers to the UE and management of the related transport and physical channels
- Management of encryption mechanisms used over the radio interface
- Mobility management as per the RRC state of the UE
- Control of handovers and measurements that the radio network performs for the UE to support RNC handover decisions
- Transport and integrity protection of signaling messages exchanged between the core network and the UE

The WCDMA RRC message consists of the following fields:

- Message type
- Identifier of the RRC transaction, used to match a request and corresponding response
- Checksum of the message
- A set of other fields as defined for the specific type of RRC message

6.1.6 Signaling Protocols between UE and Core Network

The following four signaling protocols between the UE and UMTS core network are defined in 3GPP TS 24.008 [31]:

- MM protocol for mobility management
- GMM protocol for GPRS mobility management. In this conte, GPRS covers also the WCDMA packet switched services, rather than only the traditional GPRS service
- CM (or CC) protocol for circuit switched connection and supplementary service management
- SM protocol for packet switched session management

The messages defined in TS 24.008 [31] have two fields commonly used, the values of which determine the rest of the message structure:

- Protocol discriminator (PD), identifying one of the protocols as above
- Message type, identifying the type of the specific message within the context of the protocol identified by the PD field

6.1.6.1   Mobility Management Protocol

Mobility management (MM) protocol is used between the UE and circuit switched core network for mutual authentication and tracking the location and state of the UE. MM messages are carried by two underlying protocols: RANAP between the MSC/VLR and RNC, and RRC is used between the RNC and UE.

The tasks of the MM protocol can be categorized under three areas:

MM connection management procedures:

- Establishing and releasing MM connections. An MM connection is created by the UE or MSC/VLR with call or connection initiation message of the CM protocol.
- Transport of CM protocol messages between the UE and MSC/VLR.

MM common procedures, which are used over an existing MM connection:

- Query of the UE identifiers IMSI and IMEI
- Mutual authentication and exchange of encryption keys between the UE and network
- Change of the TMSI identifier allocated to the UE
- IMSI detach when the UE disconnects from the UMTS service

MM specific procedures for location management:

- Sending a location update message to the network core when the UE has moved from one location area to another
- Sending regular location update messages
- IMSI attach when UE starts using the UMTS service

6.1.6.2   Connection Management Protocol

Connection management (CM) protocol is used to establish and release circuit switched connections between the UE and MSC/VLR, as well as for call- related supplementary service management. The CM protocol has been developed from the call management protocol used in the GSM system. The CM messages are transported over the MM protocol between the MSC/VLR and the UE.

The CM protocol has the following tasks:

- Establishing and releasing circuit switched calls
- Managing the states of the supplementary services, following the approach as defined in 3GPP TS 24.010 [32]
- Cooperation with the RANAP protocol for setting up and tearing down radio access bearers for circuit switched connections

6.1.6.3   GPRS Mobility Management Protocol

GPRS Mobility Management (GMM) protocol is used between the UE and packet switched core network for mutual authentication and tracking UE location and state. Like the MM protocol, also GMM uses RANAP and RRC protocols as carriers between the SGSN and UE.

The tasks of the GMM protocol are similar to the MM protocol, which is used for the circuit switched domain. GMM is developed from the MM protocol. The GMM protocol supports combined location updates where one single message is used to update the location of the UE for both the packet and circuit switched domains. The combined procedure requires interconnection of the SGSN and MSC/VLR over the Gs interface so that the SGSN is able to pass the state and location updates to the MSC/VLR in the CS domain.

The tasks of the GMM protocol can be categorized in a way similar to MM:

GMM connection management procedures:

- GPRS attach and GPRS detach. When attaching to GPRS service, the UE tells the network its current routing area and its most recent P-TMSI identifier.

GMM common procedures, which are used over an existing GMM connection:

- Query of the UE identifiers IMSI and IMEI
- Mutual authentication and exchange of encryption keys between the UE and network
- Change of the P-TMSI identifier allocated to the UE
- Connection setup request sent by the UE

GMM specific procedures for location management:

- Sending a routing area update message to a network core when the UE has moved from one routing area to another
- Sending regular routing area update messages

6.1.6.4   Session Management Protocol

Session management (SM) protocol is used to manage packet switched connections between the network core and the UE. SM messages are transported over the GMM protocol between the SGSN and UE.

SM protocol has the following tasks:

- Creating and deleting PDP contexts
- Changing the QoS parameters for an existing PDP context

6.1.7   Protocols of UTRAN Radio and Core Networks

6.1.7.1   Link and Network Layers

The links between the UMTS network equipment were initially built with E1, PDH, or SDH trunks. One E1 link or a group of them can be used for connecting the NodeB to the RNC. The capacity of a NodeB is bigger than GSM BTS, for which a single E1 is sufficient. PDH and SDH trunks are often used in the high-capacity links between the RNC and core network. The protocol stacks used on top of the trunk technology depend on the interface of the UMTS network reference model.

The connections between the HLR database and other types of equipment that communicate with the HLR are based on the SS7 protocol stack as within the GSM systems. MTP2 and MTP3 protocols are used on the link and network layers of the SS7 stack toward the HLR.

UTRAN Iu is the interface between the RNC and UMTS core network elements MSC/VLR and SGSN. Data transport services of the Iu are defined in 3GPP TS 25.414 [33]. Two options have been specified for building Iu interface network layer services:

- The ATM option, where the ATM protocol is used on top of the E1 and SDH physical and link layers to build permanent virtual end-to-end data circuits. Two different adaptation layer options can be used within the ATM PVC to carry SCCP signaling messages:
  - SSCOP, SSCF-NNI, and MTP3B protocol
  - IP, SCTP, and M3UA protocol
- The IP option, where an RTP/UDP/IP protocol stack is used above the physical and link layers to transport various types of protocols, such as FP, over the Iu or Iub interfaces. Technology options for the link and physical layers are not specified by 3GPP but are left as the task of network vendors and operators. Ethernet over fiber or microwave radio are the typical options deployed.

The original UMTS design only had the ATM, but Rel-4 and Rel-5 specifications aimed at replacing ATM with IP, as it was seen that IP would be the dominant core network protocol in the future [34]. This approach is known as the **bearer-independent core network (BICN)**, which gradually became the dominant option for UMTS deployments. One major challenge with BICN was synchronization. E1 and SDH protocols carry the reference clock signal used for synchronizing the RNC and NodeB, but that does not apply to IP over Ethernet. Synchronization is essential for minimizing buffer delays in the RNC. This issue was solved with a time alignment function, which was added to the Iu UP protocol.

Setting up signaling connections within the core network, between different networks and between the core network and RNC, is done with the SCCP protocol of SS7 stack. The SCCP protocol uses the global title addresses for message routing. Globally unique address space is used for interoperator communication, as when sending messages between the MSC/VLR of the visited operator and the HLR of the user's home network.

6.1.7.2 Iu User Plane Protocol

Iu user plane (Iu UP) protocol is used to transport circuit and packet switched user data over Iu interfaces between the SGSN or MSC and RNC nodes. Transparent Iu UP mode is used to carry user data packets over Iu, so that the transported upper layer frames are not encapsulated into any additional Iu UP frames. Support mode is used when Iu UP specific support services are needed. The Iu UP protocol is also used to carry circuit switched data to/from the MSC/VLR.

The Iu UP protocol frames are carried either over an IP or ATM connection (layer 2) or a GTP-U tunnel on the higher layer of the protocol stack. The Iu UP protocol is specified in 3GPP TS 25.415 [35].

The Iu UP protocol supports the following procedures:

- Transfer of user data with both transparent and support modes
- Initialization of Iu endpoints to configure them with RAB sub-flow combinations and the related RAB sub-flow combination indicators (RFCI)
- Iu rate control of data transfer over the Iu interface
- Time alignment messages to minimize the buffer delay in RNC
- Handling of error events and frame quality classification reporting

In support mode, the Iu UP message has the structure shown in Figure 6.13:

- Frame control part
  - PDU type
  - Ack/Nack acknowledgment
  - Iu UP mode version
  - Procedure Indicator – indicating the procedure being acknowledged
- Frame checksum part
  - Header CRC
  - Payload CRC
- Frame payload part
  - Spare extension
  - PDU type specific fields

Frame Control Part				Frame Checksum Part		Frame Payload Part	
PDU type	(N)ACK	Mode version	Procedure Indicator	Header CRC	Payload CRC	Spare Extension	Other fields

Figure 6.13 Structure of the support mode Iu UP frame.

Header checksum	Type of FP message	CFN	TFI	Payload

Figure 6.14 Structure of the FP frame.

6.1.7.3 Frame Protocol

Frame protocol (FP) is used to transport packet switched user data over Iur and Iub interfaces between RNCs and NodeBs. To carry user data packets, FP transports the transport blocks transparently between the Node B and SRNC.

FP protocol messages are transported over an underlying ATM or IP network connection. The FP protocol is specified in 3GPP TS 25.435 [21] and TS 25.427 [22].

The FP protocol has the following tasks:

- Transport of transport block sets over Iub and Iur interfaces between the RNC and NodeB
- Synchronization of the transport channel to minimize any delays of downlink traffic
- Passing the radio interface parameters from the SRNC to NodeB
- Moving the outer loop power control data between the SRNC and NodeB
- Providing quality estimates from the NodeB to SNRC to help the SNRC to evaluate the best cell in a soft handover configuration
- Transport of hybrid ARQ information between the SRNC and NodeB

The FP message consists of the fields shown in Figure 6.14:

- FP header checksum
- Type of FP message
- CFN: Identifier of the radio frame from which the FP data is read or to which it shall be written
- TFI: Identifier of the transport format to be used in the radio frame sent
- Payload of the FP frame

6.1.7.4 Node B Application Protocol

Node B application protocol (NBAP) is the signaling protocol used over Iub interface between the NodeB and the controlling CRNC. The NBAP protocol frames are transported over a reliable ATM connection and encapsulated within the SSCF-UNI protocol. The NBAP protocol is specified in 3GPP TS 25.433 [36]. The NBAP protocol has the following tasks:

Common NBAP protocol procedures:

- Controlling cell configuration as NodeB logical resources and their states
- Instructing the NodeB to perform radio signal measurements and forward results to the RNC
- Providing the NodeB with SYSTEM INFORMATION messages of BCCH channel
- Managing PCH, RACH, FACH, and CPCH transport channels

NBAP protocol procedures specific to a single UE:

- Managing radio links to a specific UEs
- Managing the target levels of received UE radio signal power
- Managing dedicated and shared transport channels

The NBAP message consists of the following fields:

- An identifier telling if the message is related to common or UE specific procedures
- The type of the message

- NBAP transaction identifier, used to match a request and corresponding response
- A set of other fields, dependent on the NBAP message type

6.1.7.5 Radio Access Network Application Protocol

Radio access network application protocol (RANAP) is the signaling protocol for the Iu interface between SRNC and either MSC/VLR or SGSN nodes. The RANAP protocol is used for both the circuit switched and packet switched network domains. The RANAP protocol is specified in 3GPP TS 25.413 [37].

RANAP packets are transported over the SCCP protocol. RANAP expects the SCCP protocol to provide a reliable and errorless transport service.

The RANAP protocol has the following tasks:

- Managing radio access bearers between the network core and UE
- Moving existing RABs between RNCs during SNRC relocation
- Transporting MM, CM, and GMM protocol messages over the network section between the RNC and network core
- Forwarding paging requests from the core network to the radio network
- Managing encryption and digital signing procedures used toward the UE
- Managing error or overload cases impacting traffic between core and radio networks

The RANAP message begins with the message type field, the value of which defines the rest of the RANAP message structure.

6.1.7.6 Radio Network Subsystem Application Protocol

Radio network subsystem application protocol (RNSAP) is the signaling protocol over the Iur interface between SRNC and DRNC controllers when those are located to different RNC nodes. Like RANAP, RNSAP uses SCCP as a reliable carrier. The RNSAP protocol is specified in 3GPP TS 25.423 [38].

The RNSAP protocol has the following tasks:

- Forwarding UE messages related to location updates, paging, connection establishment, and power control between RNCs
- Managing RABs and transport channels over the Iur interface for different types of handover and RNC relocation procedures
- Forwarding measurement results and power control parameters between RNCs
- Managing SNRC relocation

RNSAP message consists of the following fields:

- Message type
- RNSAP transaction identifier, used to match a request and a corresponding response
- A set of other fields, dependent on the RNSAP message type

6.1.7.7 Mobile Application Protocol

Mobile application part (MAP) protocol as specified for GSM has been reused and enhanced also for WCDMA UMTS networks. The 3GPP TS 29.002 [39] specifies MAP protocol variants used over all 3GPP networks, including WCDMA UMTS.

6.1.7.8 GPRS Tunneling Protocol

Packet switched user data is carried between the GGSN and SNRC within **GPRS tunneling protocol for user plane (GTP-U)** tunnels. The GTP-U protocol was defined already for GPRS. The tunnels are set up for PDP contexts activated for the UEs. It is worth noting that in the original GPRS design, the GGSN and SGSN were the endpoints of the GTP tunnel. In UMTS, the tunnels are extended from the SGSN up to the RNC. This means that a UMTS SGSN does not know the cell where the UE is located, since such UE location tracking has been left for the SRNC. The SGSN knows the location of the UE by its routing area and the RNC. In GPRS, the SGSN was aware of the cell in which the UE camps. This information was delivered over the BSSGP protocol connection to the PCU [15].

GPRS tunneling protocol for control plane (GTP-C) is used to manage GTP-U tunnels over the Gn interface between the GGSN and a home SGSN and the Gp interface between the GGSN and a visited SGSN. The UE uses the SM protocol for PDP context management, and the core network uses GTP-C to manage PDP contexts and the related tunnels. The SGSN and GGSN use UDP/IP stack to transport GTP-C messages.

Different versions of GTP-C protocol have been defined, two of which have been used for WCDMA UMTS networks. GTP-C specifications can be found in 3GPP TS 29.060 [40] (v1) and 29.274 [41] (v2).

GTP-C has the following tasks:

- Managing packet data protocol (PDP) contexts, their QoS parameters, and related GTP-U tunnels
- Rerouting of tunnels at SGSN relocation caused by UE movement
- Forwarding location update messages between GSNs

The structure of the GTPv1-C message has only a few differences from GPRS GTP message. The GTPv1-C message has the following fields:

- Protocol version
- Protocol type
- An extension header flag indicating the presence of extension headers
- A sequence number flag indicating the purpose of the sequence number field
- An N-PDU number flag indicating the purpose of the N-PDU number field
- Message type
- Length of the message payload
- Tunnel endpoint ID (TEID)
- Sequence number used for GTP-C message acknowledgments and retransmissions
- Other information elements, depending on the message type

6.1.8 Radio Resource Management

The basic goal of UMTS radio resource management (RRM) is fair division of radio resources among the UEs within a cell. The main problem is how to allocate the spreading codes from the code tree when the connectivity and bitrate requests of the UEs exceed the maximum capacity of the cell. Basic approaches in UMTS RRM are admission control, cell selection and handovers, UE transmission power control, and allocation of dedicated channels only when necessary. For the UMTS WCDMA system, it is especially important to control the UE transmission power since all the UEs within the cell use the same frequencies simultaneously. One single UE with too high transmission power may block transmissions of many other UEs operating under the same WCDMA band.

Radio resource management is the responsibility of the NodeB and RNC nodes. The UMTS core network does not contribute to RRM.

6.1.8.1 UMTS Cell Search and Initial Access
After being powered up, the UMTS UE registers to a WCDMA UMTS network as follows:

1) The UE searches the primary synchronization code used by all WCDMA cells on the P-SCH channel. The UE scans over all the frequency bands used for WCDMA to find this code word. The problem is to detect the

correct moment when this code word is sent on the frequency band allocated to the network. To speed up the search, the ME can use information of the cells and networks used before the UE was switched off, as this information is stored on the USIM card equipped to the ME.

2) When the UE has recognized the primary synchronization code, it has achieved the timeslot synchronization with the cell. After that, the UE starts searching the secondary synchronization code word sequence sent on the S-SCH channel (see Section 6.1.4.6), to be able to synchronize itself to the WCDMA frames of the cell and to recognize the spreading code group allocated to the cell. WCDMA standards define 64 different secondary synchronization code word sequences (one per spreading code group) so that this search would not be too difficult.

3) After reaching the frame synchronization with the cell, the UE determines the scrambling code used by the cell. The primary scrambling code is identified through symbol-by-symbol correlation over the CPICH channel with all eight possible codes within the spreading code group identified in the previous step. When using the correct scrambling code, the UE is able to correctly decode the predefined chip sequence transmitted on the CPICH channel. The primary scrambling code is needed for the next step to decode the P-CCPCH channel.

4) The UE starts listening to BCCH traffic on the P-CCPCH channel. To decode the P-CCPCH channel, the UE uses the detected scrambling code and the constant spreading code of the P-CCPCH channel. The BCCH channel carries SYSTEM INFORMATION messages, which the base station transmits as instructed by the RNC. These broadcast messages provide the UE with various radio network parameter values, including the PLMN identifier of the network and location area code (LAC) of the cell. Based on the received PLMN ID, the UE must determine if the subscriber is entitled to use the services of that network and whether the network is a preferred network such as the home network of the subscriber. If the home network is not found, the UE searches other networks over different WCDMA network bands. The USIM application provides the ME with lists of network selection priority order and inaccessible networks. If the UE uses automatic network selection, it will register to the highest priority network found. But if the user has configured the UE to use manual network selection, the UE renders names of found networks on the user interface of the UE and lets the user pick the preferred network.

5) After selecting the network, the UE opens an RRC connection toward the cell as described in Section 6.1.8.2 and sends an initial location update message as described in Section 6.1.13.2.

6) When needed, the network may request the UE to provide different types of its identifiers, such as IMEI and IMSI, by sending an MM IDENTITY REQUEST message to the UE.

7) The UE provides requested data within an MM IDENTITY RESPONSE message.

8) The network can thereafter authenticate the subscriber as described in Section 6.1.9.2 and respond to the location update message sent in step 5.

After the UE has registered to the WCDMA network, it can later use information received over the BCCH channel to simplify further network searches. SIB block types 11 and 12 contain critical pieces of information for detecting other UMTS networks in the neighborhood, such as frequencies and cell IDs which the UE shall use for its measurements.

6.1.8.2 Opening and Releasing RRC Connections

UMTS UE may open an RRC connection with the radio network for the same reasons as a GSM/GPRS mobile station: to initiate or receive a circuit switched call, to send or receive user data, or to send a location update message.

The UE opens an RRC connection as shown in Figure 6.15:

1) The UE checks from the SIB5 block the available preamble signatures, scrambling codes, and timeslots for random access requests. The **preamble** is a predefined bit sequence which the UE uses in a random access process to request a radio connection for itself. The UE sends with low transmission power one of the available preambles to a randomly selected timeslot of the PRACH channel and starts waiting for the response. If no response is received, the UE resends the preamble with higher transmission power.

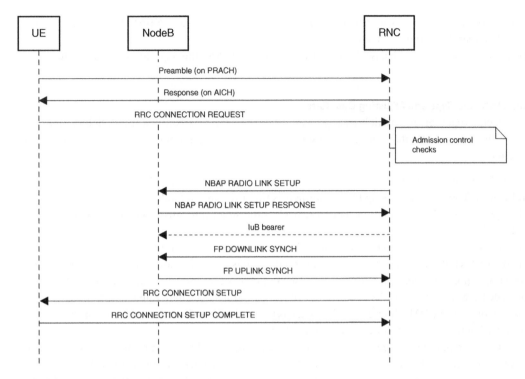

Figure 6.15 Opening UMTS RRC connection.

2) After getting a random access response (RAR), the UE sends an RRC CONNECTION REQUEST message to the RNC, describing the type of requested connection. After checking the request, the RNC estimates the expected load and interference level within the cell caused by opening the requested connection and makes an admission control decision. The RNC also decides whether to use the shared or a dedicated channel for the connection. The RNC allocates scrambling and channelization codes as well as a new S-RNTI identifier for the UE. The RNC opens an IuB bearer toward the NodeB and a radio bearer toward the UE. In the radio link setup message, the NodeB gets the scrambling codes, transport formats, and power control parameters for the new RRC connection. The IuB bearer is initialized with a downlink synch message.

3) The RNC sends an RRC CONNECTION SETUP message to the UE, describing the transport and physical channel configuration of the new connection. The UE moves to either the Cell FACH or Cell DCH state and responds to the RNC to have completed the connection setup.

To release the RRC connection, the SRNC sends an RRC CONNECTION RELEASE message to instruct the UE to release its RRC connection. After getting the response from the UE, the SRNC sends a NBAP RL DELETION message to the NodeB to tear down the Iub bearer.

For further details about the messages and parameters used in these procedures, please refer to *Online Appendix I.4.1*.

The RRC connection is released as follows:

1) The SRNC radio network controller sends an RRC CONNECTION RELEASE message to request that the UE release its RRC connection.

2) The UE responds to the SRNC with an RRC CONNECTION RELEASE COMPLETE message and stops using the RRC connection.

3) After ensuring that the UE no longer uses the RRC connection, the SRNC sends a NBAP RL DELETION message to the NodeB to tear down the Iub bearer.
4) The NodeB responds to the SNRC with a NBAP RL DELETION RESPONSE message, after which the Iu Bearer is closed.

6.1.8.3 Selection of Channel Type and Adjusting Data Rate

The RNC and UE decide together over which transport channel the user data is transmitted. Selection of the channel is based on the amount of data to be transported, the required QoS parameters, transmission rate, and burstiness of the data.

- If there is only small amount of data in small bursts, the data can be transported over the shared transport channels FACH and RACH. The benefit of using these channels is that they are available immediately without any delay caused by opening a dedicated channel. The drawback is the small transmission capacity available over these shared channels.
- If the transmission rate of the transported data varies a lot (for instance, when the data bursts should be sent with maximum data rates but there are long pauses between the bursts), the data can be transported over shared DSCH and CPCH transport channels. The benefit of these channels is that they do share one single high data rate, spreading code between multiple UEs in the time divisional manner. The data rate used by the UE can be separately adjusted for each WCDMA frame. The drawback is that the time division multiplexing of the channel resources may cause latency variations for the transmitted frames.
- If the amount of transported data is large or the data does not tolerate any long or variable latencies, the data should be transported over a dedicated DCH channel. There are two drawbacks for using dedicated channels. First of all, the signaling procedures for dedicated channel opening take time. Worse, any new dedicated channel and its channelization code take a fixed amount of transmission capacity of the cell, even if the needed data rate would vary. Using a dedicated channel also increases UE battery consumption. That is why DCH channels should be used only when necessary. The RNC releases the dedicated channels if no data is sent over in a few seconds, to save UE battery and optimize the usage of cell capacity. The data rate of the dedicated channel can also be adjusted by changing its spreading code, if the data rate needed starts to exceed or fall below to the rate supported by the spreading code allocated for the channel.

The channel type or data rate of the dedicated channel can be changed after the UE has sent an RRC MEASUREMENT REPORT to the SNRC, requesting a change to data rates. The RNC may also change the type of the transport channel after finding out the dedicated channel stayed idle for too long. The change is performed as follows:

1) The SNRC sends an RRC PHYSICAL CHANNEL RECONFIGURATION message to the UE. This message tells the UE all the necessary information about the new channel used, such as DPDCH spreading codes and the timing of applying the change.
2) The SRNC sends an NBAP RADIO LINK SETUP message to the NodeB. This message tells the NodeB to take the new transport channel into use. The NodeB responds to the SRNC with an NBAP RADIO LINK SETUP RESPONSE message.
3) When the UE has taken the new channel into use and synchronized it with the NodeB, the UE sends an RRC PHYSICAL CHANNEL RECONFIGURATION COMPLETE message back to the SRNC.

6.1.9 Security Management

Experience gained from the GSM system security was used when developing UMTS security solutions. In the UMTS system design, the following choices were made:

- Like the GSM system, the UMTS system is able to authenticate the subscriber. Additionally, in the UMTS system, the UE is able to authenticate the network by checking that the authentication vectors came from the subscriber's home network.

- Like the GSM system, the UMTS system encrypts the user's identity and data transported over radio interface between the UE and base station. The UMTS also encrypts signaling over the radio interface and may apply encryption of user data within the rest of the radio and core network.
- The GSM system relied on secret security algorithms, and the UMTS security algorithms are public. The aim is to ensure that the strengths and weaknesses of the algorithms would be understood well enough and any weak algorithms could be replaced with stronger ones, once any weaknesses would be revealed.

The UMTS security architecture is described in 3GPP TS 33.102 [42].

6.1.9.1 Security Algorithms

The UMTS security concept was designed to improve security, as compared with the GSM networks, in the following ways:

- Using mutual authentication between the UE and network as well as integrity protection of critical signaling messages to prevent rogue systems to hijack or modify sessions.
- Encrypting the traffic between the UE and RNC. In the GSM, the encryption is done only between the MS and BTS, leaving the traffic between the BTS and BSC vulnerable. This is dangerous especially when using microwave radio to connect the BTS with the BSC.
- Stronger security algorithms and longer session keys with limited lifetimes to have better protection against cryptology attacks. The UMTS security keys (CK and IK) expire with a timer to trigger reauthentication and key exchange processes.

Authentication and key agreement (AKA) is the authentication method used for UMTS. With AKA, both user authentication and data encryption are based on a secret key K allocated for the subscriber and stored both to the network authentication center AuC and the USIM application of subscriber's UICC card. This secret key is never exposed outside of these two entities. The USIM application and authentication center calculate the encryption and integrity protection keys CK and IK as well as the RES code used for user authentication based on the secret key K and a random number RAND. New RAND numbers are used and passed to the UE for each new authentication sequence.

Additionally, the network will give the UE a message authentication code (MAC) as part of the authentication sequence. With the help of the MAC code, the UE can ensure that the network has got the authentication vectors from the home network of the subscriber. To calculate the MAC, one input is the SQN value, which is incremented by both the USIM application and the AuC authentication center for every new UMTS authentication round. If the SQN sent by the network does not match with the SQN calculated by the USIM, the USIM tries to synchronize the network with the desired value. If such a synchronization attempt fails, the UE deems to be in contact with a rogue network which has not gotten the authentication vectors from the subscriber's home network.

To calculate the authentication codes and encryption keys, the UMTS uses the following algorithms, all of which use a random number RAND and the secret key K as their inputs:

- f1: algorithm to derive the MAC code from the synchronization code SQN and the other two input parameters
- f2: algorithm to derive the authentication result code RES
- f3: algorithm to derive the encryption key CK
- f4: algorithm to derive the integrity key IK
- f5: algorithm to derive the encryption key AK to encrypt the SQN parameter

Algorithms f1–f5 are operator specific. 3GPP has defined the MILENAGE example algorithm, which the operator may use if it so wished. The network and UE get the input data for these algorithms as follows:

- The secret key K is stored to the network authentication center AuC and the USIM application.
- In its authentication request, the network sends two pieces of data to the UE:
  - A random number RAND generated by the AuC for this request
  - An AUTN code, which contains the SQN code as encrypted with the AK key and the MAC code in cleartext

As its authentication response the UE sends to the network:

- The RES authentication code matching with the received RAND number

If the UE is able to match the MAC code from the network with the XMAC code calculated by the UE and the network is able to match the RES code from the UE with the XRES code calculated by AuC, the mutual authentication is successfully completed.

The data sent over the UMTS radio interface is protected as follows:

- The user data bit stream to be transported is encrypted with the f8 algorithm. Apart from the bit stream to be encrypted, f8 uses two inputs: the CK encryption key and the frame sequence number of the data protocol being encrypted.
- Signaling messages are digitally signed with the f9 algorithm, which uses two inputs: the message to be signed and the IK key.

6.1.9.2 Security Procedures

UMTS authentication and encryption key exchange can be executed when starting a circuit switched call, activating PDP context or as part of UE location update procedure. The IMSI code defined for GSM as the subscriber identifier is also used for UMTS. To avoid sending IMSI unnecessarily over the radio interface, the UMTS network uses temporary subscriber identifiers: TMSI for CS domain is specified already for GSM and P-TMSI for PS domain as specified for GPRS.

After opening an RRC connection and sending a UE INITIAL MESSAGE to the MSC/VLR or SGSN, the mutual authentication and start of encryption processes are performed as follows:

1) The MSC/VLR or SGSN sends an authentication request message to the UE. This request delivers the RAND and AUTN codes to the UE. The UE calculates the RES response, CK, and IK keys from the RAND and secret key K values. If the UE is able to authenticate the network from the AUTN code, the UE returns the calculated RES code in its response.
2) After completing the authentication, the MSC/VLR or SGSN sends a security mode command via the RNC to the UE. This sequence instructs the RNC and UE to start applying CK and IK keys for selected encryption and integrity protection algorithms.

For further details about the messages used in these procedures, please refer to *Online Appendix I.4.2.*

As a final note about UMTS security, it is worth noting that even if the UMTS security solution is better than that of GSM, it still is not perfect. The UMTS protects the traffic over the radio access network but may not provide protection of traffic between the RNC and core network. Like GSM, the UMTS uses the approach of trusting the security of the core network infrastructure, which might not always be a valid assumption. Further on, encrypting the user data is not a sufficient solution. There are providers of rogue base station devices, which are able to force UEs temporarily connecting to it and reveal identifiers like IMSI. This allows the rogue base station to track phones in its close proximity.

6.1.10 Communications Management

6.1.10.1 Mobile Originated Circuit Switched Call

After opening an RRC connection, the UE initiates a circuit switched mobile originated (MO) voice call as shown in Figure 6.16:

1) UE sends a CM SERVICE REQUEST via the RNC to the MSC/VLR. This request identifies the subscriber and the type of the service request as mobile originating call establishment. The MSC/VLR acknowledges the service request, sets up a new SCCP connection to the RNC and thereafter performs subscriber authentication and starts traffic encryption as described in Section 6.1.9.2.

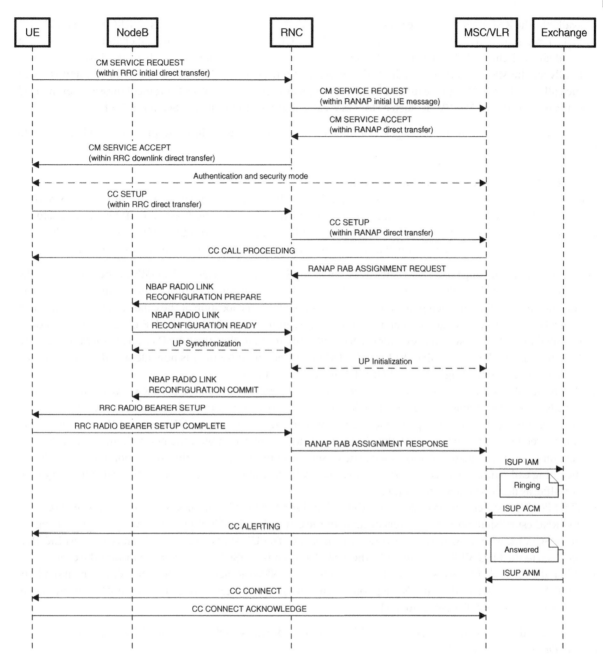

Figure 6.16 UMTS MO call setup.

2) The UE sends a CC SETUP message to the MSC/VLR. The SETUP message provides the MSC/VLR with the called telephone number, identifier of the radio bearer used, voice codecs, and QoS parameters to be used for the call. The MSC/VLR checks if the subscriber is entitled to use the requested services and returns a call proceeding the message to the UE.

3) The MSC/VLR creates new radio access bearers toward the RNC. The RNC checks if its own capacity and the capacity of the controlled base stations are sufficient to support the additional bearer. If the needed capacity is

available, the RNC then sets up the bearers with the NodeB and MSC. The RNC sends an RRC RADIO BEARER SETUP message to the UE. This RRC message tells the UE the transport and physical channel parameters of the dedicated channel. The UE takes the dedicated channel into use and responds to the RNC, which can finally tell the MSC/VLR that the radio path is prepared for the call. The MSC/VLR routes the call setup toward the callee with SS7 ISUP procedures. The call setup process continues with CC protocol messages exchanged between the UE and MSC/VLR, such as for GSM calls described in Chapter 5, Section 5.1.9.1.

For further details about the messages, parameters, and protocols used in these procedures, please refer to *Online Appendix I.4.3*.

6.1.10.2 Mobile Terminated Circuit Switched Call

Routing a mobile terminated call to the UMTS RNC is done by the CS core network in a similar way as routing calls to a BSC in the GSM system. However, the communication between the RNC and UE for MT call setup is done with the protocols defined for the UMTS network. Still, the call setup primitives of CC protocol are the same as in the GSM system. A mobile terminated call is connected to the UMTS UE as shown in Figure 6.17:

1) The telephone exchange that serves the caller sends an ISUP IAM message to the GMSC center located in the home network of the callee. The task of the GMSC is to find out the current location of the called UE and continue routing the call attempt toward it. The GMSC retrieves the location of the callee from the HLR, which contacts the MSC/VLR to get a mobile station roaming number (MSRN) for the UE of callee. The MSRN is returned to the GMSC, which uses it for the SS7 call setup routing. Now the GMSC is able to forward the ISUP IAM message to the correct MSC/VLR. In the IAM message, the subscriber is now identified with the MSRN number instead of the originally used MSISDN telephone number.

2) When the MSC/VLR receives the IAM message, it starts paging the UE and sends a RANAP paging request to the RNC. The RNC sends paging messages in those cells which belong to the location area where the UE has most recently sent a location update message or only in the cell where the UE camps, if the RNC knows it. After receiving the paging message, the UE opens an RRC connection with the network unless an RRC connection is already available. The UE then sends a service request to tell that the connection is available for an incoming call. The service request triggers the MSC/VLR to authenticate the subscriber and start encryption of traffic, as described in Section 6.1.9.2.

3) The MSC/VLR selects the voice codec for the call and sends a CC SETUP message to the UE. The MSC/VLR and RNC establish radio access bearers needed for the call as described in Section 6.1.10.1. The UE responds to the MSC/VLR with a CC CALL CONFIRMED message. The UE starts ringing and sends a CC ALERTING message to the MSC/VLR. Consequently, the MSC/VLR sends an ISUP ACM message toward the caller.

4) When the subscriber answers the call, the UE sends a CC ANSWER message to the MSC/VLR, which sends an ISUP ANM message to the caller's telephone exchange. To get the call connected, the MSC/VLR ultimately sends a CC CONNECT message to the UE.

For further details about the messages used in these procedures as well as the related RRC operations, please refer to *Online Appendix I.4.4*.

6.1.10.3 Circuit Switched Call Release

The UE closes a circuit switched call as shown in Figure 6.18:

1) The UE sends a CC DISCONNECT message to the MSC/VLR, which responds to the UE with a CC RELEASE message. To complete the release handshake, the UE sends a CC RELEASE COMPLETE message to the MSC/VLR. At this point, the call is released and the UE has stopped transmission of voice media.

2) To release the call the leg to the remote end of the MSC/VLR sends an ISUP REL message to the GMSC and gets an ISUP RLC message as a response.

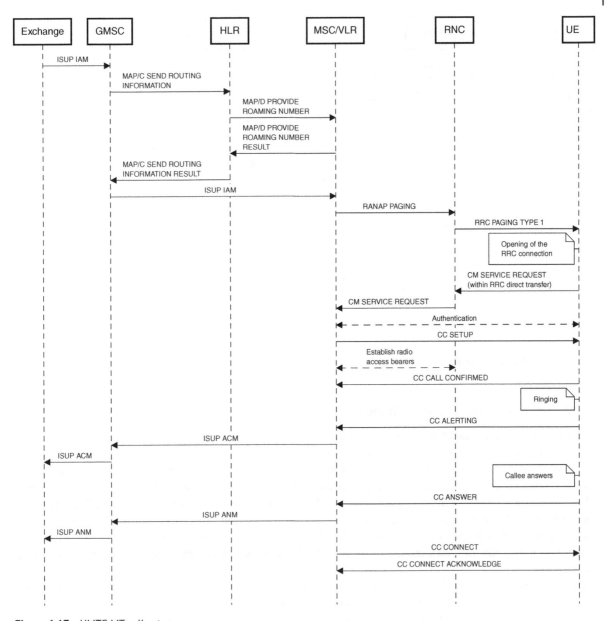

Figure 6.17 UMTS MT call setup.

3) To tear down the local radio access bearer, the MSC/VLR sends a RANAP RAB ASSIGNMENT REQUEST RELEASE message to the RNC. However, if the intention is to close all the dedicated channels for the UE, the MSC/VLR can instead send a RANAP IU RELEASE COMMAND.

4) The RNC sends an RRC RADIO BEARER (Release) message to the UE, which responds with a RADIO BEARER RELEASE COMPLETE message. This sequence is used to release either the radio bearer used for voice or all the radio bearers, including the RRC connection used for signaling.

5) The RNC returns either a RANAP RAB ASSIGNMENT RESPONSE (Release) or an IU RELEASE COMPLETE message to the MSC/VLR.

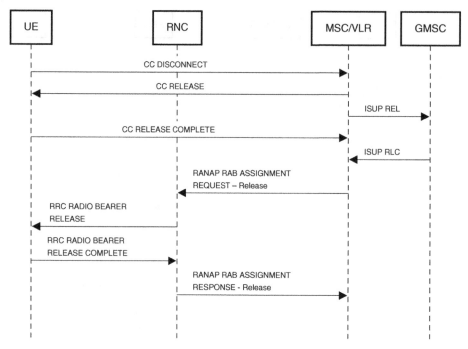

Figure 6.18 UMTS call release.

6.1.11 Voice and Message Communications

6.1.11.1 Voice Encoding for UMTS Circuit Switched Calls

Voice is digitally encoded in UMTS networks using narrowband AMR (adaptive multi-rate) and AMR wide-band (AMR-WB) codecs. The AMR technology provides the UE with a single codec capable of encoding voice different ways for different available data rates. The narrowband AMR was introduced already within GSM, but wideband AMR was added to UMTS for improved voice quality. The narrowband AMR codec supports eight different encoding methods for data rates between 12.2–4.75 kbps. One of the supported AMR encoding methods is compatible with the GSM EFR codec, another with the US-TDMA IS-641 codec, and third one with the Japanese PDC-EFR codec. This means that the AMR codec can be used also in networks which only support such legacy encoding methods.

The network and UE may choose the specific encoding method and data rate every 20 milliseconds. The selection is done based on the bandwidth of the voice samples and the available transmission capacity. By adjusting the data rate dynamically with AMR, the following benefits can be achieved:

- If the cell load and radio conditions between the UE and NodeB allow using high-enough bitrate with a spreading factor such as 128, the network is able to provide the best quality voice signal.
- When a higher spreading factor such as 256 or a more robust channel coding method is applied to the voice channel, the AMR encoding can be changed to support the lower available bitrate. This may happen due to decreased radio channel quality or an increase of the cell load, for instance, to support a higher number of simultaneous calls in the cell.

The operation of the AMR codec is specified in 3GPP TS 26.071 [43] and 26.090 [44]. The AMR codec splits the voice stream to 20 millisecond frames. Voice signal is sampled in the frequency of 8000 times per second so that every AMR frame contains information from 160 samples. Each sample is encoded with the code excited linear prediction coder (CELP) technique so that the transmitted frame contains CELP parameters needed to support decoding. The set of parameters is the same for each of the codec modes, but the parameter lengths depend on the codec bitrate. With higher bitrates, a higher number of bits can be used to describe a parameter value, and the better voice quality can be achieved. CELP parameters are divided into three classes depending on their significance. Forward error correction mechanisms of different strengths are used to protect each of the parameter classes. AMR supports voice activity detection (VAD) technique and comfort noise sample generation when the call participant does not speak. VAD helps the UE to minimize its power consumption and interference on the radio channel.

The wideband AMR-WB codec takes samples with a higher frequency than AMR codec. The sampling rate used for AMR-WB is 16 kHz. The traditional PCM 8 kHz sampling rate used for AMR is able to capture sound frequencies of 3400 Hz and below. By doubling the sampling rate of AMR-WB to 16 kHz, the captured audio frequency range is up to 7000 Hz. That is why the perceived voice quality of AMR-WB is much better than what can be achieved with narrowband encoding techniques. Like AMR, also AMR-WB provides different ways of encoding the samples, eventually producing a data stream with bitrate between 23.85–6.6 kbits. It is worth noting that the high voice quality provided by AMR-WB can be only used for a call between AMR-WB capable mobile terminals. If any of the call parties relies on the narrowband fixed-line telephone network or uses a mobile station with narrowband capabilities only, the voice quality stays at the narrowband level. In such a case, either the AMR-WB codec can be used in one of the supported narrowband modes or the media gateway (MGW) may perform transcoding between the wideband and narrowband encodings used.

6.1.11.2 Short Message
The short message service of UMTS is essentially similar to that of GSM.

6.1.12 Packet Data Connections

6.1.12.1 Quality of Service Classes and Parameters
The UMTS UE activates packet data protocol (PDP) contexts just as a GPRS terminal does. At PDP context activation, the UE defines quality of service (QoS) level needed for the packet data flow. The following **QoS classes** are defined in 3GPP TS 23.107 [10]:

- Conversational: Class of service used for symmetric bidirectional real-time data exchange, such as a voice or video call. A conversational service shall guarantee a short and constant delay and a defined minimum transfer rate.
- Streaming: Class of service used for unidirectional or asymmetric data exchange, such as transmission of streamed music or video to the UE. A streaming service shall guarantee a defined minimum transfer rate, but variation of the transfer delay exceeds the strict limits of the conversational class. The UE may buffer received data before playing it out to hide any fluctuations of the transfer delay.
- Interactive: Class of service used for asymmetric, non-real-time and bursty data exchange, such as browsing of Web pages. Interactive service shall guarantee a defined average transfer rate and a reasonable maximum delay.
- Background: Class of service used for transporting data without any response time requirements, like for downloading emails or files in the background. There are no strict data rate or latency requirements defined for the best effort background service.

In addition to the service class, the UE gives other UMTS QoS parameters to describe the characteristics of its data flow:

- The maximum data rate
- The average data rate

- The maximum latency of data transfer through the UMTS network
- The maximum error ratio of the data transferred
- The maximum size of the data packet transferred over the connection
- Whether the destination shall receive data packets in correct order
- The priority of the data flow compared to other flows in case of congestion

To find out how these parameters are used in the four UMTS QoS classes, please study table 3 of 3GPP TS 23.107 [10].

The QoS classes and parameters are used by UMTS network elements for many purposes:

- **Admission control** of new data flows, taking into account the total transmission capacity needed for the existing data flows.
- Configuration of radio access bearers, so that the allocated transmission capacity satisfies the requested criteria.
- Prioritization of different data flows in congestion cases. The stricter real-time requirements a data flow has, the higher priority the flow has for getting its fair share of the total transmission capacity at the network bottleneck.

The QoS model of UMTS system is application centric and covers only the UMTS radio interface. If a networked application running in the UE would need a certain level of QoS for its data flows, it has to provide the right set of QoS parameter values for the UE to activate the PDP context accordingly. To do that, the application developer should understand well both the UMTS QoS model and the data transport requirements of the application. Further, the developer should figure out how the application programming interfaces would support setup of QoS parameters when opening connections and exchanging data. Unfortunately, this turned out to be just too difficult so that the application- centric UMTS QoS model has not been widely used. The complexity of the UMTS QoS concept has also prohibited its implementation for real-time IMS Voice over IP support, which has later on become an important element of the 4G services even if not used over 3G networks.

The circuit switched domain of UMTS takes care of the strict QoS requirements of voice calls. The circuit switched model natively provides constant data rates and fixed latencies. For the packet switched domain, the de facto approach is the "best effort," where every data packet is treated equally regardless of the nature of the related data flows.

6.1.12.2   Packet Data Protocol (PDP) Context Activation by UE

After opening an RRC connection, the UE activates a packet data context as shown in Figure 6.19:

1) The UE sends a GMM SERVICE REQUEST via the RNC to the SGSN. This request identifies the subscriber and the type of the service request as signaling. The SGSN acknowledges the service request, sets up a new SCCP connection to the RNC, and thereafter performs subscriber authentication. Finally, the RNC starts traffic encryption as described in Section 6.1.9.2.

2) The UE sends an SM ACTIVATE PDP CONTEXT REQUEST message to the SGSN. The SM message describes the PDP context parameters such as the type of packet data protocol, APN name of the access point, QoS parameters, NSAPI identifier, and the PDP address which the UE would like to have. The SGSN checks if the subscriber is entitled to use the requested service and selects the GGSN based on the requested APN and the type of packet data protocol. The SGSN asks the selected GGSN to create a new primary or secondary PDP context. The latter is needed if a primary context already exists and the request has unique QoS requirements. The GGSN creates a GTP protocol tunnel toward the RNC and allocates an IP address for the UE, to be used for the new PDP context. The GGSN thereafter responds to the SGSN.

3) If a new radio access bearer is needed, the SGSN asks the RNC to check if RAN has extra capacity to support the additional bearer. If the needed capacity is available, the RNC then sets up the bearers with the NodeB. Thereafter, the RNC sends an RRC RADIO BEARER SETUP message to the UE. This RRC message tells the UE the transport and physical channel parameters of the dedicated channel. The UE takes the

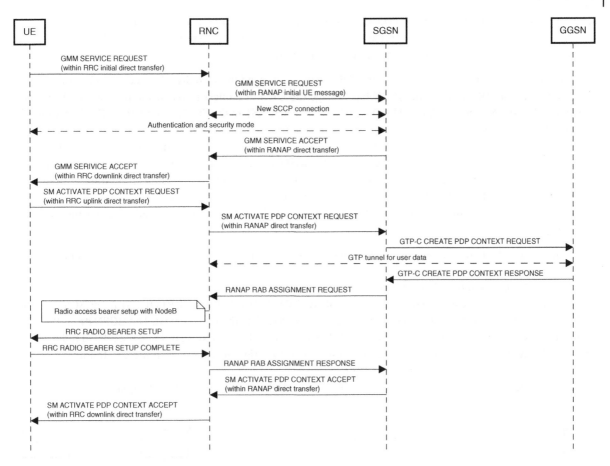

Figure 6.19 UMTS PDP context activation.

dedicated channel into use and responds to the RNC, which can finally tell the SGSN that the radio access bearer is now completed and is ready for transporting user data. The SGSN can now acknowledge the PDP context request to the UE, and packet data transfer can start over the new context.

For further details about the messages used in these procedures and secondary PDP context management, please refer to *Online Appendix I.4.5*.

The PDP context activation procedure as describes creates a radio bearer toward the UE for the PDP context. The radio bearer and its dedicated channel stay there whether or not there is any active transfer of user data, as long as the UE stays in the RRC Cell DCH state. For the idle periods, the RNC allocates a channelization code with large spreading factor and low bitrate for the dedicated channel. This is still wasteful if the cell has many UEs with PDP contexts not currently being used. To solve this problem, from Rel-5 onwards PDP contexts rely on the high-speed shared channel rather than channels dedicated for individual UEs. This allows the UEs to stay in the RRC Cell FACH state. The shared channel has only a single channelization code, which is shared among all the UEs relying on the channel.

After the PDP context has been activated, the UE may also move to one of the RRC PCH states whenever there is no user data to be transmitted. In that case, the radio bearers are dropped. When new data becomes available, the UE uses the service request procedure to return to the connected mode and get radio bearers opened for the data transmission. In that case, the SERVICE REQUEST message indicates the requested service type as data.

6.1.12.3 Packet Data Protocol Context Activation by the Gateway GPRS Support Node (GGSN)

After attaching to the UMTS packet service but yet without an active PDP context, the UE with a fixed public IP address can receive a data packet from the network as follows:

1) When the GGSN receives data packets toward the IP destination address fixedly allocated to the UE and the UE does not have a PDP context, the GGSN has to page the UE and request it to activate the context. To trigger paging, the GGSN sends an SM PDU NOTIFICATION REQUEST to the SGSN serving the UE. After responding to this message, the SGSN initiates the paging procedure.
2) The UE responds to paging with a GMM SERVICE REQUEST message, which indicates the service type as a paging response. This service request may cause the SGSN to trigger authentication as when the UE activates the context. The SGSN then requests the UE to initiate the PDP context activation procedure with an SM ACTIVATE PDP CONTEXT REQUEST message, as described in Section 6.1.12.2. After the PDP context has been activated and bearers created, the GGSN can eventually forward the data packets from the network to the UE.

For further details about the messages used in these procedures, please refer to *Online Appendix I.4.6.*

6.1.12.4 Packet Data Protocol Context Deactivation

To deactivate its PDP context, the UE sends an SM DEACTIVATE PDP CONTEXT REQUEST to the SGSN via the RNC, as shown in Figure 6.20. The SGSN requests the GGSN to delete the PDP context. The GGSN drops the GTP tunnel and responds to the SGSN, after which the SGSN together with the RNC tear down the related radio access and radio bearers. In the end, the SGSN sends a response to the original request from the UE, to inform the UE about the completion of PDP context deactivation.

For further details about the messages used in these procedures, please refer to *Online Appendix I.4.7.*

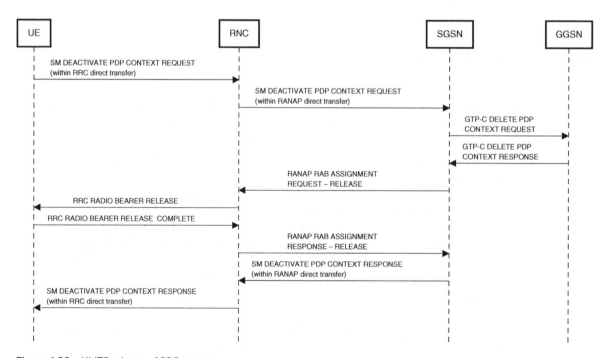

Figure 6.20 UMTS release of PDP context.

6.1.13 Mobility Management

The GSM core network, specifically its HLR and VLR databases, takes care of storing the location of a GSM mobile station. In the UMTS network, the storage of the UE location is organized in a hierarchical way. The RNC maintains location and state information of the UEs under its span of control while the HLR, MSC/VLR, and SGSN together maintain the UE location status over the whole UMTS radio network.

The accuracy of location information stored in the network depends on the state of the connectivity between the UE and the network. UE location is stored to the RNC with the accuracy derived from the RRC state of the UE:

- Idle: The RNC does not know the location of the UE since the UE has no connection to the radio network. The network core knows where the UE has reported its location most recently before becoming idle.
- Cell DCH: When the UE has a dedicated channel, the RNC tracks its location in the accuracy of cell, based on the location messages sent by the UE and the cell handovers performed under the control of the RNC.
- Cell FACH: When the UE uses shared channels, the RNC tracks its location in the accuracy of the cell, based on the location messages sent by the UE.
- Cell PCH: The RNC knows the cell where the UE has announced its location to the RNC. In this state, the UE tells the RNC always about its cell reselections so that the RNC is able to page the UE within one cell only.
- URA PCH: When the UE moves fast and sends data only infrequently so that it does not need a dedicated channel, the RNC tracks its location in the accuracy of the UTRAN registration area. The UE sends location update messages only when changing its registration area.

The MSC/VLR of the core network CS domain tracks the UE location according to the following mobility management (MM) states:

- MM detached: The UE is switched off and its location is not known.
- MM idle: The UE is registered to UMTS and attached with the MSC. The UE does not have a circuit switched connection and the MSC knows its current location area.
- MM connected: The UE has a circuit switched connection and the MSC knows its SRNC.

The SGSN of the core network PS domain tracks the UE location according to the following packet mobility management (PMM) states:

- PMM detached: The UE is switched off, does not have PDP context, and its location is not known.
- PMM idle: The UE is attached to the UMTS packet service and may or may not have a PDP context. The UE is in the RRC Idle state, so there is no active signaling connection from the SGSN to the UE. The SGSN uses the most recently reported routing area to reach the UE.
- PMM connected: The UE has an active PDP context and a signaling connection. The UE may be in any RRC state, except RRC Idle. The SGSN knows the location of the UE as its serving SRNC via which the traffic flows to/from the UE.

6.1.13.1 Cell Reselection

UEs in Cell PCH and URA PCH states are not engaged any active data transfer and reselect the camped cell by themselves. The cell selection decision of a UE is based on two pieces of information which the UE gets by listening to neighboring cells: strength of radio signals received and cell control parameters provided within SIB3 and SIB4 blocks. Among the accessible WCDMA cells, the UE tries to use the one with which it would have the best quality radio connection. On the other hand, the radio network provides the UE with hysteresis parameters to prevent the UE from excessively moving back and forth between the cells, when the UE is located on the edge of at least two cells, The UE is allowed to change the camped cell when the difference of signal strength between two cells exceeds a given hysteresis threshold continuously over a given period.

The UE may update the RNC with its current location with two types of RRC messages sent over the RACH transport channel:

- In the cell PCH state, the UE sends an RRC CELL UPDATE message to tell the network the ID of the camped cell.
- In the URA PCH state, the UE sends an RRC URA UPDATE message to tell the network the ID of UTRAN registration area (URA) as found from the SIB2 block

Both of these message types can be sent as regular updates or after changing the cell and URA. The UE may also send a cell update when it responds to paging. After receiving these messages over RACH, the network sends its response to the UE over the FACH transport channel.

6.1.13.2 Location Area Update for CS Domain

After having changed the cell, the UE may find from the system information of the new cell that the location area was changed as well. In this case, the UE shall update its location for the circuit switched network domain.

After opening an RRC connection, the UE sends the MM location update message as shown in Figure 6.21:

1) The UE sends a MM LOCATION UPDATE REQUEST message to the new MSC/VLR. This message tells the subscriber identity, the current location area of the UE, and the old location area from where the UE came to the new area. The MSC/VLR fetches subscriber authentication data from the HLR database, authenticates the subscriber, and starts encryption of traffic, as described in Section 6.1.9.2.

2) After successful authentication, the MSC/VLR informs the HLR about the new location of the UE. The HLR checks if the subscriber has the right to use the network to which the new MSC/VLR belongs and returns the result to the new MSC/VLR. If the HLR accepts the new network, the HLR stores the new location area and serving MSC/VLR to its subscriber location database and requests the old MSC to remove all data related to the

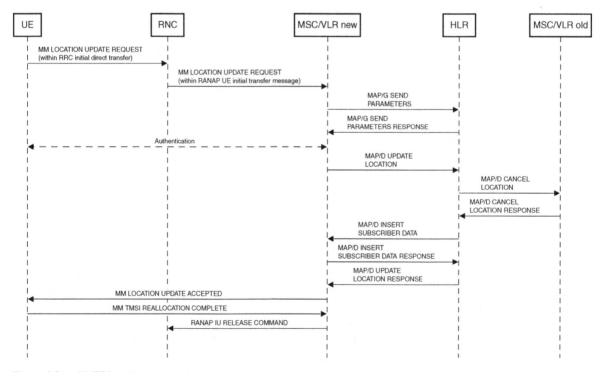

Figure 6.21 UMTS location area update.

subscriber from its VLR database. To the new MSC/VLR, the HLR sends necessary subscriber security details, as needed to authenticate the user.

3) The MSC/VLR returns an MM LOCATION UPDATE ACCEPTED message to the UE, along with the new TMSI identifier allocated for the subscriber. As the location update is now completed, the new MSC/VLR releases the signaling connection toward the UE.

For further details about the messages used in these procedures, please refer to *Online Appendix I.4.8.*

The MAP protocol messages used for the location update process are the same as used for the GSM system, since the circuit switched core network of WCDMA UMTS and GSM share the same architecture.

6.1.13.3 Routing Area Update for PS Domain

After changing the cell, the UE may find that the routing area was also changed. In this case, the UE shall update its routing area for the packet switched network domain after at first opening an RRC connection. The whole process of a routing area update is very similar to the location area update described in the previous chapter but with the following differences:

- While the UE and the HLR exchange location area update messages with the MSC/VLR of the CS network core, they exchange routing area update messages instead with the SGSN of the PS network core.
- The sequence of MAP protocol message exchange between the HLR and MSC/VLR for location area updates is reused between the HLR and SGSN for routing area updates.
- The UE uses MM protocol for location area updates, but for routing area updates it uses the GMM protocol. Within the RRC INITIAL DIRECT TRANSFER message the UE sends a GMM ROUTING AREA UPDATE REQUEST message, which the SGSN acknowledges in the end of the procedure with a GMM ROUTING AREA UPDATE ACCEPTED message.
- Instead of allocating a new TMSI identifier to the UE, the routing area update process allocates a new P-TMSI of the packet switched domain to the UE. After getting a new P-TMSI identifier, the UE responds to the SGSN with a GMM P-TMSI REALLOCATION COMPLETE message.
- The routing area update sequence has one extra step: After receiving a RANAP UE INTITIAL MESSAGE, the new SGSN sends a GTP-C SGSN CONTEXT REQUEST to the old SGSN to retrieve the PDP context data from it. The old SGSN returns the data within a GTP-C SGSN CONTEXT RESPONSE message.

The UE may also send a combined location and routing area update message to the SGSN node, if the SGSN is interconnected with an MSC/VLR over the Gs interface. In this case, the SGSN will forward the location area update to the MSC/VLR.

6.1.13.4 Handover in Radio Resource Control Connected State

When the UE is connected to a cell with a dedicated channel in a radio resource control (RRC) connected state, it no longer control its own cell reselections. Instead, the serving cell can only be changed via a handover controlled by the SRNC. Handover decisions are made for similar reasons as for GSM, such as the degradation of the radio connection between the UE and base station or the load of the cell becoming too high. Handovers can be performed in various ways:

- The UE is handed over to another UMTS WCDMA cell of the same network without changing the frequency band.
- The UE is handed over to another UMTS WCDMA cell of the same network so that the new cell uses a different frequency band than the old one. Such a change may happen when a UE in a moving vehicle changes the connection from a microcell to a macrocell operating on different bands.
- The UE is handed over to a cell that uses a different radio technology and frequency band than the currently used network. Such an **inter-RAT handover** may be caused by the UE leaving the WCDMA network coverage

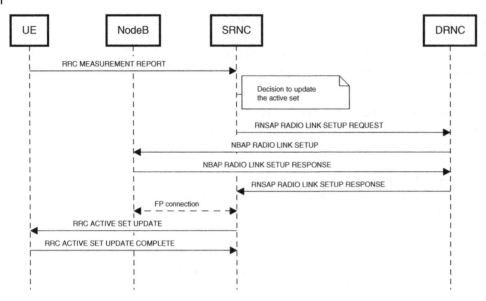

Figure 6.22 UMTS active set update for soft handover.

area or by the WCDMA network load exceeding the network capacity. Handover scenarios between UMTS and GSM are described in 3GPP TS 23.009 [45].

The measurement and handover decision processes are performed in the following way, see Figure 6.22:

1) The NodeB measures the error ratio and power loss on the active radio connection.
2) The UE measures the error ratio of its connection, the received pilot signal code power (RSCP) and EcNo, which means the energy per chip received from the pilot channel divided by the total power received (RSSI) over the same frequency. Additionally, the UE may measure the strength of signals received from adjacent cells which use either WCDMA or another radio technology. The UE leans the neighboring cells from SIB11 and SIB12 blocks broadcast in the current cell; thus, the UE does not need to make its measurements blindly.
3) The UE delivers its measurement results to the SRNC within an RRC MEASUREMENT REPORT message. The NodeB also sends its own measurement values to the RNC.
4) The SRNC compares any measurement values received from the UE and NodeB to make decisions about updating the active set of NodeB or performing inter-RAT handovers.

After making the handover decision, the SRNC performs soft handover as shown in Figure 6.22:

1) The SRNC which manages the radio access bearer toward the core network activates a new radio bearer toward the UE. If the new radio connection is created via a NodeB controlled by another radio network controller (DRNC), the SRNC sends a RNSAP RADIO LINK SETUP REQUEST message to the DRNC. The DRNC sets up radio bearers for the NodeB which was added into the active set and responds to the SNRC to have completed the radio link setup procedure.
2) The SRNC synchronizes an FP protocol connection for transporting user data between the SRNC and the new NodeB added to the active set. In the end, the SNRC informs the UE about the NodeB as a new member of the active set.

If the SRNC decides to remove a NodeB from the active set due to a degraded connection with the UE, removal is done as follows: At first, the SRNC sends an RRC ACTIVE SET UPDATE message to the UE to tell that a radio bearer has been removed from the active set. Thereafter, the SNRC either requests the DRNC to release the related radio bearers from the dropped NodeB or the SNRC releases those by itself, the NodeB is under its direct control.

After a number of active set updates, the current SNRC may not directly control any of the NodeBs in the active set. This causes an **SNRC relocation** where one of the current DNRCs will get the SNRC role for the active set. The core network shall be engaged to the SNRC relocation, since the endpoint RNC of the radio access bearer (RAB) is now changed. The SNRC relocation is performed as follows:

1) The old SRNC sends a RANAP RELOCATION REQUIRED message to the MSC/VLR and/or SGSN of the core network to inform them about the affected SRNC and UE. The MSC/VLR or SGSN moves the RAB by sending a RANAP RELOCATION REQUEST message to the new RNC and a RANAP RELOCATION COMMAND message to the old SRNC. The latter triggers transporting the user data toward the UE via the new SRNC.
2) When the new SRNC starts receiving user data from the old SRNC, it informs the MSC/VLR or SGSN about the transfer status. After that, the responsibility of the RRC connection toward the UE is moved to the new SRNC and the original RAB with the old SRNC is released.

For further details about the messages used in any of these procedures, please refer to *Online Appendix I.4.9*.

When an inter-RAT handover is performed from the WCDMA to the GSM network, the process is rather similar to SNRC relocation from the SNRC perspective. The SRNC communicates with the core network with RANAP RELOCATION messages, as described above. However, the core network communicates with the GSM base station controller with HANDOVER messages of the BSSMAP protocol.

6.2 High-Speed Packet Access

6.2.1 General

After UMTS specifications were completed for 3GPP Rel-4, it turned out that with expanded consumption of data, focus had to be shifted from voice to data use cases. The basic Rel-4 UMTS solution did not support high enough data rates, and there was room for improvement also for latencies. The enhancements of WCDMA technology in 3GPP releases 5, 6, and 7 had the goal to support high-bitrate, low latency data connections over UMTS networks.

High-speed packet access (HSPA) is a set of techniques used to increase data rates and cell total capacity for WCDMA packet switched data [46]. HSPA covers two different complementary technologies:

- **High-speed downlink packet access (HSDPA)** is able to increase the downlink peak data rates up to 14 Mbps (Rel-5) or 28 Mbps (Rel-7). Since HSDPA connections use shared channel, such peak transmission rates can be used for a single UE only over short periods. The achievable average rates depend on the other HSDPA UEs within the cell and their data flows. Furthermore, the achievable data rate depends on the UE support for optional HSDPA capabilities as well as from the distance between the UE and NodeB. By deploying HSDPA, it is possible to increase the total data transport capacity of a cell by 50–100% compared to the basic WCDMA cell capacity. This extra capacity is divided between the HSDPA and WCDMA users of the cell based on cell-specific parameter settings. HSDPA was introduced in 3GPP Rel-5 specifications and further enhanced in later 3GPP releases.
- **High-speed uplink packet access (HSUPA),** also referred as the **enhanced dedicated channel (E-DCH),** is able to increase the uplink peak data rates up to 3–4 Mbps (Rel-6) or 11 Mbps (Rel-7). HSUPA was introduced the first time in 3GPP Rel-6 specifications.

While HSPA technologies can reach bitrates of multiple Mbps, the actually reached bitrates depend on many conditions, such as supported capabilities and radio conditions between the network and UE as well as the number of HSPA users in the cell [15].

HSPA specifications are embedded to a number of technical specifications within TS 25 standardization series of 3GPP. The most important specifications dedicated for HSPA are as follows:

- TS 25.308 [47]: HSDPA overall description
- TS 25.319 [48]: HSUPA overall description
- TS 25.214 [17]: HSPA physical layer procedures
- TS 25.931 [11]: Examples of HSPA signaling procedures

3GPP HSPA standards have been developed and published within subsequent standards releases:

- 3GPP Rel-5 focus areas were the usage of packet switched protocols in a core network and HSDPA techniques to support higher downlink data rates.
- 3GPP Rel-6 defined the HSUPA technology to provide higher uplink data rates.
- 3GPP Rel-7 to Rel-9 specification releases provided several improvements to both HSDPA and HSUPA, aiming at further improving the data rates and minimizing current consumption of the mobile station that uses always-on data connections. Always-on connections are needed for various kinds of instant messaging and speech applications developed on top of IP protocols. Those applications need always-on data connection toward IP networks to be always reachable for incoming messages or calls. HSPA supports such connections with a Rel-7 feature set referred to as continuous packet connectivity (CPC) and Rel-8 fast dormancy.
- Releases from Rel-10 onwards provided various further enhancements, such as multi-carrier HSDPA, MIMO support, and transmit diversity.

Since there are many optional HSPA features specified over multiple 3GPP releases, a mechanism is needed for the UE to announce which of the features the UE supports. The UE lists the supported features into the RRC CONNECTION SETUP COMPLETE message sent in the end of the RRC connection opening sequence. This allows the network to adjust its operation against to what the UE is able to support.

6.2.2 High-Speed Downlink Packet Access

As smartphones evolved to support extensive use of the Internet, use cases like Web browsing and video streaming became dominant. To provide a good Internet experience, the data connection has to support quick downloading of Web pages and smooth streaming of videos. Technically, this means the following requirements:

- The supported bitrates shall be high enough to allow the UE to quickly download large data files (such as digital images) needed to render the Web page.
- The **round-trip delay (RTD)** should be as short as possible, because Web access means a number of access rounds over the network. At first, the browser uses DNS queries to convert the domain name within the Web URL to an IP address from which to fetch the data. Thereafter the page is retrieved with a number of HTTP queries, each fetching some elements of the page to be rendered. The longer it takes to receive responses for any of these queries, the longer it takes to get the Web page shown.
- The RTD should also be as constant as possible, to prevent TCP protocol connections (as used by HTTP) to apply congestion avoidance which slows down the effective data rate. TCP interprets the increasing RTD as a sign of congestion and slows down transmission to mitigate it (see Chapter 3, Section 3.3.5 for further details). However, RTD can also increase due to bit errors on unreliable radio links causing packet retransmissions. In such cases, the TCP congestion avoidance is an inefficient reaction.

This chapter introduces the original Rel-5 design of HSDPA while the more advanced methods added in later 3GPP releases are visited in Section 6.2.4. HSDPA applies the same basic approaches to increase WCDMA data rates as EGPRS did for GPRS. The HSDPA technology is based on the WCDMA radio interface; however, HSDPA user data is transported from the base station to UEs via the shared HS-DSCH transport channel and the corresponding HS-PDSCH physical channel. Further information about HSDPA physical layer and signaling procedures can be found from TS 25.858 [49] and TS 25.877 [50].

HSDPA uses the following new mechanisms to increase bitrates and decrease latencies:

- New modulation and channel coding methods to increase data rates. HSDPA uses **link adaptation** with which the base station selects the best modulation and line coding method that fit the current radio conditions with the UE. While WCDMA relies on QPSK modulation, HSDPA Rel-5 introduced 16-QAM (quadrature amplitude modulation), which has 16 combinations of amplitude and phase to carry 4 bits per symbol. The NodeB chooses the modulation method based on the estimates received from the UE and the ratio between positive and negative L1 acknowledgments from the UE. 16-QAM can be used on the new shared HS-DSCH transport channel when the UE is close enough to NodeB and has a strong radio connection with it. The NodeB is able to change the used modulation and coding scheme once per TTI, when also the HS-DSCH channel allocation may be switched between UEs.

- Extended multicode operation is also used to increase data rates. The HS-DSCH channel can use 15 different channelization codes on parallel. When a single HSDPA UE may simultaneously receive 5, 10, or 15 code words from the HS-DSCH channel, it is possible to multiplex one to three UEs to the HS-DSCH channel at the same time, using code division multiplexing. In a longer period, the number of UEs sharing the HS-DSCH channel can of course be much bigger, when taking the TDM multiplexing of the channel into account.

- HSDPA latencies (RTD) were decreased by moving all lower layer processing from the RNC to the NodeB. This also meant dropping the soft handover support for HSDPA. An HSDPA UE is always connected to one NodeB at a time, which streamlines the packet retransmission process. With the WCDMA DCH channel, any retransmission of a corrupted data packet is decided by the RNC, which selects the best version of uplink packets received from the NodeBs in the active set. Forwarding the packets and acknowledgments between the RNC and NodeBs increases the processing latency and makes data buffers longer with high data rates. Long buffers consume memory and increase end-to-end latency due to waiting time. HSDPA does not use soft handover to minimize buffering and to eliminate unnecessary processing steps. With HSDPA, the fast retransmissions are under control of the NodeB, simply based on the received L1 acknowledgments. In this way, the retransmissions can be sent without extra delay.

- Another approach for decreasing latencies and optimizing the cell capacity was to improve scheduling and decrease the minimum TTI time. HSDPA TTI length is only 2 milliseconds, which is shorter than the minimum UMTS TTI of 10 ms. The selection of the UEs scheduled for HS-DSCH is the responsibility of the NodeB, while the RNC has the responsibility of scheduling UEs for the traditional DSCH transport channel. As the NodeB is in direct control of the radio interface, it has the lowest latency for scheduling and rate adaptation decisions. By putting the NodeB to the decision maker seat, it is possible to maximize the overall capacity of the cell, still providing a fair share of the capacity to all UEs camping in the cell. The precise algorithm for how to divide the capacity between the UEs within the cell is not standardized, but the network vendors may select suitable algorithms by themselves. With proportional fair scheduling, NodeBs may allocate radio resources based on signal quality fluctuations toward different UEs within the cell so that minimal capacity would be lost for retransmissions or too low levels of QAM.

- In order to make the round-trip time even shorter and more predictable, HSDPA MAC protocol supports a method known as L1 **hybrid automatic repeat request (HARQ).** HARQ combines forward error correction and message retransmission, which is why it is called a hybrid method. L1 HARQ processes are run between UE and NodeB, so that no extra delay is caused by involving RNC to the process. With the short HSDPA TTI, the HARQ retransmissions can be completed in less than 10 milliseconds, which is a huge improvement compared to the typical 100 millisecond UMTS RLC frame retransmission time [15]. HSDPA uses RLC retransmissions as a secondary method in cases where the L1 level method does not work, such as in a cell handover. HARQ does not use a sliding window, but Stop and Go (see *Online Appendix A.6.3.1*) where acknowledgments are sent for every frame within 5 milliseconds from receiving the frame. To avoid a corrupted frame to stop transmissions completely, up to eight parallel HARQ processes may be running between the UE and NodeB. HARQ supports two types of retransmission approaches: soft combining and incremental redundancy (IR). In **soft combining** (or **chase combining**), the retransmitted packet is a copy of the original one, allowing the L1 receiver to combine signal energy from those. In **incremental redundancy,** the retransmission provides the receiver with the convolutional code punctured differently than for the original data packet. When a retransmission is pending, the UE and NodeB shall buffer frames received from other HARQ processes to be able to pass them to upper layer protocols in the correct order.

With these enhancements, HSDPA could achieve 14 Mbps downlink data rate and **round-trip delay (RTD)** of 100 milliseconds, being roughly half of Rel-4 UMTS round-trip delay. Usage of link adaptation and scheduling over the shared channel also have an impact to HSDPA power control. Instead of applying the fast power control, the NodeB reacts to signal quality fluctuations to the HSDPA UE with picking the right moment to schedule transmission to the UE and, when necessary, adapting the link with a more robust type of modulation.

3GPP Rel-5 specifications define 12 different HSDPA UE categories. The following UE characteristics are specified per category:

- Whether or not the UE supports 16-QAM
- How many HS-DSCH spreading codes and parallel HS-PDSCH channels the UE is able to simultaneously use
- Whether the UE is able to support continuous HS-DSCH reception or just one or two TTIs out of three consecutive ones

A WCDMA NodeB that supports HSDPA shall also support WCDMA UEs without HSDPA support. Consequently, the base station has to divide its resources between HSDPA and basic WCDMA services. The division is dynamically controlled by the RNC, based on the demand for circuit switched, traditional, and high-speed data services. To change the allocation of NodeB power and channelization codes between HSDPA and WCDMA services, the RNC sends NBAP PHYSICAL SHARED CHANNEL RECONFIGURATION messages to the NodeBs.

The HSDPA user plane protocol stack is shown in Figure 6.23. For HSDPA, a new MAC-hs entity was introduced to handle traffic of the new high-speed transport channel. Its relationship to other MAC entities is as follows:

- MAC-d forwards DTCH user data traffic and DCCH signaling either to the DCH transport channel or the HS-DSCH channel via MAC-hs. Note that when using HSDPA, the UE is still allocated a slow DCH channel for control purposes, like for RRC protocol. The MAC-d protocol entity runs within the SRNC and UE.
- MAC-c/sh in the CRNC may also use either the DSCH or HS-DSCH transport channel, the latter via MAC-hs. The MAC-c/sh protocol entity runs within the CRNC and UE.

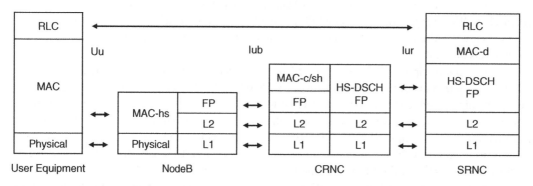

Figure 6.23 Protocol architecture of HSDPA. *Source:* Adapted from 3GPP TS 25.308 [47].

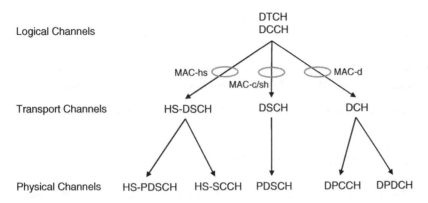

Figure 6.24 Mapping between logical, transport, and physical channels with HSDPA.

- MAC-hs maps traffic from MAC-d or MAC-c/sh to the HS-DSCH transport channel. The MAC-hs protocol entity runs within the NodeB and UE.

Since the NodeB buffers data sent over the HS-DSCH, there is a need for a flow control mechanism between the NodeB and RNC. This mechanism takes care of avoiding buffer overflows at the NodeB.

HSDPA uses three new physical channels against the HS-DSCH transport channel, as shown in Figure 6.24:

- High-speed physical downlink shared channel (HS-PDSCH) is used for high-speed, low latency user data transport. Up to three UEs can be multiplexed to the channel with code multiplexing, and in TDM dimension the channel can be allocated to a new UE every TTI of 2 milliseconds. As an HSDPA UE sends L1 acknowledgments for each TTI, the shorter TTI decreases latencies caused by both scheduling and retransmissions. A cell may have one or multiple HS-PDSCH channels. Since the HS-PDSCH channel uses fixed spreading factor 16, the maximum number of these channels per cell is 15 when no other user data channels are provided.
- High-speed shared control channel (HS-SCCH) provides the UE with parameters needed to decode and process traffic from HS-PDSCH and HS-SCCH physical channels, which carry data from the HS-DSCH transport channel. As this signaling data is critical, high spreading factor 128 is used on HS-SCCH. Each UE multiplexed to the HS-DSCH channel has its own HS-SCCH channel. The three timeslots of HS-SCCH channel contain the following information:
  - Which HS-PDSCH channel (if any) will carry HS-DSCH data for the UE in its next timeslots.
  - Which channelization codes, modulation types, and transport formats the UE should use to decode the next three timeslots of HS-DSCH channel.
  - Whether the next transport block from the HS-DSCH channel is a new one or a retransmission by HARQ. In the latter case, a HARQ process number is given so that the UE is able to match the retransmission with the corresponding original block.
- High-speed dedicated physical control channel (HS-DPCCH) is an uplink channel for the UE to send L1 acknowledgments (ACK or NACK) and downlink channel quality indicator (CQI) to the NodeB. While the NodeB uses L1 acknowledgments for HARQ retransmission control, it uses CQI to decide the combination of modulation method, transport block size, and channelization codes used for transmitting HSDPA data blocks to the UE. The NodeB may use CQI reports also for its scheduling decisions, to schedule transmission toward UEs which at the moment have favorable channel conditions. The spreading factor of 256 is used for the HS-DPCCH channel.

The UE with HSDPA support still uses UMTS dedicated channels, such as DCCH for RRC messages and DTCH for uplink packet switched data (unless the UE uses HSUPA for the uplink). DTCH may be also used bidirectionally for any voice calls while data is downloaded over the HS-DSCH channel [15].

For the RRC state model, HSDPA has a new meaning for Cell DCH state. The UE no longer gets a dedicated channel for user data transfer, but HSDPA uses Cell DCH state for the dedicated downlink control channel still being used for signaling.

HSDPA channel coding is simpler than WCDMA channel coding due to the short TTI and mapping the transport blocks to only one HS-DSCH physical channel. HSDPA channel coding has the following properties:

- The bits sent after adding the CRC checksum are scrambled to avoid long sequences of bit strings containing only 1 or 0 bits.
- There is no need for combining blocks from multiple transport channels to one physical channel.
- HSDPA uses only turbo coding. After turbo coding, each transport block is given to the HARQ process responsible of transmitting the block to the UE.
- The blocks are interleaved only within the 2 ms TTI to keep latencies short, and they are sent using the channelization codes allocated for the UE.

As mentioned earlier, soft handover is never used for HSDPA. The connection to the old cell is always dropped when a HSDPA connection is moved to a new cell. With HSDPA, the hard handover procedure is performed as follows:

1) The UE measures the pilot signal received from nearby WCDMA base stations and sends the measured values to the SRNC with an RRC MEASUREMENT REPORT message.
2) The SRNC sends a NBAP RADIO LINK RECONFIGURATION PREPARE message to both the new and old base stations. Among various handover aspects, this message determines the timing of the hard handover.
3) The SRNC synchronizes the FP protocol connection for user data between itself and the new base station.
4) The SRNC sends an RRC PHYSICAL CHANNEL RECONFIGURATION message to the UE. This message tells the UE all the information it needs to be able to take the HSDPA channel into use with the new base station.
5) When the time comes to perform the handover, the old HSDPA base station stops sending data to the UE. The UE starts listening to the new cell and sends it CQI values, with which the new NodeB can choose optimal channel configuration, if it supports HSDPA.
6) When the new base station has received the CQI value from the UE, it selects the modulation method and other parameters of HS-DSCH channel for the UE and starts sending buffered user data packets. If the handover is done between different sectors of a single base station, the base station may pass the buffered data from a sector to another within the MAC protocol layer. In other cases, RLC protocol takes care of resending those packets which the UE did not acknowledge before performing the handover.

6.2.3 High-Speed Uplink Packet Access

Web browsing was the major use case to drive HSDPA improvements for downstream bandwidth, but it did not have a similar impact for uplink requirements. The traffic pattern of Web browsing is very asymmetric so that the uplink traffic is only a small fraction of the downlink traffic. Applications like video calls and social networking cause demand for uplink traffic. The traffic pattern of a video call is symmetric, and those who publish social media contents may use uplink for uploading big video or still picture files taken with mobile phones. 3GPP specified high-speed uplink packet access (HSUPA) to improve uplink performance. This section introduces the original Rel-6 design of HSUPA; the more advanced methods added in later 3GPP releases are visited in the next Section 6.2.4.

HSUPA Rel-6 increased the maximum uplink bitrate from a few hundred kbps to 5.7 Mbps. This is achieved with the new dedicated E-DCH transport channel for which small spreading factors between 2 to 64 can be flexibly used. To increase data rates even higher, the E-DCH channel is able to use up to four channelization codes in parallel, effectively using multiple physical channels for transporting uplink data. The highest bitrate is reached with a multicode channel using two channelization codes with SF = 2 and another two with SF = 4. Since the transmission power of a single UE is much lower compared to the base station, HSUPA was originally designed to use the same QPSK modulation method like the WCDMA DCH channel. Consequently, no link adaptation was needed. This was changed from Rel-7 onwards as more advanced QAM schemes were brought to HSUPA to increase data rates further. To decrease uplink latencies, HSUPA uses same methods as the HSDPA downlink: HARQ retransmission processes and short 2 ms TTI.

The biggest difference between HSUPA and HSDPA is that while HSDPA uses a shared channel with link adaptation, HSUPA still uses dedicated channels with UE power control. There are a few benefits for that. First of all, with dedicated channels there is no need for time to synchronize the UEs within the cell, which would be necessary for them to use a shared uplink channel. No timing advance had to be introduced for HSUPA. Secondly, with dedicated channels the UEs do not need to wait for an uplink slot, which minimizes latencies. Finally, HSUPA can still use soft handover to minimize UE transmission power. While a NodeB using HSDPA can allocate its total transmission power to one or a few UEs at a time, in HSUPA the received power always comes from the whole UE population connected to the cell. HSUPA uses power control and soft handover to keep the overall interference level of the cell below planned limits.

HSUPA operates as follows:

- In HSUPA, uplink user data is transmitted via a dedicated E-DCH transport channel. The E-DCH traffic is transported over the physical E-DPDCH data and E-DPCCH control channels. While a WCDMA UE may have multiple DCH channels for its data flows, the UE may use only one single E-DCH channel at a time. If there are multiple data flows to be transported, the MAC protocol multiplexes those to the single E-DCH transport channel. The E-DCH channel may be mapped to multiple physical E-DPDCH channels, each of which has its own spreading code.

- The HSUPA signaling process for setting up the E-DCH channel for the UE is quite similar to establishing a UMTS DCH channel. At the initial setup, the RNC informs the UE within an RRC PHYSICAL CHANNEL RECONFIGURATION message about the HARQ processes which are enabled and the transport format combination set (TFCS) which the UE may use for the E-DCH channel. The UE picks a specific TFC based on the scheduling grants later received from the NodeB. The SNRC may also configure some data flows as non-scheduled so that the UE can send data any time up to a configured number of bits to keep the QoS requirements for the flow.

- E-DCH uses HARQ retransmission processes with either soft combining or incremental redundancy. The E-DCH channel is able to use a maximum of four different HARQ processes on parallel. Since retransmission of a certain transport block is done at exactly a known time, there is no need to send the HARQ process number over the E-DCH channel. It is enough for the UE to tell for each transport block whether or not it is a retransmission. The UE makes the retransmission decisions based on L1 acknowledgments from the NodeB. Since the soft handover is used for E-DCH, the UE stops the retransmissions after getting at least one acknowledgment from any of the NodeBs in the active set.

- E-DCH scheduling is done at the NodeB against the bandwidth requests that the UE sends over the E-DCH channel. The bandwidth request (SI message of MAC-e protocol) describes the amount and priority of data buffered at the UE and an estimate of the available UE transmit power. For its scheduling decisions, the NodeB also uses QoS related information from the SNRC. The precise method of allocating the transmission power and capacity between the UEs is not standardized, but the network vendors may select any method appropriate. In its absolute scheduling grants, the NodeB tells the maximum power ratio between HSUPA physical data and control channels to be used, which effectively limits the UE transmit power available for the E-DPDCH data channel. Based on the granted and available transmit power, the UE selects one of the transport format combinations to use on E-DPDCH. The TFC used then determines the bitrate of the E-DCH transport channel.

- The NodeB may also send relative grants over the E-RGCH channel to modify the current data rate. Relative grants can be given once every TTI to increase or decrease the power level step-by-step. When adjusting its transmission power accordingly, the UE may need to reselect the TFC to keep the signal quality acceptable. This in turn affects the available bitrate, like for absolute grants.

3GPP Rel-6 specifications define six different HSUPA UE categories. The following UE characteristics are specified per category:

- The number of parallel E-DPDCH channels and the spreading factors used for them
- Whether the UE supports 2 ms uplink TTI in addition to the 10 ms TTI

In addition to the E-DCH transport channel, it is possible for the UE to have a maximum 64 kbits DCH channel for circuit switched data, RRC protocol messages, and the downlink power control of DPCCH channel and pilot signal. When selecting the transport block size for the E-DCH channel, the UE must take also into account the power needed for the DCH channel. Since transport of circuit switched data requires maintaining constant data rate on the DCH, the power needed for that is automatically subtracted from the total power that the NodeB allows the UE to use for the E-DCH channel.

The HSUPA uses the following physical channels, as shown in Figure 6.25:

Uplink dedicated channels:

- Enhanced dedicated physical data channel (E-DPDCH) is the high-speed dedicated uplink user data channel, used also for RRC signaling. The TTI of E-DPDCH channel is either 10 or 2 ms. The shorter 2 ms TTI is used only with higher data rates when the UE is close to the NodeB to minimize the latencies caused by retransmissions. One single encoded transport block is sent over the E-DPDCH per TTI. The smallest spreading factor used on the E-DPDCH is two for high data rates and the channel can use up to four spreading codes in parallel. Maximum data rate of E-DPDCH is between 15 kbps–5.76 Mbps.
- Enhanced dedicated physical control channel (E-DPCCH) carries signaling needed by the NodeB to receive data from the E-DPDCH channel. Fixed spreading factor (256) is used on the E-DPCCH to send the following 10-channel coded data bits within one 2 ms TTI:
  - The E-DCH transport format combination indicator (E-TFCI) tells the size of the transport block from which the NodeB is able to deduce the number of simultaneously used E-DPDCH channels and the spreading factors applied.
  - The retransmission sequence number (RSN) bits tell the sequence number of HARQ retransmission attempts of a transport block sent over E-DPDCH.
  - A happy bit tells if the data rate provided by the NodeB for the E-DCH is sufficient to deplete the UE's data buffers within the next N x TTIs, where N is a value given by the network.

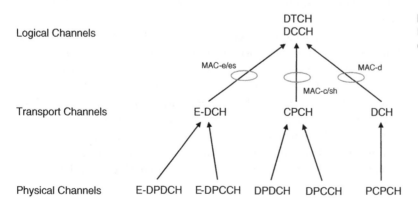

Figure 6.25 Mapping between logical, transport, and physical uplink channels with HSUPA.

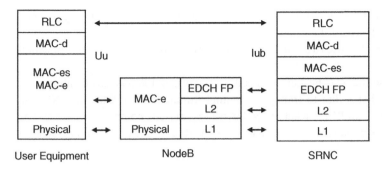

Figure 6.26 Protocol architecture of HSUPA. *Source:* Adapted from 3GPP TS 25.319 [48].

Downlink channels:

- E-DCH absolute grant channel (E-AGCH) is used by the NodeB to inform the UE about the E-DCH allowed maximum transmission power in terms of power ration between E-DPDCH and E-DPCCH. In the soft handover, only one of the NodeBs is using the E-AGCH channel at a time.
- E-DCH relative grant channel (E-RGCH) is used by the NodeB to tell the UE to either increase or decrease its transmission power by one step. In the soft handover, each NodeB in the active set may send its own power decrease command over the E-RGCH channel, but only the NodeB which uses E-AGCH may send a power increase command on the E-RGCH channel.
- E-DCH HARQ indicator channel (E-HICH) is used for sending L1 acknowledgment for HARQ data packets received from the E-DPDCH channel.

The HSUPA channel coding works like the HSDPA except that the HSUPA does not use scrambling. For the HSUPA, the following MAC entities are added below MAC-d, as can be seen in Figure 6.26:

- MAC-e protocol runs in the UE and NodeB and takes care of most of the MAC layer tasks, such as scheduling, selection of the transport format based on the available transmission power, as well as the HARQ retransmission processes. The UE may send a scheduling information (SI) message of MAC-e protocol either regularly or when running out of E-DCH buffers. The SI message tells NodeB:
  - The total amount of data in the E-DCH buffers of the UE
  - The total amount of data in the buffer of the highest priority logical channel of the UE
  - The priority level of the highest priority logical channel
  - The UE power headroom (UPH) is the ratio between UE maximum power and the power currently used to transmit a pilot bit over the DPCCH channel
- MAC-es protocol runs in the UE and RNC and takes care of reordering the data packets received from different NodeBs over soft handover. Due to the retransmissions done in MAC-e layer and the NodeB processing delays, it is possible that one of the NodeBs in the active set still provides the RNC with a transport block sent earlier than those transport blocks that the RNC already gets from other NodeBs.

6.2.4 High-Speed Packet Access Advanced

New HSPA enhancements were introduced between 3GPP Rel-7 and Rel-11 to increase data rates and decrease transmission latencies and UE power consumption. HSPA advanced (or HSPA+) enhancements to achieve higher data rates were as follows:

- New modulation methods: 64-QAM for HSDPA and 16-QAM for HSUPA to be used with very good signal conditions when the signal-to-noise ratio is high.

- MIMO technology using multiple parallel antennas to transmit two or four parallel streams of data as generated by a single channel coding block. See TS 25.214 [17] for further details.
- Dual carrier HSDPA where the NodeB and UE use two 5 MHz WCDMA carriers for downlink communication instead of one carrier only.
- Support of large RLC packets to decrease the RLC header overhead. In the original Rel-99 specifications, the size of RLC packets was limited to 40 bytes due to transport channel limitations between the RNC and NodeB. This restriction was lifted to enable an RLC packet of 1500 bytes to carry one complete IP packet. The MAC-hs layer running in the NodeB was enhanced to be able to segment RLC packets to smaller transport blocks to be sent over the radio interface.

With these enhancements (mainly new modulation methods and MIMO), it became possible to achieve maximum data rates of 42 Mbps for HSDPA and 11.5 Mbps for HSUPA. In Rel-7 and Rel-8, a total of 12 new UE categories were defined for different combinations of 64-QAM, MIMO, and dual carrier HSDPA support. In Rel-11, there were already 38 HSDPA UE categories. For HSUPA, one category was added for 16-QAM support in Rel-7 and five other categories were taken into use by Rel-11.

New MAC protocol entities were introduced as follows:

- The HSDPA provided MAC-ehs as an alternative option to an original MAC-hs design. The difference between these designs is that MAC-ehs supports dual stream transmission, logical channel demultiplexing with logical channel ID, and segmentation of MAC messages.
- The HSUPA provided MAC-i/is entities as alternative option to original MAC-e/es design. The difference between these designs is that MAC-i/is supports message segmentation, protection of messages from the MAC-c entity, and a new mechanism to determine if the UE uses a common or dedicated channel.

The HSPA network architecture was also aligned in Rel-7 toward the architecture to be used for the next generation LTE technology. The new architecture supported merging of the RNC functionality to the NodeB and dropping altogether the soft handover support. Additionally, support was introduced for GTP-U tunnels between the NodeB/RNC and GGSN, so that the SGSN become a signaling node instead of staying on the user data path. Reducing the number of network elements on the user data path decreased the latency of packet switched data processing on HSPA network.

Rel-7 **continuous packet connectivity (CPC)** supports always-on IP connections with features designed to save the power of the UE and make it easier for the UE to move between states of being idle and actively transporting data:

- Usage of discontinuous transmission and reception (DTX/DRX) on the HS-SCCH and UL-DPCCH control channels when there is no user data to be transmitted, to minimize UE power consumption. With DTX/DRX, the UE sends an uplink and listens for downlink only periodically, not continuously.
- Transmit power is saved also by reducing the number of CQI reports sent and using smaller transmit power on the UL-DPCCH control channel. The latter is achieved by using more redundant bits to protect TPC commands when no user data is sent.
- Using HSDPA logical channel HS-DSCH (mapped to physical channel HS-PDSCH) instead of the FACH logical channel (as mapped to S-CCPCH physical channel) for the UE in the enhanced Cell FACH state and instead of PCH in the enhanced Cell/URA PCH states. This enhancement brings various benefits. At first, the achievable data rates in Cell FACH state increase from a tiny 32 kbps to an astonishing 42 Mbps. With higher Cell FACH data rates, transition to Cell DCH state may be avoided, but when the transition to Cell DCH is needed to send a large amount of data, it can be done quite quickly.

Rel-8 **fast dormancy** was yet another method to decrease UE power consumption and the network signaling load for always-on IP connections. Fast dormancy means the way for the network to modify or drop the RRC connection toward the UE without releasing its PDP context. While that was possible already with earlier 3GPP UMTS releases, it was not until Rel-8 when a fast dormancy procedure was officially introduced into the specifications to improve the way for handling the idle RRC connections. Before Rel-8, the UE becoming idle would send a signaling connection release indicator to the network, which initiates the RRC connection release. With Rel-8 fast dormancy, the UE could add a new cause, "PS data session end", to the signaling connection release indicator. That would trigger the network to change the RRC state to FACH or PCH rather than to drop the RRC connection. Thereafter, the UE would start a timer between 5 seconds and 2 minutes to wait for new data to appear for the session. If the timer expires without data connection becoming active, the UE would send another signaling connection release indicator with the cause, "UE requests PS data session end", causing the RRC connection to be released. If the data connection would reactivate while the inactivity timer is still running, the RRC connection state can be changed back to DCH with less effort than when opening a new RRC connection.

A new method was defined to transport CS voice over packet-based HSPA channels. **CS voice over HSPA** uses the traditional CS voice signaling but transports the voice media as VoIP packets over high-speed data channels. The VoIP media stream endpoints are the UE and media gateway, which forwards the voice media in the traditional CS manner toward the rest of the network. This method would have the benefit of increasing spectral efficiency and minimizing UE power compression. VoIP media allows the UE to use HSPA DTX/DRX even with a continuous voice stream as VoIP packets are sent only once every 20 ms. Despite of its benefits, CS voice over HSPA has not been widely deployed, largely due to lack of real-time QoS support within the UTRAN networks.

6.3 Questions

1   What is the size of WCDMA system bandwidth?

2   What kind of roles can an RNC have?

3   What is the meaning of "bearer" in UMTS?

4   What are the purposes of WCDMA channelization and scrambling codes?

5   To which purposes is the RRC protocol used for?

6   What has the term "active set" to do with soft handover?

7   What is a Rake receiver?

8   What is the purpose of WCDMA compressed mode?

9   Why does UMTS use fast power control?

10   How is WB-AMR codec better than the GSM full rate codec?

11   How does "incremental redundancy" differ from "chase combining"?

12   In which ways does HSDPA decrease latencies compared to UMTS?

13   Why are shared channels not used for HSUPA?

References

1 ITU-R Recommendation M.687 International Mobile Telecommunications-2000 (IMT-2000).

2 ITU-R Recommendation M.816 Framework for services supported on International Mobile Telecommunications-2000 (IMT-2000).

3 ITU-R Recommendation M.1034 Requirements for the Radio Interface(s) for International Mobile Telecommunications-2000 (IMT-2000).

4 ITU-R Recommendation M.1079 Performance and Quality of Service Requirements for International Mobile Telecommunications-2000 (IMT-2000) Access Networks.

5 Holma, H. and Toskala, A. (2004). *WCDMA for UMTS: Radio Access for Third Generation Mobile Communications*. West Sussex: Wiley.

6 IEEE 802.16e-2005 Air Interface for Broadband Wireless Access Systems.

7 Anttalainen, T. and Jääskeläinen, V. (2015). *Introduction to Communications Networks*. Norwood: Artech House.

8 3GPP TS 23.205 Bearer-independent circuit-switched core network; Stage 2.

9 3GPP TS 25.401 UTRAN overall description.

10 3GPP TS 23.107 Quality of Service (QoS) concept and architecture.

11 3GPP TS 25.931 UTRAN functions, examples on signalling procedures.

12 3GPP TS 25.301 Radio interface protocol architecture.

13 3GPP TS 25.201 Physical layer – general description.

14 3GPP TS 25.213 Spreading and modulation (FDD).

15 Sauter, M. (2021). *From GSM to LTE-Advanced pro and 5G : An Introduction to Mobile Networks and Mobile Broadband*. West Sussex: Wiley.

16 3GPP TS 25.215 Physical layer; measurements (FDD).

17 3GPP TS 25.214 Physical layer procedures (FDD).

18 IETF RFC 1661 The Point-to-Point Protocol (PPP).

19 IETF RFC 1662 PPP in HDLC-like Framing.

20 3GPP TS 25.434 UTRAN Iub interface data transport and transport signalling for common transport channel data streams.

21 3GPP TS 25.435 UTRAN Iub interface user plane protocols for common transport channel data streams.

22 3GPP TS 25.427 UTRAN Iub/Iur interface user plane protocol for DCH data streams.

23 3GPP TS 25.211 Physical channels and mapping of transport channels onto physical channels (FDD).

24 3GPP TS 25.212 Multiplexing and channel coding (FDD).

25 3GPP TS 25.321 Medium access control (MAC) protocol specification.

26 3GPP TS 25.322 Radio Link Control (RLC) protocol specification.

27 3GPP TS 25.323 Packet data convergence protocol (PDCP) specification.

28 IETF RFC 2507 IP Header Compression.

29 IETF RFC 3095 Robust Header Compression (ROHC): Framework and four profiles: RTP, UDP, ESP, and uncompressed.

30 3GPP TS 25.331 Radio resource control (RRC); Protocol specification.

31 3GPP TS 24.008 Mobile radio interface Layer 3 specification; Core network protocols; Stage 3.

32 3GPP TS 24.010 Mobile radio interface layer 3; Supplementary services specification; General aspects.

33 3GPP TS 25.414 UTRAN Iu interface data transport and transport signalling.

34 Kaaranen, H., Ahtiainen, A., Laitinen, L. et al. (2005). *UMTS Networks – Architecture, Mobility and Services*. West Sussex: Wiley.

35 3GPP TS 25.415 UTRAN Iu interface user plane protocols.

36 3GPP TS 25.433 UTRAN Iub interface Node B application part (NBAP) signalling.

37 3GPP TS 25.413 UTRAN Iu interface radio access network application part (RANAP) signalling.

38 3GPP TS 25.423 UTRAN Iur interface radio network subsystem application part (RNSAP) signalling.

39 3GPP TS 29.002 mobile application part (MAP) specification.

40 3GPP TS 29.060 General packet radio service (GPRS); GPRS tunnelling protocol (GTP) across the Gn and Gp interface.

41 3GPP TS 29.274 3GPP Evolved packet system (EPS); Evolved general packet radio service (GPRS) tunnelling protocol for control plane (GTPv2-C); Stage 3.

42 3GPP TS 33.102 3G security; Security architecture.

43 3GPP TS 26.071 Mandatory speech CODEC speech processing functions; AMR speech Codec; General description.

44 3GPP TS 26.090 Mandatory speech Codec speech processing functions; Adaptive multi-rate (AMR) speech codec;Transcoding functions.

45 3GPP TS 23.009 Handover procedures.

46 Holma, H. and Toskala, A. (2006). *HSDPA/HSUPA for UMTS: High Speed Radio Access for Mobile Communications*. West Sussex: Wiley.

47 3GPP TS 25.308 High speed downlink packet access (HSDPA); Overall description; Stage 2.

48 3GPP TS 25.319 Enhanced uplink; Overall description; Stage 2.

49 3GPP TS 25.858 Physical layer aspects of UTRA high speed downlink packet access.

50 3GPP TS 25.877 High speed downlink packet access (HSDPA) - Iub/Iur protocol aspects.

7

Fourth Generation

Fourth-generation cellular technology has been designed to support data communications. It does not have any support for the circuit switched domain since voice appeared to become a data application among the others. Spectral efficiency and flexibility of resource allocation were enhanced by selecting orthogonal frequency division multiple access (OFDMA) as the air interface multiplexing method.

7.1 LTE and SAE

7.1.1 Standardization of Fourth-Generation Cellular Systems

The volume of mobile data traffic experienced constant growth since the initial deployment of wideband code division multiple access (WCDMA) networks. The growth was initially boosted by the 3G USB modems, which could be plugged into laptops, but was later spearheaded by the Internet data consumption with touch screen mobile phones. From the year 2008 onwards, mobile phones with large touch screens and Internet access capability became data platforms for various networked applications. Smartphones were no longer only used for voice calls and short messages, but they supported computer-like use cases such as email and Internet-browsing. Use cases expanded to watching videos, listening to audio books, or attending remote meetings or video conferences. High-speed packet access (HSPA) design since 3GPP Rel-5 provided improved support for those use cases, but consumers become even hungrier for higher data rates, and due to the growth of Internet traffic the network capacity had to be increased.

While the data consumption increased, the same did not happen to voice. That meant voice traffic was becoming only a small fraction of the total traffic supported by the networks. At the same time, it became evident that Voice over IP (VoIP) technology was becoming mature enough to support mobile voice calls over PS rather than the CS domain. In their long-term vision of network aspects [1] International Telecommunications Union (ITU)-T anticipated the wireless networks to provide access to fully IP-based core network. In 3GPP, this led to an important decision to drop circuit switched voice support from the new 4G system design, to be able to optimize it just for data.

In ITU-R recommendation M.1645 [2] the following goals were defined for the fourth-generation (4G) cellular technology, called IMT-Advanced by ITU:

- The system shall support maximum data rates of 1 Gbit/s when the mobile terminal is stationary or moves slowly and 100 Mbps when the terminal is in a moving vehicle.

Converged Communications: Evolution from Telephony to 5G Mobile Internet, First Edition. Erkki Koivusalo.
© 2023 The Institute of Electrical and Electronics Engineers, Inc. Published 2023 by John Wiley & Sons, Inc.
Companion website: www.wiley.com/go/koivusalo/convergedcommunications

- The system shall support a wide range of symmetrical, asymmetrical, and unidirectional services with different quality of service (QoS) levels. It was expected that there would be service convergence bringing all types of services to packet domain.
- The system shall comply with the IMT-Advanced spectrum and spectral efficiency requirements as specified by ITU-R.

3GPP started the standardization work against IMT-Advanced requirements in 2004 [3]. The basic goals for the **long-term evolution (LTE)** standard were defined as follows:

- LTE system should be optimized for IP-based data traffic, and it shall be compatible with the IMS (3GPP IP Multimedia System) core network. The system would not natively support circuit switched services.
- LTE system should support higher data rates than the HSPA system. The goal was to support 100 Mbps downlink and 50 Mbps uplink transmission speeds.
- LTE connection should have lower latencies than a HSPA connection. The goal was to have the latency (round-trip delay) between the terminal and the network below 10 ms. Additionally, setting up a new connection should not take more than 300 ms.
- LTE system transmission capacity should be larger than the capacity of a comparable HSPA system. The spectral efficiency target of LTE was two to four times higher than that of HSPA Rel-6 system.
- LTE system bandwidth should be scalable between 1.5 and 20 MHz.
- LTE system architecture shall be simpler and flatter than that of the HSPA system.
- LTE system should minimize the power consumption of the terminal.
- LTE system should support data connections for mobile users and also seamless roaming to other radio technologies (such as Global System for Mobile Communications [GSM], WCDMA, or CDMA2000) when the terminal exits the coverage of the LTE network
- LTE system security should be at least on par with the legacy cellular systems.

When specification work of the fourth-generation cellular technology was started, the effort was focused on 3GPP, making it finally a truly global forum for cellular system standardization. WCDMA UMTS became a stepping stone toward 4G systems. Further development on CDMA-based systems for American markets were discontinued and activities of 3GPP2 were eventually closed in 2013, after 4G LTE was rolled out globally. The WLAN-based WiMAX standard was also aligned with LTE from 2014 onwards. After so many years with competing cellular standards tracks, the world was ready for truly global deployment of a single new 4G cellular standard.

LTE is defined in 3GPP standard series TS 36, from Rel-8 onwards. TS 36.300 [4] provides a detailed overview of LTE architecture, functionality, and protocols. 3GPP LTE standards have been developed and published as subsequent standards releases: Those major features, which are covered in this book, were defined in the following releases:

- 3GPP Rel-8 introduced LTE radio and the new system architecture evolution (SAE) network architecture.
- 3GPP Rel-9 enhanced LTE with UMTS interoperability.
- 3GPP Rel-10 was the first release to specify **Long-Term Evolution Advanced (LTE-A)** features to meet IMT Advanced 4G requirements and introduced, for instance, carrier aggregation support, LTE relay nodes, and inter-cell interference coordination (ICIC).
- 3GPP Rel-11 continued specification of LTE relay nodes, introduced new LTE bands, enhanced carrier aggregation, and ICIC support.
- 3GPP Rel-12 introduced new carrier aggregation and multiple-input, multiple- output (MIMO) configurations as well as enhanced small cell support.
- 3GPP Rel-13 provided support for using LTE in unlicensed bands, machine type communications (MTC), and additional multiantenna configurations.

- 3GPP Rel-14 and Rel-15 introduced various enhancements for areas such as mission critical data, TV services, and Internet of Things (IoT) support while MTC were enhanced. Solutions were created to support further separation of user and control plane functions, to enhance network capacity, and minimize latencies due to processing delays.

With LTE Rel-8 the maximum downlink data rate was 400 Mbps and uplink data rate 75 Mbps. Higher data rates up to a few gigabits per second downlink can be reached with a number of additional features, such as carrier aggregation, MIMO, and higher-order QAM modulation methods introduced for **LTE Advanced** within later 3GPP standard releases from Rel-10 onwards. The research phase of LTE Advanced in 3GPP covered various topics and proposals, but the main LTE-A improvements finalized into 3GPP specifications were eventually as follows:

- Carrier aggregation (see Section 7.1.2).
- MIMO multiantenna configurations (see Section 7.1.4.6).
- Support for higher-order QAM such as 128-QAM and 256-QAM, as mentioned in Section 7.1.4.1.
- LTE-M and NB-IoT (see Section 7.1.4.10).
- ICIC where neighboring cells coordinate their scheduling decisions to reduce interference (see Chapter 4, Section 4.1.4).
- Support for heterogeneous network planning to mix large and small cells was increased by introduction of relay nodes (RNs). Relay nodes are low-power base stations to provide enhanced coverage and capacity at cell edges and hot spots.
- Offloading data from cellular network to fixed data network with LTE/WiFi interworking, where LTE RAN either assists or controls the offloading scenarios.

Good overview of LTE Advanced can be found from ITU-R recommendation M.2012 [5].

7.1.2 Frequency Bands Used for LTE

The LTE specification supports both frequency division duplex (FDD) and time division duplex (TDD) methods for separating uplink and downlink transmissions. FDD uses separate subbands for uplink and downlink and a guard band between those. TDD shares a single band between uplink and downlink with time-division. TDD uses guard period of 5 or 10 ms between uplink and downlink subframes of a complete LTE frame. Special TDD subframes carry both uplink and downlink traffic in different slots and a guard slot between them. TDD networks use an asymmetric configuration where the number of downlink subframes is bigger than uplink subframes.

The initial LTE Rel-8 specification defined 27 radio bands for the LTE system [6]. There were 18 FDD bands and eight TDD bands. Every specified radio band has its own bandwidth. Any FDD band consists of two subbands of same width between 10 and 75 MHz. The width of any TDD band is between 16 and 60 MHz.

- Global bands
  - Frequency range 824–960 MHz has three FDD bands: 5, 6, and 8
  - Frequency range 1428–1500 MHz has one FDD band: 11
  - Frequency range 1710–2170 MHz has six FDD bands: 1–4, 9–10
  - Frequency range 1880–1920 MHz has two TDD bands: 33 and 39
  - Frequency range 2010–2025 MHz has one TDD band: 34
  - Frequency range 2300–2400 MHz has one TDD band: 40
  - Frequency range 2500–2690 MHz has one FDD band: 7
  - Frequency range 2570–2620 MHz has one TDD band: 38

- USA
  - Frequency range 700–800 MHz has four FDD bands: 12–14, 17
  - Frequency range 1850–1990 MHz has three TDD bands: 35–37
- Japan
  - Frequency range 815–890 MHz has two FDD bands: 18 and 19

Additional radio bands have been defined for LTE after Rel-8. These additional bands are introduced mainly to support refarming of bands for LTE when ramping down legacy cellular networks, which have relied on those bands. Majority of the LTE deployments rely on FDD, while TDD is used widely in China and the United States, but also in some parts of Europe [7].

The mobile terminal may simultaneously use multiple LTE radio bands to increase the data throughput from what a single band could provide. LTE specifications [4] define two ways for such multi-band operation:

- **Carrier aggregation (CA)** where multiple component carriers (CC) of same radio technology from a single eNodeB are combined as a wide band carrier. Traffic is split to the CC in the Medium Access Control (MAC) protocol layer. In carrier aggregation, one of the CC may be used as the primary one used to transport RRC control signaling and hybrid automatic repeat request (HARQ) feedback, while the other secondary carriers are only used for user data. LTE CA was introduced in 3GPP Rel-10 with support of three CA band combinations. In Rel-11 the number of combinations were increased to 30 and the maximum supported CA bandwidth was 40 MHz. After Rel-11, even more CA band combinations were specified up to 100 MHz total bandwidth with five CC. See 3GPP TS 36.71x specifications for further details. Carrier aggregation is typically used for downlink where multiple carriers can be aggregated. In uplink, most often one carrier is sufficient, but the specs support aggregating two carriers. Note that carrier aggregation can be used also for increasing the capacity of the cell, even if the UEs would be scheduled to use only a single carrier any time.
- **Dual connectivity (DC)** where a number of CC from different eNodeBs are used on parallel. One of the eNodeBs is the master and the other secondary. Traffic is split to components already at the packet data convergence protocol (PDCP) layer. As every eNodeB runs its own MAC entity, every CC has its own scheduling, HARQ processes, and control data path. LTE DC was introduced in 3GPP Rel-12, but compared to carrier aggregation, DC has not been widely deployed.

7.1.3 Architecture and Services of LTE Systems

7.1.3.1 LTE Services

As described earlier, LTE is data-only system, optimized for packet switched data flows. Essentially, the LTE radio technology is a vehicle to provide a mobile device with high-speed always-on IP connectivity [8]. 3GPP decided not to build CS support into LTE in order to have best support for packet switched IP data. With LTE, the traditional CS voice services can be offered in two ways:

- Circuit switched fallback (CSFB), where the UE is redirected to another radio technology such as UMTS or GSM for CS calls.
- IP multimedia subsystem (IMS) Voice over LTE (VoLTE), which is a packet-based voice service running on top of the LTE packet connection. For detailed description of VoLTE, please refer to Chapter 9 and specifically its Section 9.4.

Even without any support for CS domain, LTE system stays complex to support seamless IP data connections regardless of the location and mobility state of the LTE terminal and the other endpoint of the IP data flow. Additional complexity comes from the QoS differentiation between various types of data flows. LTE system shall be able to ensure smooth rendering of interactive conversational or streamed media regardless of the presence of other parallel data flows or radio path fluctuations caused by moving terminals.

7.1.3.2 LTE and SAE System Architecture

LTE cellular system **evolved packet system (EPS)** consists of the following three subsystems:

- **User equipment (UE):** LTE terminal and its universal integrated circuit card (UICC) card with a USIM application.
- **Evolved UMTS Terrestrial Radio Access Network (E-UTRAN):** LTE radio network built from a number of eNodeB (or eNB) base stations. E-UTRAN architecture and functions are described in 3GPP TS 36.401 [9].
- **Evolved Packet Core (EPC):** LTE network core which consists of mobility management entity (MME), serving gateway (S-GW), packet data network gateway (P-GW), and policy and charging rule function (PCRF) network elements. Detailed description of EPC architecture and functionality can be found from 3GPP TS 23.401 [10].

In parallel to specifying a new 4G LTE air interface, 3GPP revised the architecture of the cellular network. 3GPP called this re-architecture work as **system architecture evolution (SAE)** aiming at flatter and simpler network architecture. SAE goals were to reduce the number of network element types, minimize transmission delays, increase the capacity of the network, and simplify the protocol stacks needed for controlling the network functions. SAE had an impact also on the advanced HSPA network architecture.

One of the SAE goals was to flatten the network architecture. In WCDMA UMTS radio network (UTRAN), the radio network controller (RNC) is an essential network element to support WCDMA soft handovers. With LTE, no soft handovers were to be supported and the decision was made to have the base station control functionality distributed to the base stations rather than having a centralized RNC element. Features earlier located to the RNC – such as management of radio channels, UE mobility management, packet retransmissions – were merged as part of the new eNodeB base station concept. LTE base stations operate in a rather independent way and mutually coordinate the handovers between two base stations.

Also, EPC core has been streamlined compared to legacy cellular systems. First of all, no network elements are needed for the circuit switched domain, but also the packet switched domain has been reorganized. In the traditional general radio packet service (GPRS) architecture, both the Serving GPRS support node (SGSN) and Gateway GPRS support node (GGSN) have roles for both the network control and transport of the user data. The SGSN is focused on serving the UE and tracking its location. The GGSN manages packet data contexts toward external networks and the GTP tunnels. The user data path goes through both the GSN nodes. The SAE architecture aimed at separating network elements for two planes: user plane and control plane. This was according to ITU-T vision statement Q.1702 [1], which suggested separation of transport and control functions of the network. The task of the **user plane** is to transport the user data between UE and external packet data network. In EPC, the SAE GW is responsible for user plane functions. SAE GW can be divided logically (and often also physically) into two parts: S-GW supporting connections to UEs and P-GW supporting connections toward different external packet networks. **Control plane** manages the functions of user plane and the UE. In SAE architecture, control plane functions are mainly centralized to a new network element called mobility management entity (MME), which does not participate in any user data transport. Essentially, with the new architecture the MME performs the signaling and control tasks which were part of SGSN responsibilities in GPRS, while the user data transport tasks of GSN elements are carried out by SAE GW nodes.

Separation of the control plane and user plane functionality had an impact on design of network elements. Earlier radio technologies (GSM, GPRS, and UMTS) relied on custom-designed hardware for various network elements. Separating signaling and user data transport to different elements allowed the vendors to rely on standard hardware. Signaling elements, such as MME, could be implemented on top of a standard server platform while SAE gateways could rely on a standard IP router hardware platform. Only the software layers would be specific to LTE and SAE. The separation was, however, not completed with the basic SAE architecture. The LTE eNodeB takes care of radio resource management and the MME of mobility management, but the session management is done by MME, S-GW, and P-GW in tight cooperation. Separation of user and control plane functions for session management was done as part of LTE-A from Rel-14 onwards, paving the way to the

5G network architecture. The S-GW and P-GW were split to separate control plane and user plane functions at Rel-14 as described in 3GPP TS 23.214 [11].

To support seamless packet data handovers between different radio technologies, the LTE subscriber has packet data connections toward external networks always via the SAE GW, regardless of which type of radio technology the UE uses. The LTE UE sets up EPS bearer contexts with the P-GW, which allocates the UE with an IP address. That address is seamlessly used even when the UE switches from LTE to HSPA, WCDMA, or GPRS networks. When the UE changes its radio access technology, the S-GW reconnects user data flows from the eNodeB to the SGSN node, which starts to serve the UE. To support connectivity to the EPC core, the SGSN nodes are enhanced with two new interfaces: S4 interface support toward S-GW and S3 interface support toward MME.

In UTRAN and GERAN networks, the connections between different types of network elements are fixed and hierarchical. A base station is connected to a single base station controller (BSC) or RNC, which is connected to one SGSN. The SAE architecture breaks such fixed relations. In a geographical area, there are multiple eNodeB base stations, S-GW gateways, and MME nodes, which dynamically and flexibly form logical connections between each other. One UE always has only a single combination of serving eNodeB, MME, and S-GW (except when handover is being performed). When considering the whole UE population camping in a cell, an eNodeB may be controlled by multiple MMEs and carry user data from/to multiple S-GWs serving different UEs. It is worth noting that while there is only one single serving S-GW for the UE, the UE may be connected to multiple P-GWs at a time, to be connected to multiple external data networks.

The LTE cellular system consists of the following elements, as shown in Figures 7.1 and 7.2:

- **UE,** which consists of LTE mobile equipment equipped with a UICC card:
  - Mobile equipment (ME) is an LTE terminal such as a smartphone or LTE data modem connected to a computer. As the processor technology evolved, in the typical phone the application processor consists of

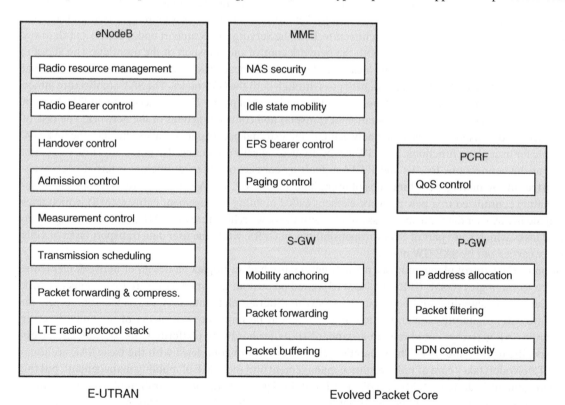

Figure 7.1 Functional split between eNodeB and elements of EPC. *Source:* Adapted from 3GPP TS 36.300 [4].

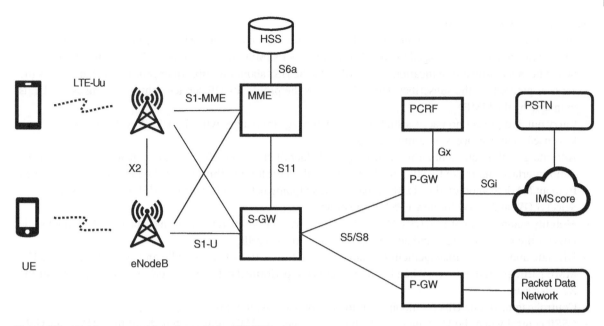

Figure 7.2 Architecture and interfaces of the LTE/SAE system.

multiple cores where different pieces of software can be run simultaneously to increase the total processing power of the device.

– UICC is a smart card which contains UMTS subscriber data and related algorithms.

• **Evolved Node B (eNodeB** or **eNB)** is an LTE (or E-UTRAN) base station, which might contain multiple radio transmitters, receivers, and antennas serving the LTE cell. In addition to those, the eNodeB has a computer system managing eNodeB functionality and the communication link from the eNodeB to the EPC. The LTE RAN has no external RAN network element (like UMTS RNC) to coordinate functions of multiple eNodeBs. Responsibility area of the eNodeB is similar to the UTRAN NodeB and RNC responsibilities combined. This architecture was chosen to minimize latencies and improve network scalability for bursty packet-switched traffic patterns, involving lots of signaling for activating and deactivating connections. For user data flows the eNodeB operates as a bridge of open systems interconnection (OSI) model layer 2 as it passes user data between the link layer of radio interface and the communications link toward the EPC. The eNodeB has the following tasks:

– Modulation and multiplexing at the LTE air interface.
– Management of LTE physical radio channels and their frame structures, channel coding, synchronization, transmission rate adaptation, and transmission scheduling.
– Radio resource and interference management, admission control, and adjustment of the transmission power with different algorithms to ensure the stability of the radio connections and the QoS for all UEs currently served.
– Management of radio bearers with the RRC protocol.
– Handover control for UEs in the active RRC state.
– Connect the user plane packet data flows between the UE and EPC.
– Compress IP packet headers to save the radio transmission capacity. Header compression is especially important for small VoIP real-time protocol (RTP) packets.
– Encrypt data transmitted over radio interface.
– Route control plane messages between the UE and MME.

- **MME** has the following tasks:
  - Interact with the home subscriber server (HSS) to authenticate the subscriber when the UE attaches to the LTE service in the area managed by the MME. Subscriber's IMSI identifier tells the MME which HSS to contact. After successful authentication, the MME asks the HSS about the subscriber profile which describes the services granted for the subscriber. The MME stores the profile for UE service control as long as the UE is managed by the MME.
  - Carry out and protect non-access stratum (NAS) message exchanges with the UE. Provide the eNodeB with keys used to protect access stratum messages.
  - Allocate global unique temporary identity (GUTI) identifier for the UE to avoid transmitting IMSI on the radio interface and make tracking of the UE more difficult for any third parties. When the UE is switched from an MME to another, the old MME passes the mapping between GUTI and IMSI to the new MME.
  - Select S-GW and P-GW for new EPS bearer contexts.
  - Mobility management of LTE UEs. MME keeps HSS aware of the current location of the UE. MME tracks idle UEs for their tracking areas and active UEs for their camped cells.
  - UE state and resource management for handovers and transitions between idle and active RRC states. The MME participates handovers when the handover is not performed between two eNodeBs connected over the X2 interface.
  - Control paging of idle UEs for mobile terminated connection attempts.
  - CSFB control when the UE which is attached to LTE gets an MT circuit switched call from UMTS or GSM core network.
- **S-GW** has the following tasks:
  - Forward user packet data within GTP tunnels between the P-GW and eNodeB. In SAE architecture GTP tunnels are set up under control of the MME and P-GW but the tunnels are routed to the eNodeB via S-GW.
  - Buffer downlink packets and trigger paging of idle mode UEs.
  - Mediate the connection requests between the MME and P-GW. With LTE, the initiative for EPS bearer context activation may come either from the UE or network. In case the UE requests activation of a packet data connection, the S-GW will pass the request from the MME to the P-GW. In the other case, when the activation request is initiated by the network, the S-GW will pass the request from the P-GW to MME. These three nodes must be synchronized for any new connections since the GWs take care of GTP tunnel creation to the eNodeB and the MME controls the eNodeB for related radio bearer configuration.
  - Anchor the connection when the UE performs handover between two eNodeB or two radio technologies. As requested by the MME responsible for the handover, the S-GW opens a GTP tunnel to the target eNodeB or SGSN toward which the UE is moving. After completing the handover, the S-GW tears down the GTP tunnel toward the old eNodeB, which no longer serves the UE. The handover may also cause the S-GW to be changed and in such cases P-GW stays as the anchor point.
- **Packet Data Network Gateway (P-GW** or **PDN-GW)** has the following tasks:
  - Connect the EPS bearers from the UE to external packet data networks or the IMS core network of the operator. Manage GTP tunnels for EPS bearers based on requests received either from the S-GW or PCRF nodes.
  - Allocate IP addresses for the UE with help of a DHCP server, external IP network, or with IPv6 autoconfiguration. The UE gets an IP address whenever it attaches to the LTE network and may get additional IP addresses when being connected to new packet data networks.
  - Forward user data packets between the S-GW and external IP network and filter packets with rules specified within a traffic flow template (TFT) specified for the data flow. Filtering of packets is done to enforce the policy received from the PCRF, specifying the QoS parameters of the data flow.
  - Collect charging information about the services used or volume of data passed over the connections, according to the user's service profile available from the HSS.

- Anchor the connection when an idle UE switches its cell so that the S-GW is changed or the UE roams between 3GPP and non-3GPP accesses.
- **PCRF** has the following tasks:
  - Control QoS policies of connections created for UEs. The PCRF gets connection requests either from the P-GW or a Call State Control Function (CSCF) of the IMS core network. The PCRF checks the applicable policy and sends policy control (PCC) rules to the P-GW, which sets up its packet data filtering and charging collection configuration accordingly.
- **Evolved Packet Data Gateway (ePDG)** has the following tasks:
  - Terminate the secure tunnel from the UE when an untrusted non-3GPP access network is used by UE to connect the EPC. In this way, users could connect for instance to 3GPP IMS service, to use VoWiFi service (operator supported IP voice over WLAN; see Chapter 9, Section 9.6). The ePDG connects the user data flows to the P-GW from where external packet data network can be reached.
- **HSS** is responsible of subscriber authentication, subscription data management, and UE location management like for UMTS.

Since the LTE E-UTRAN and EPC network architectures are quite different from GERAN and UTRAN network architectures, the LTE/SAE network has its completely own list of reference points, a subset of which is shown in Figure 7.2:

- LTE-Uu: Radio interface between LTE UE and eNodeB
- X2: Interface between two eNodeB base stations
- S1-U: Interface between eNodeB base station and S-GW gateway
- S1-MME: Interface between eNodeB base station and MME node
- S2b: Interface between ePDG and P-GW gateway
- S3: Interface between MME and the SGSN supporting UTRAN or GERAN network
- S4: Interface between S-GW and the SGSN supporting UTRAN or GERAN network
- S5/S8: Interface between S-GW and P-GW gateways. When roaming abroad and using IP home routing, this interface is run over international IP roaming exchange (IPX) network.
- S6a: Interface between MME and HSS
- S9: Interface between the PCRFs of home and visited networks
- S10: Interface between two MME nodes
- S11: Interface between S-GW gateway and MME node
- S12: Interface between S-GW gateway and the RNC of UTRAN network
- Gx: Interface between P-GW gateway and PCRF
- SGs: Interface between MME and MSC/VRL node supporting GERAN or UTRAN
- SGi: interface between P-GW gateway and service domain such as IMS

LTE extends the radio network temporary identity (RNTI) concept introduced in UMTS with a number of new RNTI types, which can be used to identify one UE, a group of UEs, or all UEs camping in a cell. Some examples are provided here, and the complete list can be found from 3GPP TS 36.321 [12]:

- Cell RNTI (C-RNTI) is used as for UMTS to identify a specific UE which has an RRC connection.
- System Information RNTI (SI-RNTI) is a well-known fixed value used by the UE for decoding system information.
- Paging RNTI (P-RNTI) is a well-known fixed value used by the UE for decoding paging messages.
- Random access RNTI (RA-RNTI) is used to encode random access response (RAR) message. The RA-RNTI identifies the resource element within which the random access preamble was sent by the UE.

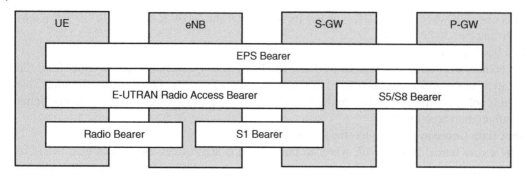

Figure 7.3 Bearer architecture of LTE system. *Source:* Adapted from 3GPP TS 36.300 [4].

7.1.3.3 LTE Bearer Model

Data connections with known QoS behavior between different LTE network elements are modeled as bearers of different types with a model similar to UMTS networks. LTE QoS and bearer models are described in 3GPP TS 23.401 [10] and TS 36.300 [4].

The model is hierarchical so that bearers on the higher layer are formed from a chain of bearers of the underlying layer.

The lowest-level LTE bearers, as shown in Figure 7.3, are as follows:

- Radio bearer is the air interface connection between the UE and eNodeB. The protocol stack of the radio bearer consists of LTE physical layer, MAC, radio link control (RLC), and PDCP protocols. The radio bearers can be divided as signaling bearers and data bearers.
- S1 bearer is a connection between the eNodeB and S-GW.
- S5/S8 bearer is a connection between the S-GW and P-GW.

Radio access bearer is a central component of the model, providing the data connection for the UE over the radio access network to the EPC core network.

The whole connection between the UE and P-GW, built from the parts described here, is called **EPS bearer** (E-UTRAN Packet System Bearer). The EPS bearer carries a single user data flow with specific QoS requirements.

7.1.3.4 LTE Protocol Stack Architecture

The protocol architecture of LTE has its roots in UMTS protocol architecture from which the LTE protocol suite has been evolved. The protocols used on the radio interface have the same names in these two systems. The name of an individual protocol identifies the purpose of the protocol and its location in the protocol stack. LTE radio protocol architecture is described in 3GPP TS 36.300 [4] and EPC protocol architecture in TS 23.401 [10].

LTE has two protocol stacks for the following purposes:

- User plane protocols for transporting the user data streams (see Figure 7.4)
- Control plane protocols used for signaling between LTE network elements (see Figure 7.5)

As Figures 7.4 and 7.5 show, the LTE protocols can be categorized as follows:

- Data transmission protocols which are used to transport either user data or signaling messages over an interface between two system elements. These physical and link layer protocols are common between user and control plane. On the radio interface, the common protocols are LTE PHY (physical layer), LTE MAC, and RLC (the link layer). Within E-UTRAN and EPC, the networking layer uses common IP/UDP protocol pair to carry both user data and signaling, except S1 and X2 interfaces, where the reliable SCTP protocol is used instead of UDP for

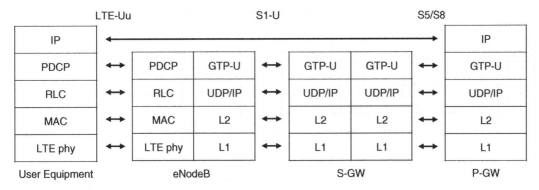

Figure 7.4 LTE user plane protocols.

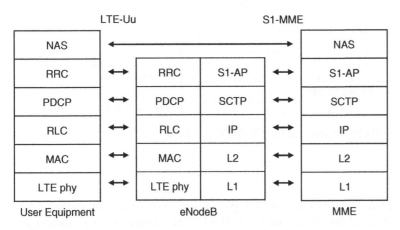

Figure 7.5 LTE control plane signaling protocols.

signaling. Underlying link and physical layers have not been specified, but typical deployments rely on Ethernet over fiber or microwave radio. For practical deployments, the S1 and X2 links are often not separate point-to-point links. Instead, both S1 and X2 may use a single transmission link from the eNodeB to a local aggregation IP router, which eventually routes the IP packets either to other eNodeBs, an MME, or S-GW [7].

- The radio network control protocols used by the MME and S-GW toward the UE and eNodeB. These protocols are used for radio access bearer management. The MME uses RRC protocol over the LTE-Uu interface toward the UE. Both the MME and S-GW use S1AP protocol over the S1 interfaces toward the eNodeB. The eNodeBs use X2AP protocol over the X2 interface for handover control.
- Mobility and security management protocols between the MME and UE. The MME uses NAS EPS mobility management (EMM) protocol over the LTE-Uu interface toward the UE for registration and location management, authentication, and encryption control.
- User data session management and transport protocols between various network elements contributing to the data sessions. The MME uses NAS EPS session management (ESM) protocol over the LTE-Uu interface toward the UE for packet data context management. Usage of the ESM protocol is not always necessary, for instance, in cases when the UE negotiates about the connections with the CSCF servers of IMS core network, which will then request the EPC network to provide the needed data connections for the UE. The user data path from the P-GW to eNodeB consists of GTP-U tunnels managed by the P-GW, S-GW, and eNodeB with the GTP-C protocol. On the radio interface, the eNodeB uses PDCP protocol to carry the user packet data to the UE.

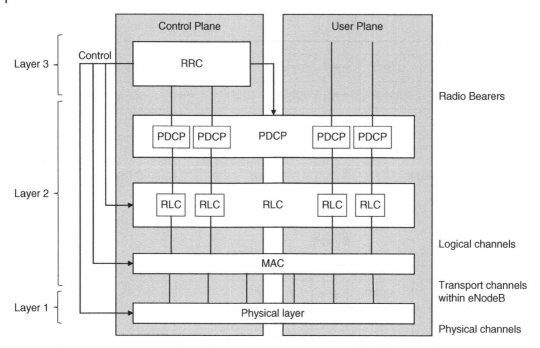

Figure 7.6 LTE radio protocol stack and channels. *Source:* Adapted from 3GPP TS 36.201 [13].

7.1.3.5 LTE Radio Channel Architecture

LTE logical channels are used for grouping data flows of various types for transport within the radio bearer. The concept of a logical channel is like defined for UMTS and GSM. The LTE MAC protocol maps data flows from logical channels to the transport channels used toward the LTE physical layer. The physical layer maps received transport blocks to the physical channels of the LTE radio interface. In UTRAN network, the transport channels are used between the RNC and NodeB so that transport blocks are carried by the Iu FP protocol over the physical link between these nodes. In LTE, both MAC and physical layers are located within the eNodeB so that there is no inherent reason (protocol connection between two nodes) for keeping the transport channel layer in the model. Nevertheless, the decision was made by 3GPP to stay with the three-level channel architecture and related terminology of LTE MAC and physical layers are inherited from the WCDMA design. See Figure 7.6 for the LTE radio protocol design, which is quite similar to the WCDMA design depicted in Figure 6.6.

The types of LTE logical, transport, and physical channels and mappings between them are introduced in the upcoming chapters of the book. Detailed specification of those channels and the mappings between them can be found from 3GPP TS 36.300 [4].

7.1.4 LTE Radio Interface

LTE uses **OFDMA** multiplexing for downlink and **single carrier frequency division multiple access (SC-FDMA)** multiplexing for uplink. Multiplexing is used for separate transmission streams to/from different UEs and the various data flows of each individual UE. General description of the LTE physical layer can be found from 3GPP TS 36.201 [13]. This specification contains also a summary of other 3GPP specifications related to the LTE physical layer.

7.1.4.1 OFDMA and QAM for LTE Downlink

With OFDMA, the deployed LTE bandwidth is divided into narrowband OFDM **subcarriers**, each of which is modulated independently. As described in *Online Appendix A.9.1*, OFDM subcarriers are orthogonal, which means that the harmonic frequencies of the subcarriers are minimized at the fundamental frequencies of other subcarriers (see Figure 7.7) to avoid excessive inter-subcarrier interference [3]. Orthogonality is achieved with suitable selection of subcarrier spacing and advanced filtering methods.

A single data flow is carried over multiple OFDM subcarriers instead of one single carrier. OFDMA is a multi-user version of OFDM where blocks of subcarriers are allocated for multiple data flows of one or multiple users of the OFDMA system.

LTE uses a fixed 15 kHz **subcarrier spacing,** which means there is 15 kHz difference between the fundamental frequencies of each subcarrier, as shown in Figure 7.7. On the other hand, the bandwidth of an LTE OFDMA subcarrier is 15 kHz. Compared to 200 kHz GSM channels, LTE subcarriers are quite narrow. The operation of LTE downlink OFDMA follows [3]:

1) The bitstream to be transmitted over a group of subcarriers are modulated with quadrature amplitude modulation (QAM) to generate a stream of QAM symbols. Each **QAM symbol** is represented by a well-defined change in the amplitude-phase space of the subcarrier. Deepening on the number of possible changes, a symbol can represent all combinations of a certain number of consecutive bits. LTE Rel-8 supports QPSK, 16-QAM or 64-QAM modulation methods where one symbol represents two, four, or six data bits, respectively. In later 3GPP LTE releases, support for 128-QAM carrying 7 data bits and 256-QAM carrying 8 data bits per symbol were added.

2) The stream of QAM symbols is distributed to the parallel OFDMA subcarriers of the allocated block of subcarriers. When multiple subcarriers are used to carry the data stream, each of the subcarriers will transport only a fraction of the QAM symbols. Consequently, the time between successive QAM symbols for one subcarrier is longer than it would have been when using only a single carrier to transport all the symbols. This is the major benefit of OFDM modulation. The changes to the amplitude and phase of subcarriers are much more infrequent than what would be needed for modulating a single carrier to achieve the same data rate. Data rate of an LTE connection can be increased or decreased by changing the number of subcarriers allocated for the data connection, rather than modifying the frequency of transmitting QAM symbols. With OFDMA, the transmission time of QAM symbols on each subcarrier is constant 66 667 µs, so that the bandwidth of the modulated subcarrier does not depend on the data rate of the link.

3) OFDMA could allow the system to select subcarriers for transporting one LTE radio connection without any limitations within the available LTE radio band. The subcarriers could be allocated in a contiguous manner so that only adjacent subcarriers could be used for the connection. Subcarriers could also be allocated discontinuously so that between the subcarriers allocated for the connection there would be subcarriers allocated for other connections. The LTE system design ended up a hybrid approach to keep the radio resource allocation algorithms and signaling messages simple enough. In LTE, subcarriers are allocated for connections as groups

Figure 7.7 Spectrum of an LTE subcarrier and the orthogonality of OFDM subcarriers.

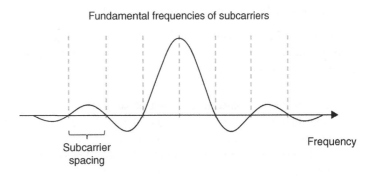

of 12 adjacent subcarriers. The allocation is done for a period of 0.5 ms, known as a **slot**. The basic unit of LTE radio resource allocation is a **resource block**, consisting of one block of 12 subcarriers within the half a millisecond slot. In LTE, the length of transmission time interval (TTI) is 1 millisecond, thus two resource blocks are transmitted within the LTE TTI. This enables a scheduler to use frequency diversity by scheduling the resource blocks of one TTI to different blocks of subcarriers. If a connection is allocated with multiple resource blocks simultaneously, those resource blocks may or may not be adjacent in the subcarrier space. The modulation and symbol type are selected separately for each resource block by the frequency domain scheduler.

4) The stream of the QAM symbols allocated for each subcarrier is given as input to the subcarrier-specific branch of an inverse fast Fourier transformer (IFFT). The IFFT transformer minimizes the harmonic frequency components spanning outside of the subcarrier band and combines the modulated parallel subcarriers into a single radio signal for the transmitter. This signal does not have a regular form as it is composed of a combination of signals which have their own the amplitudes and phases on their own subcarrier frequencies. The waveform created from a set of parallel QAM symbols within a symbol period of 66 667 μs is also known as an **OFDM symbol**.

5) A guard period is added between each OFDM symbol to reduce the impact of intersymbol interference. ISI is caused by multipath propagation, which changes signal latency when the user is moving. Signal transmission cannot be paused during the guard period since that would disrupt operation of Fourier transformation at the receiver. To support continuous transmission over the guard period, a **cyclic prefix** is added between each OFDM symbol. The cyclic prefix is a copy taken from the end of the OFDM symbol waveform and added to the front of the symbol. The receiver can ignore the cyclic prefix part of the signal when it interprets the symbols received. LTE supports normal and extended cyclic prefixes, the latter of which could be used for large cells with long transmission times.

As a summary, the OFDMA method has the following advantages compared to the other options:

- Higher bitrates can be achieved by using wider system bandwidth without increasing the symbol rate. With OFDMA, the wider bandwidth means just a larger number of narrowband subcarriers, each with a constant symbol rate. Increasing the WDCMA system bandwidth from 5 MHz would have meant a higher chip rate. The impact of multipath fading (see *Online Appendix A.3.3*) would have become more significant for the shorter chips [7]. With OFDMA, the LTE system bandwidth could be grown to 20 MHz without similar unwanted impact to the OFDM symbols.

- The system bandwidth can be easily adjusted to fit the available radio band by configuring the system for a suitable number of subcarriers. This is very convenient, for instance, when refarming frequencies from a GSM or WCDMA network ramped down to a new LTE system. In such cases, the LTE system bandwidth may be as low as 1.25 MHz.

- OFDMA subcarriers can be used to provide an optimal radio path for a single UE, based on the UE-specific radio conditions on different parts of the system bandwidth. The LTE scheduler is able to pick those subcarriers of the available radio band that have the best quality for the UE. Link adaptation described in the next chapter can then select the modulation method for each resource block to have a good match with the current radio conditions on the corresponding subcarriers [3].

The matrix of LTE subcarriers and consecutive OFDM symbols sent over them is often represented as the LTE **resource grid**, shown in Figure 7.8. One cell of the grid is called a **resource element,** which carries a single QAM symbol over one subcarrier.

7.1.4.2 Downlink Reference Signals and Link Adaptation

Link adaptation is the process where the UE and eNodeB work together to pick the best **modulation and coding scheme (MCS)** for the transmitted radio blocks. MCS is a combination of used QAM modulation type and coding ratio applied by puncturing the channel coded signal before transmission. To support downlink channel estimation, MCS selection, and demodulation, the eNodeB adds predefined **cell-specific reference signals (CRS)** to

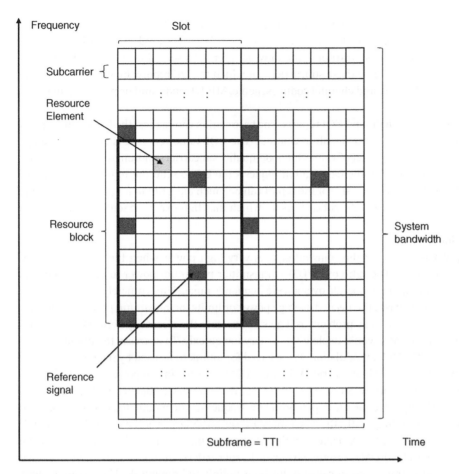

Figure 7.8 LTE resource grid. *Source:* Adapted from 3GPP TS 36.211 [14].

the **reference symbol** resource element positions of the resource grid (see Figure 7.8). CRS signals are added to every sixth subcarrier and every seventh symbol of them (except to narrow LTE bands, which may have fewer of them to spare capacity). These reference symbols do not transport any user data but have constant contents and transmission power. The receiver of the UE uses reference symbols for the following purposes: coherent demodulation, frame synchronization, comparison of signals from different cells for handover decisions, and estimation of the quality of the radio channels on different subcarriers. The UE uses received reference symbols to figure out how much does the amplitude and phase of the radio signal experience distortion on different parts of the frequency space and how much the channel has noise. The receiver can thereafter partially compensate the subcarrier-specific distortions using an equalizer.

To support link adaptation, the UE reports the results of its reference symbol measurements to the eNodeB, which uses the reported data to select the best available subcarriers for the UE and the optimal modulation and channel coding methods for them. The UE provides the eNodeB with estimates about how the data could be encoded in different parts of the LTE radio band. Estimates are based on the observations made by the UE from the downlink reference symbols. The report sent by the UE contains the indices of MCS combinations to be used for each part of the frequency band. The **channel quality indicator (CQI)** index indicates the best MCS at which the block error rate of the channel being analyzed does not exceed 10%. The CQI has 16 different values, each mapped to one MCS combination. Additionally, the report can contain parameters related to the selection of **transmission mode** for using multiple antenna ports (see Section 7.1.4.6).

The eNodeB uses the received estimates for selecting subcarriers, the MCS method, and the multiantenna configuration for each radio block. Since the base station will choose the subcarriers used for each UE, the CQI reports enable the eNodeB to distribute the subcarriers optimally between UEs served in the cell. The eNodeB tries to allocate resource blocks for the UE over those subcarriers for which the UE reports highest CQI. The second goal is the final selection of modulation and channel coding scheme, MIMO mode, and its precoding matrices for the resource block.

The eNodeB determines when the UE shall send CQI reports. The base station can request the UE to send reports periodically (once in 2–160 ms) over the shared signaling channel. The CQI value reported is either an average calculated over the whole LTE radio band or a set of values for good-quality subcarriers chosen by the UE. Another option is that instead of periodical reporting, the base station requests the UE to send the report once for every resource block transmitted. In the latter case, the size of the report can be much bigger than for periodic reporting. For per-block reporting, the eNodeB may select the subcarriers for which the UE is expect to send CQI reports.

7.1.4.3 SC-FDMA for LTE Uplink

The main drawback of OFDMA is that the radio transmitter uses very high power when the power peaks for multiple subcarriers simultaneously [8]. The ratio of OFDMA transmitter maximum power to average power is high. OFDMA requires the power amplifier to operate over a wide frequency range. Unfortunately, such amplifiers are not very power efficient, which increases the needed average power or limits the maximum power and achievable distance. For the eNodeB, this is not a huge problem as it gets power from the electricity network. When LTE was designed, it was believed that OFDMA would deplete UE battery too quickly. To decrease the uplink peak to average power ratio, single carrier SC-FDMA modulation was selected for uplink to be used by the UE.

The SC-FDMA modulation uses only one single carrier. With SC-FDMA, the maximum power is determined by the amplitude of this single modulated carrier. Since the uplink data rate cannot be adjusted by the number of subcarriers allocated, the SC-FDMA uplink data rate depends on how frequently the modulated QAM symbols are transmitted over the carrier. The more frequently those symbols are sent, the larger is the needed radio bandwidth. The bandwidth allocated to an SC-FDMA signal is defined to be a multiple of 180 kHz, which is the same amount of bandwidth used by one LTE OFDMA resource block of 12 subcarriers. In this way, the same resource block size can be used for allocating transmission capacity for both uplink and downlink. The resource block allocation is done twice at TTI for both SC-FDMA an OFDMA. Actually, the uplink uses similar LTE resource grid and resource block structure as the downlink, even if the uplink transmission is not done over narrowband subcarriers. The difference comes from the way the inverse Fourier transformation is used for generating the waveform to be transmitted. To put it simply, in SC-FDMA the parallel symbol streams are combined to modulate the single carrier.

The SC-FDMA uses normal and inverse Fourier transformations to digitally shape the stream of QAM symbols so that the bandwidth of the symbol stream can be fitted as precisely as possible to the radio band allocated, keeping the peak power lower than what OFDMA would require. Such shaping also saves bandwidth as no unused guard bands are needed between SC-FDMA carriers used by different UEs to keep the inter-UE interference low. Consequently, spectral efficiency of SC-FDMA is better than that of QAM modulation. The symbol frequency is higher on SC-FDMA carrier than on OFDMA subcarriers. On SC-FDMA, the cyclic prefix is used only once per a group of symbols rather than for every symbol. The SC-FDMA receiver must be able to process the intersymbol interference, which is a major drawback of SC-FDMA.

Reference symbols are sent within the SC-FDMA transmission, just as with OFDMA. The UE sends uplink **sounding reference signals (SRS)** to support the eNodeB to estimate the uplink channel, perform UE power control, and link adaptation. The eNodeB uses SRS to control selection of the uplink SC-FDMA carriers and the modulation method used on them. The UE is given a UE-specific SRS configuration where the bandwidth used for SRS is N×4 resource blocks. SRS is sent as the last symbol of the SC-FDMA frame when requested by eNodeB. Since the bandwidth needed for SRS symbol transmission is rather high, eNodeB activates SRS only when needed. The UE may send the SRS as one shot over the whole bandwidth or use frequency hopping so that the SRS is transmitted over different subcarriers for successive subframes. The first approach has a low delay while the latter approach conserves UE power.

The UE also sends **demodulation reference symbols (DM-RS)** to help the eNodeB to demodulate the SC-FDMA signal. The DM-RS of SC-FDMA are located in the middle of each half a millisecond slot. Additionally, the LTE cell may have a cell-specific DM-RS configuration for multiantenna use cases, communicated to UEs within the system information. Use of DM-RS is described further in Section 7.1.4.6.

7.1.4.4 Power Control and DRX

In WCDMA, power control is essential to prevent UEs near the base station to block transmission from UEs farther away. Such near-far effect is a consequence of all WCDMA UEs in a cell to use the same radio band simultaneously. Since both OFDMA and SC-FDMA allocate the UE with a dedicated subband for each resource block, near-far problem does not apply to LTE. This is why LTE does not need fast power control, and the LTE power control mechanism is much simpler to the one used in WCDMA. LTE power control procedures are specified in 3GPPTS 36.213 [15].

The LTE uplink power control has two goals:

- To conserve UE battery
- To avoid the receiver of the eNodeB to operate in too wide a range of received power

The purpose of LTE power control is to keep the power per Hertz as received from the UE in an acceptable range. The measure of power per Hertz is known as **power spectral density (PSD)**. When the UE increases its uplink data rate with SC-FDMA, both the transmission bandwidth and total transmission power increase. Since the increased total power is distributed over a wider frequency band, the PSD may stay constant. Before activating its uplink, the UE estimates the needed transmission power by measuring reference signals received from the eNodeB. After the UE has activated the uplink transmission, the eNodeB may send power control commands to the UE within the **downlink control information (DCI)** messages sent over the physical downlink control channel (PDCCH). These commands ask the UE to either increase or decrease its transmission power in 1 dB steps. For making bigger changes to maintain the connection, increase by 3 dB is also supported.

To conserve battery, the LTE UE uses discontinuous reception (DRX) in addition to power control in the RRC connected state. With DRX, the UE may pause listening to the eNodeB for certain periods. After processing the scheduled uplink or downlink resource blocks, the UE may enter a short DRX cycle. Within the short DRX period the UE saves power and does not listen to eNodeB scheduling assignments. After the DRX cycle is over, the UE resumes its normal operation and checks if further resource blocks have been allocated for it. If not, the UE may enter the short DRX cycle again. After running the short DRX cycle a number of times without any transmissions, the UE may enter a long DRX cycle. Except for its extended length, the long DRX cycle is processed the same way as the short one. When no transmissions are scheduled for the UE over a few tens of seconds the inactivity timer expires and the UE moves from long DRX to the RRC Idle state. The inactivity timer length is a network specific parameter.

7.1.4.5 Scheduling

LTE uses shared channels to provide transmission capacity for the UEs. Resources of the shared channel are allocated between UEs and their data flows by eNodeB, which runs the MAC scheduling algorithms. Centralizing scheduling to the eNodeB has the following advantages:

- The eNodeB is in a good position to ensure proper QoS handling between different data flows and to prevent network overload.
- Scheduling is done as close to the air interface as possible to avoid any extra delays for reacting to changing radio conditions and for processing the bandwidth grant requests from UEs.

Scheduling means allocation of LTE resource blocks for both uplink and downlink. The eNodeB informs UEs about the scheduling decisions with **DCI** messages sent over the PDCCH control channel. Each DCI message has a target UE for which the resources are allocated. The DCI message tells the UE which resource blocks of the LTE frame the UE is expected to use, the MCS type for them, as well as the multiantenna method used. Different types of DCI messages are specified for different resource configurations. See 3GPP TS 36.212 [16] for details about different DCI formats and detailed contents of DCI message for each format. eNodeB may also need to adjust the size of PDCCH channel used for sending DCI messages, depending on the number of UEs being scheduled. The size of the PDCCH control region is told over the PCFICH channel. The descriptions of the referred LTE physical channels and LTE frame structure used for scheduling can be found in Section 7.1.4.7.

For its scheduling decisions, the eNodeB monitors its downlink data buffers and receives buffer status reports from the UEs within MAC headers. The eNodeB must also take any random access request into account for its uplink resource grants. The eNodeB considers also the QoS requirements of the data flows. Equipped with all that information, the eNodeB may apply one of the following scheduling methods:

- **Dynamic scheduling:** Dynamic scheduling is able to reallocate the resource blocks between UEs once per TTI, which is 1 ms for LTE. When using dynamic scheduling, the symbols in the beginning of every downlink 1 ms subframe are used for the PDCCH channel to DCI messages rather than user data. When the UE receives a DCI message within a subframe for uplink allocation, the allocated uplink resource block is in the fourth subframe after the DCI, thus the UE has 4 ms to prepare itself for transmission. In the same way, the HARQ acknowledgments of MAC protocol (see 7.1.5.1) are scheduled to be sent 4 ms after the resource block to be acknowledged has arrived to the UE.

- **Semi-persistent scheduling:** When using semi-persistent scheduling, one or multiple subframes per every 10 ms LTE frame are semi-permanently allocated for one UE. Such semi-permanent allocation is configured to the UE with the RRC protocol. The benefit of this scheme is that the whole downlink subframe can be used for transporting user data as no transmission resources are needed for the PDCCH channel. Semi-permanent scheduling can be used if the UE transmits periodically but not continuously, like VoIP packets sent once in 20 ms. The UE can be configured by RRC signaling to use semi-persistent scheduling for a dedicated connection opened for IMS voice media. For semi-persistent scheduling, the PDCCH channel is used only to activate the pre-configured scheduling for a new voice burst when the user starts speaking. Semi-persistent scheduling comes with a new approach for ensuring sufficient level of transmission quality. When using dynamic scheduling, any corrupted or lost packets would be retransmitted with HARQ processes of the MAC protocol. Such HARQ retransmissions would, however, increase latency, which is problematic for real-time voice. Instead, when the uplink quality suffers at cell edges, TTI bundling is used for voice instead of HARQ. With TTI bundling, the same voice packet is sent in multiple subframes in a row to increase the likelihood of eNodeB receiving the packet correctly.

The specific algorithms for allocating the transmission resources between UEs have not been standardized but are specific to network vendors. However, there are some generic principles based on which the scheduling algorithms work. Scheduling algorithms have two dimensions based on the physical structure of LTE radio [3]:

- **Time domain scheduling** is used to select the subset of UEs, which will be granted resource blocks for each of the TTIs. The selection is based on knowledge of which UEs have data to receive or send. The eNodeB already knows the downlink data which is there in its own buffers. The NodeB asks the UEs to send buffer status reports about any uplink data waiting in the buffers of the UE. Time domain scheduling must pick only those UEs that are not in DXR sleep state during the TTI as DRX prevents UEs to detect any scheduling grants given on the PDCCH. Only a limited number of UEs can be selected to avoid running out of PDCCH channel resources for sending DCI messages.

- **Frequency domain scheduling** is used to select the frequencies and related subcarriers for the resource blocks allocated from this TTI to the UE. Frequency domain scheduling must also consider any pending HARQ retransmissions in addition to any new data waiting to be sent. Frequency domain scheduling takes into account the CQI reports sent by the UEs to use the spectrally most efficient MCS combinations for allocated resource blocks.

Time domain scheduling is the primary method to satisfy the QoS requirements of all different data flows being scheduled, and frequency domain scheduling maximizes the transmission capacity of the cell. To balance capacity and fairness, proportional fair scheduling takes the radio conditions into account and prefers such UEs that momentarily have favorable conditions, while making sure that other UEs are not starved. When the radio channel is good, high-order MCS method with high bitrate can be used for the resource block. For the scheduled resource blocks, the air interface capacity is utilized the best way when approximately 10% of packets are retransmitted by HARQ. Lack of retransmissions indicates that the selected MCS may be too low and capacity could be increased by selecting higher-order MCS for the subcarriers [7].

Schedulers may also take inter-cell interference avoidance into account. Adjacent LTE cells may use the same system frequency band. Any excessive interference is then prevented on subcarrier level with a static or dynamically adjusted configuration taking adjacent cells into account. Subcarriers may either be divided statically between the cells or the cells may dynamically allocate different subcarriers for UEs on the cell edge.

7.1.4.6 Multiantenna Methods and UE Categories

Usage of multiple antennas to transmit multiple streams in parallel over the same frequency band (as described in 7.1.4.6) is an essential part of LTE radio technology. Multi-antenna MIMO is the area of radio technology under very rapid development since the early 2000s. With improved antenna technology, it has become possible to increase the number of antennas in both the UE and eNodeB. 3GPP LTE specifications have been evolved to take advantage of the antenna technology progress. New specification releases after the initial LTE release Rel-8 have defined various multi-antenna configurations and corresponding modifications to the LTE physical layer, including resource mapping and channel quality reporting. LTE multi-antenna methods are specified in 3GPP TS 36.213 [15].

The concept of the antenna port was introduced previously in the initial 3GPP Rel-8 LTE release to support simple multi-antenna configurations. An **antenna port** consists of the transmission from one antenna with built-in reference signals used by the receiver to reliably demodulate the radio signal received from the port. As described in Section 7.1.4.2, the original LTE approach was to have **CRS** reference symbols in predefined positions of the LTE resource grid. Demodulation of physical downlink channels was done with help of CRS reference signals. As Rel-8 supported four antenna ports per eNodeB, one cell could have up to four CRS signals, one for each eNodeB transmit antenna deployed. When using multiantenna transmission, a set of antennas is used to transmit OFDM symbols toward the UE. The cell-specific reference symbols are an exception as they are sent only from one single antenna while the other antennas are pausing their transmissions. When reference signals are sent to the UEs only from one source, it stays simple enough for the UEs to synchronize themselves to the LTE frame structure and estimate distortions of the LTE signal on different subcarriers.

The number of supported antenna ports has grown over 3GPP releases. Rel-9 supported configurations with 8 antenna ports, Rel-10 up to 14 and Rel-12 up to 46 antenna ports. Not all those ports can be used simultaneously, but each 3GPP release defined antenna port combinations that the UE and eNodeB could use. The maximum number of simultaneously used ports is up to 32. The large number of antenna ports caused a new challenge for the reference signal design, as every antenna port would need its own set of reference signals. It was no longer feasible to add new reference symbols to fixed LTE resource grid positions as that would waste transmission capacity. To support expansion of MIMO configurations and the number of used antennas, two new types of dynamically used reference signals were introduced: **DM-RS** and **channel state information reference signals (CSI-RS)**.

Instead of having fixed positions in the LTE resource grid, these new reference signal types are sent along with the data scheduled for both uplink and downlink. This essentially makes the reference signals UE specific in both directions. As the CSI-RS configuration is determined dynamically per UE, the configuration is provided to the UE over the RRC protocol.

As described in 7.1.4.2, LTE Rel-8 eNodeB performs frequency domain scheduling and link adaptation based on the **CQI** reports received from the UEs. Rel-8 UE derives its CQI reports by measuring the CRS. UEs supporting 3GPP releases later than Rel-8 do advanced CQI reporting for channel estimates derived from measuring both the legacy CRS and the new CSI reference signals (CSI-RS). Depending on the number of antennas used, either 1, 2, 4, or 8 CSR reference signals can be mapped to the LTE resource grid. A A total of 40 resource elements of the grid have been reserved for CSI-RS signals. The downlink CSI-RS signal toward a UE may occupy a maximum of eight resource elements to support up to 32 antenna ports. This sounds counterintuitive, as it means four port-specific reference signals per a single resource element. The solution is to use simple code division multiplexing for distinguishing port-specific reference signals transmitted within the same subcarrier and timeslot. Each of the reference signals use its own short code word called orthogonal cover code (OCC). Code division multiplexing is also used to separate DM-RS signals of different UEs in multiuser MIMO.

Multi-antenna methods can be used to increase cell capacity, transmitted bitrates, or robustness of the transmissions against errors. The LTE air interface supports the following types of multi-antenna methods:

- **Spatial multiplexing,** where downlink bitrates are increased by splitting the data stream as segments to be transmitted with different antennas. Each of the antennas uses the same set of subcarriers. The bitrate is increased by transmitting multiple different bit streams in parallel, separated by the phase difference caused by different distance between the UE and each antenna transmitter at the eNodeB. As spatial multiplexing makes it possible for the different antenna pairs to establish independent communication paths in the same frequency space, it has been suggested to add a new coefficient to the Shannon theorem to represent the impact of those independent paths to achievable data rates. The value of this coefficient is proportional to the number of transmit and receive antennas.
- **Transmit or receive diversity** is used to increase reliability of the transmission. In transmit diversity, the same data stream is sent from different antennas using different encodings. In receive diversity, the same transmission is captured by different antennas that have different probabilities of receiving the signal correctly.
- **MIMO,** where bitrates and/or the reliability of the transmission are increased by using multiple transmitter antennas and receiver antennas in both ends of the connection (see Chapter 4, Section 4.1.3). MIMO can also be used for beamforming to increase cell coverage and capacity by focusing signal power to narrow directional beams. The beams are created with an array of transmitter antennas with which the phases of signals are adjusted in such a way that the signals amplify each other in the desired direction and cancel each other out on other directions. The UE uses one beam at any time and may switch from a beam to another when moving.
- **Virtual or multi-user MIMO,** where the uplink capacity of the cell is increased by allocating the same subcarrier for two single-antenna UEs so that one eNodeB receiver antenna is used per UE. With virtual MIMO, the eNodeB allocates the UEs with different timeslots for transmitting their reference signals. With help of the reference signals received, the eNodeB selects the best receiver antenna for each UE. The eNodeB essentially uses spatial demultiplexing for separating the transmissions of the UEs.

Multi-antenna transmission is used for shared channels when allowed by the channel conditions. LTE supports dynamic selection of the applied multi-antenna mode. When the UE moves in vehicle and the channel conditions continuously change, the network may change the multi-antenna mode used with the UE any time. For the end user point of view, at one moment additional throughput provided with spatial multiplexing may be preferable, while additional reliability provided with transmit diversity would be better at another time. Fluctuating channel conditions may also cause fading for certain transmission paths, effectively preventing usage of some antenna ports for either spatial multiplexing or transmit diversity. In such a case, the default single-stream operation can be applied.

To support the eNodeB with link adaptation and selection of the multiantenna mode, the UE gives the eNodeB feedback about downlink channel conditions as **channel state information (CSI)** reports. CSI was specified in 3GPP Rel-10 to support advanced multi-antenna configurations. The CSI report sent by the UE covers a combination of channel quality index (CQI), rank indicator (RI), and precoding matrix indicator (PMI), the last of which is described in Section 7.1.4.9. To instruct the eNodeB about the best multiantenna configuration, UE reports a rank indicator parameter to the eNodeB. **Rank indicator (RI)** is the number of independent communication channels in the MIMO configuration used, as perceived by the UE. In 2×2 MIMO, typically, the rank indicator is 2, but if the signal from one of the transmitter antennas fades out, the rank indicator drops to 1.

LTE specifications introduce a number of **transmission modes (TM)** to describe the specific ways for using multi-antenna features on the shared physical downlink shared channel (PDSCH) channel [15]. The eNodeB tells the chosen transmission mode to the UE with the RRC signaling. Thereafter, the applied DCI message format implicitly tells the UE whether to apply either spatial multiplexing or transmit diversity option available for the chosen transmission mode. DCI message also tells the UE the number of ports to be used. The choice of the transmission mode is done for a bit longer term with the RRC signaling while the specific option within the mode can be changed at any time with the frequently sent DCI messages. The options available for different transmission modes are as follows:

- The number of available antenna ports
  - TM1, TM5 use one single port only
  - TM2, TM3, TM4, TM6, TM7, and TM8 may use one, two or four ports
  - TM9 and TM10 are able to use up to eight ports
- Closed loop spatial multiplexing: TM4, TM6, TM8-10
- Transmit diversity: TM2, TM3, TM7-10
- Multi-user MIMO: TM5

Different types of transmission modes have been introduced in the successive 3GPP releases as follows:

- Rel-8 LTE supported transmission modes 1–7 with one to four antenna ports. TM7 was the first transmission mode to support DM-RS reference signals.
- Rel-10 introduced transmission modes 8–9, CSI reporting process and increased maximum number of antenna ports to eight.
- Rel-12 introduced transmission mode 10.

3GPP TS 36.306 [17] defines LTE **UE categories** which describe the UE capabilities for modulation and multi-antenna methods. The eNodeB gets the UE category at the LTE attach, after which the eNodeB knows:

- Multiantenna configuration of the UE. Category 1 UE has only one single antenna, category 2–4 UEs two antennas, category 5–12 UEs 2–4 antennas, and category 8 UE even eight antennas.
- Whether or not the UE supports 64-QAM or 256-QAM modulation. The 64-QAM option is supported from category 6 onwards, while 256-QAM is supported with categories 11 and 12.
- UE maximum data rates: 10–600 Mbps downlink and 5–100 Mbps uplink, except category 8 UEs where maximum theoretical bitrates are as high as 3000 Mbps downlink and 1500 Mbps uplink.

The first 12 UE categories were introduced in 3GPP Rel-11 or earlier. In those releases, the uplink and downlink categories were coupled so that the single UE category number indicated both the uplink and downlink capabilities of the UE. From Rel-12 onwards the uplink and downlink categories were decoupled so that a UE could have different combinations of uplink and downlink categories.

To be able to select the right set of features for a UE, the network needs to know both the UE category and which optional capabilities the UE supports. The UE tells its capabilities in the initial access sequence with the UE capability exchange, as explained later in Section 7.1.9.1.

7.1.4.7 Frame Structure and Physical Channels

The frame structure of the LTE radio interface uses 10 ms frames transmitted over the subcarriers. Each frame is divided into 10 subframes, 1 ms each, as shown in Figure 7.9. The subframe consists of two adjacent slots so that there are 20 slots in the frame. Seven or six OFDM symbols are transmitted in each slot, depending on the length of the OFDM symbol cyclic prefix.

The LTE TTI is 1 ms. This means that the LTE subframe is the unit of radio resource allocation in the time dimension. Combined with frequency dimension, the allocation is done as a resource block, consisting of subframes on 12 adjacent subcarriers. The eNodeB can select the transmission mode and bitrate per each resource block, which is scheduled for a UE in the cell. The LTE TTI is significantly shorter than the WCDMA TTI of 10 ms and even shorter than the HSPA TTI of 2 ms. The short TTI brings LTE with the following benefits: shorter delays for allocating radio resources or adjusting bit rates and decreased latencies.

The LTE system uses a slightly simpler structure of physical channels compared to the WCDMA system. Physical WCDMA channels are multiplexed with channelization code division multiplexing, but physical LTE channels use time and frequency division multiplexing. In the LTE space of subcarriers and resource blocks, the physical channels are located into predefined resource element positions. Signaling data is transmitted over the control channels in the first few resource elements of each subframe while the rest of the resource elements of the subframe are used for user data shared channels, as shown in Figure 7.10. The number of resource elements used for control channels is adjusted dynamically per resource block. Reference signals are scattered to fixed

Figure 7.9 LTE frame structure.

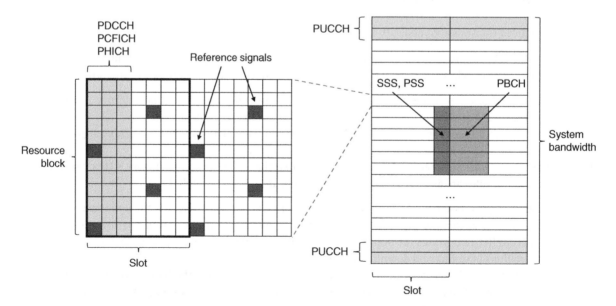

Figure 7.10 Mapping of LTE physical channels into LTE resource grid.

positions within the resource block so that they would properly represent a fair sample of subcarrier frequencies over the time domain within the block.

The physical channels of LTE radio interface are:

1) Downlink channels
   - Primary synchronization channel (PSCH) to carry primary synchronization signals (PSS) used by the UE to gain slot synchronization and frequency offset with the cell. PSS signals are sent once in four successive LTE frames in the middle of the LTE system bandwidth.
   - Secondary synchronization channel (SSCH) to carry secondary synchronization signals (SSS) used by the UE to gain frame boundaries, learn cyclic prefix length, detect if the cell uses FDD or TDD, and finally to decode physical cell ID (as a combination of PSS and SSS). Like PSS, also SSS signals are sent once in four successive LTE frames.
   - Physical broadcast channel (PBCH) carries critical network parameters for cell access within the master information block (MIB) broadcast in the cell. The PBCH channel can always be found in a fixed location of the LTE resource grid so that the UEs could find it easily without any further information from the eNodeB. Dedicated resource elements are reserved for the PBCH from resource blocks that occupy 72 subcarriers in the middle of the LTE system bandwidth. On those subcarriers, three symbols are used for the PBCH in every fourth frame, right after transmitting PSS and SSS symbols.
   - Physical downlink control channel (PDCCH) carries DCI messages to inform UEs about the resource blocks allocated to them in both uplink and downlink. In addition to telling the location of allocated resource blocks in the LTE resource grid, DCI messages tell the UE various other details related to the resource block such as MCS, HARQ process parameters, uplink transmission power levels, and multi-antenna options. The LTE uses an interesting scheme for encoding destination UEs of DCI messages sent over the shared PDCCH channel as control channel elements (CCE). Within the DCI message there is a CRC checksum calculated from the message and the unique C-RNTI identifier of the destination UE. If the UE is able to calculate the checksum correctly from a received DCI message and its own C-RNTI, it knows to be the correct destination for the message. The PDCCH channel is transmitted within 1–3 ODFMA symbols from the beginning of every subframe on each subcarrier.
   - Physical control format indicator channel (PCFICH) is used to tell how many OFDMA symbols have been reserved from the subframe for the PDCCH channel. The size of the PDCCH depends on the current usage profile of the LTE radio network. If the network is used mainly for a small number of high-speed connections, most of the resources can be used for data, and PDCCH needs only one OFDMA symbol per subframe. On the other hand, if the network is mainly used for low bandwidth VoIP calls, more resources are needed for the PDCCH channel so that the network would be able to manage the multiple simultaneous VoIP connections to large numbers of UEs.
   - Physical downlink shared channel (PDSCH) is the shared channel used for most of the LTE downlink traffic, such as data or higher layer signaling, including system information and paging messages for UEs in an idle state. The paging messages are transported within the PDSCH channel on a special resource block reserved for this purpose. All the idle UEs must wake up to listen for the paging resource block. These special resource blocks are not mapped to LTE frame structure in a fixed way, but the eNodeB tells their location to UEs within the signaling sent over the PDCCH channel.
   - Physical HARQ Indicator Channel (PHICH) carries the eNodeB HARQ acknowledgments for data packets sent by the UE.
   - Physical multicast channel (PMCH) carries downlink multicast messages.
2) Uplink channels
   - Physical random access channel (PRACH) is used by the UEs to send random access messages. PRACH capacity can be adjusted so that the channel may have one resource block in a period between 1 and 20 ms, as defined within the PDCCH signaling.

- Physical uplink control channel (PUCCH) is used by the UE for signaling when the UE does not have allocation on physical uplink shared channel (PUSCH) channel. PUCCH may be used for HARQ acknowledgments, user data scheduling requests, and CQI reports. Different PUCCH message formats have been defined for those purposes. The eNodeB informs UEs about the available PUCCH resource blocks and the formats to be used on each of them. In that way, UE is able to use the right blocks for specific types of its messages. PUCCH resource blocks are located to the lowest- and highest-frequency subcarriers of the used LTE band. Multiple UEs may use the same PUCCH resource block simultaneously since code division multiplexing is used to separate transmissions from different UEs to the same block.
- Physical uplink shared channel (PUSCH): shared physical channel via which the UEs can send user data or signaling.

7.1.4.8 Logical and Transport Channels

The set of logical channels provided by the LTE MAC is evolved from those used for WCDMA system:

- Broadcast control channel (BCCH) is a downlink channel to broadcast MIB for all the UEs camping on the cell. The MIB enables the UEs to access the shared SCH channels.
- Paging control channel (PCCH) is a downlink channel on which the eNodeB sends paging messages to notify UEs about mobile terminated data.
- Common control channel (CCCH) is a bidirectional channel shared by all UEs camping on the cell. Any message sent by the eNodeB on the CCCH has only one destination UE. The CCCH channel is used to reach UEs that are not in the active RRC mode.
- Dedicated control channel (DCCH) is a bidirectional signaling channel dedicated for a single UE.
- Dedicated traffic channel (DTCH) is a bidirectional user data channel dedicated for a single UE.
- Multicast control channel (MCCH) is a downlink channel used to transport signaling messages to control usage of the MTCH channel.
- Multicast traffic channel (MTCH) is a downlink channel used to transport data simultaneously to a group of UEs camping on the cell.

The following transport channels have been specified for LTE [4]:

1) Downlink channels:
   - Broadcast channel (BCH) is used to carry the BCCH logical data flow, which is information about PDSCH decoding, broadcast to all UEs camping on the cell. The eNodeB forwards any messages from the BCH transport channel to the PBCH physical channel. The BCH channel of a WCDMA cell carries all the SYSTEM INFORMATION messages, but in LTE the BCH channel is used only for MASTER INFORMATION and SYSTEM INFORMATION BLOCK 1 (SIB1) messages. The LTE BCH carries only a minimal set of parameters, which are necessary for the UE to decode the PDSCH channel on which the rest of LTE SYSTEM INFORMATION messages are transported. MASTER INFORMATION message tells the UE the band used, configuration of the PHICH channel, and the SFN number of the frame.
   - Paging channel (PCH) is used to carry the PCCH logical data flow consisting of paging messages sent to UEs to move them from RRC_IDLE to RRC_CONNECTED state. The base station forwards any messages from the PCH transport channel to the PDSCH physical channel.
   - Downlink shared channel (DL-SCH) is used to carry user data, SYSTEM INFORMATION messages, and other types of signaling to the UE. The eNodeB uses the RRC protocol to allocate DL-SCH resources for the UEs in the RRC_CONNECTED state.
   - Multicast channel (MCH) is used for multicasting certain user data streams (like IP-TV or radio) simultaneously to multiple UEs.

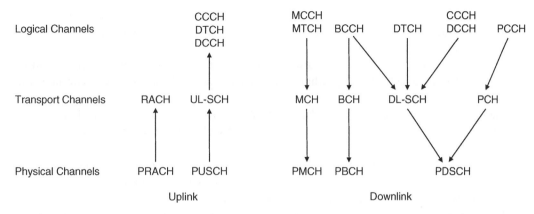

Figure 7.11 Mapping between LTE logical, transport, and physical channels. *Source:* Adapted from 3GPP TS 36.300 [4] and TS 36.321 [12].

2) Uplink channels.
- Random access channel (RACH) is used to carry random access requests when the UE wants to get uplink grant(s) to send user data or a location update message. The base station forwards messages from the PRACH physical channel to the RACH transport channel.
- Uplink shared channel (UL-SCH) is shared by multiple UEs for sending user data or RRC signaling toward the network. The UE must be in RRC_CONNECTED state to use UL-SCH channel and get transport resource allocation for it.

Mapping of transport blocks to physical channels of radio interface is done at eNodeB and UE physical layer, as shown in Figure 7.11.

7.1.4.9 LTE Transmitter Design

The processing chain of an LTE transmitter takes care of channel coding and modulation. The transmitter takes transport blocks as its input and generates the signals for the antennas as output, as shown in the block diagram of downlink transmitter within Figure 7.12. The LTE transmitter operates as follows:

1) Transport blocks are mapped to two flows, to be encoded as code word 1 and code word 2, respectively. Processing of both the flows follows steps as below.
2) CRC code is added to the end of every transport block for bit error detection.

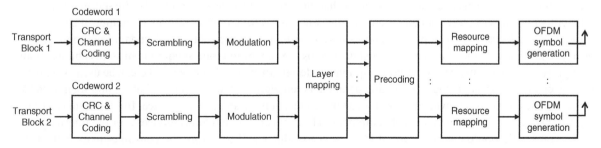

Figure 7.12 LTE downlink transmitter design.

3) The set of consecutive transport blocks are segmented to code blocks. At first, the transport blocks are concatenated to a bit sequence, which is split to code blocks of sizes suitable for the encoder. Another CRC checksum is added to each code block.

4) Code blocks are encoded for transmission with a forward error correction method such as Viterbi or Turbo coding. The encoded blocks are padded with padding bits to ensure that all the blocks have a constant size before interleaving.

5) Rate matching adjusts the transmission bit rate to the bit rate of the physical channel. Rate matching is done by either removing or repeating some of the encoded bits or disabling the transmission for a certain period (DTX).

6) The stream of encoded blocks is concatenated as a continuous bitstream of encoded code words. The bit-streams of code words 1 and 2 are scrambled to break any long sequences of zeros or ones.

7) QAM symbols are derived from the bit sequences by the modulators.

8) The generated QAM symbols are distributed to a number of layers, each feeding a subset of symbols to the precoder, which eventually allocates the symbols to 1 – N antenna ports used for transmission. The precoder uses a precoding matrix selected within the CSI process.

9) Data from antenna port(s) is mapped to the LTE resource grid. This mapping is done separately for each physical antenna.

10) OFDM signals are generated for every physical antenna used for transmission.

The eNodeB may use either open loop or closed loop process for selecting the downlink **precoding matrix**. In the open loop case, the eNodeB selects the matrix and tells the index of the matrix to the UE in the DCI message. In the closed loop case, downlink precoding matrix is selected by the eNodeB based on a recommendation received from the UE. The UE picks one of the predefined matrices and sends its index to eNodeB as **PMI** parameter. The PMI is sent along with the CQI and RI parameters within the CSI process. Closed loop spatial multiplexing is a feature of certain transmission modes used on the PDSCH channel. The UE demodulates the transmission from the eNodeB with help of downlink reference signals and combines transmissions with the precoding matrix used by eNodeB.

The UE processes its uplink data quite the same way, but there are some minor differences in the details. The major difference is that the channel coded user data and signaling streams are combined to be transmitted with SC-FDMA. In the downlink where OFDMA is used, the eNodeB divides the resource elements of different subcarriers within the resource blocks between those that carry user data and those that carry signaling symbols.

7.1.4.10 Support for Machine Type Communication and Internet of Things

As described in Section 7.1.1, the original targets of LTE came from ITU-R requirements for IMT-Advanced systems. LTE was designed to be a broadband cellular data system to provide high bitrates, low latencies, and high spectral efficiency. The focus was to support throughput-hungry end users, but different kind of needs for cellular communications have risen from another direction known as **machine type communication (MTC)**. MTC means communication between machines rather than humans. Typical use cases of MTC are sensor networks of different sorts. Modern pieces of equipment and machines often have built-in sensors to monitor the status and activity. In the 2000s, many equipment vendors have found it useful to remotely collect data from these sensors over a cellular data connection. The collected data helps the vendor to monitor equipment status and locations, plan preventive maintenance, and improve equipment design. In some other cases, MTC can be used also to control the equipment by adjusting its operating parameters remotely. The new home appliances and wearables like smart watches could also provide data over a cellular connection toward a cloud service. **IoT** is a term to mean a big number of such intelligent pieces of equipment to be connected with (cloud) services over the Internet.

The communication patterns of typical MTC use cases differ radically from the consumer mass market use cases [7]. MTC UEs perform communication periodically, but not very often. The needed bitrates are typically low, but the number of UEs per cell may be high. The cellular signal quality might be low if the equipment is located in some sort of industrial environment, a basement of a building or far at the sea. Further on, the expected power consumption should be low, especially for any battery-operated devices without regular maintenance support. The cost of both the communication and the cellular modem within the devices should be as low as possible. Until recent years, GSM has been the technology to support such low-cost narrowband communications. But operators have become interested on refarming GSM bands for LTE and ramping down their GSM networks to optimize cost and maximize the value of their scarce radio spectrum. This raised a need to support MTC and IoT use cases also with LTE systems.

While 3GPP has used much effort to scale up LTE with solutions such as MIMO and carrier aggregation, MTC use cases required just the opposite. In order to scale down LTE to support narrowband, low bitrate and low-power connections, the following approaches have been used:

- Use single-antenna configuration.
- Define a system that could rely on narrower bands than the full LTE 20 MHz system bandwidth.
- Specify additional support for UEs, which stay a very long time in power save mode (PSM) and listen to paging rather infrequently.

As mentioned in Section 7.1.4.6, 3GPP has defined a broad range of UE categories. The lowest LTE UE category specified already in 3GPP Rel-8 was Category 1 (or Cat-1), which supports 10 Mbps bitrates and a single-antenna configuration. To support MTC and IoT use cases, 3GPP has defined the following low-end single-antenna UE categories together with the related modification to the air interface and system features to support them:

- Category 0 specified in 3GPP Rel-12 with 1 Mbps maximum bitrate. Cat-0 UEs may support half-duplex mode so that the device is not able to receive and transmit simultaneously, but must switch between those. To save power, Cat-0 devices may stay in RRC Idle state without monitoring paging for days or weeks, and the network side timers must take that into account.
- Category M1 specified in 3GPP Rel-13 with 1 Mbps maximum bitrate. Cat-M1 UEs need only support maximum 1.4 MHz bandwidth, instead of the LTE common 20 MHz system bandwidth. For LTE air interface this means a major change due to the physical control channel configuration. The PUCCH uplink control channel is located to the few outermost subcarriers of system bandwidth, and the PDCCH downlink control channel occupies a few resource elements of all the other subcarriers for each slot. For Cat-M1 devices, additional control channels have been specified, available within the 1.4 MHz bandwidth allocated to Cat-1 UEs from the full LTE system bandwidth.
- Category NB1 specified in 3GPP Rel-13 with approximately 200 to 300 kbps maximum bitrate and improved support for power saving and low signal quality conditions. Cat-NB1 UEs support only a 180 kHz wide NB-IoT channel, which consists of 12 adjacent 15 kHz subcarriers like a regular LTE resource block. The LTE signaling protocol suite is used to manage Cat-NB1 radio resources and security. Otherwise, Cat-NB1 UEs have a stripped support of LTE features to simplify the modem to a bare minimum.

From these options, Category NB1 deserves a closer look. The 180 kHz NB-IoT channels still align with LTE OFDM resource grid. Like LTE, also NB-IoT uses OFDM in downlink and SC-FDMA in uplink. Only the low-order QPSK and BPSK modulation methods and half-duplex mode are used for both simplicity and low-quality signal support, since there is no need to support high bitrates. The NB-IoT channel frame structure is the same as the LTE frame structure.

In the LTE system there are three options for locating the NB-IoT 180 kHz channels:

- Carry NB-IoT channel within LTE system bandwidth so that no regular LTE resource blocks could be allocated to the subcarriers assigned to the NB-IoT channel.
- Carry NB-IoT channel on the guard bands in both ends of LTE system bandwidth.
- Carry NB-IoT channel outside of LTE band, like within a GSM band.

For the first option, the LTE cell may carry one or multiple NB-IoT channels within the LTE band. One of these channels is the anchor carrier, which delivers all NB system information and control channels. Other channels are used as per downlink and uplink resource grants given within the anchor channel. At any time, the Cat-NB1 UE uses only one single NB-IoT channel, which in RRC Idle state must always be the anchor channel. Each NB-IoT channel is permanently mapped to certain 12 subcarriers.

As mentioned earlier in the context of Cat-M1, LTE control channels use resources of every subcarrier of the system bandwidth. Like the Cat-M1 channel, also the NB-IoT channel uses its own control channel setup, as well as the positioning of synchronization and reference signals within the available 12 subcarriers of the anchor channel. The downlink control channel does not use symbols from every resource block but instead occupies a complete resource block once over a number of subframes. Exact positioning of the control and RACHs is given within NB system information blocks (NB SIB).

For user data, the NB-IoT uses narrowband physical shared channels, NPDSCH for downlink and NPUSCH for uplink. Resource allocation for these shared channels is done as follows:

- Downlink resources are allocated as resource blocks of 12 subcarriers times seven symbols like for regular LTE resource blocks. Such a block occupies the full NB-IoT channel in the duration of one slot; thus, only one Cat-NB1 UE may receive a downlink signal over the NB-IoT channel at a time. Because of this, NB-IoT downlink assignment may grant the UE with multiple subframes with a delay of several subframes from getting the assignment.
- Uplink resources are assigned per UE as 1, 3, 6, or 12 parallel subcarriers per slot. In uplink there may be up to 12 Cat-NB1 UEs transmitting simultaneously over the NB-IoT channel. Within the allocated subcarriers, the UE may either use its whole 15 kHz bandwidth or focus its transmission power to narrower 3.75 kHz band in order to improve signal reach in very weak signal conditions.

To reduce the overhead with user data transmission, NB-IoT uses the following approaches:

- RRC connections are not automatically released but only suspended so that they could be resumed without extra signaling, such as authentication or RRC reconfiguration. If the UE has no data to be sent, it may request immediate release of the RRC connection with release assistance indication (RAI) carried in NAS or AS signaling
- An option has been defined for the UE to piggyback IP user data packets to the ESM signaling management messages. With this option, MME extracts the uplink user data packet and forwards it to the S-GW serving the UE or gets the downlink user data packet from S-GW to be forwarded to UE along with ESM message. With the early data transmission (EDT) option, the data can be piggybacked to Msg3 to Msg5 of the random access procedure.

The LTE MTC support has been designed for UEs, which sent very little data with minimum power consumption. The expected communication pattern for the UE is to send a few tens of user data bytes very infrequently, like once an hour, day, or week. Between successive transmission bursts, Cat-M1 and Cat-NB1 UEs may agree with the network to enter extended idle mode discontinuous reception (eDRX) mode to save battery. In eDRX mode, the UE stops listening to control channels or paging over a period of a maximum 43 minutes, as agreed with the network. Cat-NB1 UEs support also PSM, where the device may wake up once over many days to send a tracking area update and possibly uplink data.

For LTE network NB-IoT, system support means the following requirements:

- The cell should provide one or multiple narrowband NB-IoT channels.
- Any downlink data may need to be buffered by S-GW for a long period over which the UE is in PSM and does not listen to pages.
- A cell or single sector should support tens of thousands of Cat-NB1 UEs simultaneously, but with only a few of them being active at any moment.
- The overhead of a single Cat-NB1 UE becoming active to send small amounts of user data should be minimized.

In practice, the power saving features of Cat-M1 and Cat-NB1 have turned out to be problematic. Buffering of downlink data is not commonly supported. Any centralized control of the IoT devices is difficult as the network has no way to wake up the UEs from the sleep states.

Cat-NB1 devices do not support handovers of any kind, whether Intra-LTE or Inter-system. This means Cat-NB1 UE is an LTE-only UE, which can change the serving cell only by going to RRC Idle state and performing cell reselection. This is feasible for stationary devices, which transmit only small amounts of data quite infrequently. To simplify the system even further, the Cat-NB1 UE does not support channel measurements.

7.1.5 Protocols Used between UE and LTE Radio Network

7.1.5.1 MAC Protocol

LTE Medium Access Control (MAC) protocol takes care of mapping the logical channels to the transport channels. LTE MAC applies the HARQ retransmission technique introduced already for HSPA (see Chapter 6, Section 6.2.2). LTE MAC protocol is defined in 3GPP TS 36.321 [12].

The following functions have been defined for LTE MAC protocol:

- Multiplexing and demultiplexing traffic between different logical channels and DL-SCH and UL-SCH transport channels.
- Passing RLC protocol messages to the physical layer within transport blocks and selecting the transport formats for those blocks, to be used at the physical layer.
- Correcting errors at link level with HARQ retransmission processes. LTE HARQ uses soft combining and incremental redundancy processes similar to HSPA HARQ. The LTE MAC layer uses eight parallel HARQ processes to ensure that new user data can be transmitted over those HARQ processes that are not currently processing negative acknowledgments and HARQ retransmissions. LTE HARQ acknowledgments are sent for each 1 ms subframe in the fourth subframe after the reception. In this way, the HARQ retransmission time 5 ms is half of that supported by HSDPA HARQ. Note that this applies to FDD networks, while timing may be different with TDD deployments having asymmetric configuration of uplink versus downlink subframes [7].
- Scheduling and setting priorities between different logical information flows to allocate transmission capacity between different UEs and data streams. Priority setting is needed for those logical flows that are transmitted on the DL-SCH and UL-SCH transport channels as those channels support multiple flows for various purposes and UEs.
- Random access control.
- Reporting of the transmitted data volume to the RRC layer, which manages the radio resources.
- Activating and deactivating CC within the band combination configured with the RRC signaling.

The most important difference between WCDMA MAC and LTE MAC protocols is that the LTE MAC does not take care of encrypting the data. In LTE protocol, architecture encryption has been left to protocol layers above MAC. While WCDMA MAC was divided into multiple entities, LTE MAC has only one type of entity for the UE and another for the eNodeB.

LTE MAC frame consists of three parts:

- The MAC header consists of a number of subheaders, each of which is related to a MAC control element or service data unit (SDU) field within the MAC frame. A subheader describes the type of related MAC control elements or the logical channel ID and length of the related SDU.
- MAC control elements:
  - Downlink: Timing advance to be used by the UE, an instruction for the UE to use discontinuous reception (DTX), and a UE contention resolution identity for random access process.
  - Uplink: The C-RNTI identifier of the UE, a buffer report about the amount of UE data queued for transmission, and a power headroom report about the difference between UE maximum and currently used transmission power levels.
- MAC SDUs for the user data within the MAC frame.

LTE MAC subheaders consist of the following fields:

- LCID:
  - Logical channel of the SDU; or
  - MAC control entity of the MAC control element
- Format of the SDU length field as 7 or 15 bits
- Length of the SDU measured in bytes.

7.1.5.2 RLC Protocol

LTE Radio Link Control (RLC) protocol provides a link level connection (radio bearer or signaling radio bearer) between the UE and eNodeB. LTE RLC protocol is defined in 3GPP TS 36.322 [18].

The LTE RLC protocol has the same tasks as the WCDMA RLC protocol, except that LTE data encryption is done by the PDCP protocol. LTE RLC takes care of segmentation and reassembly of upper layer data as well as RLC packet acknowledgments and retransmission.

Like WCDMA RLC, also the LTE RLC protocol provides three different service types: transparent mode (TM), acknowledged mode (AM), and unacknowledged mode (UM). RLC message structure depends on the RLC mode:

- In TM, the frame has no RLC header fields, but only user data.
- In UM, the frame has the following fields:
  - Segment information: whether the frame contains a complete SDU or a segment of it and whether the segment is the first or last one for the SDU
  - Sequence number of the RLC frame
  - Segment offset: the position of the segment within the SDU
  - User data
- In AM, the frame has the following fields:
  - D/C: frame type as control or user data frame
  - Polling bit: request for the receiver to acknowledge the frame
  - Sequence number of the RLC frame
  - Segment offset: the position of the segment within the SDU
  - User data

RLC acknowledgments are carried within RLC status messages.

7.1.5.3 PDCP Protocol

LTE Packet Data Convergence Protocol (PDCP) protocol is used to transport both user data and signaling between the eNodeB and UE. The WCDMA PCDP protocol is used only for user data, but the LTE PDCP transports also RRC signaling messages. PDCP is run on top of the LTE RLC protocol. LTE PDCP protocol is defined in 3GPP TS 36.323 [19].

The LTE PDCP protocol has the following tasks:

- Compression of TCP/IP headers to minimize the overhead data transmitted over the radio interface. For that purpose, PDCP uses robust header compression (RoHC) as specified by IETF in RFC 3095 [20] and 5795 [21]. IP header compression is especially important for the small VoIP packets where combined header size is significant (up to two-thirds) compared to the size of carried voice media packet [7]. Header compression is useful also for NB-IoT use cases.
- Encryption of all messages and integrity protection of the signaling messages.
- Reliable transmission of data. When the UE performs a handover, PDCP ensures that no packets are lost or duplicated and the packets are delivered in the correct order.

LTE PDCP has two message types, data and control messages, as indicated by the first bit of the message. The structure of PDCP message depends on message type.

- Data messages carry user data together with a PDCP message sequence number.
- Control messages can be used as status reports or for RoHC feedback. The PDU Type field of the control message determines which other fields the message contains.

7.1.5.4 RRC Protocol

LTE Radio Resource Control (RRC) protocol is used to manage the radio resources at the air interface. Thus, RRC is an access stratum protocol. LTE RRC messages are carried over the LTE PDCP protocol. LTE RRC protocol is defined in 3GPP TS 36.331 [22].

LTE RRC state model is much simpler than the WCDMA UMTS RRC model. The LTE state model has only two mobility management states:

- RRC_IDLE: The idle UE is camping in an LTE cell, has announced its location, and listens to the control messages sent over shared channels of the cell. Additionally, the UE measures the radio signals sent by the neighboring cells so that the UE can autonomously select the cell with the best radio signal quality. The UE does not have a dedicated channel allocated for itself and initiates communication only for announcing a new tracking area caused by a cell reselection.
- RRC_CONNECTED: The UE is connected to an eNodeB via a dedicated channel and may send data any time. The UE reports radio signal quality estimates of different LTE subcarriers to the eNodeB. The UE may continue measuring the radio signals of neighboring cells, but handover decisions are done by the eNodeB.

The LTE RRC protocol has the following tasks:

- Broadcasting MASTER and SYSTEM INFORMATION messages of the BCCH logical channel. LTE system information covers system information blocks (SIB) 1–12, which provide the following pieces of information to the UEs camping in the cell:
  - SIB1: cell and network identity, access, and scheduling related parameters
  - SIB2: common and shared channel configuration
  - SIB3-8: intra- and inter-ratio access technologies (RAT) cell reselection parameters
  - SIB9: name of small home eNodeB
  - SIB10-12: mobile alerts, earthquake and tsunami warnings
- Paging UEs to deliver mobile terminated data or to notify UE about changes in system information.
- Managing the RRC state of the UE, creating and releasing RRC connections.
- Managing radio bearers and corresponding transport and physical channels.
- Configuring the UE with CC for carrier aggregation and dual connectivity.
- Managing the UE radio measurement and reporting activities.
- Cell reselection and handover procedures.
- Managing encryption functionality of PDCP protocol.
- Transporting NAS protocol messages between eNodeB and UE.

The LTE RRC message has the following fields:

- Message type
- RRC transaction ID used to match the response with related request
- A set of other fields as defined for the specific type of RRC message

7.1.6 Signaling Protocols between UE and Core Network

7.1.6.1 NAS Protocols EMM and ESM

LTE non-access stratum (NAS) protocols **EPS Mobility Management (EMM)** and **EPS Session Management (ESM)** are used between the UE and MME for mobility and session management. LTE NAS protocols are defined in 3GPP TS 24.301 [23].

Tasks of the EMM protocol:

- Attaching the UE to LTE service and detaching it from the service
- Authenticating the subscriber and controlling NAS encryption and integrity protection
- Tracking area updates
- Paging and service request procedures used to reach the UE, establish new connections, and reserve radio resources for the UE

Tasks of the ESM protocol:

- Activation, modification, and deactivation of **EPS bearer contexts**, which is the term used in LTE for packet data protocol contexts
- Management of Packet Data Network connectivity and bearer resource allocation

In certain cases, like the LTE attach procedure, there is a need to send EMM and ESM messages at the same time. For those cases, the EMM message encapsulates the ESM message into the ESM message container field.

NAS signaling messages are encrypted and integrity protected over the NAS signaling connection, after the subscriber has been authenticated and NAS security is enabled with EMM procedures. NAS protocol messages start with the Protocol Discriminator field, the value of which determines the composition of other information elements within the message.

7.1.7 Protocols of LTE Radio and Core Networks

7.1.7.1 S1AP Protocol

S1 Application Protocol (S1AP) is used by the MME to manage functionality of the eNodeB and UE. LTE S1AP protocol is defined in 3GPP TS 36.413 [24].

S1AP provides two different types of services:

- UE-associated services for managing specific UEs
- Non-UE-associated services, which are related to an eNodeB and the whole S1 interface

The S1AP protocol is used for the following purposes:

- Paging control
- Managing radio access bearers and contexts for UEs
- Intra-LTE handover control between eNodeBs without an X2 connection
- Inter-RAT handover control
- Transporting NAS protocol messages between the eNodeB and MME
- General eNodeB and S1 interface management procedures such as configuration updates or S1 setup

The S1AP message starts always with the Message Type field, the value of which determines the composition of other information elements within the message. S1AP protocol messages are carried between the eNodeB and MME over the SCTP protocol, which provides reliable IP-based transport services.

7.1.7.2 X2AP Protocol

X2 Application Protocol (X2AP) is used over the X2 interface between two eNodeBs for inter-eNodeB handover control and cell load management. As eNodeBs work without centralized control, the X2 interface has an important role for coordinating actions between eNodeBs. LTE X2AP protocol is defined in 3GPP TS 36.423 [25].

The X2AP protocol is used for the following purposes:

- Setup of the X2 connection between two eNodeBs
- UE handover control
- Load and interference management between eNodeBs

At first, two eNodeBs must discover each other to be able to establish the X2 interface between them. The discovery may happen by either the network management system or a UE telling the eNodeB about other cells in neighborhood. The measurement reports sent by UEs contain the physical cell IDs of the measured cells. When the eNodeB finds a new physical cell ID from such a report, the eNodeB asks the UE to read and report the global cell ID of that cell. When the eNodeB knows the global cell ID, it can contact the eNodeB of that cell to set up the X2 connection with it. This mechanism is known as **automatic neighbor relation (ANR)**.

The X2 connection enables eNodeBs to exchange information about their load levels and deliver overload notifications. X2AP can also be used for inter-cell interference avoidance by coordinating transmission power allocation between subcarriers used for UEs at the edge of both cells. This is useful, as neighbor LTE cells use the same carrier frequencies.

The X2AP message starts always with the Message Type field, the value of which determines the composition of other information elements within the message. X2AP relies on the underlying SCTP transport protocol.

7.1.7.3 GTP-C Protocol

GTP Control Plane (GTP-C) signaling protocol is used in most of EPC interfaces, such as:

- S3 between MME and SGSN
- S4 between S-GW and SGSN
- S5/S8 between S-GW and P-GW
- S10 between two MMEs
- S11 between MME and S-GW

GTP-C is specified in 3GPP TS 29.274 [26]. The GTP-C protocol provides support for many functional areas, such as:

- Path and GTP tunnel/bearer management to support user data transport for the UE
- Mobility management and UE location tracking
- CSFB to support CS calls over non-LTE radio technologies
- Error handling and recovery over the interface supported by GTP-C

The GTP-C message has the following fields:

- GTP-C Protocol version of the message
- Piggybacking bit indicating if there is another GTP-C message present in the end of this message
- TEID flag indicating if the header contains the Tunnel endpoint ID (TEID)
- Message type
- Message length
- GTP tunnel endpoint ID (TEID) identifying a specific GTP-U tunnel
- Sequence number of the message
- Other protocol fields (Information Elements) as defined for the message type

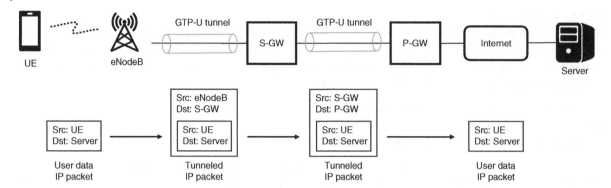

Figure 7.13 GTP-U tunneling.

7.1.7.4 GTP-U Protocol

GTP User Plane (GTP-U) protocol is used between the eNodeB, S-GW and P-GW to transport user data and signaling messages to/from the UE. GTP-U for LTE is specified in 3GPP TS 29.281 [27].

The messages of GTP-U belong to tunnels, which encapsulate the transported messages. GTP-U messages are carried over UDP/IP protocol. Since the GTP-U messages might also contain user data in UDP/IP format, GTP-U provides encapsulation to carry UDP/IP user data flows over the UDP/IP transport infrastructure of the network. In such a nested structure, the IP addresses used within the inner IP packets are the source and destination address of the user data flow. The IP addresses of the outer data flow are the IP addresses of GTP-U tunnel endpoints. See Figure 7.13 for details.

The GTP-U protocol supports the following functions:

- Tunnel and path management including error handling and sending echo requests to the other end to check if the tunnel works correctly.
- Transporting user data or signaling messages in the tunnel within G-PDU messages. Packet retransmissions and reordering for reliable data transfer.

The GTP-U message has the following fields:

- GTP-U Protocol version of the message
- Protocol type
- Extension header flag indicating the presence of extension headers
- Sequence number flag indicating the presence of sequence number
- N-PDU flag indicating the presence of N-PDU number
- Message type.
- Message length
- TEID of the GTP-U tunnel
- Sequence number (optional)
- N-PDU number (optional) used in some UTRAN or GERAN related handover and routing area update procedures
- Next extension header type (optional)
- Extension headers (optional).
- T-PDU payload, which contains either user data or the signaling message transported within the tunnel

7.1.7.5 Diameter Protocol

Diameter protocol is used for communicating with the HSS. Diameter is specified in IETF RFC 6733 [28]. Additionally, IETF RFC 3589 [29] specifies Diameter command code range reserved for 3GPP purposes.

Diameter is enhanced by 3GPP technical specifications such as TS 29.212 [30], 29.214 [31], 29.272 [32], and TS 29.336 [33].

Diameter supports the following functions related to the HSS:

- User authentication and service authorization
- Management of user location and subscription records

The Diameter message has the following fields:

- Protocol version
- Message length
- Command code
- Application, hop-by-hop, and end-to-end ID
- Other fields (Attribute Value Pairs) as defined for the command code

7.1.8 Protocols Used between EPC and UTRAN or GERAN Networks

7.1.8.1 SGsAP Protocol

SG Application Protocol (SGsAP) is used between the MME and MSC/VRL to support traditional circuit switched services and short messages. SGsAP protocol is defined in 3GPP TS 29.118 [34].

The SGsAP protocol supports the following inter-RAT functions:

- Updating MSC/VLR about the location and status of the UE camping in an LTE cell
- Forwarding short messages between the UE and the circuit switched core network
- Initiating CSFB or inter-RAT handover to move the UE from the LTE cell to a UTRAN or GERAN cell to use CS or PS services there

The SGsAP protocol messages start with Message Type field, the value of which determines the composition of other information elements within the message. SGsAP messages are transported over the underlying SCTP protocol.

7.1.9 Radio Resource Management

Radio resource management means eNodeB functions for allocating LTE radio blocks among the UE population camping on the LTE cell and managing various parameters affecting the transmission within those blocks. Radio resource management has the following goals:

- Efficient utilization of the radio frequencies assigned to the LTE network to maximize the network capacity
- Fair allocation of radio resources within the UE population to satisfy the transmission needs of the UEs
- Minimization of inter-cell interference between the same subcarriers in adjacent cells

7.1.9.1 LTE Initial Access

In the initial access procedure, the UE performs the following tasks:

- Finds LTE cells, selects the network, and performs cell access
- Opens RRC connection for exchanging signaling messages with the network
- Attaches to the LTE service and authenticates itself
- Announces its optional capabilities and gets an initial default bearer for user data

In the RRC_IDLE state, the UE scans LTE radio frequencies to find a suitable cell. For cell selection, the UE applies PLMN selection priorities stored to the USIM application. The scans are not limited to LTE radio frequencies. The UE may also scan frequencies allocated for other RAT. Once a suitable LTE cell is found, the UE attaches to the LTE network to camp in the cell. An LTE cell is identified by its cell ID. Actually, LTE has two cell ID types. The short physical cell ID has one of the 504 available values and is transmitted over the SCH synchronization channels of the cell. The physical cell ID is used to distinguish different cells without yet decoding the shared downlink channels. In addition to the physical cell ID, the eNodeB has a longer globally unique LTE cell ID, which the UE learns from system information messages of the cell.

The physical cell ID is the combination of two synchronization signal values:

- The **PSS** has three possible values
- The **SSS** has 168 possible values

The eNodeB uses the following scheme for transmitting its synchronization signals, as shown in Figure 7.10:

- In the frequency domain, the synchronization signals are transmitted in six radio blocks in the middle of the LTE system bandwidth.
- In the time domain, synchronization signals are sent twice in every frame in the blocks mentioned above. In those blocks, the LTE frames have two fixed positions for synchronization signals, in the end of the first and eleventh slots of the frame.
  - PSS and SSS are sent in two adjacent OFDMA symbols in the end of the corresponding slot.

When the UE has successfully acquired both the synchronization signals, it has synchronized itself to the LTE frame structure and decoded the physical cell ID of the cell.

The LTE UE initial access procedure is performed as shown in Figure 7.14:

1) After being powered on, the UE tries to find the synchronization signals from the middle of radio bands allocated for LTE. When the UE has recognized the PSS on an LTE band, it has reached frame synchronization with the cell. Thereafter, the UE can complete the synchronization process by reading the SSS, which completes the physical cell ID to the UE.
2) Next, the UE reads the MIB from the PBCH channel, which is transmitted on the same 1.08 MHz radio band as the synchronization signals. Master information parameters enable the UE to access the PDCCH channel and eventually locate system information block 1 (SIB1). The UE starts listening to the PDSCH channel to read other system information blocks sent periodically by the eNodeB. The SIB1 contains parameters related to network selection and cell access as well as scheduling. From SIB2, the UE finds the configuration of other LTE physical channels in the cell and values of various timers and other constants. The UE performs network selection process as explained in Chapter 6, Section 6.1.8.1 for UMTS.
3) The UE opens an RRC connection as described in Section 7.1.9.2. Within the connection opening process, the UE sends an EMM ATTACH REQUEST with an ESM PDN CONNECTIVITY REQUEST message to the MME. In the attach request, the UE tells the address of the MME with which the UE was previously registered and the PLMN identifier of the preferred operator. In the PDN connectivity request the UE may specify the APN via which the UE wants to be connected to certain packet data network (PDN). If the APN parameter is omitted, the network will instead use the default APN given within the subscriber profile of the user.
4) The MME performs subscriber authentication and starts NAS level encryption and integrity protection as described in Section 7.1.10. The MME thereafter updates the location of the UE to the HSS and gets back subscriber data, such as the default APN and QoS parameters to be used for the default EPS bearer. The MME proceeds with the initial default EPS bearer creation as described in Section 7.1.11.3. In the end, the MME sends an EMM ATTACH ACCEPT message to the eNodeB and instructs the UE to activate the default EPS bearer.
5) The eNodeB may query the UE for its LTE capabilities by sending an RRC UE CAPABILITY ENQUIRY message to the UE. With this message, the eNodeB may request information for various UE capabilities, such as the UE category (see Section 7.1.4.6), supported carrier aggregation band combinations, and other radio

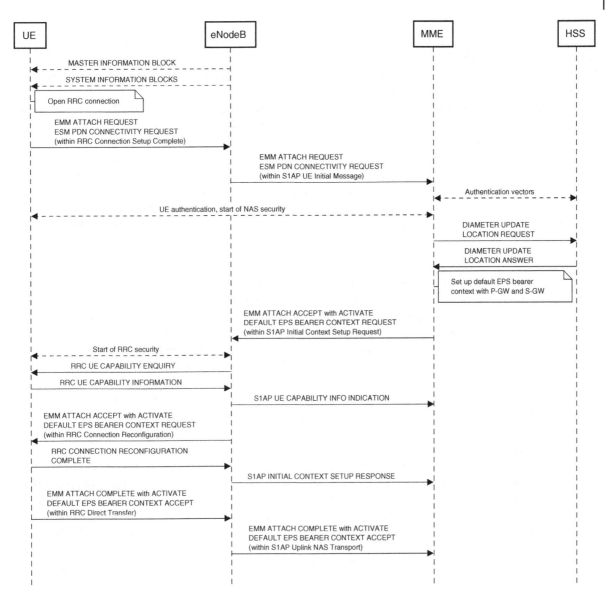

Figure 7.14 LTE UE initial access.

technologies like support for GSM, UMTS, or 5G NR EN-DC. The eNodeB forwards the returned UE capabilities to the MME.

6) The eNodeB forwards EMM messages received from the MME to the UE within an RRC CONNECTION RECONFIGURATION message and starts AS level encryption and integrity protection for the PDCP protocol as described in Section 7.1.10.2. Finally, both the eNodeB and UE provide their responses to the MME about the completion of the attach and initial default EPS bearer setup.

For further details about the system information blocks and other messages used in these procedures, please refer to *Online Appendix I.5.1.*

7.1.9.2 Opening RRC Connection

To be able to communicate, the UE needs a resource grant on a shared channel. An RRC connection is needed between the UE and eNodeB to manage resource grants and other aspects of communication. The UE enters RRC_CONNECTED state with the random access and contention resolution process. The random access process of LTE resembles closely the UTRAN random access process.

The UE may initiate random access either on its own or after being paged by the network. Paging may be done due to mobile terminated data, CS call attempt, or even for changes in system information. To page a UE, the eNodeB sends an RRC PAGING message to it over the PDSCH channel. At this point, the UE does not yet have a resource grant on the PDSCH, but in the RRC_IDLE state the UE must listen to paging slots of the channel. The UE may apply discontinuous reception (DRX) for its listening process. With DRX the UE regularly wakes up to listen to paging slots, as defined in the DRX control parameters of the SIB2 block. Paging message is sent in cells of the tracking area where the UE has most recently sent a tracking area update (TAU) message. The paging message contains either the IMSI identifier of the subscriber or the S-TMSI identifier, which the MME has given to the UE.

RRC connection is opened with random access as shown in Figure 7.15:

1) The UE, which is in the RRC_IDLE state, reads SIB2 system information message from the PDSCH channel. From the SIB2 block, the UE learns configuration or the PRACH channel and random access preamble sequences used in the cell. The preamble is a predefined bit sequence used for random access. The UE selects randomly one of the preambles and PRACH channel timeslots for sending the preamble. If the UE does not receive a RAR from the eNodeB, the UE shall send the preamble once again with higher transmission power.

2) The eNodeB responds to a received preamble with a MAC RAR message. The RAR message contains the original preamble sequence against which the RAR is sent and various parameters which the UE needs to use on the UL-SCH transport channel. After receiving both the RAR message and an PUSCH uplink grant in a DCI message, the UE uses the grant to send an RRC CONNECTION REQUEST message to the eNodeB. This RRC message informs the eNodeB about the purpose (signaling, user data, voice call, etc.) for which the UE requests to have an RRC connection.

3) The eNodeB evaluates the received connection request to decide if it is able to open a new connection or not. In this admission control process, the eNodeB checks the availability of radio resources, taking into account all its existing RRC connections and the bursty nature of LTE packet data traffic. The eNodeB sends an RRC CONNECTION SETUP message to the UE. This RRC message informs the UE about various aspects and parameters related to the new RRC connection, such as configuration parameters of LTE radio interface

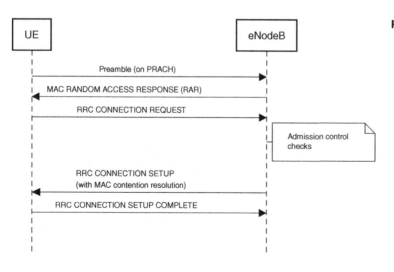

Figure 7.15 Opening LTE RRC connection.

protocols, logical and physical channels to be used, an initial UE power control command, cell measurement, and CQI reporting configuration. The UE enters the RRC_CONNECTED state and responds to the eNodeB to confirm that the RRC connection is now open.

For further details about the messages used in these procedures and the LTE contention resolution process, please refer to *Online Appendix I.5.2*.

7.1.9.3 Releasing the RRC Connection

The RRC connection is released as follows:

1) The eNodeB sends an RRC CONNECTION RELEASE message to the UE in order to request the UE to release the connection. This message tells the UE the priority order of radio technologies and frequency bands, which the UE may use in future.
2) The UE responds to the eNodeB with an RRC CONNECTION RELEASE COMPLETE message. After sending the response, the UE stops using the radio connection. The eNodeB can now release the radio bearers, which were used for the connection.

The RRC CONNECTION RELEASE message may also contain a redirection order with which the eNodeB instructs the UE to reselect another UTRAN or GERAN cell. The redirection process can be used instead of handover when no active data transfer is going on and the radio connection between the UE and eNodeB becomes too weak. After being synchronized with the new cell under other radio technology, the UE performs location and routing updates.

7.1.10 Security Management

UTRAN and LTE systems share the same Authentication and Key Agreement (AKA) method, which was originally specified for UMTS. While sharing the authentication solution, UTRAN and LTE have different designs for protecting traffic over the air interface. In UTRAN networks, the RLC protocol encrypts the traffic, except to the RLC transparent mode where encryption is done by UTRAN MAC. In LTE, encryption is done on higher layers of the protocol stack with protocols specific to the context as follows:

- **Access-Stratum (AS) security:** PDCP protocol encrypts and integrity protects RRC signaling messages and encrypts the user data between UE and eNodeB.
- **Non-Access-Stratum (NAS) security:** NAS protocol messages are encrypted and integrity protected by the NAS signaling connection between the UE and MME.

LTE security architecture is described in 3GPP TS 33.401 [35].

7.1.10.1 Authentication

The LTE network may authenticate the user and change the encryption keys as part of the UE initial attach, tracking area update and EPS bearer context activation procedures. When the UE has opened an RRC connection and sent an EMM ATTACH REQUEST to the MME, mutual authentication is performed between the user and network as shown in Figure 7.16:

1) After checking the UE security capabilities from the NAS EMM ATTACH REQUEST message, MME sends a DIAMETER AUTHENTICATION INFORMATION REQUEST to the HSS in order to get the authentication vectors for the subscriber. The HSS returns multiple authentication vectors, which the MME stores for any further reauthentication rounds. The MME selects one of the vectors and sends an authentication request to the UE. This message contains the serving network identity, RAND and AUTN numbers, as well as the Key Selection Identifier KSI_{ASME}, which will later be used to identify the primary key generated from this vector.

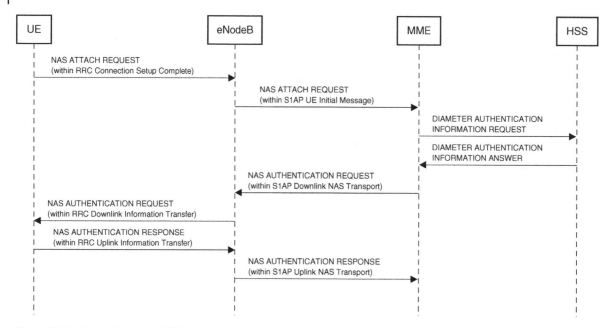

Figure 7.16 Authentication in LTE.

2) The UE authenticates the network based on the serving network identity, AUTN number, and internally stored SQN sequence number corresponding to the AUTN. Thereafter, the UE calculates a RES response based on the RAND number and the shared secret available on the UICC card and sends the response message to the MME.

3) The MME compares the received RES value to the expected XRES value received from the HSS. Authentication is successful when there is a match.

For further details about the messages used in these procedures, please refer to *Online Appendix I.5.3*.

7.1.10.2 Encryption and Integrity Protection

All the LTE security algorithms related to authentication, data encryption, and integrity protection are based on secret keys shared between the HSS server and the USIM application within the UICC card of the UE. In this respect, the architecture is similar to the one used in UTRAN. For key derivation, LTE has its own mechanisms supporting separation of AS and NAS security. LTE uses AKA algorithm at the HSS and UE to calculate the K_{ASME} primary key from which the other keys are derived by the MME and UE as shown in Figure 7.17:

- K_{NASint} key used for NAS integrity protection
- K_{NASenc} key used for NAS encryption
- K_{eNB} sent by the MME to the eNodeB, from which the eNodeB derives the AS keys
  - K_{RRCint} key used for RRC message integrity protection
  - K_{RRCenc} key used for RRC message encryption
  - K_{UPenc} key used for user data encryption

When enabling traffic protection, the MME sends key selection Indicator KSI_{ASME} to the UE. This parameter is an index to the K_{ASME} key instance to be used for key derivation.

Encryption and integrity protection is started separately for NAS and AS within the UE processes of LTE attach and security context creation. The security mode procedure is performed as shown in Figure 7.17:

1) After the successful subscriber authentication, the MME sends an EMM SECURITY MODE COMMAND message to the UE in order to activate NAS security mechanisms. The message is integrity protected with the NAS

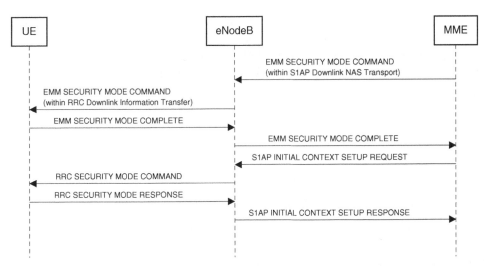

Figure 7.17 Starting LTE encryption and integrity protection.

integrity protection key. From this message, the UE learns the algorithms to be used for encryption and integrity protection as well as the key selection Indicator KSI_{ASME}. The UE derives the NAS integrity protection and encryption keys with help of KSI_{ASME} and the given algorithms. The UE starts NAS encryption and sends an integrity protected and encrypted response to the MME.

2) When accepting the ATTACH request from the UE, the MME sends an S1AP INITIAL CONTEXT SETUP REQUEST to the eNodeB to initiate security context setup at RRC level. This message contains the K_{eNB} key, from which the eNodeB derives AS integrity protection and encryption keys. Thereafter, the eNodeB sends an RRC security mode command to the UE, in order to tell the AS security algorithms and the K_{eNB} key used by the UE to derive AS encryption and integrity protection keys. After taking AS security into use, the UE responds to the eNodeB.

For further details about the messages and security keys used in these procedures, please refer to *Online Appendix I.5.4.*

7.1.11 Packet Data Connections

7.1.11.1 Quality of Service

As described in Chapter 6, Section 6.1.12.1, the application centric UTRAN QoS model found little use. A paradigm shift from the application centric to the network centric QoS control was taken for LTE. In the UTRAN architecture the application running in the UE was in charge of managing QoS parameters of its connections, but in the LTE architecture it is the network which takes the charge. The LTE RAN just needs to know the communications service used and either the average or peak data rate needed for it. When using standardized services, such as VoLTE calls, the network derives the needed parameters from the call setup signaling, without any extra support from the UE. In other cases, the UE may need to indicate the network with the type of the communication service and needed data rates, when setting up the new data flows with specific QoS requirements.

In the context of LTE QoS, the type of communication service is given as the value of **QoS class identifier (QCI)** of new data flow. Rel-8 LTE system design introduced nine **QCI classes**, with predefined maximum acceptable latencies and error ratios. Further classes have been introduced in the later releases. Some of the classes support also a guaranteed data rate. The QCI class is chosen for a new connection based on the related data flow characteristics. QCI classes have a priority order which is used for packet scheduling decisions during a congestion. Highest priority QCI classes are used for signaling and real-time voice or video calls. Lower priority classes are used for services such as streaming video or audio as well as Internet browsing. Background services like email or file downloading may use the lowest-priority best effort QCI class.

The basic LTE Rel-8 QoS classes as defined in table 6.1.7-A of 3GPP TS 23.203 [36] are as follows:

- QCI1: Conversational voice
- QCI2: Conversational video
- QCI3: Real-time gaming
- QCI4: Non-conversational video
- QCI5: IMS signaling
- QCI6 – QCI9: Voice and video streaming, interactive gaming

The following LTE QoS parameters complement the chosen QCI class:

- Guaranteed bitrate (GBR): The average transmission rate guaranteed for the data flow, used for GBR QCI classes 1–4.
- Maximum bitrate (MBR): The peak transmission rate provided for the data flow.
- Allocation and retention priority (ARP): The priority level followed by the admission control. The ARP priority level is used to constrain creation of additional data flows when the network load is high enough and limits the available bandwidth that can be used for data flows with the lowest QCI classes.

The QoS parameters of the data flow are delivered during the packet data context setup or modification procedure to the PCRF network element, which manages QoS policies for EPS bearer contexts. The PCRF then orchestrates the rest of the network to support data flows according to the agreed policies. See 3GPP TS 23.203 [36] for further details.

7.1.11.2   EPS Bearers

To provide UEs with connectivity to a packet data network (PDN), the EPC sets up EPS bearers for data flows. An **EPS bearer** is a combination of the following items used for transporting a data flow:

- An IP address allocated to the UE to be used toward the connected PDN, which is identified by its access point name (APN)
- QoS policies agreed for the data flow, to be enforced by the network
- A traffic filter template (TFT) used by the P-GW and UE for mapping and filtering packets of different flows for the EPS bearers
- GTP-U tunnels via which packets are transported between and among the eNodeB, S-GW, and P-GW nodes
- Radio bearers via which packets are transported between the eNodeB and UE

When the UE requests to be connected to an APN, the network will open a **default EPS bearer** toward the related PDN. The UE is given with an IP address and can thereafter exchange packet data with the PDN. The default EPS bearer has a non-GBR QCI class. If some of the data flows toward the PDN it would need specific QoS processing; at that point, the UE itself or an IMS CSCF (see Chapter 9, Section 9.3.2.2) may request the EPC to set up additional **dedicated EPS bearers** toward the same (or a different) PDN. The QoS parameters given by the UE or CSCF for a dedicated EPS bearer define the QoS characteristics of the specific data flow. The PCRF screens the QoS request against operator policies and the subscriber profile. After making the policy control decision, the PCRF instructs the P-GW to set up the EPS bearer and its traffic flow filtering rules accordingly.

7.1.11.3   Initial Default EPS Bearer Opening at LTE Initial Access

As data transport is the core service provided by the LTE network, the network opens an **initial default EPS bearer** to a packet data network (PDN) for the UE whenever the UE attaches to LTE. Either the UE or the network may choose the destination PDN for the initial default EPS bearer. In the former case, the UE sends a packet data network connectivity request toward the desired **Internet access point (IAP)** along with the LTE attach request. In the latter case, the UE omits the APN name from its PDN connectivity request and the network will then connect the UE toward the predefined subscriber-specific default PDN.

The initial default EPS bearer is opened as shown in Figure 7.18:

1) As described in Section 7.1.9.1, the UE sends an ESM PDN CONNECTIVITY REQUEST message within an EMM ATTACH REQUEST message to the MME when attaching to the LTE service. If the PDN connectivity request contains an APN, it will be used for the initial default EPS bearer. Otherwise, the MME uses subscriber's default APN as retrieved from the HSS within the subscriber records while performing subscriber authentication.

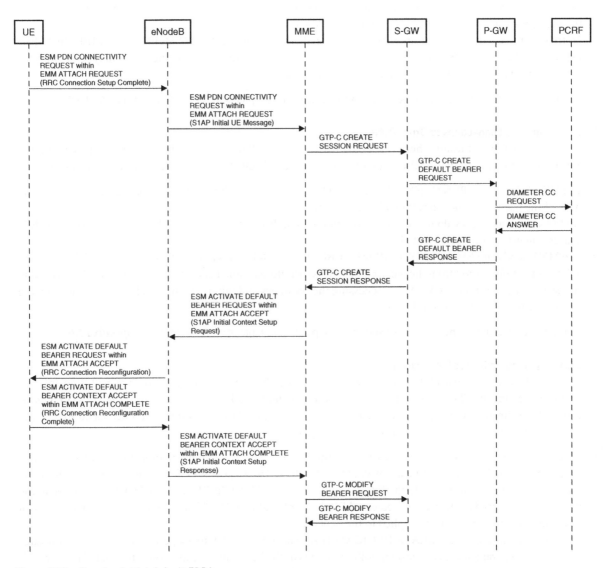

Figure 7.18 Opening initial default EPS bearer.

2) The MME sends a GTP-C CREATE SESSION REQUEST to the S-GW to initiate creation of the default EPS bearer for the UE. From this message, the S-GW learns APN and QoS parameters requested for the bearer and the address of the P-GW supporting the related PDN. The S-GW instructs the P-GW to create the default bearer for the requested APN and QCI class. The P-GW contacts the PCRF to check if the request can be accepted and to get the related policy control and charging rules. The P-GW responds to the S-GW and proceeds with setting up GTP-U tunnels between and among the P-GW, S-GW, and eNodeB. For the EPC core network, the new EPS bearer is ready for downlink packet data forwarding. The S-GW reports this status to the MME.

3) To prepare the radio access network part for the EPS bearer, the MME sends an EMM ATTACH ACCEPT and an ESM ACTIVATE DEFAULT BEARER REQUEST message to the UE. When the PDN connection supports IPv6, the ESM message contains a temporary IPv6 interface identifier, which the UE should use together with the link local prefix for the IPv6 address autoconfiguration process described in Chapter 3, Section 3.4.2.1. When forwarding the EMM and ESM messages to the UE, the eNodeB uses the information from these messages to build the radio bearers to the UE and map them to the EPS bearers toward the S-GW. After processing the messages, the UE responds to the eNodeB and MME.

4) To finalize the creation of the initial default bearer, the MME sends a GTP-C MODIFY BEARER REQUEST message to the S-GW to tell that the connection is now open up to the UE. Now the S-GW may send any buffered data packets it has received from the PDN to the UE.

For further details about the messages used in these procedures, please refer to *Online Appendix I.5.5.*

7.1.11.4   Opening Connections to Other PDNs

If the UE has data flows toward other packet data networks, the UE may request the network to open additional default EPS bearers toward those APNs. Opening of a new default EPS bearer is done as shown in Figure 7.19:

1) The UE sends an ESM PDN CONNECTIVITY REQUEST message to the MME. The ESM message contains the APN of the packet data network for which connectivity is requested. The MME, S-GW, and P-GW will cooperate the same way as they do for setting up the initial default EPS bearer at the UE attach. For each new EPS bearer, a new GTP tunnel is created.

2) The MME sends an ESM ACTIVATE DEFAULT BEARER CONTEXT REQUEST message to the UE. The eNodeB uses the information from the received messages to build the radio bearer to the UE and map it to the radio access bearer toward the S-GW. After processing the messages and taking the new radio bearers into use, the UE responds to the eNodeB and MME.

For further details about the messages used in these procedures, please refer to *Online Appendix I.5.6.*

7.1.11.5   Opening Dedicated EPS Bearers

Dedicated EPS bearers are used for data flows which have specific QoS requirements, such as real-time, interactive, or streamed media. The PCRF gets a request for a new **dedicated bearer** either from the UE via P-GW or from the IP Multimedia System (IMS) core network elements. The following example of a dedicated EPS bearer opening follows the latter of these two scenarios, see Figure 7.20:

1) When an MT VoLTE call is established toward the UE, the IMS P-CSCF element sends a DIAMETER AA REQUEST to the PCRF, requesting a new data flow for the UE with specific QoS parameters. After authorizing the request, the PCRF responds to the P-CSCF and sends a DIAMETER RE-AUTH REQUEST message to the P-GW about the need to open a new dedicated EPS bearer toward the UE. This message contains policy control rules with QoS parameters to be applied to the new bearer.

2) The P-GW sends a GTP-C CREATE DEDICATED BEARER REQUEST message to the S-GW. This message describes QoS parameters to be used for both directions of the bearer and a traffic filter template (TFT), which shall be used to identify the uplink user data packets for this bearer. The S-GW proceeds to create GTP-U

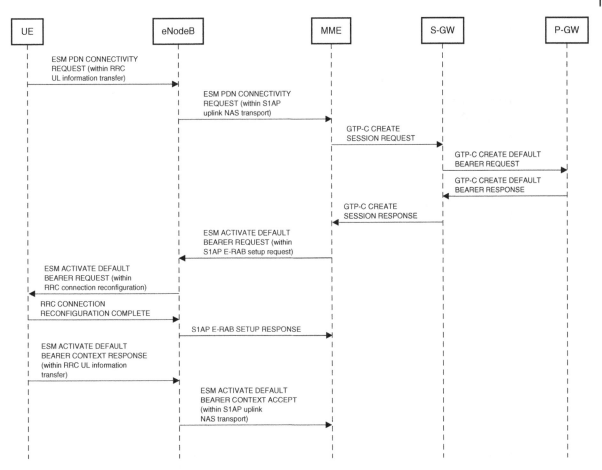

Figure 7.19 Opening additional default EPS bearers.

tunnels for the dedicated bearer toward the P-GW. Additionally, the S-GW contacts the MME to get the new radio access bearers opened toward the UE.

3) The MME sends an ESM ACTIVATE DEDICATED EPS BEARER CONTEXT REQUEST to the UE via the eNodeB and uses S1AP protocol to instruct the eNodeB to create the needed radio bearers with RRC procedures. The eNodeB forwards the ESM message to the UE within an RRC message, which instructs the UE to apply semi-persistent scheduling for the voice packets sent over the dedicated bearer. The ESM message contains the traffic filter template with which the UE is able to find out which packets should be transported over the new dedicated bearer. After processing the RRC message, the UE responds to the eNodeB, which further responds to the MME. After the UE has processed the ESM message, it responds also to the MME.

4) After receiving responses from both the eNodeB and UE, the MME can finally inform the S-GW about the successful creation of radio access bearers for the new dedicated EPS bearer. The S-GW then enables the related GTP-U tunnels to transport data packets to/from the UE and acknowledges the original request from the P-GW.

For further details about the messages used in these procedures, please refer to *Online Appendix I.5.7*.

7.1.11.6 User Data Transport
If the UE is in RRC_IDLE state, it shall at first open an RRC connection before any user data can be transmitted. If the data arrives from network, the UE is at first paged. The paging message is sent in all cells within the

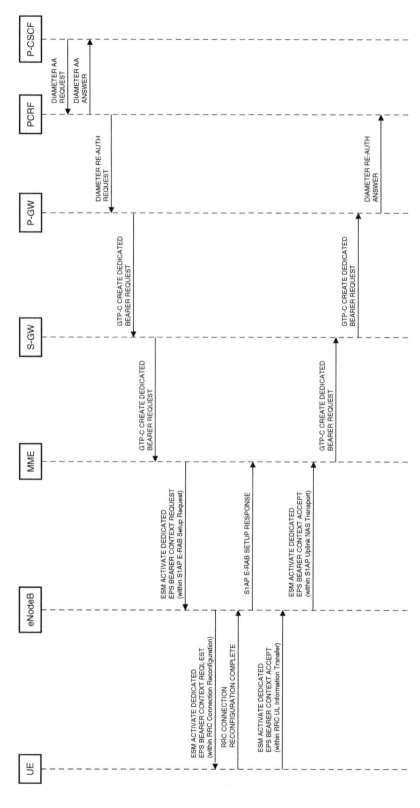

Figure 7.20 Opening dedicated EPS bearer.

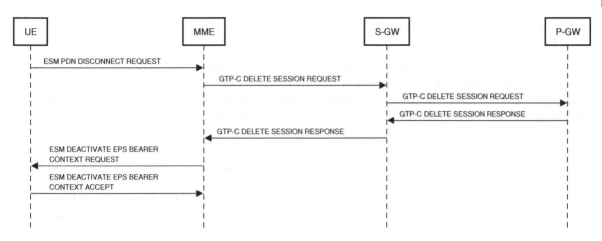

Figure 7.21 Disconnecting from a packet data network.

previously reported tracking area. If the data originates from the UE or if the UE receives a paging message, a random access procedure is performed to open an RRC connection for data transfer. After the UE has answered to paging, a GTP tunnel is established from the eNodeB to the S-GW for transferring the mobile terminated data packets over the completed EPS bearer.

After completing the random access procedure and moving to the RRC_CONNECTED state, the UE can send uplink user data over the EPS bearers as follows:

1) When sending an IP packet to the network over the RRC connection, the UE compares the destination IP address, protocol (TCP/UDP), and port numbers of the packet to the TFTs it has received within activate EPS bearer request messages. With help of the TFT, the UE determines to which EPS bearer and related radio bearer the packet belongs to.
2) The UE sends the packet to the eNodeB over the radio bearer determined with the TFT.
3) The eNodeB forwards the IP packet to the S-GW using the GTP-U tunnel set up for the EPS bearer.
4) The S-GW forwards the IP packet to the P-GW using the GTP-U tunnel set up for the EPS bearer.
5) The P-GW routes the IP packet to the destination packet data network.

7.1.11.7 Disconnecting from Packet Data Network

The UE disconnects itself from a PDN by deactivating the related EPS bearer as shown in Figure 7.21: The UE sends an ESM PDN DISCONNECT REQUEST message to the MME. The ESM message identifies the EPS default bearer to be disconnected. The MME sends a GTP-C DELETE SESSION REQUEST to the S-GW. This message identifies the EPS default bearer to be disconnected and the disconnect cause as "UE initiated PDN release." The SG-W forwards the GTP-C request to the P-GW, after which the GTP-U tunnels are torn down. After getting a response from the S-GW, the MME releases the resources reserved for the EPS bearer and tells the UE to deactivate the bearer.

For further details about the message sequence used in this process, please refer to *Online Appendix I.5.8.*

7.1.12 Mobility Management

As the LTE system does not support circuit switched services, its mobility management covers only the packet switched domain. Due to the lack of soft handover support, the RRC state model of LTE is simpler than that of UTRAN, which simplifies LTE mobility management. In the LTE network architecture, the MME takes care of mobility management. The knowledge of the UE location is maintained at the MME as follows, based on the RRC state of the UE:

- RRC_IDLE: The MME knows the location of the UE in the accuracy of an LTE tracking area. The UE performs LTE cell reselection and tracking area update procedures independently of network control.
- RRC_CONNECTED: The MME knows the location of the UE in the accuracy of an LTE cell. The UE can change the camped cell only with the handover procedure controlled by the eNodeB.

7.1.12.1   Cell Reselection and Tracking Area Update

The UE learns the current tracking area and other LTE cells in the neighborhood from the system information messages broadcast in the LTE cell. In the RRC_IDLE state, the UE monitors the quality of radio signal received from the eNodeB. If the radio signal quality with the current cell drops under a defined cell reselection threshold, the UE starts scanning to find better cells. The UE shall eventually select a cell that can provide the highest quality radio connection. In the absence of a suitable LTE cell, the UE may also select a UTRAN or a GERAN cell according to the priorities for radio technology selection. These priorities are received from the network either within a SYSTEM INFORMATION or an RRC CONNECTION RELEASE message. The UE performs a tracking area update procedure after its initial LTE attach or an LTE cell reselection, which has caused the UE to have arrived to a new tracking area, see Figure 7.22:

1) At first, the UE must move from the RRC_IDLE to the RRC_CONNECTED state as described in Section 7.1.9.2. In the end of the RRC connection opening, the UE sends an EMM TRACKING AREA UPDATE (TAU) REQUEST. The TAU message contains RRC parameters such as the selected network, the old GUTI of the UE, last visited tracking area, and the P-TMSI identifier of the UE. The eNodeB derives the identity of the MME from the parameters within the TAU message. The eNodeB forwards the EMM TAU message to the new MME and provides the MME with the tracking area identity and global cell ID of the cell, as taken from the received TAU message.
2) When the serving MME and S-GW have changed, the new MME derives the address of the old MME from the GUTI of the UE. The new MME sends a GTP-C CONTEXT REQUEST message to the old MME to retrieve the subscriber information. The old MME returns the UE context to the new MME, after which the new MME authenticates the subscriber as described in Section 7.1.10.1.
3) The new MME sends a GTP-C CREATE BEARER REQUEST message to the selected S-GW gateway. From this message, the S-GW checks bearer contexts to be created, protocols and RAT used, as well as identifiers of the MME and UE (IMSI) and addresses of the P-GW gateways supporting the bearer contexts. The S-GW contacts P-GW(s) to get it (or them) to update the bearer and GTP-U tunnels toward the new S-GW.
4) After getting response from the S-GW, the new MME sends a DIAMETER UPDATE LOCATION REQUEST message to the HSS. In this message, the MME tells its own identity and the IMSI of the UE. The update type parameter of the message makes the HSS to cancel the UE location from the old MME. After the HSS has recorded the location of the UE, it responds to the new MME. Finally, the new MME sends an EMM TAU ACCEPT message to the UE.

For further details about the messages used in these procedures, please refer to *Online Appendix I.5.9.*

7.1.12.2   Handover in RRC Connected State

The LTE UE is connected with one single eNodeB at any time while using LTE services. LTE supports only hard handovers where connection to the old eNodeB is released when the UE reconnects with another eNodeB. In the RRC_CONNECTED state, handovers between two eNodeBs are done under network control. The handover decision is done by the serving eNodeB based on the measurement results received from the UE. Handover decision may be done due to the degradation of the radio signal between the eNodeB and UE or the eNodeB running out

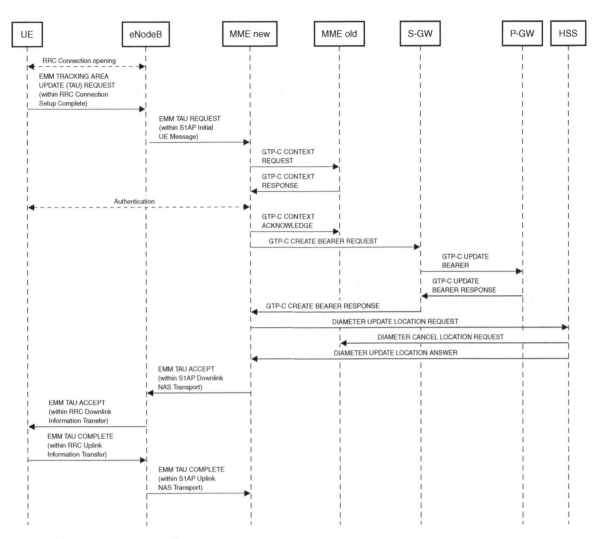

Figure 7.22 LTE tracking area update.

of capacity. Depending on the case, the eNodeB may decide to perform an intra-LTE handover between two eNodeBs or an inter-RAT handover toward another cellular radio access technology, such as HSPA or GSM.

The LTE UE measures the quality of radio signals and reports the measurement results to the eNodeB according to instructions received from the eNodeB [8]. The UE always measures the quality of the LTE signals currently used in terms of the following parameters:

- Received signal strength indication (RSSI)
- Reference signal received power (RSRP)
- Reference signal received quality (RSRQ) equaling RSRP divided by the RSSI

When the eNodeB anticipates reasons for upcoming handover, it may instruct the UE to start measuring neighboring LTE, WCDMA, and GSM cells, following the RAT selection priorities and cells available at the UE location. The measurement request is sent to the UE within an RRC CONNECTION RECONFIGURATION message, which indicates the RATs and frequencies to be measured as well as the reporting procedures to be used.

To perform measurements on different frequencies, the eNodeB grants the UE with measurement gaps, during which the UE is not expected to listen to the serving LTE cell. The eNodeB may ask the UE to send its measurement reports either with regular intervals or when the measurement results cross various thresholds defined by the eNodeB. The latter is known as event-based measurement reporting. The measurement request may also indicate when the UE can skip its current cell measurements so that it can instead measure neighboring cells.

The UE performs requested measurements and summarizes results within an RRC MEASUREMENT REPORT message sent to the eNodeB. To make the handover decision, the eNodeB considers UE measurements over different target cells and the current load levels of those target cells, which the eNodeB has received over inter-eNodeB or inter-RAT signaling.

If the handover is done between two eNodeBs connected over the X2 interface, the handover type is known as X2 handover. The X2 handover between two LTE cells is performed as shown in Figure 7.23:

1) The serving eNodeB sends an X2AP HANDOVER REQUEST message to the eNodeB of the target cell. The target eNodeB checks if it is able to take the responsibility of the packet data connections of the UE. The handover is done only to such a target eNodeB that has enough capacity to support QoS requirements of all its current radio access bearers and the new ones it would get at the handover. To prepare the handover, the target eNodeB sets up new GTP-U tunnels to the S-GW so that any uplink packets from the UE can be routed toward the PDN right after handover. In the end, the target eNodeB returns an acknowledgment to the old eNodeB. The target eNodeB describes the new radio connections for the UE within an RRC CONNECTION RECONFIGURATION message, which is embedded into the X2AP response.

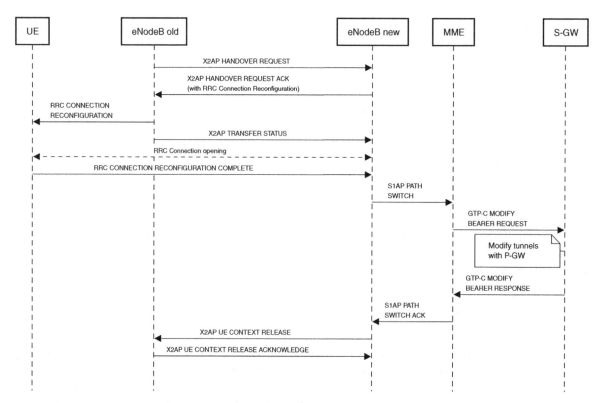

Figure 7.23 Inter-cell X2 handover in LTE.

2) The eNodeBs create a temporary GTP tunnel between each other, to be used during the handover process until the final tunnel setup is in place. The old eNodeB sends an X2AP TRANSFER STATUS message to the target eNodeB to trigger forwarding of downlink user data over the temporary GTP tunnels.

3) The old eNodeB forwards the RRC CONNECTION RECONFIGURATION message to the UE, which uses information within the message to connect itself toward the new cell. The UE synchronizes itself to the target cell and uses random access to request the target eNodeB to open an RRC connection. The UE uses the new RRC connection to send an RRC CONNECTION RECONFIGURATION COMPLETE message to the target eNodeB. User data transport can now resume over the target cell.

4) The target eNodeB sends an S1AP PATH SWITCH message to the MME, which consequently informs the S-GW about the handover. The S-GW establishes new GTP tunnels toward the target eNodeB and tears down the GTP tunnels toward the old eNodeB. After the path switch has been completed, the target eNodeB tells the old eNodeB to release its UE context. The temporary GTP tunnel between eNodeBs is also torn down.

If the two eNodeBs do not have an X2 connection, the handover is known as S1 handover since in this case the eNodeBs communicate over theS1 interface via the MME. The MME also coordinates the temporary forwarding of downlink data between eNodeBs during the handover execution. The MME is engaged also for all inter-RAT handovers. In the inter-RAT handover from E-UTRAN (LTE) to UTRAN the data path anchored to the P-GW of EPC will be routed to the RNC of the UTRAN network. A GTP-U tunnel is established in the handover between the SGSN on the UTRAN side and the S-GW on the E-UTRAN side. The tunnel creation is coordinated by the SGSN based on the request from the MME. While the handover process is ongoing, the downlink data may be forwarded from the eNodeB directly to the RNC or indirectly via the S-GW and SGSN to the RNC, according to the decision made by the MME.

Figures 7.24 and 7.25 visualize an inter-RAT packet switch handover from the LTE to the UTRAN network:

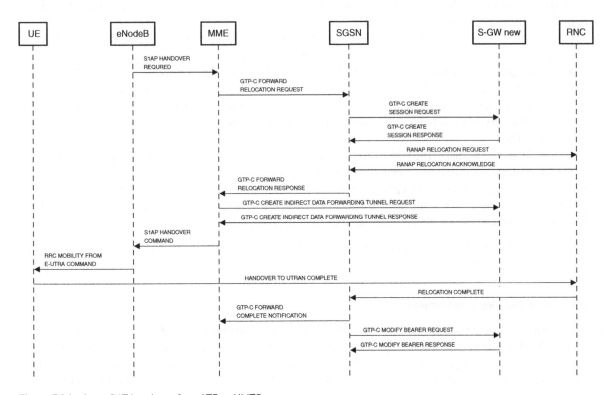

Figure 7.24 Inter-RAT handover from LTE to UMTS.

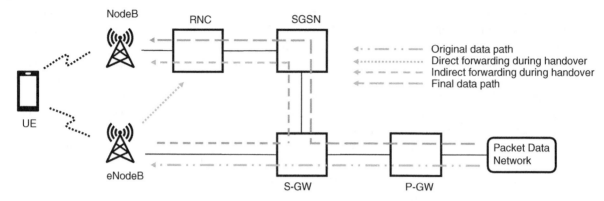

Figure 7.25 User data paths before, during, and after inter-RAT handover.

1) The eNodeB decides that inter-RAT handover will be performed for the UE and sends an S1AP HANDOVER REQURED message to the MME. This S1AP message identifies the target RNC and the UTRAN cell toward which the handover shall be done.

2) The MME sends a GTP-C FORWARD RELOCATION REQUEST message to the SGSN node of the UTRAN network. From this message, the SGSN learns the EPS bearer contexts of the UE, which shall be reestablished on the UTRAN side and the identity of the S-GW currently serving the UE. The SGSN selects an S-GW for the new user data path and sends a GTP-C CREATE SESSION REQUEST message to the S-GW for each PDN connection of the UE. After the S-GW has allocated local resources, it replies to the SGSN, which then contacts the UTRAN RNC to initiate handover preparations, such as establishing radio access bearers for the UE. After performing these activities, the SGSN sends a GTP-C response to the MME.

3) If the S-GW is changed and indirect forwarding used (rather than direct forwarding from the eNodeB to the RNC), the SGSN and MME send GTP-C CREATE INDIRECT DATA FORWARDING TUNNEL REQUEST messages to the old and a new S-GW to set up GTP-U tunnels for forwarding downlink data temporarily from the eNodeB to the RNC via the S-GW. The indirect tunnel setup is shown in Figure 7.25. After the data path is ready, the MME sends a handover command to the eNodeB, which serves the UE. After receiving the S1AP handover command, the eNodeB sends an RRC handover command to the UE and starts forwarding user data toward UTRAN.

4) After the UE has connected to the UTRAN cell, it informs the RNC about handover completion and thereafter performs location and routing area updates. The RNC informs the SGSN about UE relocation to prepare the SGSN for receiving uplink user data packets from the RNC. Thereafter, the SGSN informs the MME and completes bearer modification procedures with the S-GW and P-GW. The bearer contexts are now moved from LTE to UTRAN and the new GTP tunnel setup between the SGSN and new S-GW is completed.

For further details about the messages used in any of these procedures or indirect versus direct forwarding, please refer to *Online Appendix I.5.10*.

7.1.13 Voice and Message Communications

Since LTE networks do not support circuit switched domain at all, **circuit switched fallback (CSFB)** is used to move the LTE UE to another radio technology such as UTRAN or GERAN for receiving or initiating voice calls. Even when UTRAN networks are gradually being decommissioned, CSFB to GERAN is expected to stay as a valid use case for 4G phones or networks without VoLTE support. CSFB is specified in 3GPP TS 23.272 [37] and TS 29.118 [34].

As a packet switched network, the EPC does not have native support for short message service (SMS), which has been designed as an extension of circuit switched services. LTE radio interface signaling has a built-in SMS support between the UE and MME. The MME uses a new SG interface toward MSC/VLR to contact the short message center. This SMS delivery method is called **SMS over SG**, which is specified in 3GPP TS 23.272 [37].

To receive or initiate mobile terminated circuit switched calls or short messages over the SG interface, the LTE UE must register to the CS domain when attaching to the LTE PS service. To do that, the UE sends a special **CSFB indicator** within the EMM ATTACH REQUEST message to define the attach type as **combined EPS/IMSI attach**. After finding the CSFB indicator from the EMM message, the MME sends an SGsAP LOCATION UPDATE REQUEST over the SG interface to the MSC/VLR to register the UE to the CS domain and to update its location to the MSC/VLR. The MME derives the location area identity (LAI) at the CS side from the tracking area to which the UE has registered for the LTE network. The LAI consists of UTRAN or GERAN cells and is inherently different from the LTE tracking area. When an MT CS call arrives, the LTE network is able to page the UE within its tracking area. In the MT CSFB procedure, the UE is informed about the location area that was reported to the CS domain. But since there is no perfect match between the tracking area and the location area, the location area may have been reported incorrectly. If the UE finds out after changing the RAT that its actual location area is different than the reported one, it must immediately perform the location area update procedure. The call can be properly connected only after the location area update has been processed. This is because the CS core network may have different MSC/VLRs serving the actual location area and the other location area reported by the MME.

After performing the combined EPS/IMSI attach, the UE may receive paging requests with a reason indicator as an incoming mobile terminated CS call or an MT SMS. In the case of SMS, the UE stays in LTE and the SMS is transported to the UE within LTE signaling. In the case of a CS call, the MME instructs the UE to move to another radio technology, either with a PS handover if the UE has ongoing packet data sessions or a cell change order (CCO) if the UE has been idle. For mobile originated CS calls, the UE requests the network to trigger a PS handover or a cell change to another radio technology for initiating the voice call.

7.1.13.1 CSFB for Voice Call

The CSFB process for an MT CS call works as shown in Figure 7.26:

1) The MSC/VLR sends an SGsAP PAGING REQUEST message over the SG interface to the MME of the UE, according to the UE location as known to the MSC/VLR. When the UE is in idle mode, the MME instructs eNodeBs within the tracking area to page the UE.
2) After receiving the paging request, the UE performs random access to open an RRC connection. The UE sends to the MME an EMM EXTENDED SERVICE REQUEST with CSFB indicator to receive the MT call. The EMM message indicates the request status either as accepted or rejected depending on what the user decided to do with the call. The MME responds the UE with an ESM INITIAL CONTEXT SETUP REQUEST that also has the CSFB indicator. The ESM message tells the UE the location area to which the UE has been registered for CS calls.
3) The UE starts measuring the cells with radio technologies as instructed by the MME for the CSFB procedure. The UE reports the measurement results to the eNodeB with an RRC MEASUREMENT REPORT message. Depending on the capabilities of the new target network and state of the UE, the UE is moved to the other RAT with either the packet switched handover (PSHO) procedure, RRC connection release with redirection, or a CCO in case of GSM. Before changing the radio technology, the UE may release its LTE context. The CS call setup is completed with the chosen target radio technology specific procedures.

For further details about mobile originated CSFB, the messages used in CSFB or differences for cases when the phone is either in idle or connected mode, please refer to *Online Appendix I.5.11*.

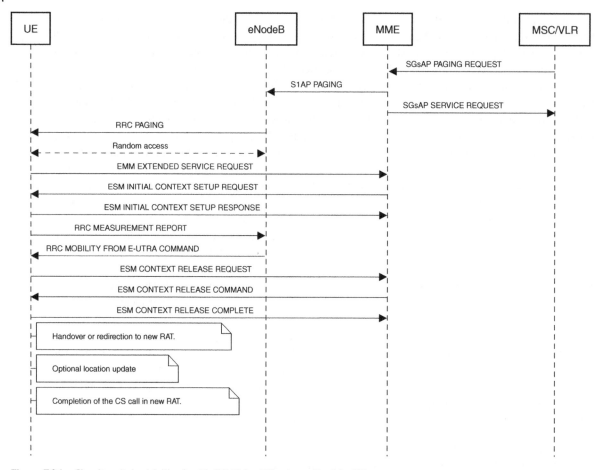

Figure 7.26 Circuit switched fallback with PSHO for MT voice call – idle UE.

After the CS call is released and if the LTE RAT has the priority, the UE may perform an LTE random access procedure to get back to the LTE service and send a new tracking update message. It is also possible for the network to use a forced cell reselection procedure to get idle UE back to the LTE service.

7.1.13.2 SMS over SG

The SMS over SG process for mobile originated short message works as shown in Figure 7.27:

1) The UE opens an RRC connection to the eNodeB and sends an RRC UPLINK INFORMATION TRANSFER message, which contains the short message (SM-TL and SM-RL protocol data units as described in Chapter 5, Section 5.1.10.2).
2) The eNodeB forwards the short message to the MME within an S1AP UPLINK NAS TRANSPORT message.
3) The MME sends the short message to the MSC/VRL within an SGsAP UPLINK UNITDATA message. The MSC/VRL forwards the message to the SMSC short message center.

The SMS over SG process for mobile terminated short message works as follows:

1) The MSC/VRL sends an SGsAP PAGING REQUEST to the MME to wake up the UE for short message reception. Consequently, the MME initiates paging with eNodeBs within the recorded tracking area of the UE.

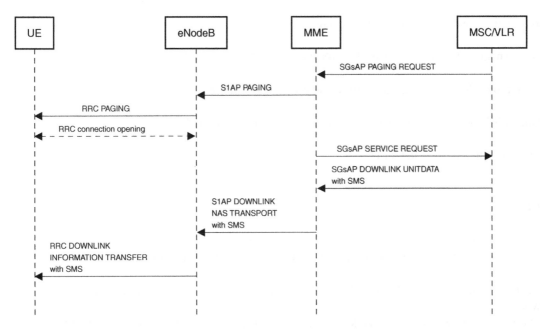

Figure 7.27 MT SMS over SG.

The paging request indicates the reason for paging to be an MT SMS. The MME informs the MSC/VLR about the paging procedure and gets back the short message sent to the UE. The MME forwards the short message to the eNodeB.

2) The UE responds to paging and opens an RRC connection with random access. Now the eNodeB is able to forward the short message to the UE and receive back the SMS delivery report.

For further details about the messages used in the MT procedure, please refer to *Online Appendix I.5.12*.

7.2 Questions

1 Why doesn't LTE support circuit switched domain?

2 How can LTE systems support voice calls?

3 What is the difference between LTE FDD and TDD systems?

4 What are the goals and main benefits of 3GPP System Architecture Evolution?

5 What is the LTE resource grid?

6 What are the main benefits of OFDMA for LTE?

7 To which purposes are reference symbols used in LTE?

8 Why does LTE use SC-FDMA instead of OFDMA on uplink?

9 How does semi-persistent scheduling work and for what is it used?

10 What are the main differences between LTE QoS and UMTS QoS concepts?

11 Which LTE protocols belong to access stratum and non-access stratum?

12 What kind of information does the UE get from LTE system information blocks?

13 How is NB-IoT different from Rel-8 LTE?

References

1 ITU-T Recommencation Q.1702 Long-term vision of network aspects for systems beyond IMT-2000.
2 ITU-R Recommendation M.1645 Framework and overall objectives of the future development of IMT-2000 and systems beyond IMT-2000.
3 Holma, H. and Toskala, A. (2009). *LTE for UMTS: OFDMA and SC-FDMA Based Radio Access.* West Sussex: Wiley.
4 3GPP TS 36.300 Evolved Universal Terrestrial Radio Access (E-UTRA) and Evolved Universal Terrestrial Radio Access Network (E-UTRAN); Overall description; Stage 2.
5 ITU-R Recommendation M.2012 Detailed specifications of the terrestrial radio interfaces of International Mobile Telecommunications Advanced (IMT-Advanced).
6 3GPP TS 36.101 Evolved Universal Terrestrial Radio Access (E-UTRA); User Equipment (UE) radio transmission and reception.
7 Sauter, M. (2021). *From GSM to LTE-Advanced Pro and 5G : An Introduction to Mobile Networks and Mobile Broadband.* West Sussex: Wiley.
8 Sesia, S., Toufik, I., and Baker, M. (2009). *LTE – The UMTS Long Term Evolution: From Theory to Practice.* West Sussex: Wiley.
9 3GPP TS 36.401 Evolved Universal Terrestrial Radio Access Network (E-UTRAN); Architecture description.
10 3GPP TS 23.401 General Packet Radio Service (GPRS) enhancements for Evolved Universal Terrestrial Radio Access Network (E-UTRAN) access.
11 3GPP TS 23.214 Architecture enhancements for control and user plane separation of EPC nodes.
12 3GPP TS 36.321 Evolved Universal Terrestrial Radio Access (E-UTRA); Medium Access Control (MAC) protocol specification.
13 3GPP TS 36.201 Evolved Universal Terrestrial Radio Access (E-UTRA); LTE physical layer; General description.
14 3GPP TS 36.211 Evolved Universal Terrestrial Radio Access (E-UTRA); Physical channels and modulation.
15 3GPP TS 36.213 Evolved Universal Terrestrial Radio Access (E-UTRA); Physical layer procedures.
16 3GPP TS 36.212 Evolved Universal Terrestrial Radio Access (E-UTRA); Multiplexing and channel coding.
17 3GPP TS 36.306 Evolved Universal Terrestrial Radio Access (E-UTRA); User Equipment (UE) radio access capabilities.
18 3GPP TS 36.322 Evolved Universal Terrestrial Radio Access (E-UTRA); Radio Link Control (RLC) protocol specification.
19 3GPP TS 36.323 Evolved Universal Terrestrial Radio Access (E-UTRA); Packet Data Convergence Protocol (PDCP) specification.
20 IETF RFC 3095 RObust Header Compression (ROHC): Framework and four profiles: RTP, UDP, ESP, and uncompressed.

21 IETF RFC 5795 The RObust Header Compression (ROHC) Framework.

22 3GPP TS 36.331 Evolved Universal Terrestrial Radio Access (E-UTRA); Radio Resource Control (RRC); Protocol specification.

23 3GPP TS 24.301 Non-Access-Stratum (NAS) protocol for Evolved Packet System (EPS); Stage 3.

24 3GPP TS 36.413 Evolved Universal Terrestrial Radio Access Network (E-UTRAN); S1 Application Protocol (S1AP).

25 3GPP TS 36.423 Evolved Universal Terrestrial Radio Access Network (E-UTRAN); X2 Application Protocol (X2AP).

26 3GPP TS 29.274 3GPP Evolved Packet System (EPS); Evolved General Packet Radio Service (GPRS) Tunnelling Protocol for Control plane (GTPv2-C); Stage 3.

27 3GPP TS 29.281 General Packet Radio System (GPRS) Tunnelling Protocol User Plane (GTPv1-U).

28 IETF RFC 6733 Diameter Base Protocol.

29 IETF RFC 3589 Diameter Command Codes for Third Generation Partnership Project (3GPP) Release 5.

30 3GPP TS 29.212 Policy and Charging Control (PCC); Reference points.

31 3GPP TS 29.214 Policy and charging control over Rx reference point.

32 3GPP TS 29.272 Evolved Packet System (EPS); Mobility Management Entity (MME) and Serving GPRS Support Node (SGSN) related interfaces based on Diameter protocol.

33 3GPP TS 29.336 Home Subscriber Server (HSS) diameter interfaces for interworking with packet data networks and applications.

34 3GPP TS 29.118 Mobility Management Entity (MME) - Visitor Location Register (VLR) SGs interface specification.

35 3GPP TS 33.401 3GPP System Architecture Evolution (SAE); Security architecture.

36 3GPP TS 23.203 Policy and charging control architecture.

37 3GPP TS 23.272 Circuit Switched (CS) fallback in Evolved Packet System (EPS); Stage 2.

8

Fifth Generation

The earlier generation cellular systems had always completely redefined the structure of the air interface, but the same does not apply to the 5G cellular system. All the earlier generation systems had their own multiplexing methods, but the 5G new radio system just extends and enhances the OFDMA-based air interface adopted already for LTE. The novelty of the 5G design was the built-in support for different use cases having different requirements for network performance, reliability, and terminal power consumption. 5G core networks were defined to be extremely scalable so that apart from network operators, other enterprises also could set up and maintain their own private 5G networks, just as they have internal data networks supported by their IT departments.

8.1 5G

8.1.1 Standardization of Fifth-Generation Cellular Systems

The drivers of earlier generation mobile system designs emerged from human centric communication patterns. Early on, the mobile telephony and messaging use cases were at first introduced and thereafter evolved as mass market consumer services. In the twenty-first century, mobile data consumption became predominant due to the ever-expanding Internet and the proliferation of networked applications used for social media, real-time gaming, or music and video streaming. Especially video streaming use cases, which require higher data rates whereas real-time gaming requires minimal latencies. The 4G LTE systems provide quite sufficient support for those use cases.

The human centric communication paradigm or consumer services alone do not create high demand for even better 5G systems. Although the spectral efficiency of LTE OFDMA system is excellent and the deployed LTE system bandwidths support high enough data rates for consumer services, what could the 5G system add on top of that? The answer comes from the connected world paradigm. **Connected world** simply means equipping very different sorts of devices with intelligence and communication capabilities. **Internet of Things (IoT)** is a term often used for a communication pattern where devices connected to the Internet communicate with each other, rather than with humans. Known IoT use cases cover both consumer and enterprise scenarios.

Organizations such as global enterprises have already started to implement various scenarios of **industrial Internet**. Enterprises have started to build solutions to collect massive amounts of sensor data from their products (such as airplanes, cranes, lifts, etc.) and send to a cloud storage where the data can be processed with help of advanced algorithms and artificial intelligence. The data analysis helps enterprises to optimize preventive maintenance activities or to design products that fit well to their typical usage patterns. The new 5G systems should support collecting such data without overloading the network or consuming too much electricity. Whereas factory automation has been a local matter due to strict real-time requirements, low-latency 5G networks could support remote rather than local control. Also, the local control at the shopfloor may benefit from

Converged Communications: Evolution from Telephony to 5G Mobile Internet, First Edition. Erkki Koivusalo.
© 2023 The Institute of Electrical and Electronics Engineers, Inc. Published 2023 by John Wiley & Sons, Inc.
Companion website: www.wiley.com/go/koivusalo/convergedcommunications

the low-latency 5G access technology. One of the promises of 5G is to support both public networks provided by network operators and non-public networks, which could be established and operated by other private enterprises for their own internal purposes. Private 5G networks may be used for various enterprise IoT and factory automation use cases. 5G systems can also support edge computing scenarios where the 5G network provides access to some local servers or resources at the edge of the network, rather than access to/over the Internet.

The **consumer centric IoT** comes with visions about intelligent homes, networked appliances with remote controls, or refrigerators able to automatically request food supplies from a grocery. While autonomous self-driving cars are already being tested, networked cars might communicate with each other and some sort of road control system. Such cars could safely drive closer to each other and increase the capacity of roads. Instead of starting to brake after detecting braking lights ahead or recognizing the speed of the car in the front decreasing, the cars on the road could simply agree on pushing the brake pedal at the same time. Naturally, such use cases would put stringent requirements to the used communication network reliability, availability, and latencies.

In 2015, ITU-R published the IMT-2020 vision statement as ITU document M.2083 [1]. The statement defined the following usage scenarios for the fifth-generation (5G) cellular technology:

- **Enhanced mobile broadband (eMBB)** to extend the user experience of mobile devices with applications like augmented reality, virtual reality, streaming ultra-high-definition video, or sharing recorded high-definition videos via social media. All these applications would require even higher bit rates than provided by LTE, and also when the user is moving fast. eMMB would mean increased network capacity requirements to support expected high volumes of 5G mobile devices. Very large numbers of simultaneous 5G users should be served at local hotspots such as shopping malls and sports arenas.
- **Ultra-reliable low-latency communication (URLLC)** to support mission critical use cases such as remote control of production machinery, remote-operated surgery, or self-driving cars, where no delays or errors over the connection would be tolerated. Furthermore, the network service availability requirements are paramount for these use cases.
- **Massive machine type communication (mMTC)** to support very large numbers of low-cost devices, such as sensors connected to a network. Sensors typically send small amounts of data infrequently. Many sensors are battery-operated with limited battery charging capabilities or maintenance support, so low power consumption is required for the radio connections.

ITU-R has derived the following technical requirements from these scenarios for the 5G cellular radio system:

- Maximum and sustainable bitrates of the 5G system shall be higher than those of LTE. To support the eMBB scenario, the peak data rates should be up to 20 Gbps downlink and 10 Gbps uplink, and 100 Mbps downlink and 50 Mbps uplink sustainable rates should be achievable at the cell edges.
- The latencies of the 5G system should be smaller than those of LTE. For the eMBB scenario, the target round-trip time is 4 ms, but for the URLLC scenario the target is as low as 1 ms.
- The area traffic capacity of the 5G system should be larger than the capacity of a comparable LTE system. Area traffic capacity is calculated as the total capacity (number of users times the bit rates per user) in an area populated by the users. The area capacity target for the eMBB scenario was set to $10 \, \text{Mbps/m}^2$.
- The spectral efficiency of 5G was supposed to be two to three times higher than that of LTE system, reaching peaks of 30 bps/Hz downlink and 15 bps/Hz uplink.
- For the eMBB use cases, 5G system should support mobility up to 500 km/h.
- The connection density target for the mMTC scenario is 1 million connected devices per square kilometer.
- Despite meeting the requirements for higher bitrates, area traffic capacity, and connection density, the energy consumption of the network should not be significantly larger than for LTE network deployments. Energy efficiency (measured as bit/J) of 5G technology shall be much better than that of LTE.

- 5G base station density must not be very much higher than that of LTE. The coverage of a 5G cell shall be large enough even if higher frequencies would be used than for LTE cells.
- 5G system should be more secure than the earlier cellular systems, to support the critical URLLC scenarios and a high number of connected devices in mMTC scenarios.
- 5G system shall support flexible allocation of radio bands, widths between 5 and 400 MHz.

To comply with these ITU-R requirements, 3GPP has specified the 5G new radio (NR) system. 5G specification work was started at 3GPP in 2016 with 5G study items [2]. In two years, various technical reports were produced to evaluate technical options. Thereafter, 3GPP carried out detailed 5G technical specification work. 5G NR specifications are covered in the new 3GPP standard series TS.38, which became available from 3GPP Rel-15 onwards. The overall 5G NR description can be found from TS 38.300 [3]. The 5G new radio system is heavily based on the solutions used for the LTE system, but with a number of extensions and evolved mechanisms to support IMT-2020 use cases.

3GPP 5G standards have been developed and published as subsequent standards releases. Major 5G features covered in this book have been defined in the following releases:

- 3GPP Rel-15 introduced the 5G NR interface and network system.
- 3GPP Rel-16 enhanced cooperation between 5G NR and 4G LTE networks, improved support for network slicing, and increased 5G bitrates with new carrier aggregation configurations and 256-QAM support. Rel-16 specifications introduced also various new capabilities for the New Radio and overall 5G system to support the URLLC scenario and non-public 5G networks.

With the Rel-16 feature set, the maximum practical downlink data rate was approximately 1 gigabit per second with 100 MHz system bandwidth. Even higher peak data rate could be achieved with a split bearer where the 5G NR UE would use multiple independent 5G bands and a non-standalone configuration where LTE would be used together with the 5G radio. Peak data rates can be achieved only when the cell uses most of its capacity toward one single UE, as the total maximum downlink capacity of a 5G cell is around 3–4 Gpbs for a 100 MHz band. The uplink peak data rates are under 100 Mbps because of many reasons, like lack of MIMO and highest order QAM modulation types, while the UE only has a single transmitter and limited amount of transmission power [4]. The new wide 5G bands are the main contributor of high 5G downlink bitrates. Provided with the same 20 MHz system bandwidth, there is no significant difference between LTE and 5G speeds, since both rely on similar OFDMA, QAM, and MIMO methods.

It is worth noting that the 5G technologies adopted by ITU-R are not limited to those specified by 3GPP. ETSI has specified DECT-2020 New Radio technology, which ITU-R has included in the scope of IMT-2020 in October 2021. Digital enhanced cordless telecommunications (DECT) is an ETSI term for short-range cordless communications over unlicensed frequencies, and DECT-2020 NR [369] was designed to support 5G URLLC and mMTC scenarios. DECT-2020 relies on mesh networking with autonomous routing and self-organizing capabilities. Terminals of a mesh network may communicate directly between each other, rather than relying on base station infrastructure. DECT-2020 NR shares many features with 3GPP 5G NR, such as OFDM combined with TDMA and FDMA, turbo coding, and Hybrid ARQ. Still, DECT-2020 NR is not compatible with the 5G NR. It remains to be seen how widely and to which purposes DECT-2020 will be deployed.

8.1.2 Frequency Bands Used for 5G NR

Two frequency ranges (FR) were allocated for 5G NR system:

- FR1 as 410–7125 MHz with 56 operating bands (Rel-17). Apart from one single band specified for the United States, all other bands of FR1 are below 6 GHz, so often FR1 is referred to as the 5G NR sub6 band.
- FR2 as 24 250–52 600 MHz with six operating bands, used especially in North America. Due to the short wavelength of radio waves on these high frequencies, FR2 is known as the 5G mm wave band.

The bandwidths used for individual 5G TDD and FDD operating bands under FR1 are between 5 and 100 MHz, and FR2 uses wider bandwidth, between 50 and 400 MHz.

5G basic approach is to use the different frequency bands as follows:

- Low bands below 2 GHz should be allocated for cells providing coverage over wide geographical areas as well as deep indoor coverage as the low frequencies are good for penetrating obstacles and have low attenuation. The challenge is that many low bands are used already for other radio technologies, such as LTE. One solution available is called **dynamic spectrum sharing (DSS),** where the same band is used to operate both LTE and 5G networks.
- Mid-bands between 2 and 6 GHz should be used for additional capacity and coverage over wide areas.
- High bands over 6 GHz should be used to provide capacity for eMBB use cases at hot spots or locations where very high data rates would be needed. The typical coverage area of a FR2 cell is between 200 and 300 m, which makes the technology suitable for open indoor areas, such as airports and exhibition halls [4].

As the low bands are narrower than the high bands, that restricts the achievable 5G bitrates to levels available with LTE over the same band. Motivation for using low bands for 5G rather than LTE does not come from high-bitrate eMMB use cases but from other 5G use cases and supporting a multi-band 5G network.

3GPP TS 38.104 [5] Rel-16 defines 23 FDD and 19 TDD operating bands for FR1 and FR2. Additionally, the spec has eight uplink-only and three downlink-only bands. While the number of FDD bands is still larger than TDD bands, the TDD technology is attractive for 5G networks. With TDD, the operator can easily support asymmetric traffic pattern where the downlink is allocated with more bandwidth than uplink. As FR2 bands are very wide, no additional band is needed for the uplink. The TDD bands are more suitable for multi-antenna operations. With TDD, the reference signals used for channel estimation are sent over the same frequencies on both directions. The drawback of TDD is the need for time synchronization between cells and networks to avoid interference from downlink signals to overlapping but weaker uplink signals. Without such synchronization and network-wide alignment of downlink and uplink periods, wasteful guard periods or guard bands would be needed for interference avoidance.

TDD networks typically divide the band between uplink and downlink traffic so that a complete frame is sent only to one direction. To minimize latencies, 5G NR applies a new **flexible duplex** approach where the same frequencies may be used in both directions even within individual frames or slots. Due to the additional interference, DM-RS reference signals are sent along with the data to support demodulator to recover the data signals properly.

8.1.3   Architecture and Services of 5G Systems

8.1.3.1   5G Services

The 5G system is a packet-based network, so essentially it provides the same services as the 4G LTE network. However, the packet services of 5G have been enhanced to support different kinds of use cases as outlined by ITU-T: eMMB, URLLC, and mMTC.

8.1.3.2   5G System Architecture

The 5G cellular system consists of the following three subsystems:

- **User equipment (UE)** is an NR terminal and its UICC card with USIM application.
- **5G New Radio (NR)** is the 5G radio access network (5G-RAN), which consists of gNB base stations. 5G-RAN architecture and functions are described in 3GPP TS 38.401 [6].
- **5G Core (5GC)** is the 5G network core with various network functions to support packet switched data flows between the UE and external packet data networks.

Overall architecture of the 5G system is specified in 3GPP TS 23.501 [7] and system procedures are defined in TS 23.502 [8]. Overall description of the interface between 5G-RAN and 5GC can be found from TS 38.410 [9].

The early 5G NR deployments have been **non-standalone NR/LTE dual connectivity** configurations (see Section 8.1.4.6), where 5G NR is essentially used as an extension of an existing LTE network. The role of 5G NR in such networks is to provide higher bitrates over new frequency bands. In the dual connectivity setup, the complete NG dual carrier DC-RAN consists of gNB NR base stations and ng-eNodeB LTE base stations. Such a RAN may be connected either to the 5G or EPC core, depending on the type of the specific deployment. Initially, the existing EPC core networks were used. To support mixing different types of core and radio access networks, a new generation LTE ng-eNodeB supports interfaces to both LTE and 5G core networks. Similarly, also 5G gNB supports both native 5G interfaces to 5G core network functions and the traditional LTE interfaces to EPC core network elements. The long-term architectural target is to build standalone NR/5GC core networks, in order to have better support for different 5G scenarios, minimize latencies for connection setup, and simplify network operations and maintenance.

With the LTE SAE architecture, 3GPP already made important steps to separate network functionality to control and user planes. The LTE MME is dedicated for control plane signaling while all user data is transported by the S-GW and P-GW nodes. However, the Rel-8 LTE split between control and user plane functions was far from complete. The S-GW and P-GW elements still embedded significant control plane functions. In the later LTE specification releases, 3GPP introduced **control and user plane separation (CUPS)** work items to divide both S-GW and P-GW to two entities, one for the user plane and another for the control plane functionality. The 5G core network architecture work finalized the split between the control and user planes via the introduction of three new major 5G network elements to redistribute functions of LTE MME and SAE-GW nodes:

- Access and mobility management function (AMF)
- Session management function (SMF)
- User plane function (UMF)

The first two of them, AMF and SMF, are control plane functions. The AMF manages mobility and network access of the UE. The SMF manages user data sessions and connectivity, with some help from the AMF. The UMF is effectively a router for user data flows and related tunnels, as set up under the control of the SMF. A 5G **packet data unit session (PDU session)** is a similar concept as LTE bearer context or GPRS packet data protocol context. 5G supports IP PDU sessions as LTE does, but also Ethernet PDU sessions to interconnect Ethernet networks in industrial or enterprise environments. In the LTE architecture, the SAE-GW was split to the P-GW toward data networks and the S-GW toward the UE. In 5G core, the SMF may dynamically allocate only one single UPF for the UE or multiple UPFs so that one of them works as PDU session anchor (like the P-GW) and the others as intermediate UPF facing toward UE (like the S-GW). To support edge computing, the UPFs may connect the UE with local servers or computing resources available at the network edge rather than to servers accessed over the Internet.

In addition to these three major elements, 5G core network design introduced a number of other network functions to provide various services (such as authentication) for their client functions. These supporting entities were specified explicitly as service-based software functions rather than network elements, to be able to run them either on dedicated hardware or within a cloud environment used to host all the necessary core network functionality. In the 5G context, the latter approach is called **network virtualization**. Such virtualization supports network scalability and deploying network functions on standard cloud platforms. Instead of only the big operators creating their 5G networks with dedicated pieces of hardware, any corporation could operate its own internal 5G network core with a cloud service provider hosting the network functions.

In the earlier generations of cellular network design up to LTE, the network interfaces between different network elements were implemented as custom protocol stacks supporting point-to-point communication between the elements. This approach is still partially used for 5G core network design, so that the 5G core is able to interact with legacy core networks of the earlier 3GPP radio technologies. The 5G core network design follows a newer inter-function communications paradigm known as **service-based architecture (SBA)**. The various functions of 5G core publish their interfaces as **Web services**, accessible over the same standard HTTPS protocol that Web browsers use to access web pages. The various 5G core functions are able to communicate with each other by publishing their service-based interfaces via a network repository function (NRF). The client functions use NRF to discover services and related Web service operations provided by the supporting functions.

In the SBA architecture, the service provider is called a producer and the client of Web service is called a consumer. Most of the 5G Web service interfaces follow the request-response model typically used for traditional 3GPP protocol messages. But some Web service interfaces follow the subscribe-notify model in which the consumer subscribes to an interface and the producer accordingly provides notifications when activities take place to trigger the service on the producer side. The naming convention for the complete interface provided by a network function is as follows: an uppercase N followed by the network function name in lowercase. For instance, AMF provides Namf interface. The names of individual Web service primitives start with the interface name followed by an underscore and the name of the primitive, such as Namf_Communication_RegistrationStatusUpdate.

This kind of service-based architecture makes it possible for a single core network to have a number of function instances of a same type, to support load balancing between the instances, and also a new 5G feature called network slicing. **Network slicing** means an approach where the core network is logically divided into slices, each of them to support different use cases such as eMBB, ULRRC, or eMTC [3]. As these use cases have different requirements for the network service, the instances of functions attached to the slices may then be optimized according to the slice-specific requirements for capacity, latencies, bitrates, and so forth, provided. The network slicing concept was defined as a systematic approach to support different 5G use cases, based on the experience gained when splitting the LTE system for broadband versus NB-IoT users.

The network slice is selected for the UE within the registration procedure of NR UE initial access sequence and renewed periodically afterwards. The network selects the slice by checking the **network slice selection assistance information (NSSAI)** identifier, which the UE sends in its registration request. The NSSAI defines both the slice/service type (SST) and the slice differentiator (SD) to identify the instance of slice in case the network has many slice instances of the same type. The NRF discovery function supports the client functions to find slice-specific instances of producer functions, to be assigned for the registered UE, as shown in Figure 8.1. It is also possible that the UE registers to multiple slices simultaneously so that the different PDU sessions and QoS flows would use services from different network slices. In such a case, the network assigns a single AMF function but may assign slice-specific SMF and UPF functions for the UE.

In addition to the core network, a network slice also has the corresponding part in the 5G radio access network. In the RAN side, the slice can be implemented with different scheduling and layer 1–2 configurations for packet data sessions opened. The 5G-RAN may also support slice specific admission and congestion control mechanisms. Different slices may also be supported by dividing the system bandwidth to bandwidth parts, each of which has its own subcarrier spacing (see Section 8.1.4.1).

The 5G cellular system consists of UEs, base stations, and core network functions. Note that the following list of functions is not complete, but provides an overview of the most important functions, shown also in Figure 8.2 and Figure 8.3. A complete list of 5G network functions and their descriptions can be found from 3GPP TS 23.501 [7].

- **Mobile equipment (ME)** is an NR terminal such as a mobile phone or NR data modem connected to a computer.
- **Universal integrated circuit card (UICC)** is a smart card, which contains subscriber data and related security algorithms, available via UICC applications such as USIM.
- **5G Node B (gNB)** is the NR base station, which has similar physical components as LTE eNodeB. The gNB typically has a big number of antennas to support various MIMO use cases, like NR beamforming described in

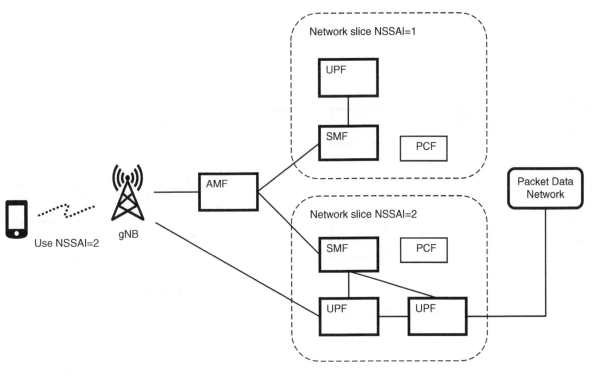

Figure 8.1 5G network slicing.

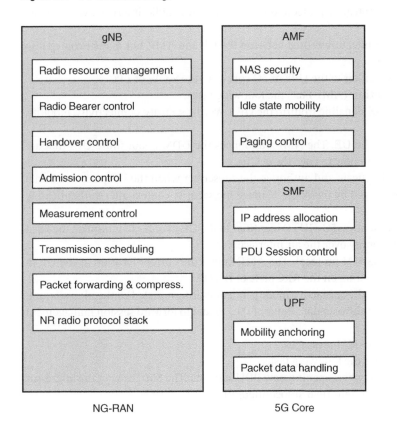

NG-RAN   5G Core

Figure 8.2 Functional split between gNB and key elements of 5GC (adapted from 3GPP TS 38.300 [3]).

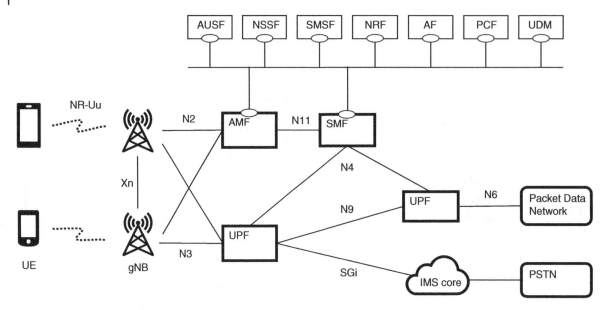

Figure 8.3 Architecture and interfaces of standalone 5G system.

Section 8.1.4.5. The list of gNB tasks is nearly identical to the task list of LTE eNodeB in Chapter 7, Section 7.1.3.2, but with the following minor differences:

- The gNB supports both encryption and integrity protection of user data over radio interface, whereas the eNodeB supports only encryption.
- The control plane messages are in 5G architecture routed between the UE and AMF, but may be routed also to MME in case gNB is connected to EPC.
- The gNB connects user plane data flows either to the UPF of 5G core or to the S-GW of EPC.
- **Access and mobility management function (AMF)** has the following tasks:
  - Serve as the signaling frontend for all NAS signaling from/to the UE, covering security, connectivity, mobility, and session management.
  - Manage the 5G NR registration state of the UE. The AMF announces the UDM about those UEs that are served by the AMF so that the UDM knows how to page those UEs.
  - Interact with the AUSF and UE to authenticate and authorize the subscriber when the UE performs initial access to the NR service in the area managed by the AMF. Manage the security context after authenticating the subscriber.
  - Allocate a 5G global unique temporary identity (5G-GUTI) identifier to the UE in order to minimize the need to transmit the permanent subscriber identity over the air interface.
  - Manage mobility of the NR UE with NAS MM protocol. The AMF knows the current location of the UE in the accuracy of cell or tracking area, depending on the RRC state of the UE.
  - Page the UE with the gNB when the UE in idle or inactive mode gets a mobile terminated connection attempt.
  - Select an SMF to manage PDU sessions for the user data flows of the UE and to forward session management messages between the UE and the SMF.
  - Encryption and integrity protection of NAS MM messages.
- **Session management function (SMF)** has the following tasks:
  - Allocate IPv4 or IPv6 addresses for the UE when opening new PDU sessions. The SMF can get the addresses from a DHCP sever, external IP network, or with IPv6 autoconfiguration.

- Select UPFs for new PDU sessions or PDU sessions moved to a new UPF during handover.
- Manage PDU sessions over UPF(s) toward destination data networks. In 3GPP NR architecture, the packet data connections are transported through GTP tunnels between the gNB and UPF(s). These tunnels are set up under the control of the SMF.
- Manage QoS of the PDU sessions. Inform UPF(s) and the gNB (via the AMF) about the QoS and policy information received from the PCF.
- Communicate with the UE for PDU session management using NAS SM protocol.
- **User plane function (UPF)** has the following tasks:
  - Forward packets between the UE and external data network, over the GTP tunnels as set up by the SMF.
  - PDU session packet inspection and QoS handling. The UPF works as policy enforcement point where the forwarded packets are treated according to the QoS parameters given for the data flow by the SMF.
  - Serve as the PDU session anchor (PSA) when the UE performs handover between two gNBs or between NR and LTE. In the handover, the SMF controls reconfiguration of tunnels between the PSA and the local UPF.
  - Collect traffic statistics and other information needed for charging the packet data services used by the UE.
- **Policy and charging function (PCF)** has the following tasks:
  - Manage policies related to UE access, mobility, and PDU session management including QoS. These policies are applied based on the user subscription parameters, which the PCF retrieves from the UDR.
  - Flow-based offline and online charging control.
- **Unified data management (UDM)** has the following tasks:
  - Support subscriber identification and access authorization in the same way as the HSS does for other 3GPP radio technologies.
  - Generate authentication and key agreement (AKA) credentials. As the secret key K of the subscriber is shared over all the 3GPP radio technologies, the UDM is typically implemented as an extension of the AuC network element.
- **Unified data repository (UDR)** has the following tasks:
  - Store the subscription data for NR subscribers to identify the services subscribed. To support subscribing services on all RATs supported by the operator the UDR uses database shared with the HSS.
- **Authenticating server function (AUSF)** has the following tasks:
  - Interact with the AMF to authenticate the user. Provides the AMF with subscriber authentication vectors as retrieved from the UDM.
- **Network repository function (NRF)** has the following tasks:
  - Registration and discovery of the Web service interfaces provided by various 5G core network functions.
  - Registration and discovery of the network function instances related to a specific network slice.
- **Network slice selection function (NSSF)** has the following tasks:
  - Select the AMF to serve the UE.
  - Assist the AMF to select the right network slice, based on the NSSAI and SD parameters received from the NR UE and the operator policy stored into the NSSF.
- **Short message service function (SMSF)** has the following tasks
  - Support transporting short messages over the NAS protocol.
- **Network exposure function (NEF)** has the following tasks:
  - Expose services and features of the 5G core in a secure way for external application functions (AF).
  - Manage packet flow descriptors (sets of IP address, protocol and port number) as provided by servers within the 5G core for external communication purposes.
- **Application function (AF)** is a function external to the 5G core, interacting with the core interfaces either directly or via the NEF. The following tasks are examples of those which an AF may perform:
  - Influence traffic routing for the application, following the policy framework.
  - Request specific QoS for a traffic flow on behalf of the UE.

- **Security edge protection proxy (SEPP)** has the following tasks:
  - Protect the communication between home and visited network.
- **Network data analytics function (NWDAF)** has the following tasks:
  - Collect data and provide analytics services for network usage.

The NR architecture allows splitting the gNB functionality to a gNB-CU (central unit) used to control the number of gNB-DUs (distributed units). The DU node takes care of functions in PHY, MAC, and RLC layers, and the CU node supports the higher layer protocols RRC, PDCP, and SDAP. While the CU can be run within a cloud relatively easily, the DU typically requires dedicated hardware environment.

While 5GC network functions communicate between each other over Web service interfaces, the 5G core network model still specifies interfaces between the functions as reference points in the traditional 3GPP manner. Interfaces with legacy LTE protocol stack are still used to support the NR/EPC non-standalone mode where 5GC network functions work together with the corresponding network elements of the LTE architecture. To support the non-standalone mode, the 5GC network functions may be co-located with corresponding LTE network elements. Typically, both the AMF and SMF may be implemented as 5G functions within the 4G MME network element.

The following reference points (see Figure 8.3) are part of 5G system design:

- NR-Uu: Radio interface between NR UE and gNB
- F1: Interface between gNB-CU and gNB-DU, if distributed gNB architecture is used
- Xn: Interface between two gNB base stations. In the EN-DC architecture (see Section 8.1.4.6) where gNBs are connected to EPC, the gNBs use X2 interface of LTE.
- N1: Interface between NR UE and AMF
- N2: Interface between gNB and AMF
- N3: Interface between gNB and UPF (possibly collocated with P-GW)
- N4: Interface between SMF and UPF (possibly collocated with P-GW)
- N5: Interface between PCF and AF
- N6: Interface between UPF and external data network
- N7: Interface between SMF and PCF (possibly collocated with PCRF)
- N8: Interface between AMF and UDM (possibly collocated with HSS)
- N9: Interface between anchor and intermediate UPF
- N10: Interface between SMF and UDM (possibly collocated with HSS)
- N11: Interface between AMF and SMF (possibly collocated with P-GW)
- N15: Interface between AMF and PCF (possibly collocated with PCRF)
- N26: Interface between AMF and LTE MME

The following different types of identifiers have been defined for the 5G subscriber and UE in 3GPP TS 23.003 [10]:

- **Subscription permanent identifier (SUPI)** is assigned for each subscriber accessing the 5G network. The SUPI retrieved from the USIM card is actually the IMSI identifier used since GSM. Devices without USIM may generate the SUPI as a network access identifier (NAI) following procedures as specified in 3GPP TS 23.003.
- **Subscription concealed identifier (SUCI)** is a global identifier that can keep the UE identity hidden. The SUCI is an encrypted version of the SUPI. The SUCI is used in the user authentication process to avoid sending the SUPI (or IMSI) as cleartext, as that would reveal the location of the subscriber. Public key of the network, as stored to the UICC card, is used for SUCI encryption. During the encryption, the UE also creates one-time public/private key pairs, enabling the UE to use the private key to derive unique SUCI instances. The network can decode those with its own private key and the one-time public key received from the UE along with the SUCI.

- **5G Global unique temporary identity (5G-GUTI)** is an identifier for the UE that does not reveal the SUPI. The 5G-GUTI is allocated for the UE by the AMF and it is used like the LTE GUTI.
- **5G Temporary mobile subscriber identity (5G-S-TMSI)** is a shortened form of 5G-GUTI which is used in radio signaling procedures to identify the user.
- **Permanent equipment identifier (PEI)** is assigned to each UE accessing the 5G network. The PEI corresponds the IMEI, which was introduced with GSM and used to block stolen phones for network access.

The 5G NR reuses and further extends the Radio Network Temporary Identity (RNTI) concept of LTE (see Chapter 7, Section 7.1.3.2) with a number of new RNTI types for single or group of UEs. Among those, I-RNTI (Inactive RNTI) is used to identify UEs in the RRC_INACTIVE state and the gNBs which suspended their RRC connections.

8.1.3.3 5G Protocol Stack Architecture

Layer structure of the 5G protocol stack is very similar to that of the LTE stack. Both 5G NR control plane and user plane stacks have the same set of protocols as corresponding LTE stacks. Individual protocols in the NR stack have mainly the same high-level responsibilities as their LTE counterparts. The protocol details have, however, been respecified for NR in order to support NR specific features. Additionally, NR introduces NR Service Data Adaptation Protocol (SDAP) as a new protocol to the top of the user plane protocol stack. SDAP is used for mapping QoS flows to NR radio bearers as described in Section 8.1.10.1.

The control and user plane protocol stack diagrams of the 5G system (Figures 8.4 and 8.5) are very close to those of LTE. The major difference comes from the different functional network element architecture so that the endpoints of protocols are specific to 5G.

8.1.3.4 5G Bearers and Radio Channels

The channel architecture of 5G is also very similar to LTE, as can be seen by comparing Figure 8.6 to Figure 7.6. 5G logical channels are mapped to transport channels, which are eventually mapped to 5G physical channels. While the 5G physical channel structure is somehow different from LTE physical channels, the 5G design mostly reuses the structure of logical and transport channels as defined for LTE.

The bearer model of 5G, shown in Figure 8.7, is also streamlined from LTE bearer model. 5G still uses the concept of radio bearers to carry data flows with similar QoS requirements. Due to the rather flat structure of 5G network, 5G model otherwise just refers to GTP tunnels between the gNB and core network UPFs. The **QoS flows** of the 5G connection model correspond with LTE end-to-end bearers.

Figure 8.4 5G packet system user plane protocols.

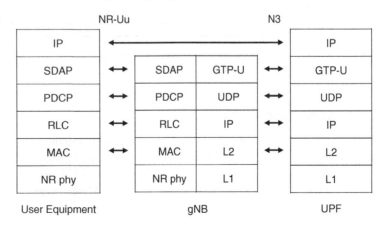

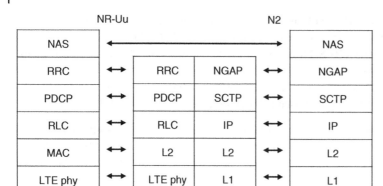

Figure 8.5 5G control plane signaling protocols.

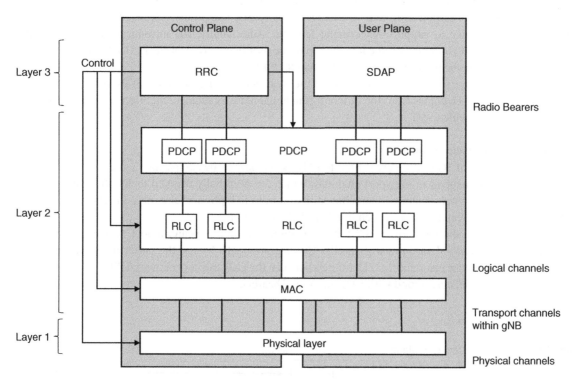

Figure 8.6 NR radio protocol stack and channels (adapted from 3GPP TS 38.300 [3]).

8.1.4 5G NR Radio Interface

The 5G New Radio (NR) physical layer design is derived from the LTE physical layer. These two radio technologies use a common multiplexing method and share rather similar high-level design of physical channels and reference signals. The NR physical layer is, however, different from LTE in certain details to support new frequency bands, enhanced resource mapping schemes, and large-scale MIMO setups. NR uses those features to achieve the capacity and latency targets set for IMT-2020 technologies.

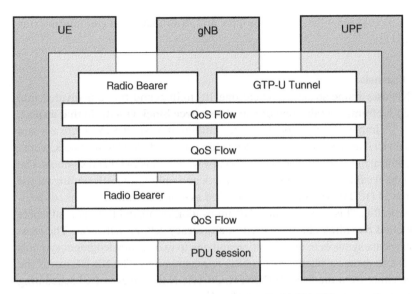

Figure 8.7 5G QoS architecture (adapted from 3GPP TS 38.300 [3]).

8.1.4.1 Modulation and Multiplexing

5G NR uses orthogonal frequency division multiple (OFDM) modulation and OFDMA multiplexing like LTE does, except that OFDMA is used for both NR downlink and uplink traffic. The 15 kHz subcarrier spacing of LTE is also the foundation of 5G subcarrier structure. However, 5G NR does not rely on only one fixed subchannel bandwidth, but uses a number of **subcarrier spacing (SCS)** options to support higher frequencies and high-speed mobility. In high velocities, the Doppler effect distorts the waveforms, so wider subcarriers are used to avoid inter-subcarrier interference. The bandwidth of an NR subcarrier is n * 15 kHz, where the value of n is either 1, 2, 4, 8, or 16. On FR1 the subcarriers can be up to 120 kHz wide; 30 kHz is a value typically used [4]. The widest subcarriers are used especially for the high-frequency bands of FR2 frequency range. One consequence of using wide subcarriers is that the number of subcarriers becomes smaller, which makes the NR radio design simpler.

The modulation options for NR OFDM symbols on the shared downlink channel are QPSK, 16-QAM, 64-QAM, and 256-QAM. The transmission time of the OFDM symbol depends on the bandwidth of the subcarrier. The wider the subcarrier is, the shorter is the OFDM symbol period, which helps 5G NR to minimize the latencies. The length of the cyclic prefix is proportional to the length of the OFDM symbol, except that with bands under 3 GHz the extended CP is used for 60 kHz SCS to support the high reliability requirements of URLLC use cases.

As described earlier, an LTE radio block consists of LTE subframes (each carrying 12–14 OFDM symbols) transmitted over 12 adjacent subcarriers within 1 millisecond. The 5G NR radio block also uses 12 adjacent subcarriers within 1 millisecond, but the number of OFDM symbols per subframe depends on the subcarrier spacing and the OFDM symbol period used.

While the LTE UE always operates on the full LTE radio band deployed, 5G NR divides the system bandwidth to **bandwidth parts (BWP),** which are contiguous subsets of subcarriers taken from the full network band. Every bandwidth part may use its own subcarrier spacing value to serve a network slice optimized for specific types of services. The 5G NR UE may use only a single BWP or up to four BWPs at any time, which helps it to balance its operating power consumption with other operation parameters. The UE may for instance receive one BWP to cover the full system bandwidth and another for a subset of it. The UE chooses one of those BWPs based on the amount of traffic being transported and the needed bitrates. The UE may also select the bandwidth part(s) based on the UE type and used network slice. The BWP mechanism was created based on the experience from extending the LTE air interface for NB-IoT support.

In other respects, NR resource grid structure is similar to LTE resource grid. NR modulation and multiplexing aspects are defined in 3GPP TS 38.211 [11].

8.1.4.2 Frame Structure and Physical Channels

For TDM and synchronization, 5G NR uses frame structure where one 10- millisecond frame is divided into 10 subframes, 1 millisecond each. Each subframe contains one **physical resource block (PRB)**. In that respect, the NR frame structure is similar to LTE frame, but the slot structure is different as shown in Figure 8.8 and Table 8.1. The LTE subframe is always divided into two slots, but 5G NR subframes are divided into 2^n slots, where n is between 0 and 4 depending on the subcarrier spacing. The 5G NR slot always contains 14 OFDM symbols. The slot structure of 5G NR TDD band can be flexibly divided between downlink and uplink slots, depending on the network configuration. Every slot of the 14 OFDM symbols can be classified as either downlink, uplink, or flexible, mixing both uplink and downlink traffic. This is very radical NR innovation compared to rather inflexible allocation mechanism between uplink and downlink channels used in any legacy cellular networks. The most typical use of NR flexible slots is to carry data symbols to one direction and one to two acknowledgment symbols to the opposite direction within one slot, to decrease latencies of HARQ retransmissions. The NR UE learns the UL/DL slot configuration of the frame either from the RRC RECONFIGURATION messages or DCI scheduling messages, while the RRC option has been deployed in early 5G networks [4].

The physical channels of 5G NR radio interface have the same names and for most of the cases also the same purposes as the LTE physical channels. However, the NR physical channels are mapped to the NR resource grid in a different way than LTE channels are mapped to the LTE resource grid. The NR physical channels also use their own data structures and indicator values, so their internal implementation differs from LTE.

The physical channels of 5G NR are as follows [11]:

1) Downlink channels
   - Physical broadcast channel (PBCH) carries NR master information block (MIB), from which the UEs learn system frame number (SFN), subcarrier spacing, and SSB offset used by the network, positions of certain reference signals and parameters for searching system information block 1 (SIB1). The PBCH also carries a few indicator bits used for frame decoding. An NR UE uses the PBCH channel for NR initial access as described in Section 8.1.8.1.

Subframe = 1 ms

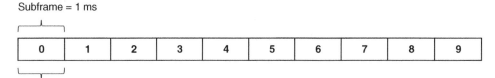

1, 2, 4, 8 or 16 slots per subframe depending on subcarrier spacing.
Each slot contains 14 OFDM symbols.

Figure 8.8 The 5G NR frame structure from gNB to UE.

Table 8.1 5G NR frame versus LTE frame.

	Symbols per slot	Symbol duration	Slot duration	Slots per subframe	Subframe duration	Subframes per frame	Frame duration
LTE	7	66.7 μs	0.5 ms	2	1 ms	10	10 ms
NR	14	Depend on subcarrier spacing			1 ms	10	10 ms

- Physical downlink control channel (PDCCH) is used to carry downlink control information (DCI) messages to tell the UEs about scheduling grants. Just like LTE, also NR has many different formats for DCI messages. The DCI messages describe which bandwidth parts, resource blocks, and formats of slots the UE shall use. The UE learns from DCI messages the modulation and coding schemes (MCS) to be used for the resource blocks. The DCI messages also provide the UE with power control commands and HARQ process mappings for transported MAC frames. Mapping of the PDCCH channel to NR resource elements is not fixed but dynamic. The dynamic PDCCH mapping is used to support PDCCH configurations specific for different bandwidth parts. The UE learns from RRC RECONFIGURATION messages the control resource set (CORESET) used to carry the PDCCH channel. CORESET consists of a set of resource blocks and one to three OFDM symbols within those. CORESET has the following internal structure used to carry DCI messages: Resource element group (REG) has 12 resource elements of one OFDM symbol within a physical resource block. Six REGs are aggregated into control channel elements (CCE). A single DCI element is carried within one or multiple CCEs.
- Physical downlink shared channel (PDSCH) is a shared physical channel over which the UEs can receive paging, data, or signaling messages from the network. The PDSCH of NR uses a single transmission scheme similar to LTE TM9, relying on the DM-RS reference signals transmitted to the UE within the PDSCH channel to support reliable demodulation of the PDSCH.

2) Uplink channels

- Physical random access channel (PRACH) is used by the UE for random access messages.
- Physical uplink control channel (PUCCH) is used by the UE to send channel state information (CSI), HARQ acknowledgments, and scheduling requests for sending uplink data over the PUSCH channel. Like for LTE, different PUCCH formats have been defined for different purposes. Formats 1–2 do not provide CSI reports, and formats 3–5 support all three types of CSI information, as described in Chapter 7, Section 7.1.4.6. Formats 0 and 2 use only one to two OFDM symbols per slot, while other formats use 4–14 symbols. Consequently, the PUCCH overhead can be minimized with formats 0 and 2 but with the cost of delay.
- Physical uplink shared channel (PUSCH) is a shared physical channel via which the UEs can send data or signaling messages. Selection and activation of NR PUSCH features like precoding type, codebooks used for the precoding matrices, and the transmission rank are done by gNB based on the sounding reference signals sent by UE. Please refer to LTE Chapter 7, Section 7.1.4.9 for further discussion about these features.

5G NR has a smaller set of physical channels compared to LTE because of the following reasons:

- LTE uses the PCFICH channel to indicate the resource (OFDMA symbol) reservations for the PDCCH channel. While the PDCCH resource mapping is dynamic also in NR, no separate physical channel is used to describe NR PDCCH configuration. Instead, at first the NR UE learns from MIB the parameters necessary for searching the SIB1 block from the PDCCH channel. Later, the RRC protocol parameter ControlResourceSetIE is used to inform the UE about the PDCCH configuration. Finally, DCI messages can be used to indicate those resource elements, which have usually been semi-statically allocated for the PDCCH, to be temporarily used for the PDSCH traffic.
- Since NR HARQ acknowledgments and multicast messages are carried over the NR downlink shared channel PDSCH, there is no need for separate PHICH and PMCH channels in NR.

NR radio transmitter design is very similar to LTE design described in Chapter 7, Section 7.1.4.9, except the following:

- NR introduces new channel coding methods used for forward error correction. Instead of using turbo coding, NR uses two channel coding methods, which due to their complexity are not described in this book:
  - Low-density parity check (LDPC) coding for both downlink and uplink data.
  - Polar codes for both downlink and uplink control information, such as DCI and UCI
- Precoding is not used for NR PDSCH but it is used for uplink PUSCH.

8.1.4.3 Scheduling

The configuration of bandwidth parts to be used by the UE and allocation of the slots for different directions is controlled by parameters, which the network configures to the 5G UE:

- Cell-specific higher layer configuration, specifying default patterns with which the UE may to divide consecutive slots between downlink, flexible and uplink.
- Network may also use RRC messages to provide the UE with a UE-specific configuration, which then overrides the default cell-specific configuration used by other UEs.
- The DCI signals sent to the UEs may also be used for ad-hoc scheduling downlink and uplink transmission for specific slots. The DCI signals are the highest priority parameters that can be used to temporarily override the other more persistent configurations.

As we already know, the transmission time interval (TTI) is the period of resource allocation used by the cellular system. In the legacy WCDMA and LTE systems, the TTI length is fixed and transmission scheduling is done once per TTI. Due to the flexible allocation of 5G NR slots between uplink and downlink, controlled by both persistent configurations and DCI signals, NR does not use any fixed TTI length. Instead, the period of NR resource allocation depends on the slot lengths as well as the usage of flexible NR slots, carrying traffic to both directions.

NR supports both the DCI-based dynamic scheduling and semi-persistent downlink scheduling methods, as used also for LTE. For uplink transmissions to complement dynamic scheduling, NR uses configured grants of two types:

- Type 1: RRC protocol is used both to provide the UE with an uplink scheduling configuration and to activate the given configuration.
- Type 2: RRC protocol is used to provide the UE with an uplink scheduling configuration, but DCI signaling is used to activate the configuration.

8.1.4.4 5G NR Reference Signals

LTE Rel-8 networks relied on cell specific reference signals in fixed positions of the LTE resource grid. Unfortunately, that approach could not be scaled for increasing number of antenna ports and support of small cells. The newer LTE transmission modes, specified in subsequent 3GPP releases, introduced DM-RS and CSI-RS reference signals for more flexible resource mapping schemes. To support scalability, these reference signals were coded with orthogonal OCC code words so that the same resource elements could be used to transport multiple reference signals simultaneously.

The fundamental design decision of 5G NR was not to use the cell specific reference signals at all, but to redistribute their tasks to other reference signals. Consequently, the NR CSI-RS are used for synchronization, RRM measurements, and CSI channel estimation while DM-RS signals alone support demodulation. NR took also a new type of **phase tracking reference signals (PT-RS)** into use on the FR2 bands to compensate phase noise, which impacts especially the high-frequency NR subcarriers.

The NR DM-RS design had to address a diverse mix of requirements, such as good channel estimation performance as well as support for a wide range of frequencies, antenna ports, low latency URLLC services, and flexible NR frame structures. Every NR resource block carries DM-RS signals. Within the resource block, the DM-RS signals occupy one to six OFDM symbol positions, over all the 12 subcarriers of the resource block. Front-loaded DM-RS signals are located to OFDM symbol positions right after the control symbols in the beginning of the resource block, as shown in Figure 8.9. Additional DM-RS signals in the other two OFDM symbol positions may be used to support high mobility use cases. NR has two DM-RS configuration types so that type 1 supports fewer antenna ports than type 2. In type 1, the 12 subcarriers of one OFDM symbol position are divided between two DM-RS signal groups, while in type 2 those subcarriers are divided between three DM-RS signal groups. One signal group may contain one to four reference signals for different antenna ports, multiplexed with OCC codes.

Figure 8.9 DM-RS type 2 configuration with front-loaded DM-RS signals only.

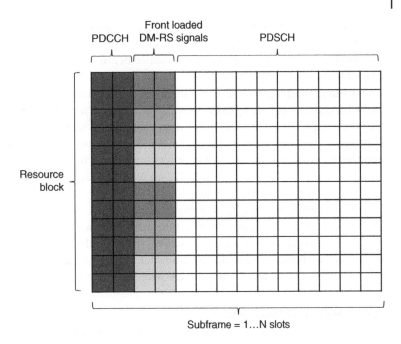

PDCCH | Front loaded DM-RS signals | PDSCH

Resource block

Subframe = 1…N slots

NR design for CSI-RS must also support different antenna port configurations up to 32 parallel ports. Mapping of CSI-RS signals to OFDM symbol positions and subcarriers of the resource block depends on the antenna port configuration. The UE learns deployed CSI-RS configuration from the SIB1 block. Configuration specific CSI-RS signal resource sets are optimized for different purposes such as CSI acquisition, time and frequency tracking, mobility measurement, and beam management. Resource configuration covers also UE CSI report settings and triggers, which are used to control when and which kind of CSI reports (channel, interface, or beam reports) the UE should send. Due to the complexity of NR and its CSI reporting options, calculating CSI reports is computationally very intensive. To avoid overload, the UE may follow priority order of different CSI reporting types and provide only highest priority reports when the UE is under high load.

The NR UE uses **sounding reference signals (SRS)** for the uplink. An NR SRS supports one, two, or four uplink antenna ports. The SRS may use up to six OFDM symbol positions in the end of the slot and it is possible to send SRS periodically or aperiodically. When the same frequencies are used for both uplink and downlink (in TDD systems), the gNB may use SRS for downlink channel estimation and uplink beam selection when scheduling uplink grants. With 5G NR carrier aggregation, the UE may be able to receive downlink traffic from a higher number of aggregated bands than what the UE can use for uplink transmission. To support downlink channel estimation over the whole aggregated band, the UE shall send SRS also on those bands that it does not use for uplink data. With SRS carrier switching, the UE pauses its uplink data transmission to transmit SRS signals on the extra downlink bands.

8.1.4.5 Beam Management
In the early LTE releases, the eNodeB had only a few antennas with which only rather wide beams could be created. The eNodeB could support a few sectors toward different directions from the antenna mast, so that each sector would be an own cell. With a rather large number of antennas, the NR gNB is able to generate much narrower beams. A single NR cell may consist of as many as 64 beams to different directions. Such beams can be generated with a **massive MIMO** configuration where the gNB has an antenna array of 32 or 64 transmit antenna elements. NR uses **beamforming** to improve coverage of cells operating in FR2 high frequencies, subject to high

path loss. Also, an FR1 NR cell may have up to eight beams to different directions [4]. Any NR UE wishing to camp in such a cell must at first scan the SSB blocks of available beams and select the best one according to the received reference signal power level. This process is called **beam sweeping**. The SSB block carries the SSB beam ID of the selected beam.

After beam sweeping, the UE shall use the PRACH channel of the selected beam for random access. From the random access attempt, the gNB learns which beam to use for scheduling any downlink transmissions toward the UE. When responding to the UE, the gNB provides the UE with a beam configuration covering the currently used beam as well as other adjacent beams for the UE to measure. The measurement results are used in the **beam recovery** process after the UE has lost or is close to losing its connection to the currently used beam and another beam of the cell must be taken into use. The quality of a beam may deteriorate due to UE mobility or some moving obstacles blocking the beam.

Beam recovery is triggered by the measurement and CSI reporting process when the UE finds the quality of the current beam to have dropped below a threshold given within the beam configuration. If the block error rate (BLER) of the currently used beam stays above the given threshold long enough, the UE reports the condition to the gNB and identifies some candidate beams for recovery. Based on the report, the gNB may switch the downlink traffic to another beam and provide the UE with a new beam configuration.

8.1.4.6 Multi-Band Coexistence

The following 5G NR use cases have been identified for parallel use of multiple carriers/bands:

- Better throughput compared to the single-band operation.
- Better network coverage, especially when using high-frequency NR bands above 3 GHz. As the higher frequencies experience high path loss and attenuate quicker than lower frequencies, the effective cell radius of the NR high-frequency cell is smaller than that of a typical LTE cell. Because of that, it is not straightforward to co-locate NR gNB base stations with LTE eNodeBs. While LTE cells would cover the whole area, coverage gaps might be left between the NR cells, impacting especially low-power uplink signals. Without a multi-band arrangement, either extra gNB sites would be needed or alternatively the system might suffer from a high amount of NR to LTE inter-RAT handovers at the edges of the NR cells.
- Balancing spectral efficiency with network latency for various eMBB, URLLC, and mMTC use cases with conflicting uplink requirements. While NR TDD bands support switching between DL/UL traffic at the level of an NR slot, a short guard period is still needed during the switch to prevent transmissions from overlapping. Frequent switches support short latencies, but the frequent guard periods decrease the spectral efficiency. One solution is to combine the NR TDD band with another always-on **supplementary uplink (SUL)** channel on a different band. The SUL channel can be used for uplink traffic and the regular NR TDD band for downlink traffic, without any need to switch traffic direction on either of the bands.

NR supports a number of multi-band coexistence approaches. Some of the approaches use multiple NR bands (**NR standalone operation**) while others combine NR bands with LTE bands (**NR non-standalone operation**):

- **NR carrier aggregation and dual connectivity (DC)** use multiple separate bidirectional NR bands for high-performance connections. These NR mechanisms are essentially similar to those of LTE.
- **NR band combined with NR SUL** channel. In this band combination a unidirectional SUL band of lower frequency is combined with a single or aggregated bidirectional NR normal band. Because the SUL band uses lower frequencies, it provides additional uplink cell coverage. The normal band is used to provide high throughput in both directions close to the cell center, while at the cell edge the SUL band is used for uplink and the normal band for downlink.
- **Dual connectivity of NR and LTE bands**, where the lower frequency LTE band is used to ensure good enough coverage at the cell edges, as shown in Figure 8.10. In the **E-UTRAN New Radio – Dual Connectivity**

Figure 8.10 Increasing the uplink coverage with supplementary uplink (adapted from 3GPP TS 38.300 [3]).

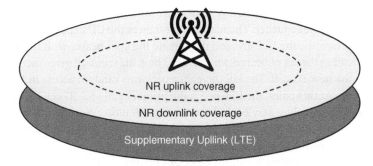

NR uplink coverage

NR downlink coverage

Supplementary Upllink (LTE)

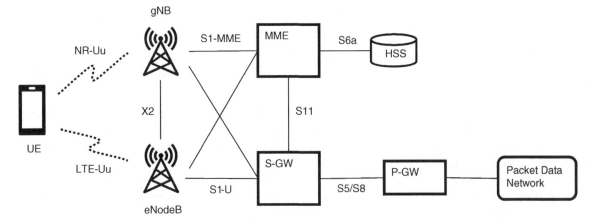

Figure 8.11 Non-standalone EN-DC NR dual connectivity with LTE supported by EPC core.

(EN-DC) mode, LTE is used as the anchor (referred to as master cell group MCG) and NR to provide extra high throughput for UEs close to the base station. EN-DC networks do not use 5G core, but the gNB base stations are connected to the EPC core shared for both LTE and NR RAN, as in Figure 8.11. 3GPP has also specified options where LTE eNodeBs could be connected to the 5G core and either LTE or 5G NR is the anchor radio technology.

- **NR/LTE dynamic spectrum sharing (DSS),** where the LTE uplink band is shared between LTE and NR. In LTE FDD, both downlink and uplink use bands with same widths. The LTE traffic pattern is typically downlink dominated, which leaves much of the available uplink capacity unused. This spare bandwidth can be used for an NR SUL uplink, which increases both the NR cell coverage and total spectral efficiency of the NR/LTE system. DSS is possible since 5G NR air interface has been specified to be quite similar to LTE air interface.

NR non-standalone configurations with NR and LTE deployments, like the one depicted in Figure 8.11, are especially important due to the wide availability of LTE networks. As LTE already occupies many of the low bands specified also for 5G NR, there is a need to use these two technologies in a complementary way for practical deployments. In such cases, LTE is typically used to provide coverage and basic support for LTE devices while 5G NR is used to boost the maximum available bitrates for 5G UEs. The first 5G network deployments have used the EN-DC architecture [4]. With EN-DC, the network is able to use either one of the radio technologies or both at the same time to deliver traffic over the Uu interface. **Split bearer** refers to a setup where UE uses both LTE and NR bearers with base stations connected over the EN-DC enhanced X2 interface. The EN-DC split bearer approach impacts a number of system processes:

- The EN-DC capable UE typically performs LTE access and announces its 5G capabilities within the capability exchange sequence. The network configures the UE to perform 5G NR measurements. After receiving the measurements, the LTE eNodeB may create the split bearer with the RRC reconfiguration procedure to add the 5G cell to the set of bearers for the UE. The RRC message gives the UE all the 5G NR parameters needed to access the new 5G cell. The UE thereafter performs random access in the 5G cell to get the additional NR bearers set up. At this point, the LTE cell is the primary cell and 5G cell the secondary cell.
- RRC signaling may be delivered to the UE either over the primary LTE connection or over both LTE and NR connections. In the latter case, the UE may either have a split signaling bearer or both the eNodeB and gNB have their own signaling bearers.
- The intra-system handover processes between cells of same radio technology are performed independently for LTE and 5G NR or alternatively the UE must perform a new 5G NR access after LTE handover.
- When using an EN-DC split bearer, typically the VoLTE service (see Chapter 9, Section 9.4) is carried over the LTE bearer using lower frequencies, which are more robust for the voice call.

The NR/LTE spectrum sharing approach has multiple implementation alternatives. Sharing the uplink band between LTE and NR could be done with static or semi-static frequency division, static time division, or dynamic FTDM based on scheduling:

- Static FDM is an extension of LTE carrier aggregation. In static FDM, two LTE bands are used so that in downlink both are aggregated for LTE, but in uplink one of the bands is used for LTE and the other for NR. Static FDM does not apply any dynamic coordination between the RATs; thus, from the UE perspective the NR uplink is like any standalone NR uplink band.
- Semi-static FDM divides subcarriers of the LTE uplink band between LTE and NR uplinks. The division is done on the granularity of resource blocks consisting of 12 subcarriers. The subcarriers on the edge of the band are used for NR while the subcarriers on the middle are allocated for LTE.
- Semi-static TDM uses a trick to allocate LTE resources for NR in time domain. Network tells LTE UEs that LTE MBSFN pattern would be used for multimedia broadcast even when no multimedia broadcast is really provided. The resource elements nominally allocated for multimedia are actually given for the NR uplink.
- Dynamic TFDM uses scheduling to allocate uplink resource blocks dynamically either to LTE or NR. Dynamic TFDM is the only spectrum sharing option that requires the LTE and NR networks to share a common uplink scheduler. One drawback of TFDM is that it may cause delay for LTE HARQ processes if uplink is scheduled for NR over a long period.

In static and semi-static approaches, the networks do not coordinate their scheduling decisions. It might happen that the UE is scheduled to use both LTE and NR uplinks at the same time. This introduces two extra problems to be solved: **spurious emissions** and the need to have tight synchronization between LTE and NR networks to align their resource grids in both frequency and time domain.

When the UE transmits simultaneously to both the LTE and NR bands, spurious emissions might take place. Spurious emissions mean unwanted signal components emitted by the transmitters, causing noise or even blocking of signals at the UE receiver. Two different mechanisms are used for controlling spurious emissions generated by two different root causes:

- **Harmonic emission:** The transmitter of the lower band emits harmonics in the bandwidth of the upper band, disturbing the UE receiver at the upper band. Harmonics emission can be mitigated with a harmonic rejection filter or coordinating the downlink resource allocation with the uplink of the same band.
- **Intermodulation:** Applying simultaneous transmissions on lower and upper bands to the shared antenna may cause unwanted emissions that interfere with the received signals. Severity of intermodulation depends on the band combination used. For those band combinations where intermodulation could cause severe issues, 3GPP specified single-UL solution where only one of the uplink bands is used at any time.

In the single-UL solution, the UE does the selection between the normal and SUL as follows:

- For initial access, the UE reads a threshold for SUL reference signal received power (RSPR SUL threshold) from NR system information messages. If the received power from the NR downlink is below this threshold, the UE is on the cell edge and shall use low-frequency SUL for random access. Otherwise, the UE shall use the normal NR uplink.
- In the connected mode, the UE reports its downlink measurements. The gNB uses the reported values to decide which of the uplink bands the UE should use. The gNB either indicates the chosen uplink band within DCI messages or gives the UE a semi-static PUSCH uplink grant configuration with RRC signaling.

When deploying spectrum sharing, NR and LTE networks shall be tightly synchronized on both frequency and time domains to maintain high spectral efficiency. Without synchronization, wasteful guard bands or guard periods are needed to avoid overlapping transmissions and mutual interference. The synchronization approach shall take the following aspects into account:

- Subcarrier frequencies and spacing shall be aligned to avoid interference or introduction of guard bands between LTE and NR uplinks. While the alignment is important for both static and semi-static FDM, it is absolutely essential for dynamic TFDM operation.
- Physical resource blocks shall be aligned in the frequency domain. This alignment supports independent scheduling on both static and semi-static FDM and shared scheduling in TFDM without introducing any guard bands at resource block boundaries.
- Channel rasters of NR and LTE bands shall be aligned for spectrum sharing. In the standalone case, the NR system bandwidth (channel raster) is N*15 kHz, and LTE systems use bands as multiples of 100 kHz. If those rasters would be applied to the shared bands, the available bandwidths would have to be multiples of 300 kHz, causing difficulties for practical deployments. For NR/LTE spectrum sharing, it has been decided that the width of the NR band shall be N*100 kHz to match the LTE channel raster.
- OFDM symbol periods shall be aligned in the time domain, to avoid causing any inter-symbol interference or introduction of guard periods for LTE/NR switching. Alignment of synchronization also helps the receiver to run fast Fourier transformations for both NR and LTE signals. For static and semi-static FDM, the time synchronization is useful to avoid inter-subcarrier interference without guard bands. Alignment of OFDM symbol periods is achieved by applying the same timing advance value for both LTE and NR uplinks of the UE.
- The timing of NR and LTE subframes and slots shall be synchronized too, to enable scheduling of complete resource blocks between RATs for the dynamic TFDM mode.

8.1.5 Protocols Used between UE and NR Radio Network

8.1.5.1 MAC Protocol
NR medium access control (MAC) protocol has same tasks as the LTE MAC, such as mapping logical channels to transport channels. NR MAC protocol is defined in 3GPP TS 38.321 [12].

The major differences between NR and LTE MAC protocols are as follows:

- NR MAC has been enhanced for the NR specific features, such as variable subcarrier spacing, bandwidth parts, beam management, and SULs.
- NR MAC supports static and semi-static scheduling. The scheduling request (SR) and buffer status report (BSR) mechanisms have been improved. Multiple SR configurations are supported to cover different bandwidth parts or serving cells for NR/LTE non-standalone operation.
- HARQ processing is largely revised for NR MAC. While LTE uses eight parallel HARQ processes, NR MAC has 16 of them. The gap between PDSCH reception and related HARQ-ACK in LTE is always 4 milliseconds, but in

NR it is configurable. The NR UE learns from DCI message when it shall return HARQ-ACK. This DCI mechanism supports the URLLC ultra-low-latency requirements as well as eMBB more relaxed requirements with smaller HARQ overhead.

– Note: NR MAC specifications define constraints for UEs about how quickly the UE should be able to return HARQ-ACK. The applied constraints depend on subcarrier spacing (which determines OFDM symbol period length) and the number of DM-RS reference signals. Depending on the case and UE capabilities, HARQ-ACK shall be sent 3–24 symbols after receiving the data over the PDSCH channel.

The NR MAC frame carries either MAC control data or user data. The frame consists of the following parts:

- MAC subheader, which encodes MAC SDU length and logical channel identifier or control elements used in the payload.
- MAC payload, which contains either MAC control elements or MAC SDUs for user data transport.

MAC control elements are used to carry various pieces of data related to the operation of MAC protocol and NR physical layer, such as:

- Contention resolution ID for random access procedure
- Timing advance and DRX commands
- Configured grants
- CSI resource sets and triggering
- SRS activation
- Recommended MAC bitrates
- Buffer status reports

8.1.5.2 RLC Protocol

NR radio link control (RLC) provides a link level connection between the UE and gNB. NR RLC protocol is defined in 3GPP TS 38.322 [13].

NR RLC protocol has the same tasks as LTE RLC protocol, but with two exceptions:

- NR RLC does not support segmentation and reassembly, because that process would cause additional latency.
- NR RLC does not perform message reordering, as that is the task of NR PDCP or protocols above PDCP.

Like LTE RLC, also NR RLC protocol provides three different types of services for transporting user data: transparent mode (TM), acknowledged mode (AM), and unacknowledged mode (UM). The structure of NR RLC frames is similar to LTE RLC.

8.1.5.3 PDCP Protocol

NR packet data convergence protocol (PDCP) is used to transport both user data and signaling between the eNodeB and UE. NR PDCP is defined in 3GPP TS 38.323 [14].

Main responsibilities of the NR PDCP are the same as LTE PDCP:

- Compression of the TCP/IP headers
- Reliable transmission of data during handovers

The NR PDCP has the following differences with LTE PDCP:

- NR PDCP may be used in a mode where it does not guarantee the order of received packets so that the reordering would be left to an upper layer protocol.
- NR PDCP supports both integrity protection and encryption while LTE PDCP supports only encryption.

- NR PDCP supports packet duplication to increase reliability of transmission over multiple parallel links. In a NR/LTE coexistence case, a single NR PDCP instance may use a bearer, which is split to both radio technologies, and interacts with underlying NR and LTE RLC protocols. A single PDCP packet can be sent to the remote end over both the NR and LTE links on parallel, causing packet duplication for the PDCP receiver.

The LTE PDCP message is either a data or control message. The structure of the PDCP message depends on its type, which is encoded to the first bit of the message.

- Data messages have PDCP sequence number and data thereafter.
- Control messages start with the PDU Type field whose value determines which other fields the message has. Control messages can be used as status reports or for robust header compression feedback.
- Sidelink messages start with the SDU Type field whose value indicates the type of layer 3 data within the message.

8.1.5.4   RRC Protocol

NR radio resource control (RRC) protocol is used to manage the radio resources at NR air interface. NR RRC protocol is defined in 3GPP TS 38.331 [15].

The NR RRC state model has the following three states:

- RRC_IDLE: The idle UE is camping in an NR cell, has announced its location, and listens to the control messages sent over the shared channels of the cell. Additionally, the UE measures the radio signals sent by the neighboring cells so that the UE can select the cell with the best radio signal quality. No dedicated channel has been allocated for the UE.
- RRC_INACTIVE: No dedicated channel has been allocated for the UE, but the UE still has the RRC context, as used earlier in the RRC_CONNECTED state. The AMF has moved the UE from the RRC_CONNECTED to the RRC_INACTIVE state with an RRC RELEASE message that had a suspend indication. The suspended UE can get back to the RRC_CONNECTED state with a resume procedure by sending an RRC RESUME REQUEST to the AMF and taking the stored context back into use. The resume procedure is quicker and requires less signaling overhead compared to the full random access procedure which the UE has to use to get out of the RRC_IDLE state.
- RRC_CONNECTED: The UE is connected to the gNB via a dedicated channel. The UE reports radio signal quality estimates of different NR subcarriers to the gNB. The UE may continue measuring the radio signals of neighboring cells, but handover decisions are done by the gNB.

In addition to the tasks that the LTE RRC protocol has, the NR RRC protocol has the following extra responsibilities:

- Support bandwidth parts, SULs, and beam management.
- On-demand delivery of system information messages SIB2–SIB9 to the UE. On-demand delivery model reduces the overhead caused by continuous SIB broadcast. If there are any changes to essential system information after UE initial access, the gNB must indicate that to UE via DCI messages so that the UE can request getting new copies of changed SIB blocks. The UE may also need to request delivery of SIB blocks when camping into a new cell. NR uses area-based system information to limit the amount of extra system information messages sent due to cell reselections. The NR SIB1 block carries the basic pieces of information of the network, such as MCC, MNC, cell ID, and tracking area code. SIB1 has also a tag of system information area. If the tag does not change between two cells, the UE does not need to read system information again.
  - SIB blocks 2–5 prepare the UE for cell and RAT reselection and handovers. These blocks announce frequencies and radio access technologies of any neighboring NR or LTE cells for the UE to measure.
  - SIB blocks 6–8 carry public warnings for natural disasters, etc.
  - SIB9 provides GPS and UTC time for the UE.

The NR RRC message has the following structure:

- Message type.
- A set of other fields as defined for the specific type of RRC message.

8.1.5.5 SDAP Protocol

NR service data adaptation protocol (SDAP) is used for QoS flow management. NR SDAP is defined in 3GPP TS 37.324 [16].

SDAP has the following tasks:

- User data transfer
- Mapping of QoS flows and data radio bearers (DRB)
- Marking QoS flow identifier (QFI) to both downlink and uplink packets

A SDAP packet may or may not have a header. When used, the structure of a SDAP header is as follows:

- Downlink packets
  - RDI: indicates whether the QoS flow to DRB mapping rule should be updated
  - RQI: indicates whether NAS should be informed about the mapping rule update
  - QFI: identifies the QoS flow
- Uplink packets
  - D/C: data or control packet
  - R: reserved bit not used for any purpose
  - QFI: identifies the QoS flow

8.1.6 Signaling Protocols between UE and Core Network

8.1.6.1 NAS Protocols MM and SM

NR non-access stratum (NAS) protocols **mobility management (MM)** and **session management (SM)** are used for mobility, connection, and session management. NR NAS protocols are defined in 3GPP TS 24.501 [17].

These two NR NAS protocols have different endpoints in a 5G core system:

- NAS MM is run between the UE and AMF for registration, mobility, and connection management.
- NAS SM is run between the UE and SMF for session management.

Tasks of the MM protocol:

- Registering the subscriber for NR service and management of registration areas
- Authenticating the subscriber
- Tracking and RAN-based notification area updates
- Paging and service request procedures used to reach the UE, establish new connections, and reserving radio resources for the UE

Tasks of the SM protocol:

- Activation, authorization, modification, and deactivation of PDU sessions
- Management of packet data network connectivity and bearer resource allocation

NAS protocol messages start with the protocol discriminator field, whose value determines the other information elements within the message.

NAS protocol messages are carried between the UE and gNB over the RRC protocol and between the gNB and AMF over the NGAP protocol. While the SMF is the endpoint for NAS-SM messages, those messages are carried to the UE via the AMF. The HTTPS protocol is used to transport NAS-SM messages between the SMF and the AMF.

8.1.7 Protocols of 5G Radio and Core Networks

8.1.7.1 NGAP Protocol

NG application protocol (NGAP) is used by the AMF to manage gNB functionality and certain UE-related gNB procedures. NGAP protocol is defined in 3GPP TS 38.413 [18].

NGAP provides two different types of services:

- UE-associated services, which are related to one single UE
- Non-UE-associated services, which are related to the whole N2 interface between the AMF and gNB

NGAP is used for the following purposes:

- Paging control
- UE context and PDU session management
- UE location management and information retrieval
- UE handover control in intra- and inter-RAT cases
- Transporting NAS protocol messages between the AMF and gNB
- NG interface management
- AMF selection, management, reallocation, and load balancing

The NGAP message starts always with the Message Type field, the value of which determines the composition of other information elements within the message.

NGAP protocol messages are carried between the gNB and AMF over a reliable SCTP protocol connection.

8.1.7.2 XnAP Protocol

Xn application protocol (XnAP) is used between two gNBs to manage UE mobility and inter gNB handovers. NR XnAP is defined in 3GPP TS 38.423 [19].

NR processes for gNB discovery and Xn interface setup between two gNBs are similar to those processes used in LTE for mutual eNodeB discovery.

XnAP is used for the following purposes:

- Setup and management of Xn connections between two gNBs.
- UE mobility management, including context transfer, handovers, and paging. For handovers the protocol supports both handover control and data transfer between gNBs.
- Dual connectivity.
- gNB resource coordination for load and interference management

The XnAP message starts always with the Message Type field, the value of which determines the composition of other information elements within the message.

8.1.7.3 PFCP Protocol

Packet forwarding control protocol (PFCP) is used by the SMF to manage PDU sessions and QoS flows. The SMF uses PFCP when communicating with UPF functions. PFCP is defined in 3GPP TS 29.244 [20] and it effectively replaces GTP-C protocol used for LTE.

PFCP has the following tasks:

- Management of PDU session contexts within UPFs
- Management of packet detection, packet forwarding, and QoS enforcement rules, which UPFs apply to the PDU sessions
- Management of GTP-U tunnels for PDU sessions toward the UE
- UPF function load control and health monitoring

The PFCP message starts always with the Message Type field, value of which determines the composition of other information elements within the message. PCFP messages are transported over UDP/IP.

8.1.8 Radio Resource Management

NR radio resource management means gNB functionality for allocating NR transmission resources dynamically between the UEs to satisfy their communication needs. Various NR/LTE coexistence scenarios increase the complexity of NR radio resource management.

8.1.8.1 Initial Access and Registration to 5G

In addition to its globally unique NR cell ID, every gNB has a physical cell ID. NR has a total of 1008 physical cell ID values, which is twice the number of LTE physical cell IDs. The NR physical cell ID is encoded into the NR primary and secondary synchronization sequences PSS and SSS.

The NR synchronization signals and the PBCH channel are carried within a special **SS/PCBH block (SSB)** of the NR resource grid. The SSB consists of four consecutive OFDM symbols on 240 adjacent subcarriers (or 20 physical resource blocks). The SSB block contains also DM-RS reference signals to support the UE to demodulate OFDM symbols of the SSB. Both the primary and secondary synchronization sequences PSS and SSS consist of 127 OFDM symbols transmitted in subsequent SSB blocks in the subcarriers at the center of the block. The PSS occupies the first and the SSS the third OFDM symbol position of the block. The PBCH channel and DM-RS signals are located to OFDM symbol positions 2–4 as shown in Figure 8.12.

Figure 8.12 NR SSB block structure.

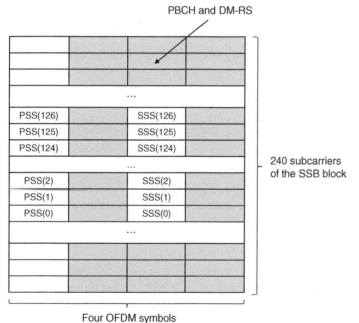

Four OFDM symbols

Location of the SSB block and its NR synchronization signals are not fixed in the subcarrier space to the middle of system bandwidth where LTE synchronization signals can be found. This is due to the variety of NR bands used as well as the NR design of bandwidth parts. When beam management is used, every NR beam carries its own SSB blocks so that the UE is able to do beam sweeping and select the best beam available for the UE. The beam selection is done with help of the CSI-RS beam management reference signals within the SSB block. The UE measures the power of CSI-RS signals as received from different beams and selects the beam with highest power.

Further on, the first subcarrier of the SSB block may not be aligned with the first subcarrier of any other NR resource block. The UE at first needs to search for NR synchronization signals to detect the SSB. After the UE has detected the SSB and read MIB from its PBCH channel, the SSB subcarrier offset of the MIB tells the difference between the first subcarrier of the SSB block versus the closest common NR resource block. In time domain, there are predefined patterns of how SSB blocks are distributed to NR frames. This is to support beam-specific SSBs in beamforming, as mentioned earlier. After selecting a detected SSB, the UE learns NR frame structure from the SSB index and the system frame number within the MIB. At this point, the UE has synchronized itself to the NR resource block structure and read the physical cell ID of the cell. With the help of SIB1 search parameters from the MIB, the UE is now equipped to find the SIB1 block from the PDSCH channel.

Due to the flexible locations of SSB blocks, the NR UE is not able to deduce the first subcarrier of the system bandwidth from the location of synchronization signals. Instead, NR SIB1 block provides the UE with the absolute frequency position of the reference resource block, starting from the lowest subcarrier of the band, referred as Point A in NR specifications. After reading the SIB1, the UE knows the exact location of NR resource grid of the band, both in time and frequency dimensions.

The NR UE registers to NR standalone service as shown in Figure 8.13:

1) After being powered on, the UE searches the radio bands allocated for NR and tries to find SSB blocks in the frequencies used for the NR SS/PBCH block raster. SSB block is found by detecting synchronization signals, from which the UE deduces the physical ID of the found cell. If the cell has multiple beams, the UE may find beam-specific SSB blocks and chooses the one with the strongest signal.

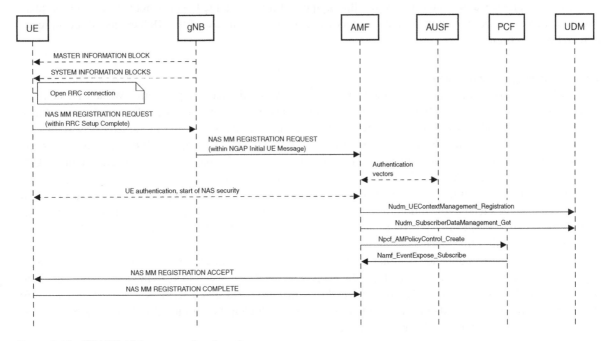

Figure 8.13 NR UE initial access and registration.

2) Next, the UE reads MASTER INFORMATION BLOCK (MIB) from the PBCH channel. The MIB provides the UE with NR system frame number, frequency offset between the SSB block and next common NR resource block, positions of DM-RS signals and resource configuration needed to find and decode the SIB1 block. SIB1 contains all the rest of the information the UE needs for initial access, like PLMN selection and RRC parameters common for all UEs, bandwidth parts, and their subcarrier spacing, configuration of downlink and normal/SULs.

3) The UE opens an RRC connection over the selected NR beam as described in Section 8.1.8.2. In the random access process, the gNB learns which downlink beam it shall use toward the UE. The UE sends an NAS MM REGISTRATION REQUEST message to the AMF and requests initial registration to a network slice identified with the NSSAI parameter. The UE provides its security capabilities to be used for authentication and identifies itself with the 5G-GUTI.

4) The gNB selects an AMF based on the NSSAI and the UE identity. The gNB forwards the NAS message to the selected AMF and provides the AMF also with the current UE location. The AMF selects an AUSF and authenticates the subscriber as described in Section 8.1.9.1.

5) The AMF selects a UDM and fetches the access, subscription, SMF selection, and SMF context data from it. The AMF selects a PCF and fetches the non-session policies applicable to the access and mobility management of the subscriber. In the end, the AMF sends an NAS MM REGISTRATION ACCEPT message to the UE.

For further details about the messages and service operations used in these procedures, please refer to *Online Appendix I.6.1*.

It is worth noting that while the LTE initial access procedure always provides the UE with an initial default access bearer, the 5G access procedure does not provide the UE with any bearers. 5G data bearers must be explicitly set up with session management procedures.

If the UE uses non-standalone mode, it may at first perform the LTE initial access procedure described in Chapter 7, Section 7.1.9.1. Thereafter, the LTE network may configure the UE with NR as a **secondary cell group (SCG)** used to provide additional capacity and throughput. At this point, the UE does not perform the complete NR initial access procedure, but instead just searches and decodes SSB blocks of the NR band for NR beam selection and synchronization.

8.1.8.2 Opening RRC Connection

The NR UE requests opening an RRC connection with the MAC random access procedure. While LTE supports random access only with contention resolution, NR supports two different types of random access procedures: **contention-based** and **contention-free access**. The contention-based access works in the same way as the LTE random access. The UE randomly picks a preamble and sends it to the PRACH channel. If two UEs used the same preamble simultaneously, the contention is detected with the MAC resolution procedure only after both the UEs try to use the same scheduled uplink resources. To support contention-free access, the gNB assigns a preamble and sends it to the UE at RRC connection opening or resume. The UE may use this preamble for its next random access, which would then be contention-free as the UE has a unique preamble that no other UE is allowed to use. Contention-based process is used for NR initial access, but after getting the RRC connection opened, the UE starts applying NR contention-free process.

The UE may initiate random access either on its own or after being paged by the network. The gNB performs paging according to the request received from the 5G core network. To reach a UE in the RRC_IDLE state, the AMF requests all gNBs within the registration area of the UE to send an RRC PAGING message to the UE. To identify the paged UE, the RRC PAGING message contains either the IMSI identifier of the subscriber or the 5G-S-TMSI identifier, which the AMF has earlier given to the UE.

Figure 8.14 Opening NR RRC connection.

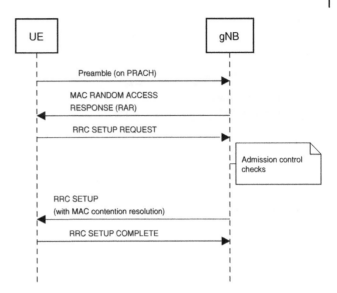

The NR UE can apply discontinuous reception (DRX) for the PDSCH channel. The UE uses DRX configuration as received from the AMF. With DRX, the UE regularly wakes up to listen to paging occasions of the NR frame, in which paging DCIs may be delivered over the PDCCH. The DCI message gives the UE with the downlink resources over which the RRC PAGING message will be delivered to the UE.

NR UE performs random access procedure as shown in Figure 8.14:

1) At first the UE checks if it is allowed to access the cell by comparing its access use case (such as SMS, emergency call, or MMTEL voice call) to the access barring information found from the MIB block. If access is allowed, the UE proceeds with the random access process, but otherwise it shall select another cell. The UE selects the uplink for random access and either picks a random preamble for contention-based access or uses the preamble received earlier from the gNB for contention-free access. The UE sends the preamble sequence over the PRACH channel.

2) The gNB returns the random access response to the UE over the PDSCH channel. This response contains an uplink grant for the UE to send an RRC SETUP REQUEST message. The RRC request contains either the 5G-S-TMSI or a random identity that the UE has chosen for contention resolution (unless contention-free access is used) and the establish cause about the purpose of the connection such as signaling, data, or voice call.

3) The gNB makes an admission control check and responds to the UE with the RRC SETUP message, telling the UE about the master cell group, contention-free random access preamble, and signaling radio bearer to be used for further signaling messages. In the contention-based access procedure, this message closes the contention resolution as it carries the random identity of the UE in the MAC frame.

For further details about the messages and service operations and options used in these procedures, please refer to *Online Appendix I.6.2*.

8.1.8.3 Resuming RRC Connection

The RRC connection can be resumed without the random access procedure when the connection has been suspended to the RRC_INACTIVE state. The UE may initiate the RRC resume process either on its own or after being paged by the network. To reach a suspended UE, the AMF requests RRC PAGING messages to be sent only by those gNBs that belong to the RAN-based notification area of the UE.

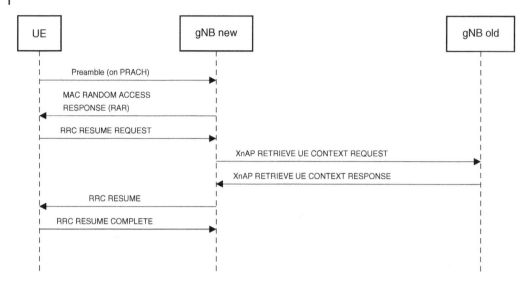

Figure 8.15 Resuming NR RRC connection.

RRC connection resume is performed as shown in Figure 8.15:

1) At first, the UE performs contention-based random access procedure as described in Section 8.1.8.2. This procedure is needed to get uplink resources for sending RRC messages. The UE sends an RRC RESUME REQUEST to the gNB over the grant acquired by the random access procedure. To send this message the UE must at first restore the signaling bearer and protocol layers under the RRC protocol to the state which was stored when the RRC connection was suspended.
2) If the serving gNB has been changed after the RRC connection was suspended, the gNB sends an XnAP RETRIEVE UE CONTEXT REQUEST to the old gNB, which returns the UE context back to new gNB. The new gNB sends an RRC RESUME message to the UE, describing the configuration of radio bearers and cell groups that the UE shall use for the RRC connection.

For further details about the messages and service operations used in these procedures, please refer to *Online Appendix I.6.3*.

8.1.8.4 Releasing the RRC Connection

The RRC release procedure can be used to get the UE to either RRC Idle or Inactive states, when no information transfer takes place. RRC release may also be used to redirect UE to another RAT. The message exchange used for RRC release is very simple. The gNB sends an RRC RELEASE message to the UE in order to request the UE to release the RRC connection. When suspending the connection, the gNB delivers the suspend configuration and information of RAN notification area in the RRC message. The UE will not send any response to the RRC RELEASE message.

8.1.8.5 Service Request

The UE initiates a service request to set up a secure connection with the AMF for sending signaling messages or to activate a PDU session for user data. The UE may trigger the service request procedure by itself or after being paged for mobile terminated SMS over NAS.

The UE-initiated service request process is done as shown in Figure 8.16: After opening the RRC connection, the UE sends a NAS MM SERVICE REQUEST message to the AMF. If the service request is invoked for sending

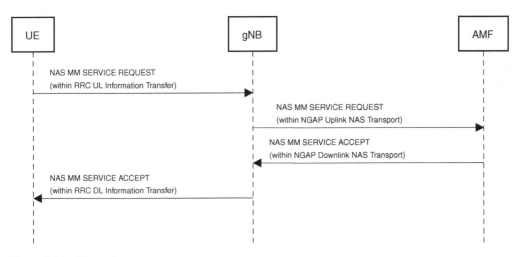

Figure 8.16 NR service request.

user data, the request contains a list of PDU sessions to be activated. The AMF will activate those sessions with procedures described in Section 8.1.10.3. Finally, the AMF will respond to the UE.

For further details about the messages, service operations and options used in these procedures, please refer to *Online Appendix I.6.4.*

8.1.8.6 Beam Recovery

At the initial access and beam selection, the UE reads system information of the cell. The system information defines conditions and thresholds for beam failure detection. While camping in a cell, the UE continues measuring the CSI-RS signals used for beam management. If the beam failure criteria are fulfilled long enough or too many times, the UE deems the currently used downlink beam to have failed. The UE tries to access another beam with the contention-free random access procedure, using the preamble allocated by the gNB to the UE for beam recovery requests.

Beam failure recovery procedures are specified in 3GPP TS 38.321 [12].

8.1.9 Security Management

8.1.9.1 Authentication and NAS Security Mode

5G uses AKA authentication and key derivation processes evolved from the LTE AKA as described in Chapter 7, Section 7.1.10.1. The 5G network may authenticate the subscriber and change the encryption keys at initial access, tracking area update, or when opening PDU sessions. One major difference in comparison with the LTE solution is that the 5G authentication is always done by the home network. In LTE, the authentication vectors are given to the visited network, which performs the authentication. 5G security architecture is described in 3GPP TS 33.501 [21].

After receiving the NAS MM REGISTRATION REQUEST message from the UE, the AMF authenticates the subscriber and switches NAS security mode ON as shown in Figure 8.17:

1) If the AMF does not have the concealed SUCI identifier of the subscriber (from its own context or from the registration request from the UE), the AMF sends a NAS MM IDENTITY REQUEST message to the UE, requesting the UE to deliver the SUCI. The UE provides the SUCI within its response.

2) The AMF selects an AUSF from which to fetch the authentication vectors for the subscriber identified by the SUCI. The AUSF itself retrieves these vectors from the UDM and returns them to the AMF. Thereafter the

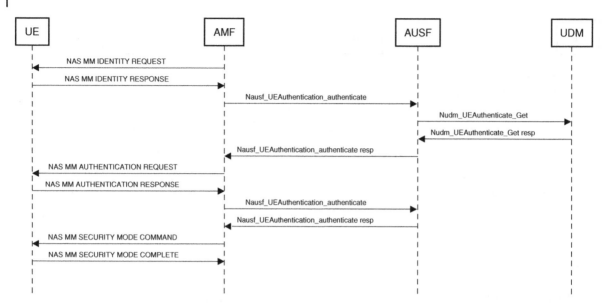

Figure 8.17 Authentication and NAS security in 5G.

AMF sends a NAS MM AUTHENTICATION REQUEST to the UE. This request contains AUTN and RAND parameters of AKA. The UE authenticates the network, calculates the AKA RES and returns authentication response to the AMF, which forwards the result to the AUSF for checking. If there is a match, authentication is completed successfully and the AUSF provides the AMF with the the non-encrypted SUPI identifier of the user.

3) To start NAS encryption the AMF sends a NAS MM SECURITY MODE COMMAND message to the UE. This message tells the UE about security algorithms to be used and parameters with which to derive the security keys. The UE starts encryption and responds to the AMF.

For further details about the messages, service operations, and security parameters used in these procedures, please refer to *Online Appendix I.6.5*.

5G supports encryption and integrity protection also with the RRC and PDCP protocols. Those AS security mechanisms are enabled between the UE and gNB with RRC SECURITY MODE procedures, which are similar to those used in LTE.

8.1.10 Packet Data Connections

8.1.10.1 Quality of Service Model

NR QoS model is network centric like the LTE model. The QoS parameters have however been redefined for 5G. Instead of the LTE QoS class identifier (QCI), 5G specifications use **5G QoS identifier (5QI)**. These mechanisms are however quite similar. Like QCI, also 5QI has a number of values which indicate the QoS requirements of the data flow. Some of the defined 5QI values support guaranteed bitrate and others do not. Every 5QI value comes with an acceptable packet error rate and delay budget. Just like LTE QCI values, 5QI values also have been assigned to different priority levels. The main difference between 5QI and QCI comes with the types of different services specified for 5G. A total of 27 different 5QI values have been defined for in TS 23.501 [7] Rel-16.

In the perspective of Quality of Service, the 5G user data flows are called **QoS flows**. Each QoS flow is identified by the **QoS flow identifier (QFI)**, which the SMF allocates for the flow when a PDU session is set up for the flow. The PCF element manages QoS properties of user data flows. At the setup phase of the QoS flow, the 5QI value and related QoS parameters are used to describe the QoS requirements of the flow. Those parameters are passed to the PCF while the PDU session is being established. After checking the parameters, the PCF gives the granted QoS properties of the flow as **PCC rules** to the SMF [22]. Thereafter, the SMF proceeds with providing the gNB with a QoS profile, the UE with QoS rules, and the UPF(s) with packet detection and QoS enforcement rules to be applied to the flow. The user data packets within the flow contain QFI identifier in the extension header of the GTP-U packet, which encapsulates the user data packet within the tunnel. Every network element that processes GTP-U packets can identify the QoS flow and its rules based on the QFI value within the packet.

The QoS profile given to the gNB contains the following QoS parameters:

- The 5QI value of the flow.
- Allocation and retention priority (ARP), which is the priority level according to which the admission control works for connections. The ARP priority level is used when the load in the network is high enough to limit creating new connections or selecting user data packets for processing, according to their QoS requirements.
- For GBR flows, the maximum bit rate (MBR), guaranteed bit rate (GBR), and maximum packet loss rate of the flow.

For the downlink packets, the UPF performs packet QFI marking and QoS enforcement based on QoS enforcement rules received from the SMF. Mapping of packets to different QoS flows and related GTP tunnels is done based in the packet detection rules received from the SMF. Packet detection relies on IP address, protocols, and port numbers of the user data packets. The UE performs classification and marking of uplink user data packets based on the QoS rules received from the SMF for the QoS flows. A QoS rule defines a set of packet filters and other QoS parameters to be applied for uplink packets. The UE also marks the packets with the QFI identifier of the flow, as given in the QoS rule.

As learned in Chapter 7, Section 7.1.11.2, EPC bearer is the LTE concept tying together a user data flow with specific QoS parameters, related packet filtering rules in the LTE network elements, and a GTP-U tunnel between them. In LTE every data flow is carried within its own tunnel specific to the EPS bearer. In 5G the tunneling structure is different:

- There is one single GTP-U tunnel between the gNB and UPF(s) carrying all the user data flows within a PDU session.
- The QoS flows of a PDU session are carried between the gNB and UE by data radio bearers (DRB). Each DRB may carry either one single or multiple QoS flows having similar QoS requirements. See Figure 8.7 in Section 8.1.3.4.

After the PDU session has been established, the tunnel setup between the gNB and UPF(s) needs no changes for additional user data flows even when the flows had different QoS requirements. It is sufficient that the SMF updates the QoS profiles and rules to those network elements to enforce correct handling of packets transported over the single GTP-U tunnel. Different DSCP markings can still be used within the same tunnel to provide differentiated QoS treatment for flows belonging to different 5QIs.

8.1.10.2 Creating 5G PDU Session

The UE may any time request creation of a new PDU session to an external data network identified by the **data network name (DNN)**. 5G system elements take care of the following activities in the PDU session establishment procedure depicted in Figure 8.18.

1) The UE sends a NAS SM PDU SESSION ESTABLISHMENT REQUEST to the AMF for a new PDU session. In this message, the UE describes the requested data flows and the network slice.
2) The AMF checks if the subscriber is entitled to use the requested slice. The AMF selects an SMF based on the DNN and the accepted NSSAI slice identifier from the UE. The AMF forwards the PDU session establishment request to the selected SMF.

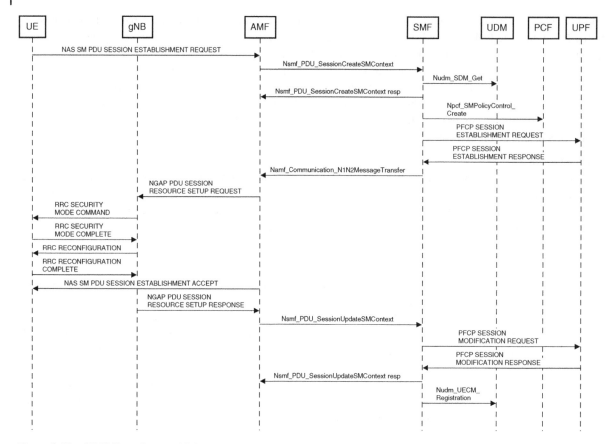

Figure 8.18 5G PDU session establishment.

3) The SMF authorizes the PDU session establishment request based on subscriber-specific authorizations and policies as retrieved from the UDM. The SMF selects a PCF to get dynamic QoS control and charging policies for the PDU session. The SMF creates a session policy association with the selected PCF, which returns the PCC rules derived from the subscriber data records and the session request received from the SMF. From these PCC rules, the SMF generates the QoS flow configuration for the PDU session. The SMF selects UPF(s) for carrying the authorized QoS flows. Thereafter, the SMF sets up GTP tunnels, QoS enforcement, and packet forwarding rules to the chosen UPF(s). The SMF allocates IPv4 or IPv6 addresses for the UE and with help of the AMF configures QoS rules and GTP tunnels for the gNB, which serves the UE.
4) The gNB makes an admission control decision whether the PDU session and QoS flows suggested by the SMF can be accepted, either fully or partially. When the session is accepted, the gNB forwards the session acceptance message from the AMF to the UE. The gNB configures and starts the encryption and integrity management procedures with the UE, for both signaling and user data. Finally, the gNB sets up the new radio bearers toward the UE for the PDU session, using RRC procedures for these tasks. Thereafter, the gNB informs the AMF and the UE about the QoS flows created for the PDU session. The AMF forwards the information to the SMF.
5) After getting the response from the gNB about accepted QoS flows, the SMF reconfigures UPF(s) accordingly and provides an IPv6 router advertisement to the UE, if the UE was granted with IPv6 access. Finally, the SMF registers to the UDM just in case the subscriber policies would be changed and the session should be updated accordingly.

6) UPF(s) start forwarding the data between the UE and the packet data network, based on the forwarding and QoS rules received from the SMF and the GTP tunnels created between UPF(s) and the gNB.

For further details about the messages and service operations used in these procedures, please refer to *Online Appendix I.6.6.*

8.1.10.3 Activating Existing PDU Sessions

Activation of an existing PDU session is done in the following cases:

- The UE changes its registration area and performs a new registration to 5G. As the change of the registration area may cause the serving AMF to be changed, the UE indicates the existing PDU sessions in the registration request. The new AMF contacts the SMF to activate those PDU sessions.
- The UE needs to send user data over the suspended PDU session and sends a service request as described in Section 8.1.8.5.

The new AMF activates PDU sessions after a registration area change as shown in Figure 8.19:

1) The AMF calls Nsmf_PDUSession_UpdateSMContext service operation of the SMF to activate those PDU sessions that the UE has listed in its NAS MM REGISTRATION REQUEST message. The SMF instance contacted is the one associated with those PDU sessions. The SMF checks which UPF(s) are currently assigned for the PDU sessions. Thereafter, the SMF selects the new set of UPF(s) for the PDU sessions, based on the new location of the UE as received from the AMF. In case the intermediate UPF was changed, the SMF reconfigures the GTP tunnel used for the PDU session, using PFCP session establishment and modification messages toward the UPFs. After reconfiguring the tunnels, the SMF responds to the AMF.

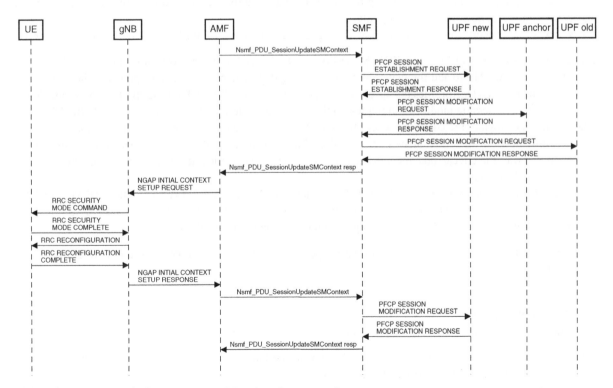

Figure 8.19 Activation of PDU sessions in 5G registration area update.

2) The AMF sends an NGAP INTIAL CONTEXT SETUP REQUEST message to the gNB. This message provides the gNB with the IP address of the UE, tunnel endpoint IDs, and other parameters needed by the gNB to set up the PDU session and its QoS flows to the UE. The gNB checks the request and makes admission control decision for the QoS flows to be activated. The gNB uses RRC procedures to start PDCP integrity protection and encryption, set up new radio bearers, and configure the UE to measure the current and secondary cells. NAS MM registration accept and complete messages are piggybacked to the RRC reconfiguration messages. In the end, the gNB responds to the AMF, providing the NAS MM registration complete message from the UE along with the response.

3) The AMF calls service operation Nsmf_PDUSession_UpdateSMContext once again to update the SMF about the downlink TEID and other parameters provided by the UE. The SMF sends PFCP session modification request(s) to UPF(s) to update the PDU session with the new tunnel endpoint ID and for any QoS flows rejected by the gNB.

For further details about the messages and service operations used in these procedures and the steps of tunnel modification, please refer to *Online Appendix I.6.7.*

8.1.10.4 PDU Session Modification

The PDU session may be modified to add new service data flows with specific QoS requirements or to drop some of the existing flows. Session modification may be initiated either by the UE, gNB, or SMF, depending on the specific scenario. For instance, if there is an ongoing IMS voice call and the remote UE wants to add a video component to it, the IMS core network brings up the need for additional data flow to the SMF, which starts driving the modification process for the PDU session. An application function (AF) can also request a new service data flow to be added or QoS parameters of an existing flow to be changed via the NEF or the PCF. In this case, the PCF triggers the change to the SMF by a PCC update.

Let's now walk through another scenario shown in Figure 8.20 where the UE initiates PDU session modification:

1) The UE sends a NAS SM PDU SESSION MODIFICATION REQUEST to the AMF for a new data flow. The AMF contacts the SMF to add the new flow to the existing PDU session. The SMF responds to the AMF and provides the updated QoS profiles for the gNB and updated QoS rules for the UE to support the new QoS flow. The AMF sends a session modification request to the gNB and session modification command to the UE. The latter message provides the UE with packet filter sets (PFS) for the new set of QoS flows. The session modification command from the AMF arrives to the UE within an RRC RECONFIGURATION message from the gNB. In this way, the UE can take the new packet filter sets into use while setting up the new radio bearers for the additional QoS flows.

2) After getting the response from the UE, the AMF contacts the SMF once again to provide the SMF with the UE location and get it to update the PDU session setup. Consequently, the SMF provides new packet detection and QoS enforcement rules to UPF(s), completing the configuration of new QoS flows.

For further details about the messages and service operations used in these procedures, please refer to *Online Appendix I.6.8.*

8.1.10.5 User Data Transport

User data is transported over the PDU session and its QoS flows as follows:

1) To transport user data over PDU session, the UE may at first have to invoke the service request procedure to activate the PDU session as described in Section 8.1.8.5.

2) When sending a user data IP packet to the network, the UE compares its destination IP address, protocol (TCP/UDP), and port numbers to the QoS rules received from the SMF when the QoS flows were set up for the PDU

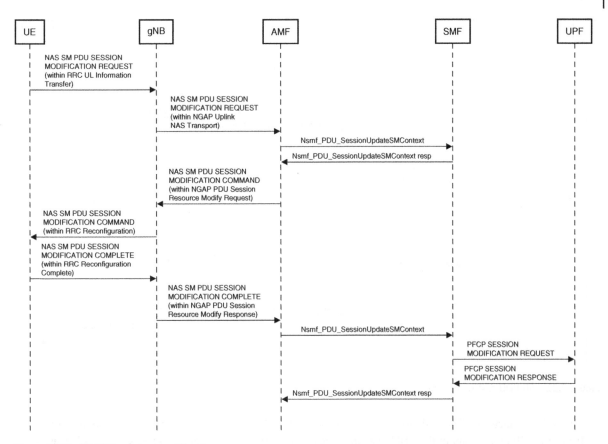

Figure 8.20 5G PDU session modification.

session. The UE uses QoS rule to determine the data radio bearer to which the packet belongs and to mark the packet with the QFI identifier of the QoS flow.

3) The UE sends the packet to the gNB over the data radio bearers of the QoS flow. The packet is carried by the PDCP protocol.

4) The gNB checks the QFI identifier of the received packet and uses it to find the matching QoS profile received from the SMF. If the data packet complies with the found QoS profile, the gNB forwards the packet to the UPF over the GTP-U tunnel. The gNB can set a QFI-specific value to the DSCP field of the inner IP packet in order to provide proper QoS treatment for the packet within the IP network that carries packets within the GTP tunnel.

5) The UPF checks the packet against the QoS enforcement rules of the flow identified by QFI of the packet. The UPF routes the packet to the destination packet data network.

8.1.10.6 Releasing PDU Session

The UE releases an existing PDU session as shown in Figure 8.21:

1) To release its PDU session, the UE sends a NAS PDU SESSION RELEASE REQUEST to the AMF. The AMF forwards the PDU session release request from the UE to the SMF, which releases the IP addresses allocated for the UE and tears down the GTP tunnel from those UPF(s) that have supported the PDU session being released. The SMF thereafter responds to the AMF, providing an updated set of QoS rules to be forwarded to the gNB and UE.

2) The AMF sends a session release command to the gNB, which forwards the session release command to the UE within an RRC message used to modify radio bearers. After releasing the data radio bearers used for the PDU session, the UE replies to the gNB and AMF.

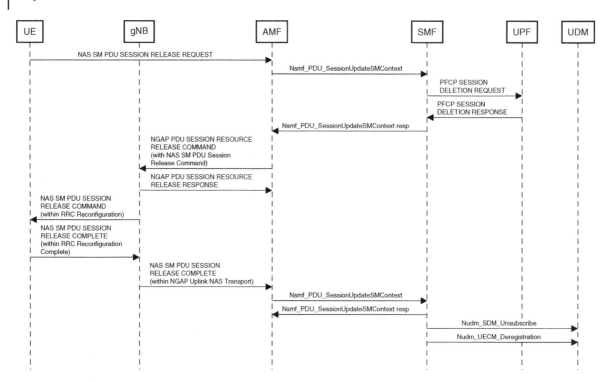

Figure 8.21 5G PDU session release.

3) The AMF contacts the SMF again to give an update about the UE location and to forward the responses received from the gNB and UE about the PDU session release. The SMF may now terminate its session policy association with the PCF and drop the session-related subscription and registration from the UDM.

For further details about the messages and service operations used in these procedures, please refer to *Online Appendix I.6.9.*

8.1.11 Mobility Management

In NR, the mobility management is the responsibility of Access and Mobility Function (AMF), which maintains the knowledge of the UE location. Location tracking accuracy depends on the RRC state of the UE as follows:

- RRC_IDLE: The AMF knows the location of the UE in the accuracy of an NR registration area which was assigned to the UE when the UE registered to the AMF. The UE takes care of the cell selection and performs a registration procedure when moving out of its current UE specific registration area.
- RRC_INACTIVE: The AMF knows the location of the UE in the accuracy of RAN-based notification area which was assigned to the UE when its RRC connection was suspended. The UE performs cell reselections such as in idle state but informs the gNB when it changes its RAN notification area using the RNA update procedure.
- RRC_CONNECTED: The AMF knows the location of the UE in the accuracy of an NR cell. The change of the cell is done with the gNB controlled handover procedure.

8.1.11.1 NR Registration in RRC Idle State

Cell selection and reselection processes of the NR UE are quite similar to LTE. One major difference is the beam measurement and selection, as the NR cell may consist of multiple beams.

In RRC_IDLE state, the UE scans NR radio bands to find a suitable cell according to the PLMN selection priorities available from the USIM application of the UICC card. After finding an acceptable cell and selecting its beam, the UE camps to the cell and registers its initial location to the AMF. If the quality of the selected beam drops below a given threshold, the UE may try out beam recovery process. If that does not help, the UE may scan for other cells to perform cell reselection.

Scanning is not limited to the currently used network and radio technology. The UE may scan over a wide range of frequencies and use various RAT specific techniques to find cells of different PLMNs and radio access technologies. To guide UEs in this process, the network assigns priorities between different radio access technologies supported by the network as well as cell reselection threshold values for the measured signal power. The network broadcasts these parameters as part of its system information. The 5G system provides cell ranking to support NR UEs for inter-RAT cell reselection decisions and making choices over available beams and SUL channels of different RATs. From the system information messages, the UEs learn also the RRC configuration to be used for measuring neighboring cells of various RATs.

Each LTE and NR cell fixedly belongs to one single tracking area. The LTE UE performs a tracking area update procedure always when moving between two tracking areas. The location update process of 5G has been redesigned and the concept of UE **registration area** was introduced on top of the tracking areas. The registration area is a set of tracking areas dynamically defined for the UE. The registration area may consist of only one or multiple tracking areas, mostly depending on the mobility pattern of the UE. In initial registration when the network does not know the mobility pattern of the UE, the registration area matches with the current tracking area. If the network later on learns that the UE moves with high speed, the network may assign the UE with a registration area, which consists of multiple tracking areas toward the trajectory of the movement.

If the NR UE moves between two tracking areas within its current registration area, it does not perform a location update. When the UE eventually moves to a tracking area outside of its registration area, the UE renews its 5G registration due to UE mobility. Within this registration process, the AMF provides the UE with a new registration area. In this way, NR uses registration areas to reduce the amount of location update messages from UEs with high mobility. The price is paid with a higher number of cells where the UE must be paged when the registration area is large. The balance is achieved by dynamic rather than static determination of registration areas, so that only UEs moving fast are given large registration areas.

Registration due to NR UE mobility is done as shown in Figure 8.22:

1) When camping to a new NR cell, the UE learns the tracking area code from the SIB1 block broadcast in the cell. If the new tracking area does not belong to the current registration area (which was given to the UE in its previous registration), the UE shall register itself to a new registration area. The UE opens or resumes an RRC connection as described in Section 8.1.8. The UE sends a NAS MM REGISTRATION REQUEST message to the AMF, indicating the registration type as registration due to UE mobility and listing the existing PDU sessions that the UE wants to activate. The UE identifies itself and the AMF used earlier and provides its security capabilities for authentication.

2) The gNB either selects a new AMF based on the NSSAI parameter or uses the old AMF. The gNB tells the UE location and forwards the NAS message to the selected AMF. The new AMF fetches the UE context from the old AMF, selects thereafter an AUSF and authenticates the subscriber as described in Section 8.1.9.1. The new AFM informs the old AMF that it has taken over the registration.

The new AMF selects a UDM and registers itself for the UE context to receive any changes to it. Since the new AMF has already gotten the subscription records from the old AMF within the context transfer, there is no need to retrieve those records from the UDM. The UDM notifies the old AMF that it no longer provides further updates about the UE to it. The new AMF may update the PCF policy association, depending on the operator local policy. After being contacted by the new AMF, the PCF subscribes the UE location and status notifications from the AMF.

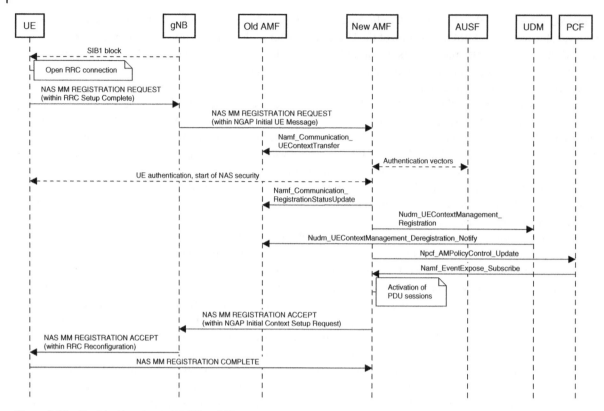

Figure 8.22 Registration due to NR UE mobility.

3) The new AMF activates the PDU sessions referred by the registration request as described in Section 8.1.10.3 and responds to the UE to have accepted the registration. The response message provides the UE with a set of tracking areas for its new registration area.

For further details about the messages and service operations used in these procedures, please refer to *Online Appendix I.6.10.*

8.1.11.2 RNA Update in RRC Inactive State
The NR UE in the RRC_INACTIVE state tracks its **RAN-based notification area (RNA)** as a set of NR cells allocated within the RRC RELEASE message, which suspended the RRC connection. The UE performs the RNA update procedure both periodically and when moving to a cell which is outside of the RNA but still within the current registration area of the UE.

The NR UE performs the RNA update procedure as shown in Figure 8.23:

1) The UE in the RRC_INACTIVE state must at first resume its RRC connection as described in Section 8.1.8.3. Within the RRC RESUME REQUEST message, the UE states the cause for resuming RRC connection as RNA update. The UE also provides the I-RNTI identifier allocated to the UE when the RRC connection was suspended.
2) If the gNB was changed since the previous RNA update, the new gNB that received the RRC RESUME REQUEST finds out the old gNB identity from the I-RNTI identifier. The new gNB sends an XnAP RETRIEVE UE CONTEXT REQUEST to the old gNB.
3) The old gNB may decide to keep the UE context or move it to the new gNB.

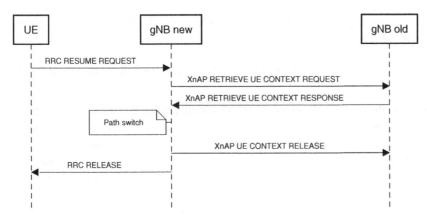

Figure 8.23 NR RAN notification area update.

- If the old gNB decides to keep the context, it sends an XnAP RETRIEVE UE CONTEXT FAILURE message to the new gNB. This message also carries an RRC RELEASE request with a suspend indication, which the new gNB shall forward to the UE. The old gNB records the UE location to the context.
- The old gNB returns the UE context to the new gNB within an XnAP RETRIEVE UE CONTEXT RESPONSE message. This causes the new gNB to send an NGAP PATH SWITCH REQUEST message to the AMF. The processing of this message is described in the Section 8.1.11.3 for the handover case. After the path switch has been completed, the new gNB sends an XnAP UE CONTEXT RELEASE message to the old gNB.
4) The new gNB sends an RRC RELEASE request to the UE to move it back to the RRC_INACTIVE state and assign the UE with a new RAN-based notification area.

8.1.11.3 Handover in RRC Connected State

In the connected mode, the NR UE may either use resources of a standalone NR cell or resources of non-standalone NR/LTE cells. In the former case, the UE may use NR normal channel for uplink or NR SUL channel, while in the latter case the SUL channel is provided by the LTE cell. Due to the RAT coexistence scenarios, there are various handover types for NR UE:

- Handover within a single RAT and system. In this case, the NR UE just changes its NR cell under a single NR access network and 5G core network.
- Handover within a single RAT but two systems. In this case, the NR UE changes its LTE cell so that one of the cells is connected to the 5G core while the other is connected to the EPC core.
- Handover between two RATs but one system. In this case, the NR UE makes inter-RAT handover between NR and LTE, both of the cells, however, are connected to a single 5G core network.
- Handover between two RATs and systems. This means that the NR UE makes a handover between an NR cell under the 5G core and an LTE cell under the EPC core network.

The specifications cover inter-RAT handovers only between 5G NR and LTE. No support has been specified for handovers from 5G NR RAN toward UTRAN or GERAN networks. This simplifies 5G design and emphasizes the need for any 5G deployments to be co-located with LTE cells [4]. The only viable alternative would be to use low enough frequencies for 5G to eliminate the need for any inter-RAT handovers.

NR handovers are done under control of the gNB. Handover decision is done based on the CSI reports from the UE. The gNB may decide to perform the handover due to radio signal degradation reported by the UE or excessive load within the cell. Handover target cell is chosen based on the UE measurements done according to the configuration received from the gNB. The target may be another NR cell or cell of another RAT. The UE does the measurements based on the measurement configuration received from the gNB over RRC protocol, in a way similar to LTE UE as described in Chapter 7, Section 7.1.12.2. The exception is that no measurements are performed toward GERAN or UTRAN cells.

The following scenario shown in Figure 8.24 describes the handover between two NR cells so that the AMF is not changed:

1) After making the handover decision, the gNB sends an XnAP HANDOVER REQUEST message to the new candidate target gNB. The XnAP message describes security keys, packet data connections, and radio access bearers of the UE subject to the handover. The target gNB evaluates if it is able to support the packet data flow of the UE. The gNB checks its current load against the QoS requirements of the new flows. If the gNB accepts the handover, it sets up a new GTP tunnel toward the intermediate UPF. The tunnel will be used to forward UPF any uplink user data that is received from the UE right after the handover. The target gNB responds to the old gNB. In its response, it provides an RRC reconfiguration message, which the old gNB forwards to the UE. The old gNB sends a status transfer message to the target gNB to start forwarding the user data flows via the temporary GTP tunnels to the new gNB.
2) The UE synchronizes itself to the target cell, performs random access with the preamble received from the target gNB, and requests the target gNB to open an RRC connection, as described in Section 8.1.8.2. After the RRC connection has been opened, the new gNB starts sending the buffered downlink user data packets to the UE while the UE does the same for uplink packets.
3) The new gNB sends a path switch request to the AMF to inform 5G core about switching the PDU sessions and their QoS flows from the old gNB to the new gNB. The AMF contacts the SMF, which manages PDU sessions

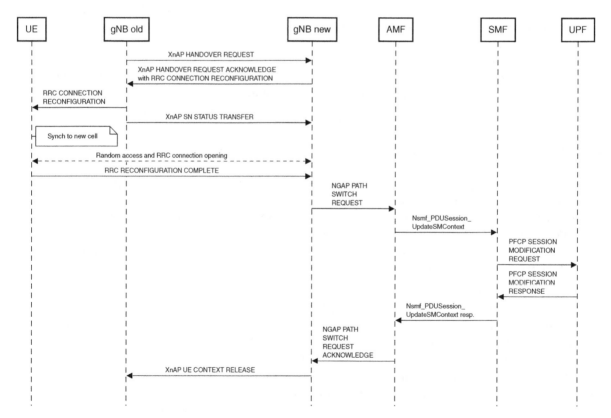

Figure 8.24 Inter-cell handover in NR.

for the UE. The SMF now determines if it has to allocate a new intermediate UPF for the new path. The SMF reconfigures GTP tunnels, packet forwarding, and QoS rules to the UPF(s), which support the new user data path toward the new gNB. UPF(s) start forwarding downlink traffic to the target gNB. The anchor UPF also sends end marker packets over the old tunnel to the old gNB, in order to inform it that the old tunnel is no longer used. The old gNB forwards end marker packets to the target gNB, so that they both know that the temporary tunnel between gNBs is no longer needed.

4) After getting an acknowledgment from the AMF, the new gNB sends a context release message to the old gNB to inform it that the handover has been completed. Now the old gNB purges the UE context and all resources reserved for the UE.

For further details about the messages and service operations used in these procedures as well as the alternatives the SMF has for the UPF reconfiguration, please refer to *Online Appendix I.6.11*.

The handover scenario as described is the hard handover case where the UE at first cuts its connection to the old gNB before it regains the connection with the new gNB. To minimize the disturbance caused by handover, the NR UE and gNBs may also apply **dual active protocol stack (DAPS) handover** for some of the data radio bearers. In DAPS handover, the UE and old gNB maintain the radio connection until the new gNB confirms the handover to have been successfully completed. Compared to the sequence as described, the following differences apply to the DAPS handover case:

- At step 5, the old gNB sends an XnAP EARLY STATUS TRANSFER message to the target gNB. In this case, the old gNB continues running PDCP and assigns the sequence numbers to every new PDCP message that the old gNB still sends toward the UE and additionally forwards to the target gNB.
- After the target gNB has received the RRC RECONFIGURATION COMPLETE message from the UE at step 8, the target gNB sends an XnAP HANDOVER SUCCESS message to the old gNB. This makes the old gNB to stop its transmissions to the UE. The old gNB now sends an XnAP SN STATUS TRANSFER message to the target gNB, causing the responsibility of allocating the PDCP sequence number to be moved to the target gNB.
- After switching its radio access bearers to the new cell, the UE may receive the same PDCP messages from the new gNB, which it has already received from the old gNB. The UE gracefully detects and ignores those duplicated messages based on PDCP message sequence numbers.

To support inter-RAT handovers between 5G NR and LTE, the 5GC and EPS elements must be able to coordinate their activities. The AMF and MME communicate with the N26 interface, but the approach for other functions is different. The SMF and UPF must be combined with P-GW functions (C and U, for control and user planes) and the UDM combined with the HSS. Main steps of the inter-RAT handover process are as follows:

- The AMF requests an LTE MME to prepare the handover. The MME instructs an S-GW and eNodeB to create contexts and bearers as necessary for the UE.
- Indirect data tunnel is created between the eNodeB and gNB to forward downlink data to the eNodeB during the handover process.
- After the MME has completed the preparation, the AMF triggers handover execution by sending a handover command to the UE via the gNB, and indirect tunnel is taken into use.
- When the UE has successfully accessed the LTE cell and GTP tunnels set up to the P-GW, the indirect data tunnel is torn down and all the resources are released on 5G side.

8.1.12 Voice and Message Communications

Voice over NR (VoNR) described in Chapter 9, Section 9.5 is the native 5G operator voice solution. The VoNR service may not be initially be available in the early 5G NR networks. The 5G UE may use EPS fallback procedures for voice calls when the network does not support VoNR. In EPS fallback, the UE is redirected to LTE service to use the **Voice over LTE (VoLTE)** service described in Chapter 9, Section 9.4.

The 5G UE may send and receive short messages with two methods:

- **SMS over IP** service, described in Chapter 9, Section 9.4.2.5
- **SMS over NAS** service as described next

SMS over NAS is similar to SMS over SG service of LTE. To support SMS service, the 5G UE may register itself to SMS over NAS service in its 5G registration procedure by including an "SMS supported" indication to its registration request. This indication makes the AMF to interact with the SMSF for the SMS support after at first completing its actions with the UDM.

- The AMF may retrieve the SMS Subscription and UE Context in SMSF records from the UDM using the Nudm_SDM_Get service operation
- If the subscriber is entitled to use the SMS service, the AMF calls the Nsmsf_SMService_Activate service operation of the SMSF.
- The SMSF retrieves SMS management records from the UDM and subscribes notifications about any upcoming changes to these records.

After registering to the SMS over NAS service the UE may receive mobile terminated short messages over NAS as shown in Figure 8.25:

1) The SMS-GMSC provides an MT SMS to the SMSF, which contacts the AMF to initiate the paging process for the UE. In the paging message, the reason is identified as MT SMS to the UE. The paged UE opens an RRC connection and performs the service request procedure to set up a connection with the AMF for uplink signaling, as described in Section 8.1.8.5.
2) After the uplink signaling communication path is opened, the AMF responds to the SMSF, which consequently provides the SMS to the AMF. The SMS is thereafter forwarded to the UE.

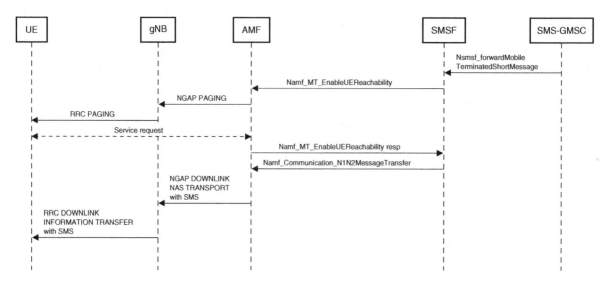

Figure 8.25 Receiving MT SMS over 5G NAS.

Omitting the details of interaction between the AMF, gNB, and UE for the lower- layer NGAP and RRC protocols, the MO SMS procedure is as follows:

1) The UE opens an RRC connection and sends a service request about the SMS to the AMF. When the RRC connection is ready, the UE sends the short message to the AMF within an EMM UL NAS TRANSPORT message.
2) The AMF forwards the SMS to the SMSF, which forwards the short message onwards to the SMS-IWMSC. The acknowledgment to the UE is sent backwards along to the same chain of network functions that processed the MO SMS itself.

For further details about the messages and service operations used in these procedures, please refer to *Online Appendix I.6.12*.

8.2 Questions

1 What are the three different usage scenarios behind the 5G system design?

2 What is meant by network virtualization?

3 What is the purpose of network slicing and how is it related to 5G NR bandwidth parts?

4 Which 5G core network functions support the same functions as the HSS of EPC?

5 What is SUCI and its purpose?

6 How does 5G TDD UE learn the allocation of 5G resource elements between uplink and downlink?

7 What are the purposes of NR beam sweeping and beam recovery processes?

8 Why does NR design introduce the concept of supplementary uplink?

9 What do the terms EN-DC network configuration and split bearer mean?

10 What are the main differences between LTE and 5G NR resource grids?

References

1 ITU-R Recommendation M. 2083 IMT Vision - "Framework and overall objectives of the future development of IMT for 2020 and beyond."
2 Lei, W., Soong, A.C.K., Jianghua, L. et al. (2020). *5G System Design – An End to End Perspective*. Cham: Springer.
3 3GPP TS 38.300 NR NR and NG-RAN Overall description; Stage-2.
4 Sauter, M. (2021). *From GSM to LTE-Advanced pro and 5G : An Introduction to Mobile Networks and Mobile Broadband*. West Sussex: Wiley.

5 3GPP TS 38.104 NR Base station (BS) radio transmission and reception.

6 3GPP TS 38.401 NG-RAN; Architecture description.

7 3GPP TS 23.501 System architecture for the 5G system (5GS).

8 3GPP TS 23.502 Procedures for the 5G system (5GS).

9 3GPP TS 38.410 NG-RAN NG general aspects and principles.

10 3GPP TS 23.003 Numbering, addressing and identification.

11 3GPP TS 38.211 NR; Physical channels and modulation.

12 3GPP TS 38.321 NR Medium access control (MAC) protocol specification.

13 "3GPP TS 38.322 NR Radio link control (RLC) protocol specification.

14 3GPP TS 38.323 NR Packet data convergence protocol (PDCP) specification.

15 3GPP TS 38.331 NR Radio resource control (RRC); protocol specification.

16 3GPP TS 37.324 Evolved universal terrestrial radio access (E-UTRA) and NR; Service data adaptation protocol (SDAP) specification.

17 3GPP TS 24.501 Non-access-stratum (NAS) protocol for 5G system (5GS); Stage 3.

18 3GPP TS 38.413 NG-RAN NG application protocol (NGAP).

19 3GPP TS 38.423 NG-RAN; Xn application protocol (XnAP).

20 3GPP TS 29.244 Interface between the control plane and the user plane nodes.

21 3GPP TS 33.501 Security architecture and procedures for 5G system.

22 3GPP TS 23.503 Policy and charging control framework for the 5G system (5GS); Stage 2.

Part IV

IP Multimedia Systems

9

Convergence

Convergence means the way of unifying earlier separate voice and data traffic. In the modern networks, voice is transported essentially as one specific type of data, having its well-defined Quality of Service requirements. Voice is an element of real-time conversation, so the latency requirements are strict. With the modern voice codecs, the needed bitrate is fairly constant and low. Voice traffic can be considered more robust against errors compared to many other types of data. IP multimedia systems are designed to support these requirements.

9.1 Voice over Internet Protocol (VoIP) and IP Multimedia

Since the share of data traffic of the total network traffic has been increasing over decades, the traditional circuit switched services have lost their earlier significance. At the time of this writing in 2022, it is estimated that the voice traffic is less than 0.5% of the total worldwide network traffic. The service providers have changed their perception about the voice services. While voice has long been the driver of telecommunication networks, data has taken the driver's seat in this century. Since 2010, voice has been seen as one specific data service rather than its own type of circuit switched service. Even though the share of voice in data volume is low, it still represents a significant share of the network operator business. It can be said that the value per bit is high in voice communication services compared to streaming. Its high value justifies operator investments to continue supporting voice services.

Consequently, packet services over the IP protocol started to replace the traditional circuit switched services in last 20 years. The following concepts and technologies have emerged in the context of packet voice and real-time multimedia services:

- **Convergence**, which means usage of a common packet-switched technology for all kinds of data- and telecommunication services. The Internet protocol (IP) is the enabler, which is able to provide end-to-end connectivity to all types of data flows, including the traditional data services (such as email and World Wide Web) and real-time services, like telephony and video services. The aim of the convergence is to unify different types of networks and provide a wider range of services than what the earlier technologies supported.
- **Voice over IP (VoIP)**, which means telephony over the IP protocol. VoIP itself is not a protocol but rather a technology supported by both standardized and proprietary protocol suites. The latter are used by various proprietary applications, such as Skype and WhatsApp. The different VoIP implementations are in many cases not compatible with each other. Either the VoIP solution is specific to an application or a gateway is used for interconnecting different VoIP and traditional circuit switched voice solutions. The task of the gateway is to do any needed conversions between the signaling and media protocols and

Converged Communications: Evolution from Telephony to 5G Mobile Internet, First Edition. Erkki Koivusalo.
© 2023 The Institute of Electrical and Electronics Engineers, Inc. Published 2023 by John Wiley & Sons, Inc.
Companion website: www.wiley.com/go/koivusalo/convergedcommunications

coding formats used in the systems connected to the gateway. In many cases, interconnections were done via the traditional PSTN service.

- **IP multimedia**, which means simultaneous real-time transport of many different types of media over the IP-based connection. Voice, video, still pictures, and text are examples of media types for an IP multimedia session. Like VoIP also, the IP multimedia solutions may be standardized ones or application specific without interoperability between different solutions. While gateways may exist for VoIP interconnection, for IP multimedia the interconnection typically is not supported at all due to the additional complexity of different systems.

Two standardized protocol suites have been defined for VoIP systems [1]:

- IP multimedia protocol stack as defined by the IETF, consisting of SIP and SDP protocols used to manage IP multimedia sessions and the RTP protocol to transport the media within the voice or multimedia session.
- H.323 [2] is the protocol stack defined by ITU-T for VoIP. H.323 was the predecessor of the IEFT multimedia protocol stack and it is no longer widely used. H.323 standardization had the goal of reusing existing standard protocols and defining new ones only when necessary. Two key protocols reused in H.323 were as follows:
  – Q.931 [3]: Signaling protocol originally defined for ISDN networks.
  – RTP: The real-time media transport protocol as defined by IETF.

VoIP and IP multimedia systems can be transported over nearly all the modern data link layer protocols, because the IP protocol itself is independent of the mechanisms used on the underlying layers of the protocol stack. This makes it possible to have many different ways for the end user to connect to the VoIP service, as demonstrated in Figure 9.1:

- Use VoIP over a fixed broadband connection, such as a DSL modem. The user can have either a dedicated VoIP phone or a VoIP application in the computer equipped with a headset.
- Use VoIP over a WiFi connection. The user can connect to VoIP service over WLAN with a laptop or mobile phone when the device supports a VoIP application. Voice over WiFi (VoWiFi), described in Section 9.6, is a standardized operator VoIP service, which relies on WLAN access.
- Use VoIP over the 4G or 5G cellular data network. Operators have deployed the IP Multimedia solution for 4G and 5G data-only networks. Such Voice over LTE (VoLTE) and Voice over NR (VoNR) solutions compliant to 3GPP standards are described in Sections 9.4 and 9.5. Both VoLTE and VoNR are based on the common 3GPP IP multimedia system (IMS) architecture.

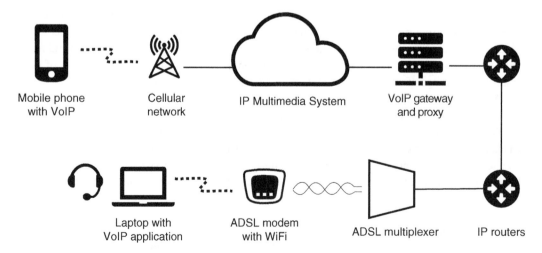

Mobile phone with VoIP Cellular network IP Multimedia System VoIP gateway and proxy

Laptop with VoIP application ADSL modem with WiFi ADSL multiplexer IP routers

Figure 9.1   End-to-end VoIP call.

In the mobile domain, the drive toward IMS services is backed by the following factors:

- The latest cellular network technologies such as 4G LTE and 5G support only the packet switched services. Such an approach was chosen to optimize those technologies for data and provide voice only as a data service. To support voice services, the network nodes in the packet switched domain must be designed for support of a large amount of small capacity VoIP connections rather than only a small number of high-speed data connections.
- The telephone operator core networks have also been transformed from the traditional circuit switched technology to the newer IP-based IMS technology. By supporting voice as a packet switched service, the network operators are able to optimize the usage of their core network link and equipment capacity and avoid investing in two separate networks.
- With help of advanced voice coding technologies, transporting packet switched voice needs less transmission capacity compared to the circuit switched voice. For instance, in theory an HSPA data network would be able to support two to three VoIP calls with the same radio bandwidth as used by the WCDMA network for a single circuit switched call. The following factors are in favor of VoIP for the spectral efficiency:
  - Packet switched link technologies are designed for very flexible real-time allocation of the transport capacity. For the VoIP stream the network can allocate the exact amount of capacity needed, whether the person speaks, is silent, or muted. As fixed data rate is used for circuit switched connections, the unused capacity is wasted when there is no voice information to be transported.
  - Compared to older technologies, the latest packet-switched radio technologies have lower latencies. A single data packet sent once every 20 milliseconds (ms) may cover the speech of the whole 20 ms period, without any noticeable delay of the speech. If transmission of the packet takes only 1 ms, other data packets can be sent during the rest of 19 ms over the same radio channel.
  - The compression of the IP packet headers improves the ratio between the transported voice media and overhead data of the packet headers. IP header compression shrinks the headers of UDP/IP packets used to carry the RTP media packets. This is important as the size of the RTP media packet is small compared to the size of regular UDP/IP headers. Without header compression the network would need to allocate 30 kbps transmission capacity to pass AMR encoded voice of 12.2 kbps. Most of the bitrate used would carry the header bytes rather than media bytes.

When deploying VoIP, the network provider must also solve the problem about how to ensure that the real-time services will be guaranteed both the capacity and short enough latencies needed for the interactive conversation in a call. There are two types of solutions for this problem:

- Over-dimensioning the network, which means providing such large total capacity that the network congestion can be avoided in all circumstances. This is, however, hard to guarantee, since the appetite for data by the network users will often eat up all the available capacity. Further on, over-dimensioning is not a viable approach for radio networks where shared bandwidth is an inherently scarce resource.
- Usage of specific QoS (Quality of Service) mechanisms to ensure that any traffic related to the real-time services will be given higher priority than for non-real-time services. Admission control mechanisms are used to ensure that the number of simultaneous calls would not exceed total capacity of the network allocated for real-time services.

9.2 SIP Systems

9.2.1 Standardization of SIP

Session initiation protocol (SIP) standardization work started in IETF in the end of the 1990s. SIP protocol structure was created by combining features from two competing protocol drafts designed earlier for session management [4]:

- Session invitation protocol was a text-based protocol with which the users were able to announce their locations to the servers, providing multimedia conferences and receiving notifications about upcoming conferences. The protocol used UDP transport and SDP payloads to describe the timing and media of the conferences.

- Simple conference invitation protocol was another text-based protocol (resembling the HTTP protocol used for Web access) with which the users were able to send invitations for bilateral or multilateral sessions. The protocol used TCP transport and had its own features to describe the media of the session.

The following features were applied to the new Session Initiation Protocol (SIP):

- SIP is text-based and its structure closely resembles HTTP.
- SIP messages can be transported over UDP or TCP protocols.
- SIP enables the users to register to the SIP multimedia service.
- SIP can be used for managing bilateral and multilateral sessions.
- The media used within the session is described using **session description protocol (SDP)**. The SDP descriptions are sent to their recipients within SIP messages.

The basic features of SIP are described in the following two IETF documents:

- RFC 2543 [5]: The first version of SIP protocol as approved on the year 1999.
- RFC 3261 [6]: The second version of SIP protocol approved on the year 2002. This document describes how the SIP user registers to the service, sets up and tears down sessions, and how the other basic SIP protocol features work.

In addition to these two specifications, IETF has published tens of other RFCs to define various extensions to the basic SIP protocol, such as:

- RFC 3262 [7]: How to provide responses in a reliable way when setting up a session
- RFC 3263 [8]: How to find out the IP address of the SIP server using DNS
- RFC 3265 [9]: SIP notification service
- RFC 3428 [10]: Instant messaging with SIP
- RFC 3515 [11]: SIP session transfer mechanism
- RFC 3856 [12]: SIP notifications for the presence service
- RFC 3903 [13]: How to pass events to SIP notification service

In addition to the SIP protocol, the IP multimedia stack uses SDP, RTP, and RTCP protocols, which IETF has originally specified prior to SIP. These protocols have been enhanced parallel to the evolution of SIP-related specifications. The *Online Appendix J* introduces all these protocols briefly and lists the IETF RFCs where the protocols have been specified.

9.2.2 Architecture and Services of SIP System

9.2.2.1 SIP Multimedia System Services

SIP multimedia system may provide the following types of services:

- Basic Voice over IP calls
- Multimedia sessions mixing various media such as voice, video, still pictures, text, and documents
- One-to-one sessions, multiparty sessions, or broadcast conferences

9.2.2.2 SIP System Architecture

The SIP VoIP system consists of the following types of elements (either devices or client and server software packages) shown in Figure 9.2:

- **SIP user agent (UA),** which is the user's SIP terminal. The UA is able to send and receive SIP protocol messages used to manage sessions and the state of the user. The user agent has two roles:
  - **User agent client (UAC),** which sends the SIP request messages and receives responses for them
  - **User agent server (UAS),** which receives SIP requests and sends SIP responses for them

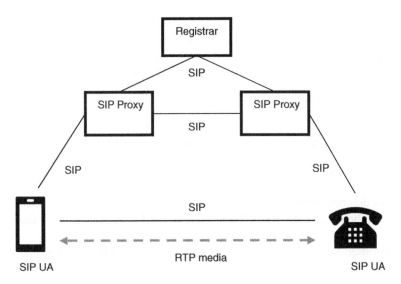

Figure 9.2 SIP VoIP system within a VoIP service provider domain.

- **SIP registrar server**, which manages registrations of SIP user agents. The registrar knows the network level addresses of those terminals, which the users currently use to access the SIP service. The address of the terminal (UA) can be given either as a public IP address or a unique instance ID of the UA, which can be used for the terminal identification. The instance ID is mapped to the private IP address of the UA by a SIP proxy, which the registrar knows to serve the user. Private IP addresses can also be directly used by registrars belonging to the same private network or when a SIP-aware gateway is used to interconnect networks. Every user of the SIP service has a **public SIP address** known by other users. The public SIP address may look like a telephone number or email address but is prefixed with a specifier to define the address type. The prefix tel.: is used for telephone numbers and the prefix sip: is used for native SIP addresses. The main task of the registrar is to store mappings between the public SIP addresses of the users and their currently valid contact (IP) addresses. Such mappings are called **Address of Records (AoR)**.
- **SIP proxy server,** which forwards SIP protocol messages between SIP user agents and other SIP servers. The proxy server is able to route the SIP message to the next server or user agent according to the SIP protocol addressing information within the messages. When a proxy receives a SIP request with the public SIP address of the destination user but without any contact IP addresses, the proxy works as follows: If the SIP address belongs to another network domain, the message is forwarded to a SIP proxy in that domain. If the SIP address belongs to the domain of the proxy, the proxy asks the local registrar to return valid contact IP address(es) for the destination user. After that point, the proxy is able to route the message to the destination UAS. If the proxy is unable to reach the UAS, it will return a response message to instruct the UAC to either give up or send its SIP request again toward another endpoint. When the SIP proxy routes a SIP request, it can either store or not store information about it for its further needs:
  - Store information about the session being created, so that the proxy is able to cancel the session if its establishment is not completed within a predefined time.
  - Store information about the ongoing SIP transaction for cases where the called SIP user has registered to the SIP service simultaneously from multiple terminals. The proxy can in such a case forward the SIP request in parallel to all those terminals. This SIP protocol mechanism is called **forking**. When one of the terminals provides a positive response, the proxy may cancel the requests sent toward the other terminals.

– Not to store any information about the forwarded SIP request. This kind of **stateless proxy** is able to route the response returned by the UAS back to the UAC by investigating the Via header of the response. The Via header of a response tells the chain of proxies via which the SIP request was routed toward the UAS. Stateless proxies can be scaled to serve a big number of SIP transactions. That is why the SIP core network servers are usually stateless.

9.2.2.3 SIP Protocol Stack and Operation

IP multimedia protocol stack defined by IETF consists of the following three complementary protocols:

- **Session initiation protocol (SIP)** is the signaling protocol for IP multimedia call management, like establishing and releasing VoIP calls.
- **Session description protocol (SDP)** is a description language used within SIP protocol messages to describe the formats of transported multimedia and the IP addresses and UDP ports to be used as the endpoints of the media path.
- **Real-time protocol (RTP)** is a protocol used to transport digitally encoded voice or real-time multimedia between the endpoints.

SIP and SDP protocols are used for call setup and management. The UA sends a REGISTER request to the registrar server to announce its contact address where the SIP user is currently reachable. The registrar stores the mapping between the contact address and the public SIP address of the user as an AoR. When trying to reach another user of the SIP system for a call, the UAC sends a SIP INVITE request toward the public SIP address of the callee. The UAC describes the available media types and endpoints within a SDP description embedded into the body of the SIP INVITE request. SIP proxies will take care of delivering the SIP INVITE request to the UAS of the callee. The UAS can be reached with help of the contact address retrieved from the AoR stored in the registrar. After getting the SIP request the UAS provides a response toward the UAC. After the user agents of caller and callee have completed the SIP INVITE transaction and created a signaling dialog between themselves, the call can be established and transport of agreed types of media started. Media is sent between the user agents as RTP packets. Media packets are transported between the endpoints which were announced in SDP descriptions. Media itself is not following the path of signaling messages, but has its own route between the endpoints, based on IP routing mechanisms. The exception to this is when media is routed via some sort of gateways to support media traversal over NATs and firewalls or when some sort of transcoding is needed between different codecs used in the endpoints. In those cases, the media will traverse via the gateway function between the endpoints.

Further details of SIP, SDP, and RTP protocols and related system procedures can be found from the *Online Appendix J*. If the reader is not familiar with the SIP protocol, it is recommended to study the appendix before going into details of 3GPP IMS, which provides various extensions on top of the "plain vanilla" SIP mechanisms described in the appendix.

9.3 3GPP IP Multimedia Subsystem

9.3.1 Standardization of IMS

The global deregulation process started in the 1980s has opened the telecommunications market for competition. The increasing competition has forced the network operators to find ways for service differentiation. While standardization aims at unifying system implementations, operators have demanded solutions with which they could introduce their own services to stay or become successful on the competitive market. In the 1980s, intelligent networks (IN) was the prevalent approach for providing differentiated telephony services. The IN allowed opera-tors to build their own enhancements on top of the shared standard call model. Intelligent networks technology applied only to the circuit switched exchanges. By the end of the 1990s it had become clear that packet switched solutions were much more flexible and allowed building applications of different kinds. It was anticipated that voice would become just one packet switched service among the others. To retain their position in voice business and keep their control on voice service provisioning and charging, operators wanted to have their own premium-level VoIP systems. In addition to VoIP, the IP protocol provided a venue toward any kind of real-time or non-real-time multimedia services, potentially deployable for the operators. The 3GPP and ETSI standardization forums addressed this demand with a new IP multimedia architecture known as 3GPP IMS.

IP multimedia subsystem (IMS) is a carrier-grade, real-time multimedia system architecture to be deployed in operator-hosted IP networks [14]. As specified by 3GPP from Rel-5 onwards, IMS systems are typically used in the context of data-only cellular mobile networks such as LTE and 5G, but IMS access over WLAN (Rel-6) and fixed DSL broadband lines (Rel-7) were also added to IMS standards by 3GPP. The ETSI project Telecommunications and Internet converged Services and Protocols for Advanced Network (TISPAN) specified a way of using IMS for new generation fixed networks [15]. The IMS specification work was started in 1999 and it has continued since then. Only limited operator-specific IMS deployments were in commercial use before 2015, which eventually became the IMS breakthrough year. Standard compliant operator IMS voice systems – VoLTE and VoWiFi – were in use by several operators in 2015. Thereafter, those IMS services have become popular in many countries along with the deployment of 4G and 5G mobile networks without circuit switched voice support.

The 3GPP standardization had the following requirements for the IP multimedia subsystem:

- IP connectivity to be able to provide a mix of packet-based multimedia services
- Access independence to be able to use IMS services over cellular network, WLAN, or fixed broadband connection to the terminal
- Quality of Service to support high-quality voice and video calls with guaranteed bitrates and bound latencies
- Interworking with other networks like independent VoIP providers or the circuit switched PSTN network
- Support for using the service when roaming in visited network
- Secure communications
- Charging support
- Flexibility for operators to develop value-added communication services

3GPP selected IETF multimedia stack as the toolkit for building the IMS protocol suite. SIP and SDP are used for IMS call control while RTP is used for the real-time media transport. Additionally, other protocols defined in IETF are used for special purposes. The Diameter protocol was chosen to communicate with the home subscriber server (HSS) and the HTTP protocol for the user equipment to communicate with IMS application servers (AS).

IMS standards have been evolved over 3GPP standards releases. This book covers these major IMS features from the four 3GPP releases where IMS foundations were laid out:

- 3GPP Rel-5 introduced IMS architecture and its basic mechanisms.
- 3GPP Rel-6 introduced IMS Phase 2 with support for IMS multi-registration, presence, messaging, and IMS interworking with other CS and PS networks.
- 3GPP Rel-7 introduced IMS Phase 3 with support for IMS multimedia telephony and conferencing services and a few additional parameters, such as ICSI and GRUU. IMS protocols and security were enhanced and aligned with IETF. Support for voice call continuity (VCC) between IMS and CS services was added as well as the IMS emergency call support.
- 3GPP Rel-8 added support for IMS service continuity and centralized services, where the IMS core could be used to provision services over all types of access networks and to support also circuit switched services. Additional IMS multimedia telephony features were specified, such as IMS supplementary services and

single-radio voice call continuity (SRVCC). Support for packet cable access was added and IMS specifications from 3GPP2 were incorporated to 3GPP standards.

Many other smaller IMS enhancements have been introduced in later 3GPP releases, such as VCC and QoS enhancements, IMS emergency callback, support for IMS over WLAN access, overload control, and restoration mechanisms. From these, the Section 9.6 describes the operation of IMS over WLAN access in the context of the VoWiFi service.

9.3.2 Architecture and Services of IMS System

9.3.2.1 IMS Services

The IMS system provides services similar to IETF multimedia stack, such as voice and video calls, instant messaging, real-time gaming, and so on. What IMS provides on top of the IETF building blocks is tight integration of IMS with different types of access networks to support Quality of Service needs for different types of media flows, support for handover scenarios, as well as the support for operator service control and charging of the services used.

9.3.2.2 IMS System Architecture

IMS system architecture, network elements, and the abstract interfaces – IMS reference points – are defined within 3GPP TS 23.228 [16].

The network element types of the IMS system, as shown in Figure 9.3, are as follows:

- **User equipment (UE)** with SIP user agent (UA) functionality. In the context of IMS, the UE is typically a mobile phone, but it may also be another type of terminal such as a laptop with a USIM card accessing the network over WiFi or cellular connection.
- **Proxy call state control function (P-CSCF)** is the first hop SIP proxy communicating directly with the UE. The P-CSCF supports the following functions:
  - Forward SIP signaling between the UE and the rest of the IMS network. What comes to the outgoing request from the UE, initial SIP REGISTER requests are forwarded to the I-CSCF and other SIP requests to the S-CSCF assigned to the user during the registration.

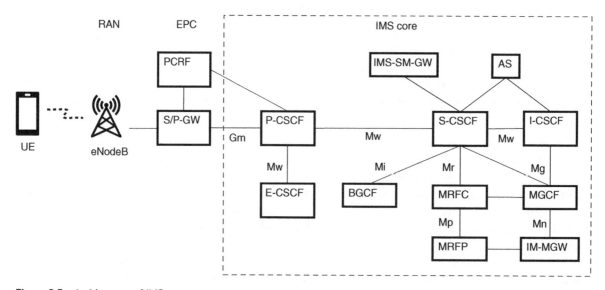

Figure 9.3 Architecture of IMS.

- Create and maintain secure IPSec connections with the UE to protect SIP signaling against modification or eavesdropping.
- Compression of SIP signaling messages sent to the UE over the air interface, if the operator applies signaling compression. Compression is done with SigComp mechanism, as specified in IETF RFCs 3320 [17], 3485 [18], and 3486 [19].
- Communicate with the PCRF to authorize usage of media streams requested for the session within the SIP INVITE transaction. Reject the request if certain required streams would not match the operator's policy or the service subscription.
- Detect and terminate any hanging sessions based on a session timer mechanism.
- When the UE uses IPv4 private address space, the P-CSCF controls the NAT functionality. When the UE communicates with external networks, the address translation is not only done for the addresses in the IP packet but also IP addresses of the UE within SIP protocol messages.

- **Serving call state control function (S-CSCF)** is another proxy in the IMS network working also as the registrar for the user. The S-CSCF supports the following functions:
  - Authenticate the user during SIP registration with IMS AKA algorithm and the related user-specific data retrieved from the Authentication Center (AuC). The AuC is a part of the HSS of a user's mobile network. To locate the correct HSS for the user, the S-CSCF may need to access another network element called subscriber locator function (SLF).
  - Store the current contact address of the registered UE.
  - Supervise registration timers and de-register the user in certain circumstances.
  - Route mobile terminating SIP traffic to the UE via the P-CSCF and mobile originating traffic to other IMS network elements (such as AS, I-CSCF, BGCF) based on the request type, destination URI, and other filter criteria.
  - Media policing and rejecting any UE terminated requests which do not satisfy the operator policy or user IMS subscription for the types of media streams offered to the UE. The S-CSCF gets the user subscription profiles from the HSS.
- **Interrogating call state control function (I-CSCF)** is a proxy whose task is to assign an S-CSCF instance for the UE in registration and thereafter route any SIP messages to that UE via the assigned S-CSCF. The I-CSCF supports the following functions:
  - When receiving a SIP REGISTER request from the UE, the I-CSCF checks from the HSS which S-CSCF capabilities the user needs. Thereafter, the I-CSCF chooses an S-CSCF instance and assigns it for the UE. Finally, the I-CSCF forwards the SIP REGISTER request to that S-CSCF.
  - When receiving another SIP request to the user from external IMS or IP multimedia networks, the I-CSCF forwards the request to the S-CSCF instance currently assigned to the user.
- **Emergency call state control function (E-CSCF)** is a proxy used only for emergency calls. The E-CSCF supports the following functions:
  - Locate the user with help of the location retrieval function (LRF) element. The network may locate the UE, for instance, with the GPS or OTDOA method of LTE. In OTDOA, the UE measures the timing of signals from three base stations. The UE tells the cell IDs and related timing differences to the network, which is then able to calculate the UE location with a triangular positioning method.
  - Select the emergency response center closest to the UE so that the center will be able to dispatch the help for the caller. The emergency center is also referred to as a public safety answering point (PSAP).
- **Application server (AS)** is a server that processes requests for certain IMS-based applications such as **IMS conferencing**. The AS server may communicate with the UE over HTTP protocol and with CSCF proxies over SIP.
- **Policy and charging rule function (PCRF)** is a server that helps the IMS network to coordinate charging and Quality of Services. The PCRF authorizes the media streams to use bearers of the mobile network. The PCRF supports the following functions:

- Inform packet network elements of the mobile core network about the media streams of the IP multimedia network. Information about the media stream types and bandwidth needs are used for setting up bearers for those streams.
- Pass the IMS charging identifier to packet network elements to enable combined charging of IMS and mobile data services.

SIP proxies and registrars as originally specified by IETF [6] are primarily SIP message routing servers. The IMS call state control functions (CSCF) support SIP proxy and registrar functionality, but in addition they do much more processing for SIP messages. Examples of such CSCF activities are:

- Setting up the needed Quality of Service support within the access network for the IMS signaling and media flows.
- Checking which services and codecs the IMS user is entitled to use and modifying the message contents accordingly.
- Checking the real identity of the IMS subscriber and encoding that to the SIP messages as routed within the trust domain.
- Providing extra information within SIP messages about the used type of access network and location of IMS user.

The following set of network elements are used to interconnect the IMS network with traditional circuit switched voice networks, IMS networks of other operators, or other non-IMS IP multimedia networks:

- **Breakout gateway control function (BGCF)** selects the breakout point from the IMS network to the CS network. If the point is in the local network, the BGCF coordinates the related actions with the local MGCF, otherwise it coordinates the actions with the BGCF of the network where breakout is done.
- **Media gateway control function (MGCF)** participates to call control of calls connected between IMS and CS networks. The MGCF is able to make conversions between SIP and SS7 call signaling protocols used in the CS network. For ISUP call control messages received from the CS network, the MGCF generates a corresponding SIP INVITE request and sends it to the local I-CSCF.
  - Signaling Gateway (SGW) function supports the MGCF. While the MGCF makes the needed conversions on the call control protocol level, the SGW converts the protocols in lower layers (network and link).
- **IP multimedia subsystem media gateway function (IM-MGW)** interconnects the media streams between IMS and CS networks when the call is connected between those. The IM-MGW terminates the media connection from one network, does the needed media protocol transcoding, and passes the media to the other network.
- **Multimedia resource function processor (MRFP)** processes the media in certain cases. The MRFP mixes streams within a multiparty conference, plays out announcements to the user, or transcodes the audio between different encoding schemes.
- **Multimedia resource function controller (MRFC)** controls the actions performed by the MRFP, based on SIP signaling received from the S-CSCF.
- **IP short message gateway (IP-SM-GW)** forwards short messages between an IMS capable UE and the circuit switched domain.
- **Interconnection border control function and transition gateway (IBCF/TrGW)** provides special protocol conversions between the networks, such as conversion between IPv4 and IPv6 addressing or conversions between different types of media codecs.

9.3.2.3 IMS User Identities and Service Profiles

Every IMS user has two types of user identities: private and public:

- **IP multimedia public identity (IMPU)** is the public SIP or TEL URI of the user, which the other users can use in their SIP requests to reach the user. The format of IMPU is like sip:user@operator.com or tel.:+123456789, where the telephone number is given in the international format. The user may have multiple public IMS user identities, related to one or multiple user profiles (like a work profile and a private profile). Having multiple IMPUs enables the user to register to a chosen subset of profiles at any time.

- **IP multimedia private identity (IMPI)** is used only internally within the IMS network to recognize a user's IMS service subscription when the user is authenticated during IMS registration. IMS UE uses IMPI only within the Authorization header of the SIP REGISTER request. The format of IMPI is "user@realm" where the realm refers to the domain of the IMS service provider.

In the typical case, the UE is able to read these identities from the Universal Integrated Circuit Card, which contains an **IP multimedia services identity module (ISIM)** application. The UICC card is the same one as used for accessing the cellular network with the USIM application. As older UICC cards do not have the ISIM application, the specifications provide a way for the IMS user to perform IMS authentication with the USIM application only. For that case, 3GPP has defined rules on how to derive the private IMPI and a temporary IMPU from the identities available on USIM. The USIM identities used are the country code and mobile network code of the operator as well as the MSISDN identifier that the operator has allocated to the user. During the registration sequence, the IMS network tells the UE any other permanent IMS public identities, as known to the network, which have been implicitly registered for the user.

The mapping between a private IMS identity and the related public identities is stored by the HSS within a user's service profiles. For each IMS Private Identity, the HSS has one or more service profiles for the user. The **service profile** determines the IMS user's service subscription with the following items:

- One or multiple IMS public identities, which will become implicitly registered when the service profile becomes registered. The service profile becomes registered when the UE sends a SIP REGISTER request for any of the IMS temporary or permanent public identities within the service profile.
- Core network service authorization, which the S-SCSF uses when checking if the user is allowed to use specific types of media streams for its multimedia sessions.
- **Initial filter criteria**, which is used to check when any Application Server must be involved for processing the signaling of the SIP service used. The filter criteria consist of trigger points and application server information. The trigger points are a set of conditions defined for the SIP request Request-Line, headers, and SDP session description, specifying when a SIP request should trigger the S-CSCF to involve an application server to process the request. The trigger points are checked by the S-CSCF as part of processing SIP INVITE requests for new sessions. Depending on the case, the application server may play different kinds of roles for SIP signaling: the AS may be a SIP UA terminating the request, SIP proxy, SIP redirect server, or third-party call control server. One example of service supported by application servers is the multi-party voice conference.

9.3.3 IMS Functions and Procedures

Detailed IMS protocol procedures are defined in 3GPP TS 24.229 [20] as complemented with various IETF RFC documents referred by TS 24.229.

9.3.3.1 IMPU Registration State Management

When the S-CSCF receives a SIP REGISTER request from the UE for a single IMS Public Identity, the S-CSCF fetches a user's service profile(s) from the HSS. From the service profile the S-CSCF finds the set of IMS public identities, which the S-CSCF now deems being registered along with the explicitly registered public identity.

The 200 OK response for the SIP REGISTER confirms the explicit registration of the IMPU that the UE used within the SIP REGISTER request. S-CSCF uses **IMS registration state event package** to inform the UE about other IMS public identities implicitly registered via the service profile.

After completing the initial registration, the IMS UE shall send a SIP SUBSCRIBE request to the S-CSCF for the registration state event package. In the SUBSCRIBE request, the user is identified with the explicitly registered IMS public identity. After the subscription has been completed, the S-CSCF sends a SIP NOTIFY request for the subscribed registration event to the UE. This initial notification tells the UE the full set of implicitly registered user identities. Those identities are carried as an XML reginfo payload within the SIP NOTIFY request.

Later, if there are any changes to the registration status of any IMS public Identities related to the subscribed identity, the S-CSCF sends a new SIP NOTIFY request to inform the UE about the status changes. The XML reginfo structure tells the current registration status of each public identity in the service profile and the reason for changing the status:

- A new IMPU has been registered (either explicitly or implicitly), what is the case after initial registration.
- The length of the registration period has been either extended (e.g., via re-registration) or shortened (e.g., because network requests user reauthentication).
- The registration has been deactivated by the network or by the user.

The UE can keep the chosen service profiles active by sending SIP REGISTER requests to explicitly register or de-register public IMS user identities.

9.3.3.2 Authenticating IMS Subscribers

The AKA authentication mechanism described in Chapter 6, Section 6.1.9.1 is used also for authenticating IMS subscribers. The same AKA algorithms and related data entities as defined for UMTS are reused also for IMS authentication. Authentication is done between the UICC card and the AuC of the network, which store the authentication master keys and exchange the response values calculated from the master key and random value generated for each authentication event.

The SIP protocol authentication mechanisms and headers as defined for HTTP digest authentication (see *Online Appendix J.1.6*) have been reused for IMS authentication. RFC 3310 [21] specifies a way to pass the IMS AKA related parameters from the S-CSCF to the UE within the WWW-Authenticate header of the 401 response and from the UE to the S-CSCF within the Authorization header of the SIP REGISTER request. The AKA RAND and AUTN values are encoded into the nonce of WWW-Authenticate header passed to the UE within the 401 response. The UE retrieves RAND and AUTN from the nonce when the WWW-Authenticate header uses AKAv1-MD5 as its authentication algorithm. The UE creates its authentication response value for the SIP Authorization header as defined in RFC 2617 [22] using the AKA RES as the password.

The IMS authentication process is done as follows:

- Both UICC and AuC store the shared secret key K of the user. They never expose K as such to any other entities.
- For an authentication event the AuC generates a random challenge RAND and derives the expected result XRES from K and RAND values.
- After receiving an initial SIP REGISTER request from the UE, the S-CSCF retrieves the RAND and XRES parameters of the user from the AuC. In this context the user is identified by the IMS private identity carried within the Authorization header of the REGISTER request.
- The S-CSCF keeps the XRES parameter and sends the RAND value to the UE in the 401 Authenticate response for the SIP REGISTER request. In that response the S-CSCF also sends a network authentication token AUTN, with which the UE can reliably authenticate the IMS network.
- The ISIM application calculates its authentication response RES from the received RAND and the key K stored into the UICC card. Thereafter the UE sends a new SIP REGISTER request to the S-CSCF. This request has an Authorization header whose response parameter value has been generated by using the RES value as the password.
- When the S-CSCF finds that the returned response value sent by the UE matches with the value the S-CSCF has itself calculated with the XRES value received from the AuC, the user is successfully authenticated.
- When processing 401 responses, the ISIM application also generates the encryption and integrity protection keys CK and IK, to be used when setting up the IPSec secure connection with the P-CSCF.

9.3.3.3   SIP Security Mechanism Agreement and Ipsec-3gpp

3GPP decided to apply IPSec to secure IMS signaling messages between the UE and P-CSCF. This Ipsec-3gpp mechanism uses both IPSec encryption and integrity protection to ensure that signaling messages are neither read or modified by unauthorized parties. Encryption is important especially for mobile IMS clients, who communicate over the radio interface and are vulnerable for eavesdropping. IPSec endpoints are the UE and P-CSCF.

The encryption and integrity protection keys CK and IK are created both at the UE and AuC during the IMA AKA authentication. Those keys are derived from the secret key shared between the AuC and ISIM application and the RAND value given by the AuC for the UE to authenticate itself. The UE generates the CK and IK keys during the authentication process. The P-CSCF gets those keys from the AuC via the S-CSCF within the SIP 401 response. The P-CSCF removes the CK and IK from the response before forwarding it to the UE so that no eavesdroppers can learn the keys from the response message.

The **IPSec security associations (SA)** are set up between the UE and P-CSCF during the initial IMS registration sequence. The term "security association" is used rather than "connection" as the IPSec SAs are used to protect SIP signaling over UDP, which is a connectionless protocol. In case the SIP signaling message is bigger than the MTU of 1300 bytes, TCP connections can temporarily be used over the same IPSec SAs.

The UE and P-CSCF may possibly support various security mechanisms for the following reasons:

- In addition to the IMS service the UE might support some Internet VoIP services relying on TLS instead of Ipsec-3gpp.
- As 3gpp IMS specs evolve, new additional security mechanism options may be added. The UE and P-CSCF must check which mechanisms both of them support and agree to use one of those.

To agree usage of Ipsec-3gpp security mechanism and UDP port numbers for the IPSec security associations, the UE and P-CSCF use the **SIP Security Mechanism Agreement** method specified in RFC 3329 [23]. The UE and P-CSCF exchange the following headers within the SIP REGISTER request and the related 401 response:

1) The UE adds a Security-Client header to its initial SIP REGISTER request. This header tells the security mechanisms supported by the UE and the IPSec client and server ports which the UE would use, if Ipsec-3gpp is the agreed mechanism.
2) The P-CSCF adds a Security-Server header to the 401 response. This header tells which security mechanisms the P-CSCF supports and the IPSec client and server ports of the P-CSCF, if Ipsec-3gpp is the agreed mechanism.
3) When the UE receives the Security-Server header it can compare its own list of security mechanisms to the list of the P-CSCF mechanisms. Among the mechanisms supported by both the endpoints, the one is chosen that has the highest usage preference recorded to the security headers. At this point, if Ipsec-3gpp is the mechanism, the related IPSec SAs are set up and used thereafter for SIP signaling between the UE and P-CSCF.
4) Finally, the UE adds both Security-Client and Security-Verify headers to its second SIP REGISTER request, which also contains the authentication credentials. The Security-Client header has the same values as before. The contents of Security-Verify header are copied from the Security-Server header received from P-CSCF. As Security-Verify header is sent reliably over the newly established security mechanism (IPSec SA), the P-CSCF can ensure that the value of its Security-Server header was passed correctly to the UE.

9.3.3.4   Resolving the Address of Local P-CSCF

In order to start communicating with the IMS network, the UE shall find out the IP address of the local P-CSCF. The P-CSCF is typically located within the home network of the user. In case of IMS roaming, the P-CSCF might also be located to the visited network of the other operator. There are a few ways how the UE might retrieve the IP address of the P-CSCF:

- The domain name of the home P-CSCF may have been stored to the configuration settings of the UE. The home P-CSCF is one within the home network.
- The IMS UE might retrieve the domain name of the home P-CSCF from the ISIM application of the UICC card, which stores various pieces of IMS access details.

- The IMS UE might use DHCP protocol to find out domain name(s) of the local P-CSCF. The related procedures are defined in IETF RFC 3315 [24], 3319 [25], 3361 [26], and 8415 [27].
- The IMS UE might use the procedures of the cellular protocol used (such as WCDMA UMTS or LTE) to find out the domain names or IP addresses of the local P-CSCF. In this case, the mobile network returns the P-CSCF address within the PDP or EPS bearer context activation or the default bearer setup procedure as used for IMS signaling. In these signaling messages, the P-CSCF address is given within a protocol configuration options (PCO) element.

After the UE has found out the domain name of a P-CSCF, it can use the DNS procedures as defined in IETF RFC 3263 [8] to map the domain name to a set of IP addresses.

9.3.3.5 Signaling Compression

SIP signaling messages might be quite long for the following reasons:

- SIP protocol is text-based. The messages of text-based protocols tend to be much longer than those of binary protocols for many reasons. Encoding one atomic value might take a number of bytes with a text-based protocol while with binary protocols 1 byte or bit in a fixed location of the protocol frame may be sufficient. Rather than a fixed frame structure, SIP uses headers whose names appear in the beginning of every line the message.
- SIP messages have bodies which carry text-based content mostly in SDP or XML formats. Especially XML bodies might be even longer than SIP headers of the messages.

When 3GPP IMS standardization was started in the early 2000s, the cellular radio interface supported quite modest data rates. It was estimated that with 3G WCDMA the transmission time of SIP signaling messages between the UE and P-CSCF might become excessive for real-time communication. To speed up the signaling, 3GPP agreed to use **signaling compression (SigComp)** to reduce the size of the signaling messages and consequent delays of the transmission. Even if running the compression and decompression algorithms in the endpoints takes time, the net effect was still estimated to be positive. SigComp is defined in IETF RFC 3320 [17] and 3322 [28]. Static compression dictionaries for compressing SIP and SDP are defined in RFC 3485 [18].

To announce SigComp support, the UE can add a "comp=SigComp" parameter to the Contact and Via header of SIP messages while the P-CSCF can add the same parameter to the Record-Route and Via headers. This parameter and its usage are defined in RFC 3486 [19]. Even if SigComp has been a mandatory feature for IMS UEs, it currently has very little value as the 4G and 5G systems have provided much higher data rates compared to 3G. Many operators do not use the SigComp option at all.

9.3.3.6 Media Negotiation

In a basic IETF multimedia system, a three-way SIP INVITE transaction completes the SDP **offer-answer media negotiation** process for the endpoints to agree to the set of media to be used for the session, as described in *Online Appendix J.2.1*. Either the caller or callee sends the offer and the other party then its answer. The answer nails down the media types and codecs used for the call. Regardless of which of the two parties starts the SDP exchange, messages of the three-way SIP INVITE transaction carry one single SDP description per direction. This model relies on the assumption that the communicating parties have a permanent access for transmission resources needed for the media streams. While such an assumption is valid for any dedicated SIP phone or workstation relying on fixed network cabling, it is not valid for mobile IMS phones relying on cellular radio access. Cellular network transmission resources are reserved only when needed, to optimize usage of scarce radio frequencies.

Because of this, IMS introduces an extended media negotiation model which consists of two or three offer-answer rounds as follows:

- In the first offer-answer round, the IMS UEs exchange proposals of media streams and codecs for the call. The offer may provide any number of media types and multiple codecs for each media type. The answer is a subset of the offer, covering those types of media and codecs that the callee UE supports and is willing to use for the call. In this round, SDP preconditions "des" attributes (see *Online Appendix J.2.2*) are used to indicate whether resource reservation is needed at either or both ends of the call. The media types are agreed in this first round.

- The second offer-answer round is an optional one. It is performed only if the answer in the first round had multiple codec options for any of the media types proposed for the call. In this second round, there is only one codec for each media stream, to nail down the codecs for the call.
- The last offer-answer round is used to update the local resource reservation status with the SDP preconditions "curr" attributes. As the resource reservation is mandatory for IMS UEs, the precondition mechanism is used to inform the other end when the radio resources have been reserved to enable transmission of media.

In the IMS call setup sequence, the SDP offers and answers are carried in the following types of SIP messages:

- The first SDP offer is carried within the initial SIP INVITE request.
- The first SDP answer is carried within the 183 response for SIP INVITE.
- The second SDP offer, if needed, is carried within a SIP PRACK request.
- The second SDP answer, if needed, is carried within the 200 OK response for PRACK.
- The last SDP offer is carried within a SIP UPDATE request.
- The last SDP answer is carried within the 200 response for UPDATE.

SIP PRCK and UPDATE request types are used because the second and third offer-answer rounds are performed while the initial INVITE transaction is not yet completed. It is worth noting that the negotiation takes place between two IMS enabled UEs. If only one of the UE uses IMS, the other endpoint of the negotiation is a gateway in the network interconnecting IMS and other domains such as the circuit switched PSTN.

9.3.3.7 Charging Support

IMS specifications define also the architecture to support charging the IMS user for the used services. The IMS charging architecture consists of the following three parts:

- **Offline charging:** Charging Collection Function (CCF) has interfaces toward all the IMS call state control function elements, application servers, and the MRFC to collect charging information from them. The CCF exchanges accounting request (ACR) messages of the Diameter protocol with the other elements. The charging data collected covers relevant IMS events and sessions per subscribers. For instance, when establishing and releasing IMS sessions, the S-CSCF sends ACR requests to the CCF for the session-based charging model.
- **Online charging:** The IMS online charging system has two major functions: Session Charging Function (SCF) which communicates with the S-CSCF and Event Charging Function (ECF) communicating with the AS and MRFC. The ECF is supported by a rating function which determines unit, price and tariff related to the service used. In addition to collecting the charging data, the online charging system is able to check the money available on the user's account (for prepaid services) and consequently either grant, deny or terminate usage of services such as voice call.
- **Combining IMS charging with GPRS data charging:** Policy and charging rule function (PCRF) is able to inform the GGSN node about the IMS charging identifier (ICID) and the P-CSCF about the GPRS charging identifier (GCID). By using these identifiers in both GPRS charging data records and IMS charging data records, the billing system can correlate these records to each other and build a consistent summary of all charging related to using different types of IMS services.

Even if the operators tend to provide monthly billed flat-rate services, there are still good reasons to collect detailed charging records. Some of the flat-rate services may have monthly caps, and exceeding those might incur extra costs for the subscriber. In roaming scenarios, the subscribers may be charged for the services used, unless roaming in the specific country is covered by the flat-rate service package. Naturally, any prepaid services require the usage of real-time online charging model.

9.3.4 IMS System Procedures

9.3.4.1 Registration and Authentication
The IMS UE registers to the IMS service as shown in Figure 9.4:

1) The UE activates a PDP context (in UTRAN) or a default EPS bearer context (in E-UTRAN) for IMS signaling to gain the IP connectivity toward the IMS core network. In absence of real-time IMS voice service in UTRAN, the UTRAN IMS service (if any) might be limited to messaging. The UE learns the IP address of the P-CSCF either from the protocol configuration options information element (PCO IE) within the response from the packet network core, retrieves it separately with DHCP procedures, reads it from ISIM application, or from the UE configuration.

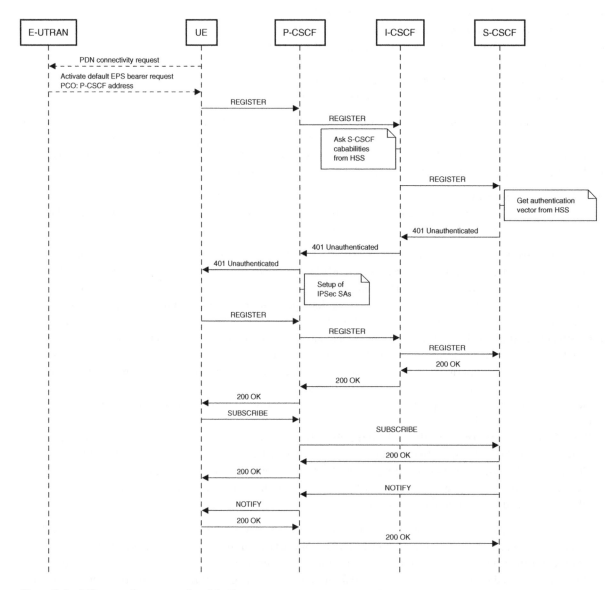

Figure 9.4 IMS proxy discovery and registration.

2) The UE composes a SIP REGISTER request for initial registration and sends it to the P-CSCF. The P-CSCF forwards the request to the I-CSCF of the user's home domain. The I-CSCF contacts the HSS to find out S-CSCF capabilities needed for the subscriber and accordingly assigns an S-CSCF for the UE. The I-CSCF forwards the SIP REGISTER request to the selected S-CSCF.

3) The S-CSCF checks the REGISTER request and uses IMS Private Identity from Authorization header to fetch the IMS authentication vectors from the HSS and its AuC function. Thereafter, the S-CSCF returns a 401 response to the UE to perform authentication. The 401 response contains the necessary set of IMS AKA parameters as well as the integrity and ciphering keys matching with the AKA credentials. Before forwarding the response to the UE, the P-CSCF removes and stores the security keys from the message so that only the IMS AKA parameters are sent over the air interface. After sending the 401 response to the UE, the P-CSCF uses the security keys to set up IPSec security associations toward the UE and starts IPSec encryption and integrity protection processes.

4) After receiving the 401 response, the UE runs IMS AKA authentication algorithms with the ISIM application of the UICC card, using the AKA values extracted from the 401 response. The UE thereafter composes a new SIP REGISTER request and adds an Authorization header to carry the IMS AKA response values, as retrieved from ISIM. The UE sends the new REGISTER request to the S-CSCF via P-CSCF over the newly established IPSec security associations. The P-CSCF decrypts the message and verifies its integrity, before forwarding it to the S-CSCF.

5) The S-CSCF checks the outcome of IMS AKA authentication. After accepting the SIP REGISTER request for the authenticated subscriber, the S-CSCF fetches the service profiles of the subscriber from the HSS and returns a 200 OK response to the UE for the successful registration. After receiving the response, the UE subscribes to the registration state event package. The S-CSCF sends a registration state notification to the UE about all IMS Public Identities, which have now become explicitly or implicitly registered.

After the UE has completed the registration, also the P-CSCF subscribes to the registration state event package of the user with the same procedures as used by the UE. In that way, the P-CSCF learns the implicitly registered user identities, which will be reachable via the registered contact address of the UE and the established IPSec security associations.

For further details about the messages used in these procedures, the DHCP option for the UE to retrieve IP address of P-CSCF and the IMS security agreement procedure used within registration, please refer to *Online Appendix J.5.1.*

9.3.4.2 Voice Call Setup

A mobile user Adam camps on his home cellular operator Avia, providing IMS services for Adam. Adam's friend Bea uses IMS services provided by her home operator Bora. When Adam calls Bea from his IMS UE A, the IMS voice call is set up as shown in Figure 9.5:

1) Adam's UE A sends a SIP INVITE request to the P-CSCF of Avia. The SIP INVITE request has Bea's IMS public identity (IMPU) both in its To header and Request URI fields. An SDP session description offer embedded into the body of the request proposes media streams and related codecs for the call. The SDP body also describes the media resource reservation status, indicating that resource reservation is mandatory at both ends of the call. At this point, no radio resources have yet been reserved at either end. The resource reservations will be done only after the media types have been agreed for the call. The P-CSCF returns a 100 Trying response to Adam's UE A to tell it to stop any resending attempts of the INVITE request.

2) The P-CSCF routes the SIP INVITE request to the S-CSCF of Avia, which finds out that the callee uses another operator, Bora. The SIP INVITE request is then forwarded to the I-CSCF of Bora. The I-CSCF forwards the request to the local S-CSCF, which is currently serving Bea. The S-CSCF fetches Bea's service profile from the HSS (of Bora) to check if Bea is entitled to use the offered services and whether any local application servers

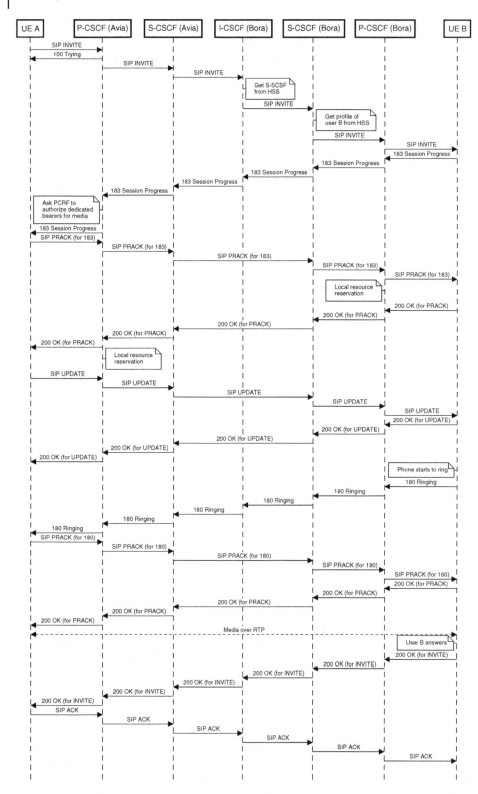

Figure 9.5 IMS voice call setup.

should be engaged to the call. Finally, the SIP INVITE request is routed to Bea's UE B via the P-CSCF, which has IPSec SAs to the UE B.

3) The UE B sends a 183 Session Progress response for the SIP INVITE request. The response contains the SDP answer of UE B, describing at least one codec for every type of media that the UE B supports, and is willing to use for the call, as a subset of those offered by UE A. The answer also indicates that the UE B has not yet reserved any resources for media. The 183 response is routed to the UE A along with the same chain of CSCFs via which the SIP INVITE request traversed to Bea. The P-CSCF intercepts the SDP description within the response to find out the types of media streams and codecs proposed for the session. If only one type of codec is proposed for any media stream, the P-CSCF may proceed with the local resource reservation since the media bandwidth needs are now known. When using LTE, the P-CSCF initiates creation of dedicated EPS bearers, while in UTRAN local SGSN policies are updated and the network starts to wait for UTRAN UE to request creation of secondary PDP context for media. The UE A sends a SIP PRACK request back to the UE B. The PRACK request tells UE B that the 183 response was received by UE A and nails down the set of media and codecs used for the call. The UE B responds to SIP PRACK with a 200 OK response.

4) The UE A gets an indication from the network that the media resources (LTE dedicated bearers or UTRAN secondary PDP context) have been reserved. Consequently, the UE A sends a SIP UPDATE request to the UE B. The precondition status in the SDP body of the request tells that the local radio resource reservation has been completed for the UE A. The precondition status in the SDP body of the 200 OK response (to SIP UPDATE) tells that the UE B has also completed reservation of radio resources for the agreed media streams.

5) After resource reservation has been completed in both ends, the UE B starts ringing to alert Bea about the incoming call. The UE B now sends a 180 Ringing response for the SIP INVITE, to indicate the ringing of the UE B. After receiving the 180 Ringing response, the UE A sends back a SIP PRACK request to tell the UE B that the response was received correctly. When the UE A receives a 200 OK response for PRACK, transporting the call media is started with the RTP protocol. When Bea picks up her phone and answers the call, the UE B sends a 200 OK response for the original SIP INVITE and the UE A returns a SIP ACK request to complete the call setup sequence.

For further details about the messages and routing procedures used in for IMS call setup, please refer to *Online Appendix J.5.2*.

It is worth noting that while this chapter describes IMS voice calls over both UTRAN and E-UTRAN networks, only the latter has been deployed. IMS voice over UTRAN HSPA did not gain wide support from operators, who believed that building the support for real-time voice data context over HSPA would not pay off the effort and cost. On the other hand, IMS voice over E-UTRAN (LTE), known as VoLTE, has become a mainstream operator voice technology.

9.4 Voice over LTE

9.4.1 Standardization of VoLTE

It was mentioned in Chapter 7, Section 7.1 that LTE only supports packet switched domain. LTE has no native telephony support so 3GPP had the plan to support both voice and real-time multimedia use cases with help of the IMS core network, tightly connected to the LTE access network.

As explained in Section 9.3.1, 3GPP specified **multimedia telephony (MMTel)** service in Rel-7 to emulate the traditional telephony services, such as voice call, messaging, and supplementary services, over the IP-enabled IMS architecture. What MMTel specifications bring on top of the plain vanilla IMS is the specific way that those telephone services would apply the basic IMS mechanisms. When MMTel services are used over LTE radio access, the service is called **Voice over LTE (VoLTE)**. VoLTE service was introduced to 3GPP Rel-8, but extensions such as VoLTE emergency calls were added in Rel-9.

The supplementary service specifications of 3GPP MMTel have originally been created within the ETSI TISPAN project. ETSI chose IMS as the base of the new generation core network technology. Additionally, TISPAN selected IETF XML Configuration Access Protocol (XCAP) for non-call-related supplementary service management, described further in Section 9.4.2.7. Later, 3GPP adopted these specifications created by ETSI and continued developing them further.

While 3GPP working groups (in tight cooperation with IETF) have been the standardization forum to specify the building blocks of VoLTE, another operator-driven organization GSM Association (GSMA) specified the minimal VoLTE service in the document GSMA PRD IR.92 [29]. IR.92 defines the way to tie 3GPP building blocks together to provide a baseline VoLTE service, which devices compliant to GSMA VoLTE are expected to support. IR.92 is a profile over the 3GPP MMTel service for building interoperable networks and user equipment. The work done in GSMA aligned the market and paved a way for nearly worldwide adoption of standard-compliant VoLTE implementations. The benefit of IR.92 is the alignment of technology over quite flexible IMS standards to provide a minimum deployable voice service. If all the operators would have promoted their own requirements for IMS-based VoIP services, the vendors could not have benefited from the economies of scale and the global travelers would have suffered from service interoperability problems.

The seeds for GSMA PRD IR.92 were laid already in 2009 when a consortium of 12 operators announced their intention to create a shared VoLTE specification. This initiative was given the name One Voice [30]. A few years later, the initial One Voce specification was adopted and extended by GSMA. The wide acceptance of GSMA PRD IR.92 also marked an end to **Voice over LTE via Generic Access (VoLGA)**, which was a competing approach to support telephony via the traditional circuit switched domain by providing mechanisms to transport the circuit switched connections over the IP-based LTE access network.

9.4.2 VoLTE System Procedures

9.4.2.1 Registering to VoLTE Service within LTE Attach

The HSS stores two different types of records used for VoLTE registration:

- EPC VoLTE subscription records with subscriber's MSISDN number, a session transfer number (STN-SR) for SRVCC handovers and APN configuration for EPS bearers.
- IMS VoLTE subscription records with subscriber's service profiles and authentication parameters. As mentioned earlier, the service profile contains a subscriber's public and private IMS user identities and the initial filter criteria for selecting IMS application servers for contributing to specific IMS services.

The VoLTE UE registers to the VoLTE service as shown in Figure 9.6:

1) The UE sends an NAS EMM ATTACH REQUEST message to become connected over LTE as described in Chapter 7, Section 7.1.9.1. When the UE supports VoLTE, it shall be assigned with an initial default EPS bearer for IMS signaling. Selection of the IMS APN may be done either by the UE or the network, if the UE omits APN from the attach request. In the latter case, the MME checks from the HSS (along with the location updating) which APN to use for the UE as default APN, after at first authenticating the user for EPC connectivity. When the HSS instructs the MME to use IMS APN, the MME sets up the initial default EPS bearer toward the IMS APN with QCI class 5 dedicated for IMS signaling. The MME proceeds with the S/P-GW to open the default EPS bearer for the UE toward the IMS APN as described in Chapter 7, Section 7.1.11.3.
2) The MME accepts the attach request received from the UE. After processing the attach accept message, the network reconfigures the RRC connection and finalizes setup of the EPS bearer as described in Chapter 7, Section 7.1.11.3. The MME sends an activate default EPS bearer request to the UE and tells it the address of P-CSCF to be used for IMS services. At this point, the UE can send a SIP REGISTER request to the P-CSCF for initial IMS communication service registration as described in Section 9.3.4.1.

For further details about the messages used in these procedures, please refer to *Online Appendix J.6.1.*

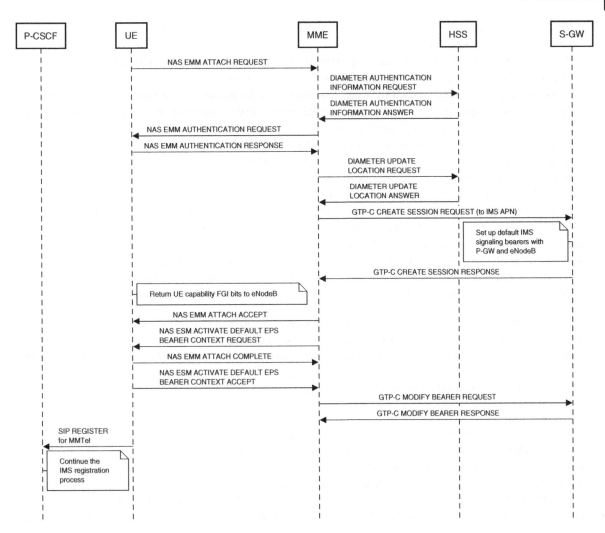

Figure 9.6 LTE attach and bearer setup for VoLTE registration.

9.4.2.2 Call Setup and Release

A VoLTE call follows the IMS call setup process described in Section 9.3.4.2, with the following enhancements:

- When composing the SIP INVITE message, the UE adds the IMS communication service identifier (ICSI) value 3gpp-service.ims.icsi.mmtel to the P-Preferred-Service, Contact and Accept-Contact headers. That is done to announce the support and preference of the voice call signaling according to the 3GPP MMTel specifications.
- The local resource reservation is done as described in Chapter 7, Section 7.1.11.5 for setting up dedicated EPS bearers for voice media. Since the pattern of voice traffic consists of small IP packets sent periodically every 20 or 40 ms, the following LTE mechanisms (described in Chapter 7, Sections 7.1.4.4, 7.1.4.5, 7.1.5.3, and 7.1.11.1) are used to support high voice quality and optimize the efficiency of the traffic:
  - Voice bearers need guaranteed bitrate provided by QCI class 1. The configured bitrate depends on the selected voice codec.
  - Semi-persistent scheduling is used to minimize the signaling overhead for scheduling the periodical resource grants.

- Robust IP header compression is used to reduce traffic overhead caused by IP headers, which are quite long compared to the encoded voice payload packets.
- Discontinuous reception (DRX) is used to optimize the UE power consumption by putting the UE receiver to sleep for the period between two periodically transmitted voice packets.
- Unacknowledged RLC mode is used since any retransmitted voice packets would anyway arrive too late to be played. The better approach is to drop any corrupted packets and instead play waveform extrapolated from the earlier packets. To support high audio quality, the radio parameters (such as transmit power, modulation, and channel coding) of the dedicated bearers are selected so that packet loss rate would not exceed 1% [31]. At the edge of the cell, TTI bundling is used to increase the reliability with additional redundancy.

- While the callee phone is ringing, the caller hears an alerting tone. When setting up a traditional circuit switched call, the alerting tone is played by the exchange over the media path opened to the phone. VoLTE has two options for generating the alerting tone. The tone may be generated locally at the UE based on SIP 180 response. As these tones are country specific, the UE should have correct versions of the tones available, to be used in different countries. The second option is that the alerting tone is generated by the network and played as early media until the SIP 200 OK response is received for the SIP INVITE request. This approach has an additional benefit that the network could also play announcements otherwise unavailable for the UE. SIP P-Early-Media header is used in 180 responses to indicate whether the early media option is used or not [32].

VoLTE implementations typically use either WB-AMR or EVS wideband audio codecs [33]. Enhanced voice services (EVS) multi-rate audio codec was specified in 3GPP Rel-12 to improve voice quality and network capacity. EVS supports as high as 20 kHz audio bandwidth with a maximum 48 kHz sampling frequency. EVS is designed to be resilient against packet loss, errors, and jitter [34]. Narrowband AMR codec can be used when the remote end does not support either of the wideband audio codecs.

The VoLTE call is released as follows:

1) The UE sends a SIP BYE request to the remote UE engaged to the call. The BYE request is routed by the SIP dialog, which is established over the CSCFs contributing to the call control.
2) When the remote P-CSCF receives the SIP BYE request, it sends a DIAMETER SESSION TERMINATION REQUEST message to its local PCRF to tear down the dedicated bearers.
3) The PCRF sends a DIAMETER SESSION TERMINATION ANSWER message back to the P-CSCF.
4) The PCRF sends a DIAMETER RE-AUTH REQUEST to the P-GW requesting it to tear down the dedicated bearers.
5) The P-GW takes care of tearing down the dedicated bearers with GTP-C procedures together with the S-GW and MME. Eventually, the P-GW sends a DIAMETER RE-AUTH ANSWER to the PCRF.
6) The remote UE sends a 200 OK response to the SIP BYE request.

9.4.2.3 Single Radio Voice Call Continuity

In the early phase of LTE deployment, the network coverage was far from ubiquitous. The current LTE networks may still have coverage gaps in sparsely populated rural areas or specific locations, such as tunnels or some corporate premises. In order to keep the voice call alive when moving out of VoLTE service coverage, 3GPP has defined the **single radio voice call continuity (SRVCC)** handover mechanism. With SRVCC, the ongoing call is transferred from the LTE packet switched domain to the circuit switched domain of another cellular radio technology. The SRVCC handover is one atomic operation, which switches the single radio modem of the UE between two radio technologies and the call between PS and CS domains. SRVCC architecture is specified in 3GPP TS 23.216 [35] for LTE RAN and TS 23.237 [36] for IMS.

The starting point for SRVCC handover is a VoLTE call managed with the SIP protocol. In the SRVCC handover, the voice media stream is handed over to UTRAN or GERAN CS side, but the UE still continues to manage the

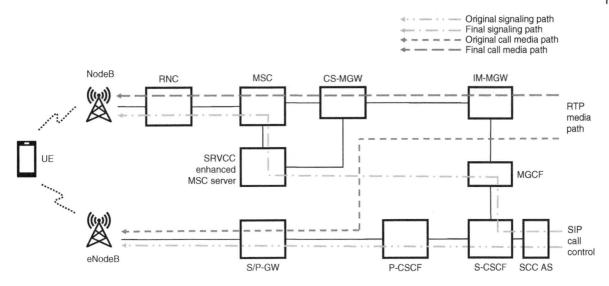

Figure 9.7 VoLTE call before and after 3GPP Rel-8 SRVCC handover.

call with the IMS core network and use the SIP protocol for that. At SRVCC handover, both signaling and media paths are changed. The IMS signaling is moved by a packet switched handover from the EPS signaling bearer to a new PDP context opened toward the IMS APN. The endpoint of the RTP voice media is transferred from the UE to a MGW node. The MGW terminates the RTP path and forwards the voice stream to/from the UE via the CS side protocols between the MGW and UE, as shown in Figure 9.7.

A few new network elements and certain enhancements to the existing elements were needed to support SRVCC. These new and enhanced elements were as follows in the 3GPP Rel-8 SRVCC architecture:

- **SRVCC enhanced MSC** supports new functionality to trigger SRVCC with the MME and control the handover together with other MSCs and radio access network elements in CS domain. A new Sv interface is defined in TS 29.280 [37] to be used between the SRVCC enhanced MSC and MME. Naturally, this interface and SRVCC procedures require additional functionality also on the MME and eNodeB side.
- **Service centralization and continuity (SCC) AS** is a new IMS application server via which VoLTE call control signaling path is routed at the call setup, to be prepared for SRVCC. The SCC AS is an anchor point for IMS signaling as it will stay on the signaling path even after the SRVCC handover. Initially, the signaling reaches the SCC AS over the P-CSCF, but after handover the signaling arrives over the MGCF and SRVCC enhanced MSC. The SCC AS informs the remote UE about any changes to the IMS session, such as new media endpoints and dropping of such media streams that are not supported at the CS side.
- **IM-MGW** terminates the RTP media path and forwards the media to the UE via CS network elements after the SRVCC handover.

The Rel-8 SRVCC handover procedure toward UTRAN network is performed as shown in Figure 9.8:

1) While being engaged to a VoLTE call, the UE detects the power of the LTE radio signal from the eNodeB to have decreased under a defined threshold. Consequently, the UE sends a measurement report to the eNodeB. When the eNodeB finds out that handover to another radio technology is needed, it sends an S1AP HANDOVER REQUIRED message to the MME.
2) The MME decides to initiate an SRVCC handover for the voice bearer and a PS handover for the IMS signaling bearer. The MME sends an SRVCC handover request to the SRVCC enhanced MSC, identifying the anchor SCC AS for the call. The SRVCC enhanced MSC sends a handover request message to the target MSC, which sends

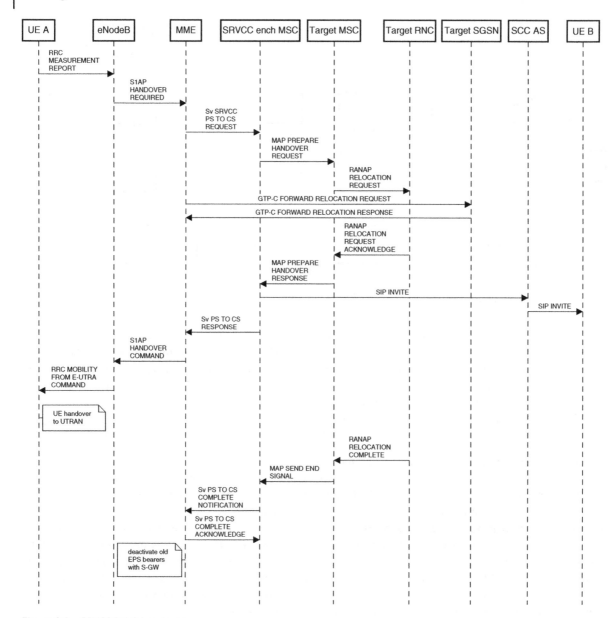

Figure 9.8 SRVCC Rel-8 handover.

a relocation request to the target RNC. While handover is being prepared on the UTRAN side, the MME sends a relocation request to the target SGSN of the core network to prepare the PS handover.

3) After getting the responses from UTRAN via the target MSC, the SRVCC enhanced MSC sends a SIP INVITE message to the SCC AS to move the IMS session from the UE to the MGW. The SCC AS itself sends a SIP INVITE message to the remote UE to get the media packets sent toward the MGW rather than the UE. The SCC AS also releases the access leg for signaling it had with the UE under the SRVCC handover. Ultimately, the SRVCC enhanced MSC responds to the MME indicating all the preparations for the SRVCC handover to have been completed in UTRAN. This response carries an embedded handover command which provides the UE with details about how to access the UTRAN cell at the handover.

4) The MME sends an S1AP handover command to the eNodeB, which forwards the embedded handover command from UTRAN to the UE within an LTE RRC message. The UE accordingly connects with the new UTRAN target cell. After the UTRAN connection has been opened, the RNC sends a relocation complete message to the target MSC. The target MSC thereafter confirms the completion of the SRVCC handover to the SRVCC enhanced MSC, which passes this status to the MME. At this point the MME deactivates the old EPS bearers toward the UE.

For further details about this sequence, please refer to *Online Appendix J.6.2.*

In 3GPP Rel-10, the SRVCC architecture was enhanced to minimize the disruption of the handover to the speech and support SRVCC also in roaming scenarios. In the renewed architecture, an access transfer anchor node function was introduced to keep the change of the media and signaling paths within the network domain of the visited operator providing the UE with access services. This change eliminates the need for the SCC AS to request the remote UE to switch its media path to the MGW. This is because the access transfer gateway always anchors the media path, before and after the SRVCC handover.

The new elements to support SRVCC according to 3GPP Rel-10 specifications are as follows:

- **Access transfer gateway (ATGW)** is the anchor endpoint of VoLTE media path for both the UE supporting SRVCC and the remote UE in the other end. After the handover, the ATGW stays as the media path endpoint for the remote UE while locally a new path is created between the ATGW and IMS-MGW, the latter of which forwards call media between IMS and CS domains.
- **Access transfer control function (ATCF)** controls the ATGW for setting up the media path and performing the handover. The ATCF is the anchor point of call signaling for both the local and remote UE. After the handover, the ATCF stays on the signaling path for the remote UE, while locally the ATGW moves the signaling path from the P-CSCF to the SRVCC enhanced MSC.

The decisions whether or not to include the ATCF to the IMS signaling path and the ATGW to media path are done at the IMS registration phase. The decisions are based on multiple conditions, such as SRVCC support of the UE, local policy, registered communication service, access network type, and whether the UE is roaming. The P-CSCF routes the SIP REGISTER request via the ATCF, which decides whether to keep itself on the path for any further IMS signaling messages. Figure 9.9 shows how the media and signaling paths are routed in the Rel-10 case.

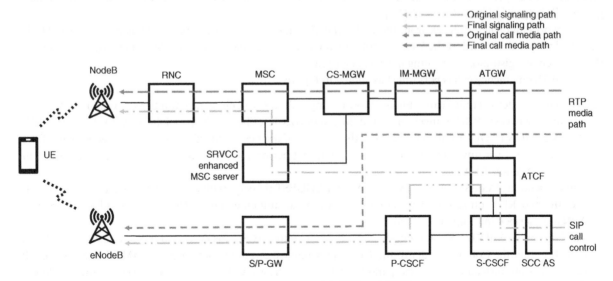

Figure 9.9 VoLTE call before and after 3GPP SRVCC handover with ATCF and ATGW.

With the Rel-10 architecture, when both ATCF and ATGW are included to the signaling and media paths, the SRVCC handover procedure is modified as follows [36, 38]:

- In step 3, the SRVCC enhanced MSC sends the SIP INVITE message to the ATCF instead of the SCC AS for new routing of both the voice media and IMS signaling during the access transfer.
- In step 3, there is no need for the SCC AS to inform the remote UE about the handover. Instead, the ATCF instructs the ATGW to switch the media path from the P-GW to the IM-MGW. The ATCF itself stops using the P-CSCF for signaling and continues signaling with the SRVCC enhanced MSC.
- Finally, the ATCF informs the SCC AS about the handover and creates a new dialog between the ATCF and SCC AS. The new dialog ensures that the new call leg toward the CS domain is not affected by possible expiration of the IMS registration of the UE. Finally, the SCC AS sends a SIP BYE request to the UE via the ATCF and P-CSCF. After receiving the BYE request, the UE relinquishes the resources it had for the packet switched call leg.

The original SRVCC Rel-8 specification covered the handover of an ongoing call from VoLTE to CS domain. Further SRVCC options were added in later releases as follows [39]:

- Rel-9 introduced the emergency access transfer function (EATF) to enable SRVCC handovers for IMS emergency calls.
- Re-10 introduced SRVCC support for voice calls in the alerting phase. SRVCC mid-call services were introduced to support SRVCC handovers for conference calls and calls on hold.
- Rel-11 introduced the reverse SRVCC from CS domain to VoLTE and SRVCC support for video calls. At the time of this writing, the reverse SRVCC is not known to be widely deployed.
- Rel-12 introduced SRVCC support for voice calls in the pre-alerting phase.

9.4.2.4 Emergency Call

VoLTE emergency call procedure differs from the normal call to guarantee preferential treatment for emergency calls and to support connecting the user to a local emergency response center known as the **public safety answering point (PSAP)**. The latter is an important requirement, especially when the user is roaming abroad far from the home network. As mentioned in Section 9.3.2.2, IMS emergency calls use a local emergency call state control function (E-CSCF) and an LRF to locate the user for dispatching rescue services. The architecture of IMS emergency calls is specified in 3GPP TS 23.167 [40].

To support emergency callback from the PSAP, the UE makes a special emergency registration, if the UE is equipped with a UICC card. If the UE does not have a UICC card, it may initiate an IMS emergency call without the emergency registration preventing the callback from the PSAP.

The VoLTE emergency call is performed as shown in Figure 9.10:

1) When the UE detects that a call is initiated toward any well-known emergency number such as 112 or 991, the UE shall open a new RRC connection toward the eNodeB as described in Chapter 7, Section 7.1.9.2. In the RRC CONNECTION REQUEST, the UE indicates the purpose of the new connection to be an emergency call. If the UE does not yet use any cellular network, it shall at first perform the LTE initial access process described in Chapter 7, Section 7.1.9.1.
2) The UE sends a NAS ESM PDN CONNECTIVITY REQUEST message with emergency indication to the MME over the new RRC connection. The network creates the default emergency EPS bearer toward an emergency APN as described in Chapter 7, Section 7.1.11.4.
3) The UE sends a SIP REGISTER request via the P-CSCF to the S-CSCF as described in Section 9.3.4.1 with the exception that the Contact header of the REGISTER request does not contain either MMTel or SMS over IP feature tags but instead has the "sos" parameter. The IP address given in the Contact header is the IP address that the UE has received for the default emergency EPS bearer.

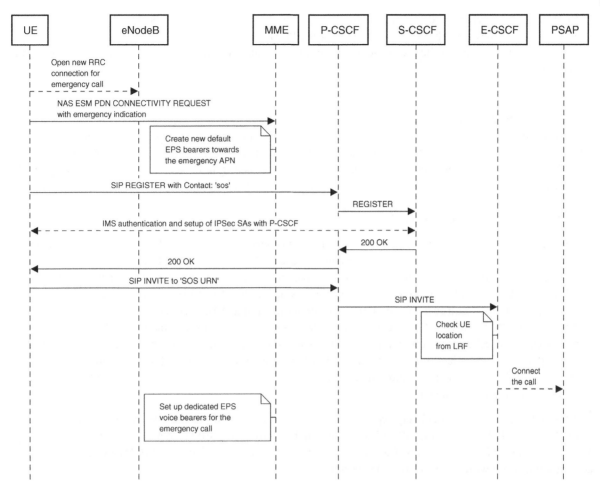

Figure 9.10 VoLTE emergency call.

4) The IMS emergency registration is completed like a normal IMS registration except that the S-CSCF does not add a Service-Route header to the SIP 200 OK response returned for the REGISTER request. For emergency calls, the S-CSCF does not want to be on the signaling path as the P-CSCF routes the call to the local E-CSCF. Registration event package is not subscribed as the emergency registration should only be valid in the duration of the emergency call. The S-CSCF consequently sets a rather short value to the expires parameter of the Contact header, but the UE may renew the registration before its expiry as long as the emergency call continues.

5) The UE sends a SIP INVITE request to the P-CSCF in order to initiate the emergency call. The Request URI and To header of the INVITE request contain an SOS URN, possibly indicating the type of the emergency service needed, such as the fire department or police. The UE may provide its initial location information as the global LTE cell ID within a P-Access-Network-Info header added to the INVITE request.

6) The P-CSCF forwards the SIP INVITE request to the local E-CSCF, which selects the PSAP toward which the call is connected either with IMS procedures or via PSTN interconnect. Which connection method the IMS core uses depends on the PSAP capabilities. The PSAP selection is done based on the location information received within the INVITE message. The P-CSCF also consults the LRF to get more precise locations of the UE to be mapped to a street address.

7) Dedicated voice bearers are set up toward the UE for the emergency call as described in Chapter 7, Section 7.1.11.5, based on the policy rules given by the PCRF.

8) The PSAP may also consult the LRF to get the location information of the UE for dispatching help to the right address.

9.4.2.5 Short Message over IP

A new **SMS over IP** service was specified for IMS to transport short messages over IMS signaling rather than the cellular network signaling. SMS over IP has the advantage that it works over any type of IMS access, such as WiFi. The structure of the transported short message is kept as originally specified for GSM to support all SMS-based interactions, such as updating SIM cards with SMS [31]. SMS over IP service is defined within 3GPP TS 24.341 [41].

The IMS UE uses SMS over IP to send short messages as shown in Figure 9.11:

1) The UE constructs the short message. As always, the SMS consists of nested SM-TL and SM-RL protocol messages. These protocols were described in Chapter 5, Section 5.1.10.2. The SM-TL message structure has the message text and the telephone number of the SMS recipient. The UE encapsulates the constructed SMS into a SIP MESSAGE request. The Request URI of the message is the SIP address of the Short Message Center (SMSC) within the circuit switched domain rather than the ultimate SMS recipient.

2) The UE sends the SIP MESSAGE request to the P-CSCF, which forwards the SIP message to the S-CSCF where the message is checked against initial filter criteria. The S-CSCF finds out that in order to reach the SMSC, the request shall be forwarded to the IP-SM-GW. The IP-SM-GW terminates the SIP transaction, takes the SMS from the SIP MESSAGE request, and sends it toward the SMSC. In the end, the IP-SM-GW returns a SIP 202 ACCEPTED response to the UE.

If the SMSC sends an SMS delivery report or a mobile terminated SMS to the UE, the report or SMS is forwarded from the IP-SM-GW to the UE within another mobile terminated SIP MESSAGE request in a similar way as described earlier for the MO SMS. The small difference is that after receiving the SIP MESSAGE request, the UE acknowledges it with a 200 OK final SIP response.

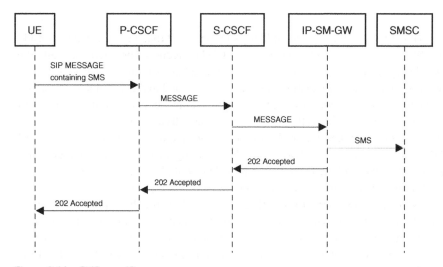

Figure 9.11 SMS over IP.

9.4.2.6 SMS Interworking

3GPP has also specified the **SMS interworking** concept where the traditional SMS protocol stack is no more used over IMS but native IME messaging methods are instead used. With this method, the IMS UE or IP-SM-GW composes a SIP MESSAGE request so that the SIP request just wraps the plain text to be rendered to the user. When the IMS UE exchanges short messages with a non-IMS UE, the IP-SM-GW terminates the traditional SMS protocols toward the circuit switched domain and uses plain SIP procedures toward the IMS UE. The term SMS interworking refers to this interworking procedure at the IP-SM-GW.

Usage of native IMS procedures for messaging between IMS-capable endpoints is defined within 3GPP TS 24.247 [42]. When both the endpoints support IMS, there is no need to use the traditional SMS protocols for message transport. Instead, short messages are transported end-to-end with IMS protocols only.

In addition to sending messages within individual standalone SIP MESSAGE requests, it is also possible to set up messaging sessions between two IMS UEs or the UE and IP-SM-GW. The session is set up with a SIP INVITE transaction between the endpoints, but the actual messages are sent between the endpoints over the **message session relay protocol (MSRP)** as specified in IETF RFC 4975 [43]. For messaging sessions, the MSRP provides several advantages over individual SIP MESSAGE requests:

- Optimized routing of MSRP messages between the endpoints. While SIP messages use SIP infrastructure (CSCFs) and routing methods on top of IP routing, MSRP messages can be sent over an end-to-end TCP connection, relying on IP routing infrastructure alone.
- MSRP supports fragmentation and reassembly of messages so that there are no limitations to the message size. It is possible to send images or even voice or video clips to the other end over the MSRP session.

The structure of the MSRP protocol message is similar to an HTTP protocol message. The MSRP message consists of three parts as follows:

1) The first line of the message, which is called the start line, contains a transaction identifier and the request or response type.
2) Header lines, which begin with the header name followed by a colon and the value of the header. The MSRP message has the following headers:
   - To-Path and From-Path with URI of a message session endpoint.
   - Message-ID.
   - Byte-Range identifies the total message size and the portion of it within the current MSRP message.
   - Content-Type defines the type of contents within the message body, such as plain text or some structured media such as a MIME encoded picture.
   - Message body, which carries the end user information.

9.4.2.7 Supplementary Services

The range of supplementary services defined in 3GPP for MMTel is wider than those specified earlier for GSM. Further on, MMTel specifications provide an extended set of conditions to trigger services such as call barring or call forwarding. The IMS MMTel supplementary services are specified in 3GPP TS 24.173 [44] umbrella specification. Detailed procedures for each of the services can be found from various technical specifications referred from TS 24.173. As GSMA PRD IR.92 [29] defines the minimum baseline for VoLTE implementation, it only refers to a subset of MMTel supplementary services. That subset is sufficient to emulate the features familiar with the users of circuit switched mobile phones.

The following VoLTE supplementary services are included with the GSMA PRD IR.92:

- Originating identification presentation (OIP), where the callee requests the identity of the caller to be rendered before answering the call.

- Originating identification restriction (OIR), where the caller instructs the network to hide the identity of the caller.
- Terminating identification presentation (TIP), where the callee requests his/her own identity to be rendered to the caller.
- Terminating identification restriction (TIR), where the callee requests his/her own identity to be hidden from the caller.
- Communications hold (HOLD) to temporarily pause the voice media.
- Conference call (CONF) to initiate a call between multiple parties.
- Communication diversion (CDIV) to forward the call to another number if the call cannot be connected to the called number.
- Call barring (CB) to prevent the call to be connected, for instance, to foreign numbers.
- Message waiting indication (MWI) to notify the user about a voice message waiting.
- Communication waiting (CW) to notify the user about an incoming call while another call is going on.
- Explicit call transfer (ECT) to transfer an ongoing call to another number. ECT was not in the scope of early versions of IR.92 but was later added to its scope.

MMTel supplementary services can be broadly categorized as the following types:

- Supplementary services (such as HOLD, CONF), which the user may activate while the call is being connected or is already connected. In MMTel, such supplementary services are controlled with SIP protocol requests. While call hold is done between two UEs, the conference call requires usage of a specific application server as a focal point setting up the conference and inviting the participants to it. A conference call uses also a media processor network element to combine voice streams from the participants.
- Supplementary services (such as MWI), where the user has to subscribe the service with SIP SUBSCRIBE request to be notified with SIP NOTIFY request when the service will be triggered.
- Supplementary services (such as OIP, OIR, TIP, TIR, CDIV, CB), which the user may activate beforehand and will be triggered by the network at the call attempt. Such MMTel supplementary services are activated with XCAP requests, and when triggered, will impact SIP call setup signaling.

XCAP is a protocol which is defined in IETF RFC 4825 [45] on top of the HTTP protocol. The UE may use XCAP request to query, configure, activate, or deactivate specific supplementary services and conditions for the network to trigger the services. An XCAP protocol request for MMTel supplementary service configuration is an HTTP protocol request that has the following special properties:

- The HTTP URI has a special format that contains identifiers of the user, operator, and the specific supplementary service being processed.
- The HTTP request body contains an XML "simservs" object providing conditions and actions for the supplementary service identified by the HTTP URI. The conditions determine when to trigger the supplementary service and actions describe the detailed behavior. For instance, the "simservs" body for call forwarding supplementary service may define a condition as "user does not answer" and action to forward the call to a given telephone number.

The UE sends XCAP HTTP requests over the Ut interface to an Authentication Proxy, which responds to any initial request with an HTTP 401 Unauthorized response. The UE then sends the XCAP HTTP request once again, now with an Authorization header that has valid credentials of the user, typically retrieved from the UICC card. This time, the Authentication Proxy forwards the XCAP HTTP request to the IMS telephony application server (TAS) responsible of the supplementary service configuration. The TAS processes the request and its "simservs" body. For HTTP GET requests the TAS returns the current state of the supplementary service to the UE. For HTTP PUT requests the TAS stores the new supplementary service state into its database. The TAS uses the stored state when it is engaged processing any SIP INVITE requests for an incoming or outgoing call, potentially triggering any of the supported supplementary services.

It is worth noting that XCAP requests are not necessarily transported over the IMS signaling bearer, but over another LTE bearer or PDP context (of GERAN or UTRAN) used to access home operator services over the Home Operator Services (HOS) APN [29]. This makes it possible to always use XCAP for supplementary service control, regardless of the type of radio access used by the UE and whether the user is in the home network with IMS or roaming abroad without using IMS [31]. Usage of the HOS APN supports fair charging in roaming scenarios as subscribers are not expected to be charged for data used to control their supplementary services.

9.4.2.8 DTMF Tones

As mentioned in Chapter 1, Section 1.1.7, DTMF tones are used to carry numbers from telephone keyboard to a remote system. Even if the DTMF mechanism was invented for analog telephony, it is still in common use for controlling voice mail systems and selecting specific services within a call center. In its original implementation, a DTMF tone was sent as in-band dual-tone audio signal, while in GSM the audio signal was replaced by signaling messages to start and stop DTMF tones. For VoLTE, a hybrid approach was chosen where the DTMF tone is carried within RTP packets (normally used to carry audio media) but as special RTP signaling events used for DTMF [33] rather than as encoded audio. In this way, DTMF tones are carried within the media path but explicitly as DTMF rather than audio which the receiver has to interpret as DTMF tones.

9.4.2.9 VoLTE Roaming

At the time of writing, VoLTE has largely stayed as a national service without support for users roaming abroad. The roaming users have to use the traditional circuit switched telephone services based on GSM or UMTS. In the future this might change, as 3GPP has specified two options to support VoLTE roaming:

- VoLTE local breakout specified in 3GPP Rel-11. Local breakout requires VoLTE support in the visited network. For Internet access, the LTE data flows of the roaming user will be connected to the P-GW in the home network, and the IMS signaling bearer of the roaming UE is connected to a P-GW in the visited network. In the local breakout scenario, the local P-CSCF within the visited network forwards IMS signaling between the UE and the S-CSCF in the home network.
- VoLTE S8 home routing is a new roaming architecture being studied in 3GPP. In S8 home routing, the IMS signaling bearer will be connected to the P-GW of the home network. The whole IMS infra, including the P-CSCF, is located to the home network. In this case the P-CSCF is not able to locally control creation of dedicated voice bearers in the visited LTE network. Instead, the dedicated bearers are set up by the home P-GW, which sends a related request to the S-GW and MME of the visited network. To support such an inter-operator procedure, the visited LTE network shall support both LTE dedicated bearers and the IPX roaming exchange. The IPX is used between operators to pass the request from the P-GW to the visited network. For SRVCC handovers, the implementation must support cooperation between the SRVCC enhanced MSC server in the visited network and the IMS infrastructure in the home network. S8 home routing appears to get increasing support in the operator community due to its apparent simplicity for the IMS infrastructure, as compared to the VoLTE local breakout architecture.

9.5 IMS Voice over 5G NR

As its name says, VoLTE was designed to operate over the LTE cellular system. The newer 5G systems are packet-only networks just like LTE networks. Consequently, IMS voice stays as the native voice call solution also for standalone 5G systems.

The mechanisms described in this book for VoLTE largely apply also to **IMS voice over 5G NR (VoNR)** systems. In some contexts, VoNR is also called Vo5G.

From the UE perspective, the differences between VoNR and VoLTE are as follows:

- In LTE, the UE declares its voice domain preference in the NAS EMM ATTACH REQUEST, but in the 5G NR system that is done with the NAS MM REGISTRATION REQUEST message. The IMS voice over PS support of the NR network is declared within the NAS MM REGISTRATION ACCEPT message of the 5G registration procedure.
- Instead of using default and dedicated LTE bearers, the IMS voice signaling and media are carried over 5G PDU sessions and QoS flows. The 5G PCF is enhanced to support VoNR voice flows.
- The UE will get the address of the P-CSCF from the SMF when setting up the PDU session and QoS flow for IMS signaling. The SMF may either utilize the Network Repository Function (NRF) to discover the a P-CSCF instance or the P-CSCF address may be just configured to the SMF.
- The SRVCC handover does not apply to VoNR as the only NR inter-RAT handover procedure is toward the LTE network where VoLTE should be available and CS domain is never supported.

Two main options exist for the operators to deploy VoNR service:

- Introduce VoNR service for those gNB base stations which are connected to the EPC core and support VoLTE service.
  - This option is not popular as instead of introducing VoNR to their networks, operators have preferred to support only VoLTE over the LTE bearers of EN-DC non-standalone configuration. In such a case, if the UE tries to set up QoS flow for voice, the network may redirect the UE toward the LTE bearer by rejecting the setup request with cause "fallback towards EPS." This mechanism in a way resembles the CSFB procedure used with early LTE networks without VoLTE support.
- Introduce VoNR service as a native service of the 5G standalone network and 5GC.

9.6 Voice over WiFi

IMS Voice over WiFi (VoWiFi) uses WLAN access toward the IMS core network, rather than LTE or 5G cellular access. For network operators, VoWiFi provides the benefit of offloading traffic from the cellular network to WLAN network. For end users, VoWiFi can also provide additional network coverage in locations where the cellular network signal is weak but WLAN network is available. GSMA has defined basic VoWiFi service in the PRD IR.51 [46] document.

To access IMS over WLAN, the UE connects to the ePDG network element, which was briefly introduced in Chapter 7, Section 7.1.3.2. The ePDG node effectively replaces S-GW and MME nodes used with LTE access option [31]. As shown in Figure 9.12, voice data packets and IMS signaling are tunneled between

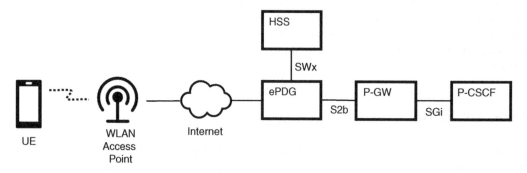

Figure 9.12 VoWiFi network architecture and interfaces.

the ePDG and P-GW nodes so that the P-GW hides the used access option from the IMS core network elements, such as the P-CSCF.

The used WLAN network may be managed by the IMS home operator or it may be any other untrusted WLAN network, such as a home WLAN. As the connection toward the ePDG may be routed over the Internet, a secure VPN tunnel is created between the UE and ePDG. This VPN tunnel uses IPSec technology as described in Chapter 3, Section 3.3.10.1. The VPN tunnel replaces LTE security mechanisms used in VoLTE to protect the traffic between the UE and LTE RAN. Inside of this VPN tunnel there is another IPSec protected IMS connection between the UE and P-CSCF as described in Section 9.3.3.3. These two nested IPSec tunnels use different IPSec configurations. IPSec-3GPP option is used between the UE and P-CSCF while IPSec with IKEv2 and EAP-AKA protocols [47] are used for the VPN tunnel between the UE and ePDG.

To connect with the ePDG, the UE must at first find the IP address of the ePDG. This is done with DNS by resolving the well-known domain name of the ePDG provided by the home operator. The domain name of the ePDG follows the standard format epdg.epc.<mnc>.<mcc>.pub.3gppnetwork.org, where <mnc> is the mobile network code and <mcc> the mobile country code of the home operator.

After acquiring the IP address of the ePDG, the UE sets up the VPN tunnel to the ePDG [47]. To create IPSec security associations for the VPN, the UE and ePDG at first negotiate the used security algorithms, IPSec modes, and security keys with IKEv2 protocol, as defined in IETF RFCs 4306 [48], 5282 [49], and 5996 [50]. IKEv2 uses the Diffie-Hellman algorithm for the key exchange. IKEv2 messages carry also subscriber identity and other critical parameters in encrypted form. The ePDG uses either certificates with EAP-AKA [51] algorithm or the 3GPP AKA mechanism to authenticate the subscriber within this initial IKEv2 sequence. In the latter case, the ePDG retrieves AKA authentication vectors from the HSS.

The UE retrieves the following IPv4 and/or IPv6 addresses from the ePDG:

- IP address of the UE used as the contact address of the UE for the SIP protocol
- IP addresses of DNS servers and P-CSCF nodes

When the VPN and its IPSec security associations are in place, the UE performs IMS registration with the IMS core network elements as described in Section 9.3.4.1. apart from the first LTE-specific step.

When no calls are in place, the UE may switch the IMS signaling bearer between WiFi and LTE accesses, to be able to initiate or answer to calls in the right access network. Depending on either user choice or operator policy, the UE may be configured to prefer either cellular or WiFi access for IMS when both types of networks are available.

- When switching to WiFi network, the UE must create the VPN tunnel to the ePDG as described earlier, apart from the fact that the UE does not get a new IP address. Instead, the IP address used for the UE over LTE will be moved to the new VPN tunnel. At the P-GW, this causes rerouting of the GTP tunnel toward the UE.
- When moving to LTE, the UE at first sets up the LTE initial default bearer, over which it sends an EMM PDN CONNECTIVITY REQUEST message. With this EMM message, the UE requests handing over the existing IMS bearers to LTE. At the P-GW this causes rerouting of the GTP tunnel toward the UE.

UEs compliant to IR.51 may hand over an ongoing VoWiFi call between WiFi and LTE access, if the network supports such handovers. The IMS core network does not even see the handover, which is done at the access networks behind the P-GW. To initiate the handover, the UE performs LTE attach with request type as handover. Consequently, the MME creates a session request with handover indication to the S-GW and P-GW. The P-GW returns the IP address of the UE as used for the ePDG access. After the LTE radio bearers have been set up for the UE, the MME instructs the S-GW and P-GWs to establish GTP tunnels to the S-GW. After the new tunnel setup is in place, the P-GW releases the old tunnels to the ePDG. The complete procedure description can be found from 3GPP TS 23.402 [52].

9.7 Questions

1 How can VoIP enhance the spectral efficiency of a cellular radio system?

2 What is the main task of the SIP registrar?

3 What is the task of the RTP protocol?

4 Please list the types of IMS call state control functions and their main responsibilities.

5 What are the two types of IMS user identities?

6 What is the purpose of the IMS security mechanism agreement procedure?

7 Why do IMS UEs have to subscribe registration state events?

8 How can the VoLTE UE learn the address of the local P-CSCF?

9 What is SRVCC?

10 How is VoWiFi different from VoLTE?

References

1 Wright, D.J. (2001). *Voice over Packet Networks*. Wiley.
2 ITU-T Recommendation H.323 Packet-based multimedia communications systems.
3 ITU-T Recommendation Q.931, ISDN user-network interface layer 3 specification for basic call control.
4 Camarillo, G. (2002). *SIP Demystified*. New York: McGraw-Hill.
5 IETF RFC 2543 SIP: Session Initiation Protocol.
6 IETF RFC 3261 SIP: Session Initiation Protocol.
7 IETF RFC 3262 Reliability of Provisional Responses in Session Initiation Protocol (SIP).
8 IETF RFC 3263 Session Initiation Protocol (SIP): Locating SIP Servers.
9 IETF RFC 3265 Session Initiation Protocol (SIP)-Specific Event Notification.
10 IETF RFC 3428 Session Initiation Protocol (SIP) Extension for Instant Messaging.
11 IETF RFC 3515 The Session Initiation Protocol (SIP) Refer Method.
12 IETF RFC 3856 A Presence Event Package for the Session Initiation Protocol (SIP).
13 IETF RFC 3903 Session Initiation Protocol (SIP) Extension for Event State Publication.
14 Poikselkä, M., Mayer, G., Khartabil, H., and Niemi, A. (2004). *The IMS IP Multimedia Concepts and Services in the Mobile Domain*. West Sussex: Wiley.
15 ETSI ES 282 001 Telecommunications and Internet Converged Services and Protocols for Advanced Networking (TISPAN); NGN Functional Architecture.
16 3GPP TS 23.228 IP Multimedia subsystem (IMS); Stage 2.
17 IETF RFC 3320 Signaling Compression (SigComp).
18 IETF RFC 3485 The Session Initiation Protocol (SIP) and Session Description Protocol (SDP) Static Dictionary for Signaling Compression (SigComp).
19 IETF RFC 3486 Compressing the Session Initiation Protocol (SIP).

20 3GPP TS 24.229 IP multimedia call control protocol based on Session Initiation Protocol (SIP) and Session Description Protocol (SDP); Stage 3.

21 IETF RFC 3310 Hypertext Transfer Protocol (HTTP) Digest Authentication Using Authentication and Key Agreement (AKA).

22 IETF RFC 2617 HTTP Authentication: Basic and Digest Access Authentication.

23 IETF RFC 3329 Security Mechanism Agreement for the Session Initiation Protocol (SIP).

24 IETF RFC 3315 Dynamic Host Configuration Protocol for IPv6 (DHCPv6).

25 IETF RFC 3319 Dynamic Host Configuration Protocol (DHCPv6) Options for Session Initiation Protocol (SIP) Servers.

26 IETF RFC 3361 Dynamic Host Configuration Protocol (DHCP-for-IPv4) Option for Session Initiation Protocol (SIP) Servers.

27 IETF RFC 8415 Dynamic Host Configuration Protocol for IPv6 (DHCPv6).

28 IETF RFC 3322 Signaling Compression (SigComp) Requirements & Assumptions.

29 GSMA PRD IR.92 IMS Profile for Voice and SMS.

30 Poikselkä, M., Holma, H., Hongisto, J. et al. (2012). *Voice over LTE (VoLTE)*. West Sussex: Wiley.

31 Sauter, M. (2021). *From GSM to LTE-Advanced pro and 5G : An Introduction to Mobile Networks and Mobile Broadband*. West Sussex: Wiley.

32 3GPP TS 24.628 Common Basic Communication procedures using IP Multimedia (IM) Core Network (CN) subsystem.

33 3GPP TS 26.114 IP Multimedia subsystem (IMS); Multimedia telephony; Media handling and interaction.

34 3GPP TS 26.441 Codec for enhanced voice services (EVS); General overview.

35 3GPP TS 23.216 Single radio voice call continuity (SRVCC); Stage 2.

36 3GPP TS 23.237 IP Multimedia Subsystem (IMS) Service Continuity; Stage 2.

37 3GPP TS 29.280 Evolved packet system (EPS); 3GPP Sv interface (MME to MSC, and SGSN to MSC) for SRVCC.

38 3GPP TS 24.237 IP multimedia subsystem (IMS) Service continuity.

39 GSMA PRD IR.64 IMS Service Centralization and Continuity Guidelines.

40 3GPP TS 23.167 IP multimedia subsystem (IMS) emergency sessions.

41 3GPP TS 24.341 Support of SMS over IP networks; Stage 3.

42 3GPP TS 24.247 Messaging service using the IP multimedia (IM) core network (CN) subsystem; Stage 3.

43 IETF RFC 4975 The Message Session Relay Protocol (MSRP).

44 3GPP TS 24.173 IMS Multimedia telephony communication service and supplementary services; Stage 3.

45 IETF RFC 4825 The Extensible Markup Language (XML) Configuration Access Protocol (XCAP).

46 GSMA PRD IR.51 IMS Profile for Voice, Video and SMS over Untrusted Wi-Fi access.

47 3GPP TS 33.402 3GPP System architecture evolution (SAE); Security aspects of non-3GPP accesses.

48 IETF RFC 4306 Internet Key Exchange (IKEv2) Protocol.

49 IETF RFC 5282 Using Authenticated Encryption Algorithms with the Encrypted Payload of the Internet Key Exchange Version 2 (IKEv2) Protocol.

50 IETF RFC 5996 Internet Key Exchange Protocol Version 2 (IKEv2).

51 IETF RFC 4187 Extensible Authentication Protocol Method for 3rd Generation Authentication and Key Agreement (EAP-AKA).

52 3GPP TS 23.402 Architecture enhancements for non-3GPP accesses.

Summary – The Transformation

During the last 40 years (1980–2020), the telecommunications industry has experienced a huge transformation, which can be characterized by the following trends:

- Shift from circuit switched telephony to packet switched data
- Shift from fixed networks toward mobile communications
- Shift from interactive communications to media consumption

This transformation was based on a number of technical and commercial drivers, such as:

- Progress of computing technology, bringing computers first to enterprises and thereafter to homes, and finally to everyone's pocket.
- Digitalization of telecommunications technologies, making it possible to increase the data rate and volume that can be transported over a single cable or fixed radio bandwidth as well as increasing the achievable distances and bringing overall cost levels down.
- Advances in various digital communication techniques, such as modulation, multiplexing, data encoding, and multi-antenna technologies, supporting ever higher bitrates, lower latencies, and bigger network capacities.
- Turning the regulated telecommunication market with a few incumbent service providers to a competitive market with multiple challenging service providers with different business models. This political change has greatly increased the investments in communications networks.
- Inventing the hyperlink, which opened a totally different approach for using the Internet via Web browsers, search engines, Web sites, pages, and links between them. This simple invention transformed the way that end users could easily find and retrieve (and later on publish) all sorts of information via the World Wide Web.
- Developing Voice over IP (VoIP) protocols capable of transporting voice over packet switched data so that it could replace the circuit switched connections for voice calls.

The shift from voice telephony to data is depicted with the two following tables S.1 and S.2. These tables provide estimates on how global communications traffic was divided between voice and data at different points of time. The given figures are very rough approximations derived from various, even conflicting sources, as the traffic volumes are very hard to measure accurately on a global level. The intention is not to provide figures with any precision but some insight into the trend of how the market has developed over the years:

Table S.1 Share of voice and data from the total network traffic.

Year	1995	2000	2005	2010	2015
Voice (%)	>99	80	20	3	1
Data (%)	<1	20	80	97	99

Converged Communications: Evolution from Telephony to 5G Mobile Internet, First Edition. Erkki Koivusalo.
© 2023 The Institute of Electrical and Electronics Engineers, Inc. Published 2023 by John Wiley & Sons, Inc.
Companion website: www.wiley.com/go/koivusalo/convergedcommunications

Table S.2 Share of voice and data from the mobile network traffic.

Year	2005	2010	2015	2020
Voice (%)	98	20	2	<1
Data (%)	2	80	98	>99

From these figures it can be seen that the volume of data traffic exceeded voice traffic in fixed networks around 2002, and in mobile networks a few years later, around 2008. Growth of the voice traffic has been very small, but data traffic has grown exponentially in the last 30 years.

During the 1980s, data communications was still a marginal business compared to the telephony. Although universities and research organizations used early Internet for their data communications, enterprises relied on slow-speed public data networks, such as X.25, for their needs. While microprocessor technology had made its breakthrough, in the telecommunications world the focus was to digitalize telephone exchanges and the trunk lines between the exchanges or toward fixed access networks. While this was being done, the rising importance of data communications was recognized. The pace of developing new types of analog modems accelerated and the supported data rates grew from 1 to 10 kbps. At the same time, the ISDN network was specified in International Telecommunications Union Telecommunication Standardization Sector (ITU-T). Narrowband Integrated Services Digital Network (ISDN) was supposed to provide 64 kbps end-to-end connections over the digitalized telephone network. To achieve higher data rates, enterprise clients could hire leased lines from telephone companies, providing a whole 2 Mbps trunk line or higher order PDH line to be used for data traffic. Actually, the core Internet routers were at that time connected over such leased lines.

On the mobile front, the first-generation analog mobile networks were specified and deployed during the 1980s. Advanced business users could enjoy telephone calls from their cars while on the road. Mobile phones were still heavy car-mounted devices, which could possibly be detached and carried outside to be used with a built-in or external battery. With the experience gathered from the first-generation mobile systems, standardization and development of the second-generation systems was kicked off.

During the 1990s, it had become clear that the data rates that could be provided by narrowband ISDN or analog modems, limited to 64 kbps, were totally insufficient. Computer technology had made great progress during the 1980s, and data volumes grew. With the emerging World Wide Web and electronic Internet email becoming commonplace in the first half of the decade, the demand for data communications started to grow rapidly. To meet with such demand, ITU-T started activities to specify Broadband ISDN, which would rely on the asynchronous transfer mode (ATM) technology to carry both circuit switched voice and packet switched data over permanent or switched virtual connections. Still, the industry was struggling with finding an optimal approach to matching the routed Internet protocol (IP) traffic to virtual links created with protocols which supported label switching. Such mapping was needed to avoid the processing complexity of IP routing at high-capacity network nodes. In the Internet engineering task force (IETF), a few different approaches were made to adapt ATM and IP with each other, but none turned out to be too successful to solving this key problem. The good news was that digital subscriber line (DSL) modem technology appeared as a good replacement for analog modems to increase data rates over existing telephone subscriber lines to the Mbps range. Asymmetric digital subscriber line (ADSL) was the most promising DSL variant, as it allowed the subscriber line to be shared with traditional analog voice and digital data by splitting the frequency ranges over the line for these two purposes. In the core and access networks, high-capacity optical SDH trunk lines of 155 Mbps and above were installed to support the needed capacity between network nodes. As the continued data growth was anticipated, actions were started in IETF and 3GPP to specify solutions for scalable Voice over IP (VoIP) systems.

On the mobile side, the success of second-generation global system for mobile communications (GSM) networks created a global boom of mobile telephony. In American markets, the IS-95 CDMA networks did the same on a bit smaller scale. The number of mobile telephone subscriptions grew exponentially in many countries, as smaller and more inexpensive mobile phones became available. While computers had started to support graphical color displays and Web browsers, mobile phone users had to rely on phones with small black-and-white screens supporting a small amount of text. Mobile phones were still true phones with some limited support for short messaging. GSM supported only circuit switched data up to 58 kbps data rates. The mobile industry had nevertheless recognized the growing need for mobile data. By the end of the decade, GPRS was standardized as the way to deliver packet switched data efficiently over the existing GSM infrastructure, increasing peak data rates up to 150 kbps.

During the 2000s, fixed Internet was booming. The first wave of Internet Web shops provided by various companies emerged. Though the early "dotcom" boom of startup companies focusing only on web sales soon turned out to be a failure, a few years later the same approach was adopted by much more successful players, such as Amazon. By the end of the decade, sales over Internet with direct deliveries to consumers had become a successful business model for a number of new companies, and was also supported by many well-established enterprises. Individuals active in the Internet started to publish their own content with blogs. In the second half of the decade, social network applications became increasingly successful even without a clear business model. With social network applications, it became possible for anyone to easily share their own content with their friends or openly to the public. The content typically consisted of some daily updates containing text and photos or some documents or videos prepared to be shared with a bigger audience. In the beginning of the decade, technology vendors had their visions of video-on-demand service with which consumers could at any time select their preferred movie and get it played over their home Internet connection. The market became ready for such use cases during the second half of the decade, when, for instance, Netflix abandoned its original business of DVD rental and started providing streamed videos instead.

Fueled with all these use cases, sales of DSL connections accelerated and various Internet service providers enjoyed the success of the consumer Internet market. The dilemma of mapping routed IP packet traffic to label switched paths was eventually solved with the introduction of multi-protocol label switching (MPLS) and its "forwarding equivalent classes." MPLS label switched routers (LSR) were able to dynamically derive the needed MPLS label switched connections and map the traffic to them by scanning the forwarded IP packets and consequent IP routing decisions applied to them. MPLS connections could be automatically created and modified on the fly, based on the perceived data routing and Quality of Service (QoS) patterns. This outcome meant the gradual end of ATM deployment and the planned ITU-T Broadband ISDN solution to use telephony-like signaling for label switched path management. While the capacity requirements increased for trunk lines, optical wavelength multiplexing (WDM) provided the solution for increasing aggregate data rates per optical single mode fiber. Different wavelength signals could be allocated to carry either Synchronous digital hierarchy (SDH) or gigabit Ethernet type of links. While SDH links were optimized to carry 64 kbps voice circuits, they could also be used for data. Where only high bitrate data support was needed, gigabit Ethernet provided a proven and cost-efficient solution.

Mobile operators tried to access the same market with the help of third-generation cellular systems, designed with native support for both packet switched data and circuit switched voice service. Compared to what had been promised, the delivered 3G data rates were disappointing to many users. While 3G WCDMA Universal mobile telecommunications system (UMTS) theoretically supported data rates up to 2 Mbps, in practice the supported data rates were around a few hundred kbps at best. It was not until the second half of the decade when High-speed packet access (HSPA) technology was deployed that the mobile data rates grew to the Mbps range. As highlighted by Table S.2, the volume of mobile data surpassed the volume of mobile voice traffic after 2007. During the standardization of fourth-generation cellular systems, a decision was made to drop support for circuit switched voice and focus only on the packet domain.

During the 2010s, the exponential growth of Internet and packet switched data volumes continued while voice traffic volume grew only modestly. The shortage of IPv4 addresses caused by the ever-expanding Internet could be overcome mainly with network address translation (NAT) technology and reusing private IPv4 address ranges within individual IPv4 networks. Transition toward IPv6 with a considerably larger address space made some gradual progress and is expected to continue in the foreseeable future. For fixed Internet access, Very high-speed digital subscriber line (VDSL) solutions capable of tens of Mbps data rates started gradually replacing slower Asymmetric digital subscriber line (ADSL) lines. Network operators increased their effort of replacing the old copper subscriber lines with underground optical fibers, reaching customer premises. IP multimedia subsystem (IMS) was at first deployed in fixed core networks when operators started to replace their international and long-distance exchanges with signaling and media gateways. Those gateways were able to convert SS7 signaling to session initiation protocol (SIP) and map 64 kbps pulse code modulation (PCM) voice to real-time protocol (RTP) streams with the help of various types of voice codecs. This move was done to optimize the backbone networks. Instead of maintaining traditional voice trunks and separate data backbones, operators could focus their investments only on the latter. As a consequence, SDH began to become outdated, and toward the end of the decade, WDM with gigabit Ethernet had become the most widely used standard approach to carry both IP data and voice traffic in the backbone networks. This development was carried further by introducing IMS to mobile networks as Voice over Long-Term Evolution (VoLTE).

The fourth-generation, packet switched Long-Term Evolution (LTE) cellular technology, was globally deployed during the first half of the decade. In their first years, LTE networks were used mainly as an alternative to fixed Internet access. Computers could be equipped with USB data dongles to provide LTE connectivity to the Internet. Operators started to sell LTE modems as an option to DSL or packet cable solutions. Mobile phone vendors were able to supply LTE phones to the market from 2012 onwards, and soon 4G LTE become the mainstream cellular radio access technology (RAT) supporting high-speed data. Mobile telephony was still done with older radio access technologies so that the LTE phone had to do circuit switched fallback (CSFB) from LTE to any RAT that was available to support circuit switched calls. By the second half of the decade, the native operator calls become available over LTE, along with VoLTE designed to support 3GPP standard operator VoIP over LTE access. In the first half of the decade, VoLTE had only a rather limited deployment (mainly by Verizon Wireless in the United States), but in 2015 VoLTE was eventually deployed by a few operators in half a dozen of European countries. In the second half of the decade, VoLTE became the global mainstream operator voice solution for LTE networks. That said, much of the voice traffic moved out of operator solutions as multiple VoIP applications by major operating system vendors and independent Internet application providers took their share of global VoIP. This development fragmented the voice market so that traditional operators have retained some of their market share and the rest is divided between a handful of other Internet players. Internet application vendors are providing both mobile and computer applications capable of voice and video calls, instant messaging, and content sharing. Traditional operators run their businesses by providing their clients with fixed and mobile Internet access solutions, the latter often bundled with either metered or flat rate monthly packages covering mobile data, messages, and some amount of voice minutes. While Internet application providers force their users to upgrade applications and even stop support of some older versions of operation systems, mobile operators continue supporting over 30-year-old GSM technology to complement VoLTE.

This book is being written in the beginning of the 2020s. The global Covid-19 pandemic has put extremely high demand on communications systems. Knowledge workers are locked in their homes, working remotely and doing video meetings with IP-based network applications. Demand for video streaming has increased, to provide some entertainment in these restricted conditions. In 2021, various industries faced problems with a worldwide lack of electronic components caused by an expanded demand of mobile phones, tablets, laptop computers, television sets, and Internet modems. As even cars need lots of electronics, not only phone manufacturers but also car manufacturers have been forced to limit their production volumes, as they cannot get sufficient quantities of microprocessors and memory components for their products. Still, in these rather unusual circumstances, the global telecommunications system has served well and enabled the high expansion of remote working, at least in the developed economies. While 5G mobile networks are being built, new opportunities open for industrial Internet and some emerging consumer use cases, such as ultra-high-definition video or augmented reality, which could be used, for instance, to support virtual traveling. In these times, when the travel industry is suffering from health-based restrictions and concerns of global climate change, the possibility to visit other countries while staying at home might become an interesting alternative.

Evolving telecommunications has a tradition of enabling activities which earlier were considered impossible or could not even be imagined. That evolution has not yet reached its end, and it remains to be seen with what its next stages will be able to provide us.

Index

Converged Communications: Evolution from Telephony to 5G Mobile Internet, First Edition. Erkki Koivusalo.
© 2023 The Institute of Electrical and Electronics Engineers, Inc. Published 2023 by John Wiley & Sons, Inc.
Companion website: www.wiley.com/go/koivusalo/convergedcommunications

Printed and bound by CPI Group (UK) Ltd, Croydon, CR0 4YY

16/04/2025

14658605-0005